# Therapeutic Proteins

# METHODS IN MOLECULAR BIOLOGY™

## John M. Walker, SERIES EDITOR

METHODS IN MOLECULAR BIOLOGY™

# Therapeutic Proteins

*Methods and Protocols*

Edited by

## C. Mark Smales

*Protein Science Group, Department of Biosciences,*
*University of Kent, Canterbury, Kent, UK*

and

## David C. James

*School of Engineering, University of Queensland, St. Lucia,*
*Queensland, Australia*

HUMANA PRESS ✳ TOTOWA, NEW JERSEY

© 2005 Humana Press Inc.
999 Riverview Drive, Suite 208
Totowa, New Jersey 07512

**www.humanapress.com**

This publication is printed on acid-free paper. ∞
ANSI Z39.48-1984 (American Standards Institute)

Permanence of Paper for Printed Library Materials.

Cover illustration: Figure 2 from Chapter 21, "Solid-State Protein Formulation: *Methodologies, Stability, and Excipient Effects,*" by Yuh-Fun Maa and Scott P. Sellers.

Cover design by Patricia F. Cleary.

For additional copies, pricing for bulk purchases, and/or information about other Humana titles, contact Humana at the above address or at any of the following numbers: Tel.: 973-256-1699; Fax: 973-256-8341; E-mail: orders@humanapr.com; or visit our Website: www.humanapress.com

**Photocopy Authorization Policy:**
Authorization to photocopy items for internal or personal use, or the internal or personal use of specific clients, is granted by Humana Press Inc., provided that the base fee of US $30.00 per copy is paid directly to the Copyright Clearance Center at 222 Rosewood Drive, Danvers, MA 01923. For those organizations that have been granted a photocopy license from the CCC, a separate system of payment has been arranged and is acceptable to Humana Press Inc. The fee code for users of the Transactional Reporting Service is: [1-58829-390-4/05 $30.00 ].

Printed in the United States of America. 10 9 8 7 6 5 4 3 2 1

Library of Congress Cataloging in Publication Data
Therapeutic proteins : methods and protocols / edited by C. Mark Smales and David C. James.
                    p. ; cm. — (Methods in molecular biology ; 308)
          Includes bibliographical references and index.
          ISBN 1-58829-390-4 (alk. paper) eISBN 1-59259-922-2
          1. Protein drugs—Laboratory manuals.    I. Smales, C. Mark. II. James, David C. (David
          Cameron), 1963-    . III. Series: Methods in molecular biology (Clifton, N.J.) ; v. 308.
          [DNLM: 1. Proteins—therapeutic use. 2. Proteins—chemical synthesis. 3. Proteins—
          isolation & purification.    QU 55 T3976 2005]
          RS431.P75T475 2005
          615'.3—dc22 2004021242

# Preface

With the recent completion of the sequencing of the human genome, it is widely anticipated that the number of potential new protein drugs and targets will escalate at an even greater rate than that observed in recent years. However, identification of a potential target is only part of the process in developing these new next generation protein-based "drugs" that are increasingly being used to treat human disease. Once a potential protein drug has been identified, the next rate-limiting step on the road to development is the production of sufficient authentic material for testing, characterization, clinical trials, and so on. If a protein drug does actually make it through this lengthy and costly process, methodology that allows the production of the protein on a scale large enough to meet demand must be implemented. Furthermore, large-scale production must not compromise the authenticity of the final product. It is also necessary to have robust methods for the purification, characterization, viral inactivation and continued testing of the authenticity of the final protein product and to be able to formulate it in a manner that retains both its biological activity and lends itself to easy administration.

*Therapeutic Proteins: Methods and Protocols* covers all aspects of protein drug production downstream of the discovery stage. This volume contains contributions from leaders in the field of therapeutic protein expression, purification, characterization, formulation, and viral inactivation. The contributors are all based at highly esteemed industrial and academic institutions from around the world and contact details are provided if researchers wish to obtain further information from the authors.

This book contains complete protocols set out in a simple step-by-step manner. It opens with an introductory chapter that discusses where therapeutic protein expression and downstream processing currently stand in terms of production, and contains thoughts on the direction of future developments from experts in the field. All other chapters contain a useful introduction describing the theory and background to the method, which is then followed by a list of all equipment and materials required to complete the protocol. The Methods section describes every step of the protocol and is cross-referenced to a Notes section that describes possible difficulties or problems that may arise, alternative methods and invaluable hints.

*Therapeutic Proteins: Methods and Protocols* includes protocols for the production of therapeutic proteins using a variety of sources, including bacterial and yeast expression systems and insect and mammalian cells. Methods for the purification of the resulting protein product are also described, as are purification protocols for the more traditional methods of preparing therapeutic proteins such as those sourced from plasma. Protocols for the characterization of therapeutic proteins throughout the production process are described, along with viral inactivation and protein formulation methods and strategies. The book contains both general methods and information and specific case studies highlighting particular expression systems, proteins of interest or characterization procedures that may be equally applicable to other systems or recombinant proteins.

A large number of people have helped to put this book together so that it ultimately provides an invaluable resource to all those working in the field of therapeutic protein production. I would especially like to thank all the contributors whom have all made many excellent suggestions, and indeed, improvements, to this book. I must also thank John Walker, the series editor, for asking me to edit this book, and for his help and advice in preparing the final product. Thanks also to those at Humana Press who have helped put this together. Finally I would like to thank my co-editor David James for all his help and advice and my family for their support.

*C. Mark Smales*

# Contents

*vii*

# Contributors

VARSHA BHAKTA • *Canadian Blood Services, Research and Development Department, Hamilton, Ontario, Canada*

JOHN R. BIRCH • *Lonza Biologics plc, Slough, UK*

ILSE BLUMENTALS • *Merck & Co., Inc., Rahway, NJ*

INEKE G. A. BOS • *Department of Immunopathology, Sanquin Research at CLB, Amsterdam, The Netherlands*

NICOLA BOSCHETTI • *R & D Virology, ZLB Behring AG, Bern, Switzerland*

SHERI BRADSHAW • *Schering Plough Research Institute, Union, NJ*

JOHN B. BRIGGS • *Department of Analytical Chemistry, Genentech Inc., South San Francisco, CA*

LEE J. BYRNE • *Department of Biosciences, University of Kent, Canterbury, Kent, UK*

MICHELE P. CALOS • *Department of Genetics, Stanford University School of Medicine, Stanford, CA*

SUSAN CANNON-CARLSON • *Schering-Plough Research Institute, Union, NJ*

JONG HYUN CHOI • *Department of Chemical and Biomolecular Engineering, Metabolic and Biomolecular Engineering National Research Laboratory, Bioinformatics Research Center, BioProcess Engineering Research Center, and Center for Ultramicrochemical Process Systems, Korea Advanced Institute of Science and Technology, Daejeon, South Korea*

LILY CHU • *Merck & Co., Inc., Rahway, NJ*

CONSTANCE CULLEN • *Schering Plough Research Institute, Union, NJ*

COLLETTE CUTLER • *Schering-Plough Research Institute, Union, NJ*

ARJO L. DE BOER • *Eukaryotic Microbiology, Groningen Biomolecular Sciences and Biotechnology Institute (GBB), University of Groningen, Haren, The Netherlands*

ERIC C. DE BRUIN • *Department of Immunopathology, Sanquin Research at CLB, Amsterdam, The Netherlands*

MARC DELORENZO • *Schering Plough Research Institute, Union, NJ*

ERWIN DUITMAN • *Eukaryotic Microbiology, Groningen Biomolecular Sciences and Biotechnology Institute (GBB), University of Groningen, Haren, The Netherlands*

ILARIA DURELLI • *Laboratory of Immunogenetics, Department of Genetics, Biology and Biochemistry, University of Torino Medical School, Turin, Italy and Research Center for Experimental Medicine (CeRMS), San Giovanni Battista Hospital, Turin, Italy*

A. N. S. ESHWARI • *Product Development Cell, National Institute of Immunology, New Delhi, India*

MARTIN FUSSENEGGER • *Institute for Chemical and BioEngineering, Swiss Federal Institute of Technology, ETH Zurich, Zurich, Switzerland*

LALIT C. GARG • *Gene Regulation Laboratory, National Institute of Immunology, New Delhi, India*

*xi*

SABINE GEISSE • *Novartis Pharma Research CT/BMP, Basel, Switzerland*

P. CLAYTON GOUGH • *Bioproduct Research and Development, Lilly Research Laboratories, Eli Lilly and Company, , Indianapolis, IN*

MICHAEL J. GRACE • *Schering Plough Research Institute, Union, NJ*

C. ERIK HACK • *Departments of Immunopathology and Clinical Chemistry, Sanquin Research at CLB, VU Medical Centre, Amsterdam, The Netherlands*

ALBERTO L. HORENSTEIN • *Laboratory of Immunogenetics, Department of Genetics, Biology and Biochemistry, University of Torino Medical School, Turin, Italy and Research Center for Experimental Medicine (CeRMS), San Giovanni Battista Hospital, Turin, Italy*

MARK J. HOWARD • *Department of Biosciences, University of Kent, Canterbury, Kent, UK*

LIHUA HUANG • *Bioproduct Research and Development, Lilly Research Laboratories, Eli Lilly and Company, Indianapolis, IN*

KEN-ICHI IZUTSU • *National Institute of Health Sciences, Tokyo, Japan*

DAVID C. JAMES • *School of Engineering, University of Queensland, St. Lucia, Queensland, Australia*

MICHAEL A. JANKOWSKI • *Department of Characterization and Analytical Development, Wyeth BioPharma, Andover, MA*

ZHENG JIN • *Institute for Cancer Research, College of Life Science and Technology, Xi'an Jiaotong University, Xi'an, Peoples Republic of China*

ANNA JOHNSTON • *CSIRO Health Science, Parkville, Melbourne, Australia*

ANDREW J. S. JONES • *Department of Analytical Chemistry, Genentech Inc., South San Francisco, CA*

MARTIN JORDAN • *Laboratory of Cellular Biotechnology, SV-IGBB-LBTC, EPFL, Lausanne, Switzerland*

RODNEY G. KECK • *Department of Analytical Chemistry, Genentech Inc., South San Francisco, CA*

MARINA KORNEYEVA • *Bayer Corporation Biological Products, Clayton, NC*

BEAT P. KRAMER • *Institute for Chemical and BioEngineering, Swiss Federal Institute of Technology, ETH Zurich, Zurich, Switzerland*

BRITTANY LARKIN • *Schering-Plough Research Institute, Union, NJ*

WENDY LAU • *Department of Analytical Chemistry, Genentech Inc., South San Francisco, CA*

SANG JUN LEE • *Department of Chemical and Biomolecular Engineering and Center for Ultramicrochemical Process Systems, Metabolic and Biomolecular Engineering National Research Laboratory, Korea Advanced Institute of Science and Technology, Daejeon, South Korea*

SANG YUP LEE • *Metabolic and Biomolecular Engineering National Research Laboratory, Department of Chemical and Biomolecular Engineering, Department of Biosystems, Bioinformatics Research Center, BioProcess Engineering Research Center and Center for Ultramicrochemical Process Systems, Korea Advanced Institute of Science and Technology, Daejeon, South Korea*

SEOJU LEE • *Neose Technologies Inc., Horsham, PA*

YAN-HUI LIU • *Schering-Plough Research Institute, Union, NJ*

SI LUSHENG • *Institute for Cancer Research, College of Life Science and Technology, Xi'an Jiaotong University, Xi'an, Peoples Republic of China*

STACEY MA • *Department of Analytical Chemistry, Genentech Inc., South San Francisco, CA*

YUH-FUN MAA • *ALZA Corporation, Mountain View, CA*

GARGI MAHESHWARI • *Merck & Co. Inc., Rahway, NJ*

FABIO MALAVASI • *Laboratory of Immunogenetics, Department of Genetics, Biology and Biochemistry, University of Torino Medical School, Turin, Italy and Research Center for Experimental Medicine (CeRMS), San Giovanni Battista Hospital, Turin, Italy*

ROSALYN J. MARCHANT • *Department of Biosciences, University of Kent, Canterbury, Kent, UK*

JOSEPH E. MCCLELLAN • *Department of Characterization and Analytical Development, Wyeth BioPharma, Andover MA*

TERESA R. MCCURDY • *Research and Development Department, Canadian Blood Services, Hamilton, Ontario, Canada*

AARON P. MILES • *Biochemical Assay Development and Quality Control, Malaria Vaccine Development Branch, National Institute of Allergy and Infectious Diseases, Rockville, MD*

CHARLES E. MITCHELL • *Bioproduct Research and Development, Lilly Research Laboratories, Eli Lilly and Company, Indianapolis, IN*

KATHY MOORHOUSE • *Department of Quality Control Clinical Development, Genentech Inc., South San Francisco, CA*

TERUHISA NAKASHIMA • *Blood Products Research Department, The Chemo-Sero-Therapeutic Research Institute, Kumamoto, Japan*

WASSIM NASHABEH • *Department of Quality Control Clinical Development, Genentech Inc., South San Francisco, CA*

KENNETH J. O'CALLAGHAN • *Department of Biosciences, University of Kent, Canterbury, Kent, UK*

YEMI ONAKUNLE • *Lonza Biologics plc, Slough, UK*

AMULYA K. PANDA • *Product Development Cell, National Institute of Immunology, New Delhi, India*

HIMAKSHI K. PATEL • *Department of Characterization and Analytical Development, Wyeth BioPharma, Andover MA*

BERNARDO PEREZ-RAMIREZ • *Scientific Director, BioFormulations Development, Genzyme, Framingham, MA*

ANDREW G. POPPLEWELL • *Celltech R&D, Slough, UK*

THOMAS J. PORTER • *Department of Characterization and Analytical Development, Wyeth BioPharma, Andover MA*

ALICE RIGGIN • *Bioproduct Research and Development, Lilly Research Laboratories, Eli Lilly and Company, Indianapolis, IN*

SCOTT ROSENTHAL • *Bayer Corporation Biological Products, Clayton, NC*

JASON C. ROUSE • *Department of Characterization and Analytical Development, Wyeth BioPharma, Andover MA*

ALLAN SAUL • *Malaria Vaccine Development Branch, National Institute of Allergy and Infectious Diseases, Rockville MD*

MUKESH SEHDEV • *Celltech R&D, Slough, UK*

SCOTT P. SELLERS • *ALZA Corporation, Mountain View, CA*

WILLIAM P. SHEFFIELD • *Department of Pathology and Molecular Medicine, McMaster University, Hamilton, Ontario, Canada*

SURINDER M. SINGH • *Product Development Cell, National Institute of Immunology, New Delhi, India*

C. MARK SMALES • *Protein Science Group, Department of Biosciences, University of Kent, Canterbury, Kent, UK*

MARIANGELA SPITALI • *Celltech R&D, Slough, UK*

JOHN J. STECKERT • *Characterization & Analytical Development, Wyeth BioPharma, Andover, MA*

BHASKAR THYAGARAJAN • *Poetic Genetics, LLC, Burlingame, CA*

KAZUHIKO TOMOKIYO • *Blood Products Research Department, The Chemo-Sero-Therapeutic Research Institute, Kumamoto, Japan*

MICK F. TUITE • *Department of Biosciences, University of Kent, Canterbury, Kent, UK*

MICHÈLE F. UNDERHILL • *Department of Biosciences, University of Kent, Canterbury, Kent, UK*

MARTEN VEENHUIS • *Eukaryotic Microbiology, Groningen Biomolecular Sciences and Biotechnology Institute (GBB), University of Groningen, Haren, The Netherlands*

MARCIO VOLOCH • *Transkaryotic Therapies Inc., Cambridge, MA*

A. NEIL C. WEIR • *Celltech R&D, Slough, UK*

FLORIAN M. WURM • *Laboratory of Cellular Biotechnology, SV-IGBB-LBTC, EPFL, Lausanne, Switzerland*

DAVID C. WYLIE • *Biotechnology Development, Schering-Plough Research Institute, Union, NJ*

LEI XIE • *Schering Plough Research Institute, Union, NJ*

WANG YILI • *Institute for Cancer Research, College of Life Science and Technology, Xi'an Jiaotong University, Xi'an, Peoples Republic of China*

LEI YU • *Bioproduct Research and Development, Lilly Research Laboratories, Eli Lilly and Company, Indianapolis, IN*

DAMING ZHU • *Biochemical Assay Development and Quality Control, Malaria Vaccine Development Branch, National Institute of Allergy and Infectious Diseases, Rockville, MD*

# 1

## Biopharmaceutical Proteins

*Opportunities and Challenges*

### John R. Birch and Yemi Onakunle

### 1. Introduction

Over the last 20 yr, there has been extraordinary growth in the biopharmaceutical industry based on the development of recombinant DNA and hybridoma technologies in the 1970s. Prior to this, the dependence on extraction from natural sources severely limited the range and quantity of proteins available for clinical use. Recombinant DNA technology made it possible to mass produce a wide range of natural and modified proteins for the first time. In addition, hybridoma technology introduced a new class of protein reagents—the monoclonal antibodies (MAbs)—that provided an alternative approach to treat many diseases.

The earliest recombinant products were replacements for existing protein products, which were extracted from natural sources, such as blood and pituitaries. Insulin was the first recombinant protein to be approved in 1982, followed by growth hormone and blood-clotting Factor VIII. In some cases, the switch to recombinant products was also motivated by safety concerns related to the natural source. In the 1980s, concern existed that growth hormone derived from human pituitaries might transmit the prion agent responsible for Creutzfeld–Jacob disease; this protein is now produced in *Escherichia coli*. Recombinant technology has also allowed the production of viral vaccines, e.g., hepatitis B, without the need to use the potentially hazardous virus in the manufacturing process.

In the second generation of products, there is a progression of new approaches to treat various diseases using recombinant proteins and MAbs. A significant proportion of applications are in the cancer field; other important areas include immune disorders and infectious diseases. The number of approved products has grown steadily. By mid-2003, 148 biopharmaceuticals were approved in the United States and Europe compared with 84 in 2000 (*1,2*; **Table 1**). Similarly, the number of products being evaluated has rapidly increased with estimates of many hundreds more in development

From: *Methods in Molecular Biology, vol. 308: Therapeutic Proteins: Methods and Protocols*
Edited by: C. M. Smales and D. C. James © Humana Press Inc., Totowa, NJ

**Table 1**
**Approved Biopharmaceutical Proteins**

Hormones

| Product | Protein | Therapeutic area | Company USA | Approval date for first indication | Production technology | Cell line/ microbial species |
|---|---|---|---|---|---|---|
| Humulin® | Human insulin | Diabetes | Eli Lilly | 1982 | Microbial | *E. coli* |
| Protropin® (somatrem) | Methionyl-human growth hormone (hGH) | hGH deficiency | Genentech | 1985 | Microbial | *E. coli* |
| Humatrope® (somatropin) hGH | | hGH deficiency | Eli Lilly | 1987 | Microbial | *E. coli* |
| Novolin® | Human insulin | Diabetes | Novo Nordisk | 1991 | Microbial | *S. cerevisiae* |
| Nutropin® (somatropin) hGH | | hGH deficiency | Genentech | 1993 | Microbial | *E. coli* |
| Norditropin® (somatropin) hGH | | hGH deficiency | Novo Nordisk | 1995 | Microbial | *E. coli* |
| Bio-Tropin® (somatropin) hGH | | hGH deficiency | Bio-Technology General | 1995 | Microbial | *E. coli* |
| Genotropin® (somatropin) hGH | | hGH deficiency | Pfizer/Pharmacia | 1995 | Microbial | *E. coli* |
| Humalog® (insulin lispro) | Human insulin analog | Diabetes | Eli Lilly | 1996 | Microbial | *E. coli* |
| Serostim® (somatropin) hGH | | AIDS-associated wasting | Serono | 1996 | Mammalian | Mouse C127 |
| Follistim® (follitropin-β) | Follicle-stimulating hormone | Infertility | Organon | 1997 | Mammalian | CHO |
| Gonal-F® (follitropin α) | Follicle-stimulating hormone | Infertility | Serono | 1997 | Mammalian | CHO |
| GlucaGen® (glucagon) | Glucagon | Hypoglycemia | Novo Nordisk | 1998 | Microbial | *S. cerevisiae* |
| Glucagon | Glucagon | Hypoglycemia | Eli Lilly | 1998 | Microbial | *E. coli* |
| Lantus® (insulin glargine) | Human insulin analog | Diabetes | Aventis | 2000 | Microbial | *E. coli* |
| NovoLog® (insulin aspart) | Human insulin analog | Diabetes | Novo Nordisk | 2000 | Microbial | *S. cerevisiae* |
| Ovidrel® (choriogonadotropin α) | Human chorionic gonadotropin | Infertility | Serono | 2000 | Mammalian | CHO |
| Natrecor® (nesiritide) | Human B-type Natriuretic peptide | Heart failure | Scios | 2001 | Microbial | *E. coli* |
| Forteo® (teriparatide) | Parathyroid hormone | Osteoporosis | Eli Lilly | 2002 | Microbial | *E. coli* |
| Somavert® (pegvisomant) | Growth hormone analog (receptor antagonist) | Acromegaly | Pfizer/Pharmacia | 2003 | Microbial | *E. coli* |

Cytokines/receptor antagonists/growth factors

| | | | | | | |
|---|---|---|---|---|---|---|
| Intron A® | Interferon α-2b | Cancer, hepatitis | Schering-Plough | 1986 | Microbial | *E. coli* |
| Roferon A® | Interferon α-2a | Cancer, hepatitis | Hoffmann-La Roche | 1986 | Microbial | *E. coli* |
| Epogen® (epoetin-α) | EPO | Anaemia | Amgen | 1989 | Mammalian | CHO |
| Procrit® (epoetin-α) | EPO | Anaemia | Ortho Biotech/J&J | 1990 | Mammalian | CHO |
| Actimmune® | Interferon γ-1b | Chronic granulomatous disease, osteopetrosis | InterMune | 1990 | Microbial | *E. coli* |
| Neupogen® (filgrastim) | G-CSF | Neutropenia | Amgen | 1991 | Microbial | *E. coli* |
| Leukine® (sargramostim) | Granulocyte/macrophage colony-stimulating factor | Cancer/bone marrow transplantation | Berlex | 1991 | Microbial | *S. cerevisiae* |
| Proleukin® (aldesleukin) | Interleukin-2 (IL-2) | Renal carcinoma | Chiron | 1992 | Microbial | *E. coli* |
| Betaseron® | Interferon β-1b | Multiple sclerosis | Berlex Inc./Chiron | 1993 | Microbial | *E. coli* |
| Avonex® | Interferon β-1a | Multiple sclerosis | Biogen Idec | 1996 | Mammalian | CHO |
| Infergen® (interferon alfacon-1) | Consensus interferon | Hepatitis C | InterMune | 1997 | Microbial | *E. coli* |
| Regranex® (becaplermin) | Platelet-derived growth factor | Diabetic ulcers | Ortho-McNeil/J&J | 1997 | Microbial | *S. cerevisiae* |
| Neumega® (oprelvekin) | IL 11 | Chemotherapy-induced thrombocytopenia | Genetics Institute/Wyeth-Ayerst | 1997 | Microbial | *E. coli* |
| Ontak® (denileukin diftitox) | IL-2/diphtheria toxin fusion protein | T-cell lymphoma | Ligand/Seragen | 1999 | Microbial | *E. coli* |
| Peg-Intron® (peginterferon alfa-2b) | Pegylated interferon α-2b | Hepatitis C | Schering-Plough | 2001 | Microbial | *E. coli* |
| Aranesp® (darbepoietin-α) | EPO analog | Anemia | Amgen | 2001 | Mammalian | CHO |
| Osteogenic protein 1 | Bone morphogenic protein-7 | Bone repair | Stryker Biotech | 2001 | Mammalian | CHO |
| Kineret® (anakinra) | IL-1 receptor antagonist (IL-1 Ra) | Rheumatoid arthritis | Amgen | 2001 | Microbial | *E. coli* |
| Neulasta® (pegfilgrastim) | Pegylated (G-CSF) | Neutropenia | Amgen | 2002 | Microbial | *E. coli* |
| Rebif® | Interferon β-1a | Multiple sclerosis | Serono | 2002 | Mammalian | CHO |
| InFuse® (dibotermin α) | Bone morphogenic protein-2 | Bone repair | Medtronic Sofamor Danek | 2002 | Mammalian | CHO |
| Pegasys® (peginterferon alfa-2a) | Pegylated interferon α-2a | Hepatitis C | Hoffmann-La Roche | 2002 | Microbial | *E. coli* |

*(continued)*

**Table 1** (*continued*)

| Product | Protein | Therapeutic area | Company USA | Approval date for first indication | Production technology | Cell line/ microbial species |
|---------|---------|------------------|-------------|------------------------------------|----------------------|------------------------------|
| | | Enzymes | | | | |
| Activase® (alteplase) | Human tPA | Thrombolysis | Genentech | 1987 | Mammalian | CHO |
| Pulmozyme® (dornase α) | Deoxyribonuclease | Cystic fibrosis | Genentech | 1993 | Mammalian | CHO |
| Cerezyme® (imiglucerase) | β-Glucocerebrosidase | Gauchers disease | Genzyme | 1994 | Mammalian | CHO |
| Retavase® (reteplase) | Modified human tPA | Thrombolysis | Centocor/J&J | 1996 | Microbial | E. coli |
| TNKase® (tenecteplase) | Modified human tPA | Thrombolysis | Genentech | 2000 | Mammalian | CHO |
| Elitek® (rasburicase) | Urate oxidase | Uric acid management | Sanofi-Synthelabo | 2002 | Microbial | S. cerevisiae |
| Aldurazyme® (laronidase) | α-L-Iduronidase | Mucopolysaccharidosis I (MPS1) | Genzyme/BioMarin | 2003 | Mammalian | CHO |
| Fabrazyme® Agalsidase β | Human α-galactosidase | Fabry disease | Genzyme | 2003 | Mammalian | CHO |
| | | Blood factors | | | | |
| Recombinate® | Factor VIII | Hemophilia | Baxter Healthcare | 1992 | Mammalian | CHO |
| Kogenate® | Factor VIII | Hemophilia | Bayer | 1993 | Mammalian | BHK |
| BeneFIX® | Factor IX | Hemophilia | Genetics Institute/ Wyeth Pharmaceuticals | 1997 | Mammalian | CHO |
| NovoSeven® | Factor VIIa | Hemophilia | Novo Nordisk | 1999 | Mammalian | BHK |
| ReFacto® (moroctocog-α) | Factor VIII | Hemophilia | Genetics Institute/ Wyeth Pharmaceuticals | 2000 | Mammalian | CHO |
| Xigris® (drotrecogin α) | Human-activated Protein C | Sepsis | Eli Lilly | 2001 | Mammalian | |
| Advate | Factor VIII | Hemophilia | Baxter Healthcare | 2003 | Mammalian | CHO |

## MAbs/fusion proteins

| Product | Target | Antibody type | Therapeutic area | Company USA | Approved date for first indication | Cell line |
|---|---|---|---|---|---|---|
| Orthoclone OKT® 3 (Muromonab) | CD3 | Murine | Transplant rejection | Ortho Biotech/J&J | 1986 | Hybridoma (not recombinant) |
| ReoPro® (abciximab) | Platelet GPIIb/IIIa receptor | Chimeric | Cardiovascular disease | Eli Lilly/Centocor/J&J | 1994 | Sp2/0 |
| Rituxan® (rituximab) | CD20 | Chimeric | Non-Hodgkin's lymphoma | Genentech/Biogen Idec | 1997 | CHO |
| Zenapax® (daclizumab) | IL-2 receptor | Humanized | Transplant rejection | Hoffman La Roche | 1997 | NS0 |
| Simulect® (basiliximab) | IL-2 receptor | Chimeric | Transplant rejection | Novartis | 1998 | NS0 |
| Synagis® (palivizumab) | Respiratory syncytial virus | Humanized | Respiratory syncytial virus infection | Medimmune | 1998 | NS0 |
| Remicade® (infliximab) | Tumor necrosis factor | Chimeric | Crohns Disease/rheumatoid arthritis | Centocor/J&J | 1998 | Sp2/0 |
| Herceptin® (trastuzumab) | HER2 receptor | Humanized | Breast cancer | Genentech | 1998 | CHO |
| Enbrel® (etanercept) | Tumor necrosis factor receptor/Ig G1 Fc Fusion protein | Fusion protein | Rheumatoid arthritis | Amgen/Wyeth Pharmaceuticals | 1998 | CHO |
| Mylotarg® (gemtuzumab ozogamicin) | CD33 used as conjugate with calicheamicin cytotoxin | Humanized | Acute myeloid leukemia | Wyeth Pharmaceuticals | 2000 | NS0 |
| Campath® (alemtuzumab) | CD52 | Humanized | Chronic lymphocytic leukemia | Berlex/Ilex Oncology | 2001 | CHO |
| Zevalin® (ibritumomab tiuxetan) | CD20 used as conjugate with Yttrium 90 | Murine | Non-Hodgkin's lymphoma | Biogen Idec | 2002 | CHO |
| Humira® (adalimumab) | Tumor necrosis factor-α | Human | Rheumatoid arthritis | Abbot Laboratories | 2002 | CHO |
| Amevive (alefacept) | Leukocyte function-associated antigen-3/IgG fusion protein | Fusion protein | Psoriasis | Biogen Idec | 2003 | CHO |
| Xolair® (omalizumab) | IgE | Humanized | Asthma | Genentech/Novartis/Tanox | 2003 | CHO |
| Bexxar® (tositumomab) | CD20 used as conjugate with iodine I 131 | Murine | Non-Hodgkin lymphoma | GSK/Corixa | 2003 | |
| Raptiva® (efalizumab) | CD11a | Humanized | Psoriasis | Genentech/Xoma | 2003 | CHO |
| Erbitux® (cetuximab) | Epidermal growth factor receptor | Chimeric | Colorectal cancer | ImClone/Bristol-Myers Squibb | 2004 | Mouse myeloma |
| Avastin® (bevacizumab) | Vascular endothelial growth factor | Humanized | Colorectal cancer | Genentech | 2004 | CHO |

*(2–5)* and predictions of 50–60 new products coming to the market in the next 5–7 yr *(4)*. A key factor driving future developments will be the genomic revolution, which is an important prelude to a better understanding of biological systems, leading to a dramatic increase in targets for drug development. MAbs represent the largest and fastest growing category of biopharmaceutical proteins, with a wide range of applications, particularly in cancer, immune disorders, and infectious disease.

The increase in numbers of products is reflected in the dramatic rise in biopharmaceutical sales. It is estimated that the annual global market grew from US $12 billion in 2000 to $30 billion in 2003 *(2)*. This success includes several blockbuster products with sales in excess of US $1 billion per year. The market is predicted to grow to US $45–50 billion by 2006 *(4)*. Biopharmaceuticals represent an increasingly important part of the overall pharmaceutical market, and continued strong growth is anticipated in the next few years *(6)*. Recently, biopharmaceuticals have represented 10–25% of all new pharmaceuticals in the United States *(4)*. Sales of MAbs have grown particularly strongly from 1% of the biopharmaceutical market in 1995 to 14% in 2001 *(4)* and to 22% by 2002 *(7)*. Sales of MAbs are projected to grow at a compound annual growth rate (CAGR) of 17.2% by 2010 and increase from $3.6 billion in 2002 to $7.9 billion in 2006 and to $12.1 billion by 2010 *(6)*.

## 2. Technologies to Make Modified Proteins

A number of licensed proteins have been engineered. It is now possible to produce not only the recombinant version of natural proteins, but also proteins that have been engineered with improved characteristics. Various approaches have been used to modify the therapeutic activity of proteins, improve their stability, or reduce the rate of clearance, including amino acid substitutions, domain removal, fusion of peptide sequences from different proteins, and glycosylation engineering. In addition to chemical modification, approx 50% of all licensed products have a modified gene sequence *(8)*.

There are several methods to modify the primary protein sequence, such as site-directed mutagenesis to change specific amino acids. Regarding insulin, analogs with altered amino acid sequences were developed that affect the speed of hormone action. The natural form of insulin tends to associate and, when formulated with zinc, exists as hexamers that slowly dissociate to give the active monomer. Insulin lispro and insulin aspart are fast-acting analogs in which specific amino acids at the C terminal were changed to reduce the self-association properties of insulin. In contrast, two products contain insulin glargine—a long-acting variant that dissociates slowly from a microcrystalline precipitate, resulting in a long-acting effect. These analogs enable better control of insulin availability based on clinical circumstances.

Another approach is to use gene shuffling to generate a large degree of sequence diversity from which to select for desirable characteristics *(9)*. Multiple rounds of "breeding" can be carried out to evolve desirable traits. Chang et al. *(10)* describe the application of gene shuffling to human interferon α, creating variants by shuffling the genes of five different parental interferons. Variants were created that were dramatically more active than the parent molecules.

The plasminogen activator Reteplase is an example of the use of domain deletion. The tissue plasminogen activator (tPA) activates plasminogen to plasmin, which dissolves fibrin blood clots. In Reteplase, the domains responsible for rapid elimination of the enzyme in vivo have been removed *(8)*. This results in an extended serum half-life, making administration more straightforward.

Another approach to modify protein characteristics has been to fuse sequences from different proteins to create a new protein with novel characteristics. One such product is Enbrel® (etanercept), a fusion protein used to treat rheumatoid arthritis. It comprises the ligand-binding domain of the tumor necrosis factor receptor coupled to the Fc portion of immunoglobulin (IgG). Amevive® (alefacept) is another example of an approved fusion protein. It is used to treat psoriasis and consists of a CD2-binding portion of the human leukocyte function-associated antigen 3 coupled to the Fc region of IgG1.

## 2.1. Antibody Humanization

Protein engineering techniques have been applied extensively to MAbs. First-generation MAbs used rodent hybridoma technology, which led to the first licensed MAb—Orthoclone OKT®-3 (murumonab-CD3)—to prevent the rejection of transplants. However, for most applications, rodent antibodies were too immunogenic, producing a human anti-mouse antibody response. This is particularly problematic if multiple dosing is required. The problem was partially resolved by using chimeric antibodies, which combined the variable region of the mouse antibody with a human constant region sequence. Further improvements came with the development of humanization techniques that made it possible to significantly reduce the rodent component of the antibody. Using transgenic mice or phage display technology *(11,12)*, it is now possible to make fully human antibodies, and the first such antibody, Humira® (adalimumab), has now been approved.

## 2.2. Engineering Process Characteristics

Protein engineering can improve the processing characteristics of a given product. An early example of this is provided by human interferon β *(13)*. Difficulties were found in expressing active interferon β in *E. coli* owing to incorrect disulfide bond formation. This problem was resolved by substituting serine for one of the cysteine residues.

## 2.3. Altered Glycosylation

Glycosylation can have a profound effect on various aspects of a protein's function, including activity, rate of clearance, and stability. There are an increasing number of examples where the glycosylation of proteins has been modified to alter the therapeutic properties of the product. Examples of licensed products are darbepoetin α (Aranesp®), a variant of erythropoietin, and the enzyme Cerezyme® (imiglucerase). Three approaches are currently used to engineer glycosylation: alteration of glycosylation sites on proteins, enzymatic modification of glycans in vitro, and manipulation of cellular glycosylation machinery.

The presence and number of terminal sialic acids in glycan structures can have a dramatic effect on the properties of glycoproteins. Elliot et al. *(14)* made analogs of rHuEPO, leptin, and MpI ligand in which they introduced new N-linked glycosylation sites that resulted in additional glycan structures. They observed greatly improved activity in vivo of all three proteins and prolonged duration of activity. The principle has been validated in the clinic for erythropoietin (EPO; darbepoietin $\alpha$). Interestingly, the approach also worked on leptin, a protein that is not naturally glycosylated.

It is also possible to consider glycan modification in vitro. The use of sialyltransferase to increase levels of terminal sialic acid on the glycans of glycoproteins and consequently increase in vivo half-life has been described *(15)*. In the case of Cerezyme®, an enzyme used to treat Gaucher's syndrome, enzyme glycosylation is modified during downstream processing. This modification facilitated the incorporation by macrophage where the enzyme deficiency occurs *(16)*. Several groups have now also described the metabolic engineering of host cell lines to alter cellular glycosylation capabilities, e.g., by overexpression of sialyl transferase or by the use of glycosylation mutants.

With antibodies, glycosylation contributes significantly to effector functions, and it is becoming apparent that subtle changes in glycan structure can profoundly affect these functions *(17)*. Several recombinant antibodies act by blocking receptors or binding soluble ligands and therefore may not require effector functions; indeed, they may be a disadvantage in some situations *(18)*. Effector functions and, particularly, antibody-dependent cellular cytotoxicity (ADCC), are thought to be very important for antibodies, such as anti-HER2 and anti-CD20, where they contribute to antitumor activity *(19,20)*. This has led to several studies conducted to improve the effector functions of antibodies by glycosylation modification with the intent to increase potency and clinical response and also to reduce treatment costs *(19)*. Umana et al. *(21)* genetically modified the glycosylation pathways of Chinese hamster ovary (CHO) cells to increase the content of bisecting GlcNAc, which improved the ADCC of antibodies produced in these cells. Shinkawa et al. *(19)* found that the absence of core fucose in antibody glycans produced by rat hybridoma YB2/0 cells significantly improved ADCC. Using a CHO variant (Lec 13), Shields et al. *(20)* also showed that the lack of fucose on human IgG1 N-linked oligosaccharide improved binding to human Fc$\gamma$RIII and increased ADCC.

## 2.4. Pegylation

Pegylation has been used to modify protein properties and, particularly, to prevent proteolytic degradation and decrease the rate of in vivo clearance *(22)*. Pegylated products that have been approved include versions of interferon [Pegasys® (peginterferon alfa-2a) and Pegintron® (peginterferon alfa-2b)], human granulocyte colony-stimulating factor (G-CSF) [Neulasta® (pegfilgrastim)] and an analog of human growth hormone [Somavert® (pegvisomant)], which acts as a hormone receptor antagonist and is used to treat acromegaly (**Table 1**). The pegylation of interferon $\alpha$ 2a has been reported to increase in vivo half-life from 9 to 77 hr, and renal clearance was reduced 100-fold when compared with the nonpegylated interferon *(22)*. Several more pegylated proteins are in development, such as pegylated fragments of antibodies *(23,24)*. Antibody fragments have a substantially increased clearance rate when compared with whole antibodies. Pegylation can restore the clearance rate to values similar to that for the whole antibody *(23,24)*. Other examples of chemical modification are the coupling of isotopes or cytotoxic drugs to MAbs, particularly in cancer applications. Two such products are Bexxar® (tositumomab), which contains $I^{131}$ coupled to a CD20 antibody, and Mylotarg® (gemtuzumab ozogamicin), a MAb linked to calicheamicin.

## 3. Production Technologies

In 2000, Walsh *(1)* reported that 34 approved products were made in *E. coli*, 14 in CHO cells, 2 in baby hamster kidney (BHK) cells, and 11 in *Saccharomyces cerevisiae*. The proportion of products made in mammalian cells has increased over time partly because of the recent increase in approved antibodies. The production hosts for licensed products are shown in **Table 1**. Polastro *(4)* estimates that mammalian cell products account for 60% of the total market. Although microbial production systems are less expensive, it has not been possible to produce all proteins in such systems. Mammalian cell culture has been the dominant choice for large complex proteins (e.g., antibodies) and for proteins where mammalian posttranslational steps (e.g., glycosylation) are required.

### 3.1. Mammalian Cell Processes

Mammalian cell culture production processes utilized to generate approval of products have been reviewed elsewhere *(25)*. CHO cells are the most commonly used for recombinant proteins. In the case of antibodies, NS0 and Sp2/0 are used along with CHO *(26,27)*. Other cell lines are in development, including a human retinal cell line *(27)*. For small-scale processes, a variety of production processes are used, such as roller bottles, hollow fiber devices, and suspension culture. In recent years, there has been increasing interest in the use of simple disposable bag systems for small-scale culture *(29)*. For large-scale (multikilogram) manufacturing, suspension culture is the method of choice *(25,30)*, and suspension reactors up to 20,000 L have been described. These systems are most commonly operated in batch or fed-batch mode, but there are several examples of perfusion processes.

In the last few years, many proteins have been developed that are used at relatively high dosage, sometimes 1000 times higher than some of the earlier biopharmaceuticals. This is especially true of MAbs where, in some cases, the demand is for multiple hundreds of kilograms per year, leading to a rapid expansion in mammalian cell capacity for pharmaceutical companies and contract manufacturers. By 2006, approx 2 million liters of mammalian cell capacity is expected worldwide—more than doubling that available in 2002.

There are several published estimates of the manufactured quantities of individual proteins *(7,31)*. Of the 957 kg of proteins manufactured in 2002, products such as cytokines, hormones, and enzymes only constituted 40 kg of the total *(7)*; the remainder consisted of MAbs and fusion proteins. They predict that the requirement for protein-based drugs will grow from 1318 kg/yr in 2003 to 6122 kg/yr in 2006. The development of products required in larger volumes is also placing increased emphasis on improving process efficiencies. The manufacturing costs of these complex biopharmaceuticals are high. One estimate *(31)* suggests that traditional chemicals cost less than $5/g, whereas proteins produced in cells may cost $100–$1000/g and have a selling price on average of $9000/g.

Mammalian cell processes are complex, and facilities are very expensive—typically $5–10 million/m³ compared with $0.5–1.0 for chemical reactors *(4)*, i.e., $200–500 million for a facility. As might be expected, the goal is to improve existing technologies and increase throughput from plants, but attention has also been directed toward evaluating alternative production technologies, e.g., transgenic plants and animals. Also, the improved design of products to increase potency or reduce the rate of clearance could lower costs by decreasing the volumes required.

### 3.2. Advances in Mammalian Cell Technology

In addition to benefiting from economies of increased reactor scale, particular attention has been focused on improving productivity of the upstream process of mammalian cell technologies and dramatic progress has been made in recent years. Although there are still widespread yields in the hundreds of mg/L range (*see* **ref.** *32* for results of an industry survey), there appears to be an increasing number of antibody processes yielding in excess of 1 g/L.

Improvements in productivity have derived from two sources: the use of highly productive cell lines and fermentation process improvements. Several efficient expression technologies have been described that, in combination with suitable screening procedures, result in highly productive cell lines. Indeed, in the case of MAbs, recombinant cell lines are often significantly more productive than hybridomas. No clear evidence suggests that one cell type is inherently more productive than another. Published production rates for MAbs are typically in the range of 20–60 pg/cell/d *(33)*, and the upper limit of the production rate for Igs is indicated to be around 60–100 pg/cell/d or 10–20% of total protein synthesis *(34)*. In highly expressing cell lines, product synthesis is probably not limited by transcription *(34)*, and considerable interest exists in identifying posttranscriptional bottlenecks, which might then be usefully manipulated.

Manipulating other aspects of cell metabolism is also focused onto improving the performance of cells in fermentation processes *(33)*. Birch et al. *(35)* described the transfection of the glutamine synthetase gene into a hybridoma to achieve glutamine independence. In this case, productivity was increased under glutamine-free conditions. Several groups have used metabolic engineering approaches to improve the efficiency of energy substrate utilization in cells. Irani et al. *(36)* described the improvement of the primary metabolism of cell cultures by introducing a new cytoplasmic pyruvate carboxylase. A poor link was found between glycolysis and the tricarboxylic acid (TCA) cycle in BHK cells, leading to glucose metabolism via oxidative glycolysis. The introduction of genes for pyruvate carboxylase led to a fourfold reduction in glucose consumption and a twofold reduction in glutamine consumption.

There was an enhanced flux of glucose into the TCA cycle. Chen et al. *(37)* manipulated the pathway of lactate accumulation in a hybridoma by partially disrupting the lactate dehydrogenase A *(LDH-A)* gene through homologous recombination. Lactate accumulation was reduced by 50%, and antibody productivity increased threefold.

Another aspect of cell physiology that has received attention is apoptosis. When cell death occurs in culture, it is usually via apoptotic pathways, and there is increasing knowledge of the molecular basis of apoptosis. This has led to a considerable effort to inhibit these pathways, consequently prolonging the duration of cell cultures *(38,39)*. The two approaches taken have been to either find culture conditions that prevent the onset of apoptosis or to introduce antiapoptotic genes (e.g., *Bcl*2) into the cell. The latter approach has given mixed results; although prolonged survival may be achieved, this does not necessarily translate into increased productivity, which may not be surprising given the complexity of apoptosis regulation. In one example *(40)*, transfection of the antiapoptosis gene *Bcl*2 into an antibody-producing cell line increased viable cell concentration by twofold and delayed the onset of apoptosis by 50 h but gave no increase in yield.

In some circumstances, it has also been possible to select natural variants of cell lines that have improved characteristics, e.g., with respect to altered nutrient requirements. Rasmussen et al. *(41)* review the isolation and application of Veggie–CHO—a CHO host cell line that grows in serum-free medium. Birch et al. *(35)* describe the isolation of a cholesterol-independent clone of NS0. Cholesterol is a problematic nutrient, especially in chemically defined media because of its insolubility. Therefore, removing its requirement simplifies the process.

In addition to developing highly productive cell lines, there has been significant success in the improvement of fermentation processes. This has typically resulted from the optimization of fermenter operating conditions, media, and feeding strategies. Physicochemical conditions, such as the pH, can have profound effects on growth and productivity. Wayte et al. *(42)* found that the pH value was a critical parameter for several cell lines. Depending on the specific cell line, effects were seen on different aspects of growth kinetics and on the specific production rate. The design of media and feeding strategies also has a strong influence on growth and productivity. In practice, several aspects of cell growth and productivity can be manipulated to increase throughput in a reactor, including specific growth rate, maximum cell concentration, culture duration, and specific production rate. It is now quite common to achieve cell concentrations in excess of $10^7$ cells/mL in fed-batch culture, an order of magnitude higher than was achieved a decade ago.

**Table 2** shows the result of several rounds of optimization for a CHO cell process producing the chimeric antibody cB72.3. This cell line was created using the glutamine synthetase (GS) gene expression system *(43)*. At the start of the optimization program, cells were adapted to grow in a chemically defined medium. There has been a general move away from media containing materials of animal origin, principally to remove the risk of introducing adventitious agents, such as prions, into the process *(44)*. This approach also simplifies purification, reducing the number of contaminating proteins to be removed, and chemical definition assists process optimization.

**Table 2**
**Effect of Cell Line and Process Improvements**
**for the Production of cB72.3 MAb in GS-CHO Cells**

| Process | Antibody (mg/L) | Fold increase |
|---|---|---|
| Original cell line | 139 | |
| Improvements to process | 585 | 4 |
| Improved cell line | 1917 | 14 |
| Further improvements to process | 4301 | 31 |

After two optimization rounds, the product concentration had been increased four-fold. A new cell line was created at this point using a variant of CHO K1, which grows spontaneously in suspension in chemically defined medium. Improved transfection and screening techniques allowed the isolation of a cell line that was threefold more productive than the original cell line, mainly owing to extended culture duration using the new cell line. Further improvement of culture conditions has increased the titer to 4.3 g/L—an improvement of 30-fold over the original process.

### 3.3. Alternative Production Technologies

Several alternative production technologies (particularly transgenic plants and animals) have been considered for the production of large-volume biopharmaceuticals, such as MAbs (45). These alternatives have the potential to substantially reduce upstream manufacturing costs. However, cell culture may only account for 30–50% of costs, whereas the remainder is attributable to product recovery and quality assurance (27). Although several products made in these systems are under clinical trial, including at least one product at a late stage (antithrombin III in goats), there are currently no approved products, and their impact in the future remains to be seen. (For reviews of the technologies and products in development, see refs. 46 and 47.) Mison and Curling (48) have given a detailed cost analysis for manufacturing proteins in plants. Chadd et al. (49) discuss the economic potential of a range of production technologies, including mammalian cell culture and transgenic plants and animals.

### 3.4. Microbial Production

A significant number of approved therapeutic protein products are already produced in the prokaryote, *E. coli*, and the eukaryote, *S. cerevisiae*. Other organisms are also being developed as production hosts, including *Pichia pastoris* and filamentous fungi. Microbial cultures are much more productive in terms of throughput per liter of reactor volume and are likely to become increasingly appealing options for protein production in the future. This is apparent in the case of the largest category of high-volume mammalian cell products, the MAbs. In some cases, particularly where the antibody is being used to block an activity or is used as a targeting agent for a drug, it may be possible to use fragments of antibodies, which can be readily expressed in microorganisms. The production, chemical modification, and uses of therapeutic antibody fragments made in *E. coli* have been described (23,24). These fragments include

Fabs and scFvs with sizes of 57 and 27 kDa compared to a whole IgG with a size of 150 kDa. It is now possible to produce even smaller "domain" antibodies that have a size of only 11–15 kDa *(50)*. Along with *E. coli*, other organisms, e.g., *P. pastoris*, may be used to produce antibody fragments *(51)*.

Notably, significant progress is being made to improve the capabilities of microorganisms to secrete large complex proteins into the periplasmic space of *E. coli* or into the culture broth with organisms, such as yeasts and filamentous fungi. The production of whole antibodies in the periplasm of *E. coli* has been reported at levels of 130–150 mg/L, and studies of one antibody indicated that the product retained the long half-life associated with antibody made in mammalian cells *(18)*. In such a system, this production could be appealing from a cost perspective, and it avoids the need for pegylation to achieve adequate half-life. Antibodies made in *E. coli* are not glycosylated, a characteristic that will impair effector functions. As mentioned above, the requirement for glycosylation and functional effector functions depends on the particular therapeutic application. The production of full-length antibody and antibody Fab fragments in the filamentous fungus, *Aspergillus niger*, has been described *(52)*. This organism is used at a very large scale for the production of industrial enzymes. However, the full-length antibody had fungal- rather than human-type glycosylation. One reason to use mammalian cell culture is because microbes do not glycosylate proteins or synthesize glycans that are dissimilar to those found in human proteins. Yet, even this hurdle is gradually being overcome. Choi et al. *(53)* describe the engineering of glycosylation pathways in yeast to enable synthesis of glycans that mimic those seen in mammalian glycoproteins. Even the general perception that glycosylation is limited to eukaryotes is proving to be false. Glycosylation pathways have been found in some bacteria and transferred to *E. coli* *(54)*. Although the glycan structures produced are not the same as in eukaryotes, it may be eventually possible to engineer relevant pathways in bacteria.

### 3.5. Chemical Synthesis

In the future, chemical synthesis may become sufficiently efficient to permit the manufacture of larger peptides and proteins. Already the 36 amino acid peptide Fuzeon (enfuvirtide), which is a virus fusion inhibitor used to treat HIV, is produced by synthesis on an industrial scale.

## 4. Conclusion

Modern molecular biology, along with progress in manufacturing technology, has made it possible to manufacture an impressive array of natural and modified proteins. It is interesting to see how the trend toward development of large-volume complex proteins, such as antibodies, is improving technology, particularly in mammalian cell culture. This, in turn, provides a stimulus for new approaches like the use of modern genomic and proteomic techniques to inform process design. Clearly, there is still significant potential to improve both upstream and downstream aspects of current technologies, while advances in other areas, including microbial systems, may provide alternative cost-effective approaches.

## References

1. Walsh, G. (2000) Biopharmaceutical benchmarks. *Nature Biotechnol.* **18**, 831–832.
2. Walsh, G. (2003) Biopharmaceutical benchmarks—2003. *Nature Biotechnol.* **21**, 865–870.
3. Andersson, R. and Mynahan, R. The protein production challenge. *In Vivo*, May 2001, 1–5.
4. Polastro, E. and Tulcinski, S. Boom time for biopharma. *Scrip*, September 2002, 45–49.
5. Gura, T. (2002) Magic bullets hit the target. *Nature* **417**, 584–586.
6. Datamonitor Report. (2002) Therapeutic Proteins.
7. UBS Report. (2003) The State of Biomanufacturing.
8. Kresse, G.-B. (2004) Custom made genetically engineered drugs. *BIOforum Europe* **1**, 38–40.
9. Stemmer, W. P. (1994) Rapid evolution of a protein in vitro by DNA shuffling. *Nature* **370**, 389–391.
10. Chang, C.-C. J., Chen, T. T., Cox, B. W., Dawes, G. N., Stemmer, P. C., Punnonen, J., and Patten, P. A. (1999) Evolution of a cytokine using DNA family shuffling. *Nat. Biotechnol.* **17**, 793–797.
11. Griffiths, D. and Duncan, A. R. (1998) Strategies for selection of antibodies by phage display. *Curr. Opin. Biotechnol.* **9**, 102–108.
12. Vaughan, T. J., Osbourn, J. K., and Tempest, P. R. (1998) Human antibodies by design. *Nat. Biotechnol.* **16**, 535–539.
13. Khosrovi, B. The production, characterization and testing of a modified recombinant human interferon beta, in *Interferon: Research, Clinical Application and Regulatory Consideration* (Zoon, K. C., Noguchi, P. D., and Liu, T.-Y., eds.) Elsevier, New York, 1984, pp. 89–99.
14. Elliott, S, Lorenzini, T., Asher, S., Aoki, K., Brankow, D., Buck, L., et al. (2003) Enhancement of therapeutic protein in vivo activities through glycoengineering. *Nat. Biotechnol.* **21**, 414–420.
15. Morrow, K. J. (2002) Glycosylation in scale-up of antibody production. *Genetic Engineering News* **22**, no. 12, p. 1.
16. Walsh, G. *Biopharmaceuticals, Biochemistry and Biotechnology*, 2nd ed. Wiley, New York, 2003.
17. Jefferis, R. Glycosylation of human IgG antibodies, relevance to therapeutic applications. *BioPharm*, September 2001, pp. 19–27.
18. Simmons, L. C., Reilly, D., Klimowski, L., Raju, T. S., Meng, G., Sims, P., et al. (2002) Expression of full-length immunoglobulins in *Escherichia coli*: rapid and efficient production of aglycosylated antibodies. *J. Immunol. Methods* **263**, 133–147.
19. Shinkawa, T., Nakamura, K., Yamane, N., Shoji-Hosaka, E., Kanda, Y., Sakurada, M., et al. (2003) The absence of fucose but not the presence of galactose or bisecting N-Acetyl-gucosamine of human IgG1 complex-type oligosaccharides shows the critical role of enhancing antibody-dependent cellular cytotoxicity. *J. Biol. Chem.* **278**, 3466–3473.
20. Shields, R. L., Lai, J., Keck, R., O'Connell, L. Y., Hong, K., Meng, Y. G., et al. (2002) Lack of fucose on human IgG1 N-linked oligosaccharide improves binding to human FcgammaRIII and antibody-dependent cellular toxicity. *J. Biol. Chem.* **277**, 26733–26740.
21. Umana, P., Jean-Mairet, J., Moudry, R., Amstutz, H., and Bailey, J. E. (1999) Engineered glycoforms of an antineuroblastoma IgG1 with optimized antibody-dependent cellular cytotoxic activity. *Nature Biotechnol.* **17**, 176–180.
22. Potera, C. (2003) Pegylation for improving polypeptide drugs. *Genetic Engineering News* **23**, no. 6, p. 58.

23. Chapman, A. P., Antoniw, P., Spitali, M., West, S., Stephens, S., and King, D. J. (1999) Therapeutic antibody fragments with prolonged in vivo half-lives. *Nature Biotechnol.* **17,** 780–783.
24. Humphreys, D. P. (2003) Production of antibodies and antibody fragments in *Escherichia coli* and a comparison of their functions, uses and modification. *Curr. Opin. Drug Discov. Devel.* **6,** 188–196.
25. Chu, L. and Robinson, D. K. (2001) Industrial choices for protein production by large-scale cell culture. *Curr. Opin. Biotechnol.* **12,** 180–187.
26. Birch, J. R. Cell products—antibodies, in *Encyclopedia of Cell Technology* (Spier, R., ed.) Wiley, New York, 2000, pp. 411–424.
27. Morrow, K. J. (2003) Cell lines for recombinant protein production. *Genetic Engineering News* **23,** p. 50.
28. Jones, D. H., van Berkel, P. H. C., Logtenberg, T., and Bout, A. (2002) PER.C6 Cell Line for Human Antibody Production. *Genetic Engineering News* **22,** no. 10, p. 50.
29. Singh, V. (1999) Disposable bioreactor for cell culture using wave-induced agitation *Cytotechnology* **30,** 149–158.
30. Birch, J. R. Suspension culture, animal cells, in *Encyclopedia of Bioprocess Technology: Fermentation, Biocatalysis and Bioseparation* (Flickinger, M. C. and Drew, S. W., eds.) Wiley, New York, 1999, pp. 2509–2516.
31. J. P. Morgan Report (2002) The State of Biologics Manufacturing: Part 2, New York.
32. Fox, S. (2002) Biopharmaceutical contract manufacturing. *Genetic Engineering News* **22,** no. 17, p. 1.
33. Andersen, D. C. and Krummen, L. (2002) Recombinant protein expression for therapeutic applications. *Curr. Opin. Biotechnol.* **13,** 117–123.
34. Reff, M. E. (1993) High level production of recombinant immunoglobulins in mammalian cells. *Curr. Opin. Biotechnol.* **4,** 573–576.
35. Birch, J. R., Boraston, R. C., Metcalfe, H., Brown, M. E., Bebbington, C. R., and Field, R. P. (1994) Selecting and designing cell lines for improved physiological characteristics. *Cytotechnology* **15,** 11–16.
36. Irani, N., Wirth, M., van den Heuvel, J., and Wagner, R. (1999) Improvement of the primary metabolism of cell cultures by introducing a new cytoplasmic pyruvate carboxylase reaction. *Biotechnol. Bioeng.* **66,** 239–246.
37. Chen, K., Liu, Q., Xie, L., Sharp, P. A., and Wang, D. I. C. (2001) Engineering of a mammalian cell line for reduction of lactate formation and high monoclonal antibody production. *Biotechnol. Bioeng.* **72,** 55–61.
38. Dickson, A. J. (1998) Apoptosis regulation and its application to biotechnology. *Tibtech* **16,** 339–342.
39. Arden, N. and Betenbaugh, M. J. (2004) Life and death in mammalian cell culture: strategies for apoptosis inhibition. *Trends Biotechnol.* **22,** 174–180.
40. Dutton, G. (2003) Reducing apoptosis: bioprocess bane or boon. *Genetic Engineering News* **23,** no. 4, p. 32.
41. Rasmussen, B., Davis, R., Thomas, J., and Reddy, P. (1998) Isolation, characterization and recombinant protein expression in Veggie—CHO: A serum-free host cell line. *Cytotechnology* **28,** 31–42.
42. Wayte, J., Boraston, R., Bland, H., Varley, J., and Brown, M. (1997) pH: Effects on growth and productivity of cell lines producing monoclonal antibodies: control in large-scale fermenters. *The Genetic Engineer and Biotechnologist* **17,** 125–132.

43. Bebbington, C. R., Renner, G., Thomson, S., King, D., Abrams, D., and Yarranton, G. T. (1992) High level expression of a recombinant antibody from myeloma cells using a glutamine synthetase gene as an amplifiable selectable marker. *Biotechnology* **10,** 169–175.
44. Lubiniecki, A. (1998) Historical reflections on cell culture engineering. *Cytotechnology* **28,** 139–145.
45. Houdebine, L.-M. (2002) Antibody manufacture in transgenic animals and comparisons with other systems. *Curr. Opin. Biotechnol.* **13,** 625–629.
46. Ma, J. K., Drake, P. M. W., and Christou, P. (2003) The production of recombinant pharmaceutical proteins in plants. *Nat. Rev. Genet.* **4,** 794–805.
47. Moskowitz, D. B. Down on the pharm. *Scrip*, March 2003, pp. 36–37.
48. Mison, D. and Curling, J. (2000) The industrial production costs of recombinant therapeutic proteins expressed in transgenic corn. *BioPharm*, May 2000, pp. 48–54.
49. Chadd, H. E. and Chamow, S. M. (2001) Therapeutic antibody expression technology. *Curr. Opin. Biotechnol.* **12,** 188–194.
50. Holt, L. J., Herring, C., Jespers, L. S., Woolven, B. P., and Tomlinson, I. M. (2003) Domain antibodies: proteins for therapy. *Trends Biotechnol.* **21,** 484–490.
51. Gurkan, C., Symeonides, S. N., and Ellar, D. J. (2004) High level production in *Pichia pastoris* of an anti-p185 HER-2 single chain antibody fragment using an alternative secretion expression vector. *Biotechnol. Appl. Biochem.* **39,** 115–122.
52. Ward, M. (2002) Expression of antibodies in *Aspergillus niger. Genetic Engineering News* **22,** no. 21, p. 48.
53. Choi, B.-K., Bobrowicz, P., Davidson, R. C., Hamilton, S. R., Kung, D. H., Li, H., et al. (2003) Use of combinatorial genetic libraries to humanize N-linked glycosylation in the yeast *Pichia pastoris. Proc. Natl. Acad. Sci. USA* **100,** 5022–5027.
54. Wacker, M., Linton, D., Hitchen, P. G., Nita-Lazar, M., Haslam, S. M., North, S. J., et al. (2002) N-linked glycosylation in *Campylobacter jejuni* and its functional transfer into *E. coli. Science* **298,** 1790–1793.

# 2

## Expression of Antibody Fragments by Periplasmic Secretion in *Escherichia coli*

**Andrew G. Popplewell, Mukesh Sehdev, Mariangela Spitali, and A. Neil C. Weir**

### 1. Introduction

Antibody-based drugs are increasingly used in the clinic, and their importance is set to escalate in the coming years as more drugs in this class progress through clinical trials. Although many such drugs utilize whole antibodies, others exploit fragments, e.g., fragment antigen binding (Fab') or single-chain fragment variable (scFv), which retain the antigen-binding specificity without the fragment crystallizable (Fc) element necessary to mediate effector functions. Antibody fragments can be advantageous for many therapeutic uses, owing to the fact that valency and half-life can be tailored through protein engineering approaches to suit the desired mechanism of action *(1)*. Furthermore, antibody fragments are more suited to expression in microbial systems, providing benefits in terms of increased scale and ease of manufacture *(2)*.

*Escherichia coli* is currently the host of choice for producing antibody fragments. For disulfide-bonded proteins (Fab' and scFv), good expression levels have been achieved via soluble production, secreting the Fab' chains into the oxidizing environment of the bacterial periplasm where assembly and disulfide bond formation can occur. Periplasmic secretion is achieved by genetically fusing the signal sequence from an *E. coli* protein onto the N-terminus of the antibody V-region sequence. Numerous systems have been developed for the overexpression of recombinant proteins and can be adapted for antibody fragment production. The method described in this chapter relies on the lactose promoter–repressor–operator system. Other systems may work equally well, and the methods given can be adapted accordingly. While moderate amounts of most, if not all, Fabs can be expressed with these methods, high-level expression may require engineering of the antibody fragment to maximize expressibility in *E. coli* (*see* **Note 1**). In the following protocols, the expression of a humanized Fab' fragment is used to exemplify the methodology. A description of small-scale production by isopropyl-β-D-thiogalactopyranoside (IPTG) induction in shake flasks

From: *Methods in Molecular Biology, vol. 308: Therapeutic Proteins: Methods and Protocols*
Edited by: C. M. Smales and D. C. James © Humana Press Inc., Totowa, NJ

is followed by a protocol for fermentation using a switch of carbon source from glycerol to lactose as the means of induction. Methods for periplasmic extraction and purification are also given.

## 2. Materials

1. Plasmids: pTTO-1 and pDNAbEng-1 (Celltech R&D).
2. *E. coli* strains: INVαF' (InVitrogen) and W3110 (ATCC).
3. Oligonucleotide primers.
4. Restriction enzymes, T4 DNA ligase and *Taq* DNA polymerase.
5. Agarose gel apparatus and DNA sequencing apparatus.
6. L-broth: 10 g/L bactotryptone, 10 g/L NaCl, 5 g/L yeast extract. Sterilize by autoclaving.
7. Transformation solution: 10 g/L bactotryptone, 10 g/L NaCl, 5 g/L yeast extract, 10 g/L MgCl$_2$, 100 g/L polyethylene glycol (PEG)4000, and 50 mL/L dimethylsulfoxide (DMSO), pH 6.5. Filter-sterilize.
8. SOC medium: 2% bactotryptone, 0.5% yeast extract, 10 m$M$ NaCl, 2.5 m$M$ KCl, 10 m$M$ MgCl$_2$, 10 m$M$ MgSO$_4$, and 20 m$M$ glucose (added from 20% stock solution after autoclaving other components).
9. Tetracycline: 7.5 mg/mL stock solution in 50% ethanol.
10. Glycerol solutions: 80% (w/w), 52.5% (w/w). Sterilize by autoclaving.
11. IPTG: 200 m$M$ stock solution, make fresh. Filter-sterilize.
12. Periplasmic extraction buffer (2X stock): 200 m$M$ Tris-HCl and 20 m$M$ EDTA, pH 7.4.
13. Coating buffer: 1.59 g/L Na$_2$CO$_3$, 2.93 g/L NaHCO$_3$, and 0.2 g/L NaN$_3$, pH 9.6.
14. Capture and reveal antibodies.
15. Phosphate-buffered saline (PBS).
16. PBS-Tween (PBST): PBS plus 0.05% (v/v) Tween-20.
17. Sample conjugate buffer: 0.1 $M$ Tris-HCl, 0.1 $M$ NaCl, 0.02% (v/v) Tween-20, and 0.2% (w/v) casein, pH 7.0.
18. Substrate solution: 10 mL sodium acetate/citrate solution (0.1 $M$, pH 6), 100 µL H$_2$O$_2$ solution (0.44% [v/v]), and 100 µL tetramethyl benzidine solution (10 mg/mL in DMSO).
19. 2.5 L Braun BiostatB batch fermenter (or equivalent) and associated pH and dissolved oxygen probes.
20. SM6 defined media: 5.2 g/L (NH$_4$)$_2$SO$_4$, 4.2 g/L NaH$_2$PO$_4$·2H$_2$O, 4.025 g/L KCl, 1.05 g/L MgSO$_4$·7H$_2$O, 5.2 g/L citric acid, 0.052 g/L CaCl$_2$·2H$_2$O, 0.0200 g/L ZnSO$_4$·7H$_2$O, 0.0275 g/L MnSO$_4$·4H$_2$O, 0.0075 g/L CuSO$_4$·5H$_2$O, 0.004 g/L CoSO$_4$·7H$_2$O, 0.1000 g/L FeCl$_3$·6H$_2$O, 0.0003 g/L H$_3$BO$_3$, 0.0003 g/L Na$_2$MoO$_4$·2H$_2$O, and 31.1 g/L glycerol. Sterilize by autoclaving.
21. Struktol (antifoam agent) stock solution: 10% (v/v). Sterilize by autoclaving.
22. Lactose solution: 40% (w/w). Sterilize by autoclaving.
23. Sartobrand P capsule (Sartorious) fitted onto a peristaltic pump and tubing.
24. Nalgene stericups: 0.45 µm and 0.22 µm (Millipore).
25. Affinity chromatography equilibration buffer: PBS.
26. Affinity chromatography elution buffer: 0.1 $M$ glycine-HCl, pH 2.7.
27. Protein G Gammabind Plus Sepharose (Amersham Biosciences).

## 3. Methods

These methods outline the (1) construction of the expression plasmid; (2) small-scale expression of the protein in *E. coli*; (3) large-scale expression by fermentation; and (4) extraction and purification of the protein.

### 3.1. Construction of the Expression Plasmid

The construction of a plasmid for expression of a Fab' fragment by translocation to the *E. coli* periplasm is described. The same principle can be applied to construct plasmids to express other antibody fragments, including scFv. However, the construction is more complicated for those fragments made up of two component polypeptides, such as the Fab'. Such "double gene" expression can be achieved by the use of separate promoters for each gene, but perhaps the simplest solution is to use a dicistronic expression method as described here. This indicates that two genes are produced from one transcript, requiring only a single promoter.

#### 3.1.1. pTTO Expression Vector

Plasmid pTTO-1 (**Fig. 1**) is a derivative of plasmids pTTQ9 *(3)* and pACYC184 *(4)*. The expression unit consists of the strong *Tac* promoter *(5)* and dual *rrnB_{t1t2}* transcriptional terminator *(6)*. The plasmid contains the *lacIq* gene *(7)*, giving constitutive expression of the lac repressor protein necessary to keep the *tac* promoter repressed. Derepression (or induction) is mediated by the addition of lactose (which is converted to allolactose, the natural inducer of the *lac* operon in *E. coli*), or IPTG, a nonhydrolyzable synthetic inducer. The plasmid also contains the origin of replication from plasmid p15A *(8)*, conferring a low-copy number, and it carries the tetracycline resistance gene. As shown in **Fig. 1**, the plasmid contains DNA encoding a portion of the signal peptide from the *E. coli* OmpA protein *(9)*. Insertion of DNA encoding an antibody fragment and the remainder of the OmpA signal sequence creates a gene encoding a protein product that will be translocated to the *E. coli* periplasm.

#### 3.1.2. Insertion of Fab' Light-Chain Gene

The light chain of a Fab' can be inserted into the pTTO-1 polylinker so that an in-frame OmpA signal peptide is created on ligation. A polymerase chain reaction (PCR) strategy is required to "build" DNA encoding the C-terminal portion of the signal peptide onto the 5' end of the light-chain gene. A suggested oligonucleotide to act as a 5' forward PCR primer is shown in **Fig. 1**. Thus, the unique *Mfe*I site within the pTTO-1 polylinker sequence is employed. For the 3' reverse primer, no additional components are necessary, apart from a restriction site for cloning into the vector *Bsi*WI site. The cloning scheme is as follows (*see* **Note 2**):

1. Use PCR to amplify the light chain, adding a 5' *Mfe*I site and DNA encoding part of the OmpA signal peptide and adding a *Bsi*WI site to the 3' end of the light-chain gene.
2. Purify the amplified DNA, digest with restriction enzymes *Mfe*I and *Bsi*WI, and repurify.
3. Prepare vector pTTO-1 by restriction with *Mfe*I and *Bsi*WI, and purify the cleaved fragment.
4. Ligate the two fragments together, and transform into an appropriate host strain (*see* **Note 3**).

This creates the light-chain intermediate plasmid. This method can be adapted for creation of a plasmid for single-gene expression (e.g., scFv), in which case the expression construct is now complete. For Fab' expression, the heavy-chain gene must also be inserted.

**A**    Restriction Map of Plasmid

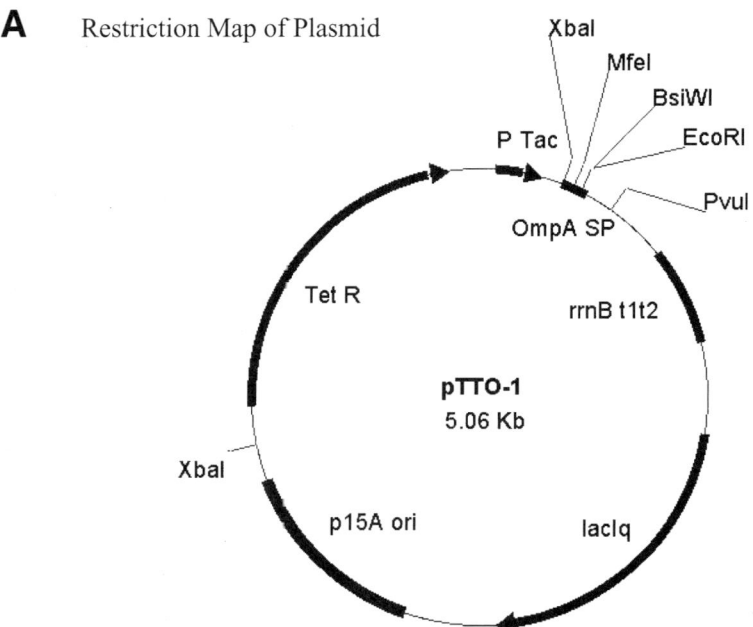

**B**    Polylinker sequence showing N-terminal portion of OmpA sequence

**C**    Proposed oligonucleotide sequence to add remainder of OmpA signal sequence to V-region sequence

```
           MfeI
NNNNNNGCAATTGCAGTGGCCTTGGCTGGTTTCGCTACCGTAGCGCAAGCTXXXXXXXXXXXXXXXXXXXXXXXXXXXX
      A  I  A  V  A  L  A  G  F  A  T  V  A  Q  A
```

Fig. 1. Expression plasmid pTTO-1. (**A**) Map of plasmid. P Tac, *tac* promoter and *lac* operator region; SP, signal peptide; rrnBt1t2, transcriptional terminator; lacIq, gene constitutively expressing lac repressor protein; p15A ori, origin of replication; Tet R, tetracycline resistance gene. (**B**) Nucleotide sequence of polylinker showing N-terminal portion of OmpA signal peptide amino acid sequence. Restriction enzyme recognition sequences are underlined. The ribosome binding site (RBS) is shown in bold. (**C**) Potential sequence of oligonucleotide to add remainder of OmpA signal peptide sequence to sequence of antibody fragment gene. N, any nucleotide; X, nucleotide identical to that in antibody fragment to be amplified.

### 3.1.3. Insertion of the Fab' Heavy-Chain Gene

To create a dicistronic message expressing both Fab' chains, the gene for the heavy chain also needs to be inserted 3' to the light-chain gene. Another signal sequence is required, in addition to a ribosome-binding site to ensure efficient translational initia-

**A** Restriction Map of Plasmid

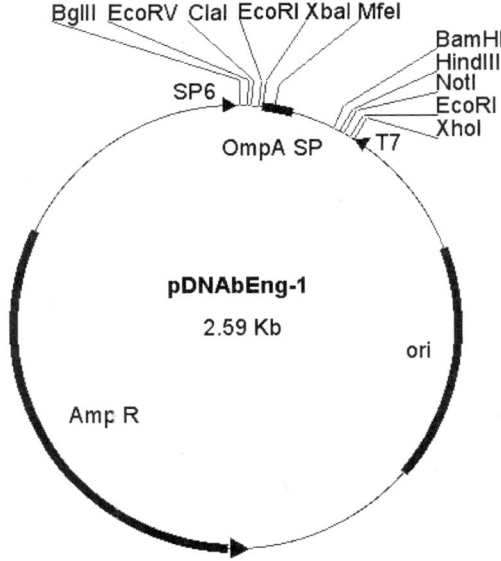

**B** Sequence of Polylinker 1

```
Bgl II  EcoRV  ClaI   EcoRI  XbaI        RBS                                              MfeI
AGATCTGATATCATCGATGAATTCTCTAGATAACGAGGCGTAAAAAATGAAAAAGACAGCTATCGCAATTGCAGTG
                                                  M   K   K   T   A   I   A   I   A   V
```

**C** Proposed oligonucleotide sequence to add remainder of OmpA signal sequence to V-region sequence

```
            MfeI
NNNNNNGCAATTGCAGTGGCCTTGGCTGGTTTCGCTACCGTAGCGCAAGCTXXXXXXXXXXXXXXXXXXXXXXXXXXX
          A   I   A   V   A   L   A   G   F   A   T   V   A   Q   A
```

**D** Sequence of Polylinker 2

```
   BamHI  HindIII   NotI   EcoRI  XhoI
GAGGATCCAAGCTTGCGGCCGCGAATTCCTCGAGGCCGGT
```

Fig. 2. Cloning intermediate plasmid pDNAbEng-1. (**A**) Map of plasmid. SP, signal peptide; ori, origin of replication; AmpR, ampicillin resistance gene; SP6, T7, promoter sites. (**B**) Nucleotide sequence of polylinker 1 showing N-terminal portion of OmpA signal peptide amino acid sequence. Restriction enzyme recognition sequences are underlined. The RBS is shown in bold. (**C**) Potential sequence of oligonucleotide to add remainder of OmpA signal peptide sequence to sequence of antibody fragment gene. N, any nucleotide; X, nucleotide identical to that in antibody fragment to be amplified. (**D**) Sequence of polylinker 2.

tion. This can be achieved by cloning the heavy chain through an intermediate vector. One such vector, termed pDNAbEng-1, is shown in **Fig. 2**. Again, this contains part of the OmpA signal peptide, requiring the use of a "forward" oligonucleotide to amplify the heavy chain and, at the same time, introduce the remaining sequences of the OmpA signal peptide. Once the heavy chain is cloned into this intermediate, it can be excised as an *Eco*RI fragment and inserted into the light-chain vector *Eco*RI site.

1. Use PCR to amplify the heavy chain, adding a 5' *Mfe*I site and DNA encoding part of the OmpA signal peptide and adding a *Hin*dIII site to the 3' end.
2. Purify the amplified DNA, digest with restriction enzymes *Mfe*I and *Hin*dIII, and repurify.
3. Prepare vector pDNAbEng-1 by digestion with restriction enzymes, *Mfe*I and *Hin*dIII, and purify the cleaved fragment.
4. Ligate the two fragments together and transform into an appropriate host strain.
5. Prepare plasmid DNA from a transformed cell.
6. Digest a preparation of DNA with *Eco*RI, run on agarose, and purify the released fragment (containing DNA encoding the heavy chain plus a ribosome-binding site and OmpA signal peptide).
7. Ligate into the light-chain intermediate plasmid cut with *Eco*RI, and transform it into an appropriate host to generate single colonies containing the Fab' double-gene vector. Colonies need to be analyzed for insertion of the heavy-chain fragment in the correct orientation (PCR colony screen or plasmid minipreparation; *see* **Note 4**).

### *3.2. Small-Scale Expression in Shake Flasks*

Competent cell preparation is discussed for the *E. coli* expression strain and transformation of the plasmid. Small-scale culture and induction of the plasmid-carrying cells is then described, together with the sampling necessary to give an induction time course—this permits an evaluation of the expression of the antibody fragment before undertaking large-scale expression studies. A simple enzyme-linked immunosorbent assay (ELISA) procedure is also described to enable Fab' quantification.

### *3.2.1. Competent Cell Preparation*

A number of host strains can be used for protein expression; ideally, several should be tried with each recombinant product. One widely used strain is W3110, which has a history of safe use in the production of biopharmaceuticals. The following method *(10)* can be used to make competent W3110 cells.

1. Inoculate 50 mL L-broth in a 500-mL Erlenmeyer flask with a single colony of W3110, and incubate with shaking (250 rpm) at 37°C.
2. Measure the optical density at 600 nm ($OD_{600}$) at intervals until a value of 0.3–0.5 is attained (*see* **Note 5**).
3. Chill the culture on ice for 5 min, then spin at more than 5000$g$ for 10 min in a centrifuge rotor precooled to 4°C.
4. Discard the supernatant, and resuspend the cells gently in an 5-mL ice-cold transformation solution.
5. The cells are now competent and can be used immediately or alternately can be frozen at −70°C in 50 µL aliquots.

### *3.2.2. Transformation of Cells with Expression Plasmid*

To transform the competent cells with expression plasmid, the following protocol is used:

1. Thaw an aliquot of competent cells slowly on ice. Add 50–100 ng plasmid DNA, and mix gently with the pipet tip. Do not pipet up and down.

2. Leave on ice for 30 min.
3. Heat-shock at 42°C for 1 min.
4. Return to ice for a further 3 min.
5. Add 250 µL SOC, and incubate at 37°C for at least 1 h.
6. Plate 100 µL and 100 µL of a 10-fold dilution of the transformation mix (to ensure that single colonies are obtained) onto fresh L-broth agar plates containing 7.5 µg/mL tetracycline.
7. Incubate plates overnight at 37°C.

### 3.2.3. Preparation of Glycerol Stock (Working Cell Stock)

This procedure describes the creation of a glycerol stock of the transformed strain.

1. Inoculate 40 mL L-broth (containing 7.5 µg/mL tetracycline) in a 0.25-L Erlenmeyer flask with a single colony of transformed cells.
2. Incubate in an orbital shaker at 30 ± 1°C, 250 rpm for 8–10 h to an $OD_{600}$ of 0.9–1.1.
3. Combine 30 mL culture aseptically with 4.5-mL sterile 80% (w/w) glycerol and divide into 1-mL aliquots. These aliquots are frozen at –70°C to form the working cell stock.

### 3.2.4. Shake Flask Expression: Induction Time Course

The following procedure permits the small-scale analysis of expression of a Fab' (or other) fragment prior to production at scale.

1. Thaw 1-mL glycerol stock and transfer to 100-mL L-broth (containing 7.5 µg/mL tetracycline) in a 1-L baffled Erlenmeyer flask.
2. Incubate in an orbital shaker at 30°C, 250 rpm overnight (12–16 h). By the following morning, the culture will be at the stationary phase (*see* **Note 6**).
3. Measure the $OD_{600}$ of the overnight culture. Inoculate three 2-L baffled Erlenmeyer flasks containing 200 mL L-broth (containing 7.5 µg/mL tetracycline) with an appropriate volume of the culture, such that the $OD_{600}$ is approx 0.1. These three flasks should be prewarmed to 30°C; incubation is continued at this temperature while growth is monitored.
4. When the culture reaches an $OD_{600}$ of approx 0.5 (about 2 h), all three flasks should be sampled (*see* **step 5**). Each culture can then be induced by addition of IPTG to 200 µ$M$ (*see* **Note 7**). The OD of the cultures should be measured and further samples taken at 0.5, 1, 2, 3, and 4 h postinduction.
5. To sample the cultures so that they are normalized for differing culture ODs, the following protocol is used:

   a. Measure the $OD_{600}$ and use the formula below to calculate the volume of culture to sample:

   $$\text{Volume (mL)} = 9/\text{culture } OD_{600}$$

   Remove this volume of culture, chill on ice for 5 min, and harvest the cells by centrifugation (>5000$g$ for 15 min).

   b. Decant the supernatant, and retain a 1-mL sample (some antibody fragments will "leak" into the culture supernatant; retention of this sample allows analysis of this). Pellets can be stored frozen at this point if desired.

   c. Resuspend the cell pellet in 300 µL 1X periplasmic extraction buffer.

   d. Remove an aliquot (50 µL) for sodium dodecyl sulfate-polyacrylamide gel electrophoresis (SDS-PAGE) analysis of total cells. Incubate the remainder at 30°C overnight (12–16 h) with shaking (*see* **Note 8**).

e. Spin the sample in a microfuge (>12,000$g$ for 10 min). Retain both the supernatant sample (for ELISA, SDS-PAGE, and Western blot), and pellet sample (for SDS-PAGE and Western blot; *see* **Note 9**).

### 3.2.5. Expression Analysis: Assembly ELISA

The following method describes an ELISA technique to quantify the amount of Fab' material present in soluble extracts (*see* **Note 10**).

1. Dilute the capture antibody to 2 µg/mL in coating buffer, add 100 µL to each well of a 96-well ELISA plate, and leave overnight at 4°C.
2. Wash the wells three times with PBST.
3. Load 100 µL sample to be quantified in each well of the first column. Include a well of purified Fab' (initially at 2 µg/mL) and a well of PBST blank.
4. Serially dilute samples twofold across the plate in a sample conjugate buffer.
5. Incubate at room temperature for 1 h with agitation.
6. Wash and dry the plates. Add 100 µL reveal antibody (diluted to 0.2 µg/mL in sample conjugate buffer).
7. Incubate at room temperature for 1 h with agitation.
8. Wash and dry the plates. Add 100 µL substrate solution. Wait 4–6 min for the color to develop.
9. Read the absorbance at $A_{630\,nm}$. Determine concentrations of antibody in the samples by comparison with a standard curve obtained using the control Fab'.

## 3.3. Fermentation

These methods apply to the production of a Fab' fragment and the recovery of this material from the *E. coli* periplasm. The steps outline the (1) preparation of preinoculum and inoculum and (2) fermentation procedure and cell harvest. The induction system employed for this section differs from that used for the shake flask cultures and is based on switching the carbon source from glycerol to lactose. Carbon source switches can be more readily controlled in the defined medium used for fermentation, and the lactose induction appears to support a more gradual rate of Fab' expression that can be sustained over a comparatively long induction phase. Samples should be taken throughout the fermentation. A method for making small-scale periplasmic extractions from these samples is described. In this way, an optimum harvest time for repeat fermentations can be determined.

### 3.3.1. Shake Flask Preinoculum (Culture 1) and Inoculum (Culture 2)

1. Thaw a vial of the working cell stock at room temperature.
2. Transfer 40 µL to a 0.25-L Erlenmeyer flask containing 40 mL L-broth and 7.5 µg/mL tetracycline (culture 1).
3. Incubate in an orbital shaker at 30°C for 8–10 h until an $OD_{600}$ of 1.0–1.5 is reached.
4. Use 20 mL to inoculate 200 mL sterile SM6 medium (supplemented with 5 µg/mL tetracycline) in a 2-L Erlenmeyer flask (culture 2).
5. Incubate in an orbital shaker at 30°C (250 rpm). Monitor the $OD_{600}$.
6. When an $OD_{600}$ of 3.0–3.5 is attained, use 0.1 L to inoculate the production fermenter.

### 3.3.2. Fermentation Procedure

1. Prepare 1 L SM6 media, transfer to a 2.5-L batch fermenter, and sterilize at 121°C for 30 min; adjust the pH to 7.0 with the addition of 50% (v/v) $NH_4OH$. Add Struktol antifoam stock solution (2.0 mL/L).
2. Inoculate with 0.1 L culture 2, and incubate at 30 ± 1°C. Maintain at pH 7.0 ± 0.2 with 50% (v/v) $NH_4OH$ and 20% (v/v) $H_2SO_4$. Maintain the DO at more than 20% using variable aeration and agitation. Monitor the $OD_{600}$.
3. At $OD_{600}$ 18–22, add 49.2 mL sterile 52.5% (w/w) glycerol solution.
4. At $OD_{600}$ 38–42, lower the incubation temperature to 25°C, and add a further 49.2 mL sterile 52.5% (w/w) glycerol solution, together with 16.7 mL sterile 0.83 $M$ $MgSO_4$ and 16.7 mL sterile 0.1 $M$ $CaCl_2$.
5. At $OD_{600}$ 58–62, add 115 mL sterile 40% (w/w) lactose solution. When the culture reaches $OD_{600}$ 75–80, glycerol will become depleted, and carbon utilization switches to lactose. This point will be marked by a reduction in oxygen utilization and represents the start of the induction phase.
6. During the induction phase, add lactose as required to maintain the concentration between 20 and 50 g/L. Continue the induction for 24–36 h.
7. To harvest, cool the fermenter to 20°C, and remove the culture. Dispense into 1-L centrifuge pots and spin at more than 5000$g$ for 1 h.
8. Decant the culture supernatant from the cell pellet, and store the pelleted cells at –20°C.
9. Take 1-mL samples throughout the fermentation. Recover cells by centrifugation (12000$g$ for 10 min). Retain both the supernatant (for subsequent analysis of extracellular Fab' leakage) and cell pellet fraction (for small-scale periplasmic extraction and subsequent analysis). Both fractions can be stored at –20°C.

### 3.3.3. Small-Scale Periplasmic Extraction

These steps describe the procedure for small-scale recovery of soluble Fab' from the *E. coli* periplasm.

1. Thaw the 1 mL *E. coli* cell pellets at room temperature, and resuspend to the original culture volume in 1X periplasmic extraction buffer.
2. Incubate at 30°C overnight (12–16 h) with constant agitation in an orbital shaker.
3. Separate the cell debris from the periplasmic extract by centrifugation (>15,000$g$ for 10 min).
4. Decant the periplasmic extract from the cell debris, and retain the extract (for assembly ELISA, SDS-PAGE, and Western blot analysis).

**Figure 3** shows the growth profile for the fermentation of W3110 expressing a humanized Fab' fragment, in combination with expression profiles of periplasmic and extracellular Fab' estimated by the ELISA method outlined in **Subheading 3.2.5.**

## 3.4. Primary Recovery and Purification

These methods outline the: (1) extraction of soluble periplasmic Fab' and (2) purification of Fab' using affinity chromatography.

### 3.4.1. Periplasmic Extraction

This section outlines the procedure for primary recovery of soluble Fab' from the periplasm of the *E. coli* using Tris and EDTA to disrupt the outer cell membrane.

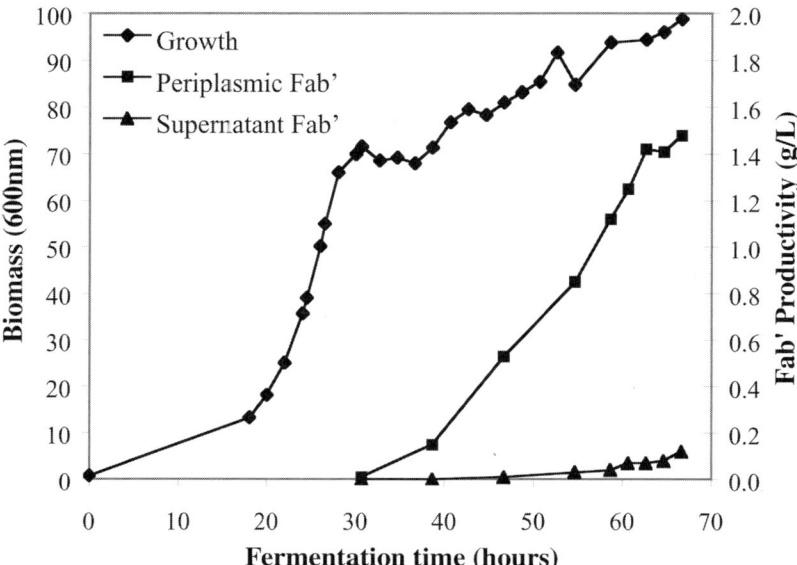

Fig. 3. Fermentation of W3110 expressing a Fab' fragment. Fermentation growth profile and humanized Fab' productivity profile. Samples were taken through the induction phase and analyzed as described in the text. The supernatant Fab' samples were taken from the culture medium. With this particular Fab', the vast majority of Fab' is retained within the periplasm.

1. If frozen, the harvested *E. coli* cell paste should be thawed at ambient temperature before use. Once thawed, the cell paste is resuspended to the original culture volume in 1X periplasmic extraction buffer.
2. The *E. coli* resuspension is incubated at 30°C overnight (12–16 h) with constant agitation by either returning to the fermenter vessel or incubation in an orbital shaker (*see* **Note 11**).
3. The cell extract is then clarified by centrifugation at more than 15,000*g* for 45 min, followed by filtration using 0.45 μm followed by 0.22-μm filters. For larger volumes, a Sartobrand P (0.45–0.22 μm) filter capsule could be used. If filtration difficulties are encountered, a prefilter capsule, e.g., Sartoclean GF (0.8–0.65 μm), could be implemented.

### 3.4.2. Affinity Chromatography

Fab' can be purified to greater than 95% using protein G chromatography, as detailed in the following steps. To avoid product loss, aim to load the column at less than 6 mg Fab' per milliliter of resin.

1. Sanitize the protein G column using two column volumes of 6 *M* guanidine-HCl, followed by two column volumes of 10% (v/v) methanol, then equilibrate in PBS for approximately five column volumes.
2. Adjust the pH of the clarified cell extract to 7.5 using 2 *M* Tris-HCl stock solution and apply to the prepared column at 75 cm/h.
3. After loading, the column is washed with PBS until the absorbance at 280 nm returns to baseline.

4. Bound Fab' is step eluted with 0.1 $M$ glycine-HCl buffer, pH 2.7. Start product collection when the absorbance at 280 nm goes above baseline, and stop collection when the absorbance returns to baseline.

5. Neutralize the column eluate using 2 $M$ Tris-HCl solution dropwise.

6. After use, the column is washed using 6 $M$ guanidine-HCl and 10% (v/v) methanol as described in **step 1**. When not in use, the column should be stored in 20% ethanol.

7. Filter-sterilize the neutralized column eluate through a 0.22-μm filter, and store at 2–8°C (*see* **Notes 12** and **13**).

**Figure 4A** shows the protein G chromatography profile, whereas **Fig. 4B** shows SDS-PAGE analysis of the affinity purification.

## 4. Notes

1. This can include removal of rare codons, removal of unfavorable residues or even wholesale "grafting" of the complementarity determining regions (CDRs) onto optimized frameworks. The balance of expression of the two chains may also need to be optimized. Such engineering is beyond the scope of this chapter.

2. Because of space limitations, the precise details of these molecular biology techniques are not included.

3. Strain W3110, which is used for expression studies, is *Eco*KI restriction-competent, i.e., it will restrict DNA that is not protected by adenine methylation at the appropriate sites. To be transformed into this strain, DNA should be prepared in a host that is competent in *Eco*KI methylation. This phenotype is conferred by the *hsdRMS* alleles, and any host strain used for subcloning work should not carry a *hsdS* allele, which abolishes both restriction and methylation (r⁻m⁻). We recommend strains carrying *hsdR* alleles (r⁻m⁺), such as JM109 or INVαF'.

4. The copy number of the pTTO-1 vector is low and can result in poor yields from plasmid preparations. We routinely use 10-mL L-broth cultures for minipreparations (Qiagen) and up to 500-mL L-broth cultures for maximum preparations (Qiagen).

5. Measurements of OD$_{600}$ need to be in the linear range of the curve of cell number vs OD; in practice, this means values in the range of 0.05–0.35. Where the density of the culture is too high, it should be diluted in L-broth so that it falls within this range.

6. In an ideal experiment, the cells would not be allowed to enter the stationary phase, but they would be subcultured in the late log phase of growth. However, this is rarely practical, and allowing the cultures to reach the stationary phase is unlikely to have deleterious effects.

7. An additional flask of culture to which IPTG is not added can be used as a control to highlight changes in the growth profile from induction and fragment production. It is possible to vary the strength of the induction by varying the IPTG concentration added. Although 200 μ$M$ will give a strong induction, for some antibody fragments this can result in the accumulation of insoluble protein. In such cases, the use of 50 μ$M$ IPTG can be beneficial in "slowing" the induction, allowing more protein to fold correctly and remain soluble. With thorough sampling and analysis, any insolubility issues can be monitored, and adjustments to inducer concentration can be considered.

8. This extended incubation period works well for most antibody fragments because they are generally very resistant to proteolysis. Western blot analysis of samples pre- and postincubation will provide a check for this.

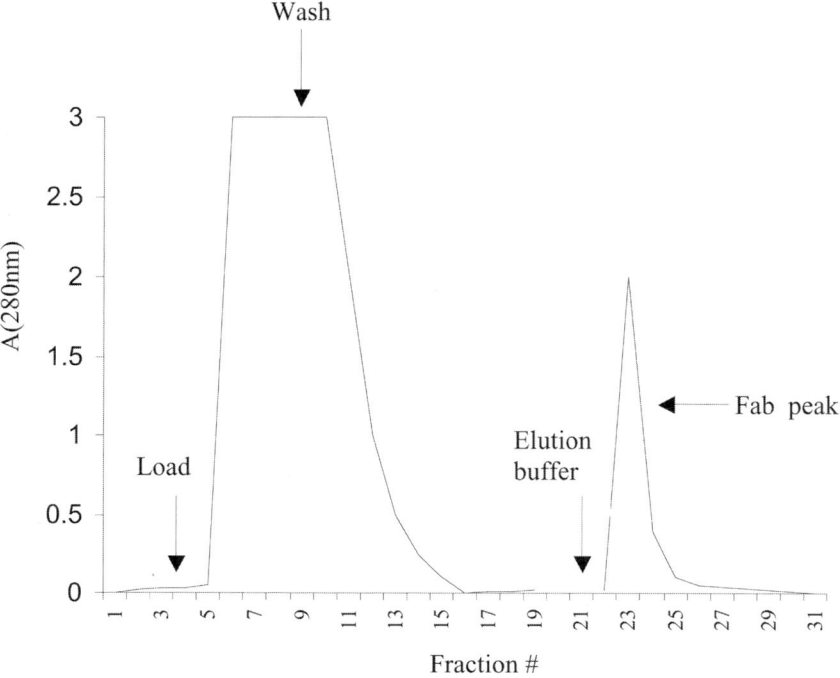

**A** Protein G chromatography profile

**B** SDS-PAGE analysis of Fab purification using Protein G chromatography

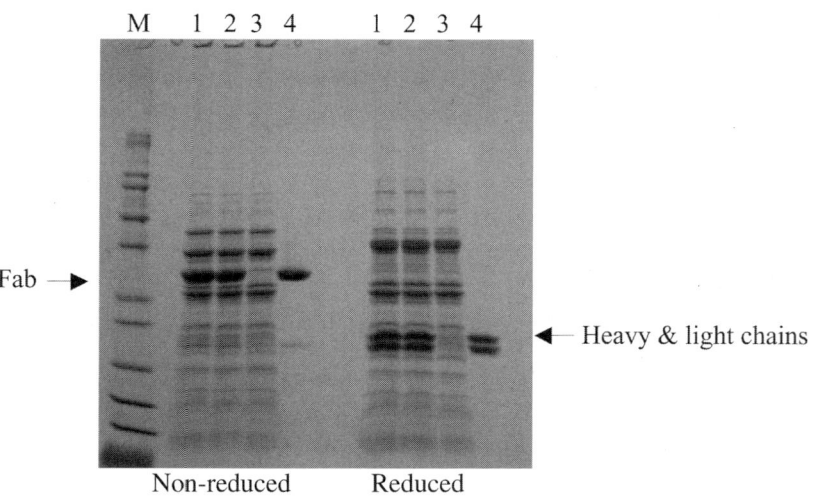

Fig. 4. Fab' purification. (A) Elution profile of protein (A280) against the fraction number, showing the elution of a sharp protein peak after application of elution buffer. (B) SDS-PAGE analysis of Fab' purification. Coomassie blue-stained reducing and nonreducing 4–20% Tris-Glycine SDS-PAGE gel. M, Mark 12 markers (Novex); 1, cell extract, 5 µL; 2, column load, 5 µL; 3, column flow-through, 10 µL; 4, column eluate, 2 µg protein.

9. SDS-PAGE and Western blot analysis will give useful information about the integrity of the expressed protein as well as an indication of the level of expression. It is also advisable to analyze medium supernatant, whole cell, periplasmic extract, and cellular fractions to provide information about periplasmic retention and any propensity to form insoluble aggregates.

10. The capture and reveal antibodies used will vary according to the nature of the antibody fragment being expressed. For a humanized Fab' fragment with human constant regions, we use anti-human Fd 6045 (The Binding Site) as the capture antibody and anti-human C-κ GD12-peroxidase (The Binding Site) as the reveal antibody. Amounts to use may have to be fine-tuned for other antibody pairings.

11. The purity of the final product may be improved by performing the periplasmic extraction using elevated temperature, e.g., 60°C instead of 30°C *(11)*. The heat treatment will ensure the removal of any incorrectly folded nondisulfide Fab' and truncated Fab' species, as well as ensuring the removal of some host cell proteins. Some Fab' constructs may be unstable at 60°C, resulting in loss of product. In this case, the temperature of extraction may need to be reduced to 55°C or 50°C.

12. If high-purity material is required, the affinity chromatography fractionated Fab' can be further processed using anion-exchange chromatography.

13. In most cases, the final purified product will contain some diFab'. If diFab' removal is necessary, it can be reduced to mono-Fab' by incubation in 5 m$M$ β-mercaptoethylamine at 37°C for 30 min. To avoid over-reduction, resulting in the breaking of the disulfide bonds between the heavy and light chains, it is important to remove the reductant immediately after the incubation period. β-mercaptoethylamine can be removed by ultrafiltration or for volumes less than 1 mL by implementation of a desalting column, such as a PD10 (Amersham). Once the Fab' has been reduced, the unpaired hinge thiol can be capped with *N*-ethylmaleimide to avoid oxidation and redimerization. The reduction reaction will also remove any adducts, e.g., glutathione, on the hinge thiol introduced during biosynthesis. Alternatively, any diFab' species could be fractionated from the Fab' using size exclusion chromatography.

## Acknowledgments

We would like to acknowledge the contribution of our many colleagues at Celltech, both past and present, who have helped in the development of these methods.

## References

1. Weir, A. N. C., Nesbitt, A., Chapman, A. P., Popplewell, A. G., Antoniw, P., and Lawson, A. D. G. (2002) Formatting antibody fragments to mediate specific therapeutic functions. *Biochem. Soc. Trans.* **30,** 512–516.

2. Humphreys, D. P. (2003) Production of antibodies and antibody fragments in Escherichia coli and a comparison of their functions, uses and modification. *Curr. Opin. Drug Discov. Devel.* **6,** 188–196.

3. Stark, M. J. (1987) Multicopy expression vectors carrying the lac repressor gene for regulated high-level expression of genes in Escherichia coli. *Gene* **51,** 255–267.

4. Chang, A. C. and Cohen, S. N. (1978) Construction and characterization of amplifiable multicopy DNA cloning vehicles derived from the P15A cryptic miniplasmid. *J. Bact.* **134,** 1141–1156.

5. de Boer, H. A., Comstock, L. J., and Vasser, M. (1983) The tac promoter: a functional hybrid derived from the trp and lac promoters. *Proc. Natl. Acad. Sci. USA* **80,** 21–25.

6. Brosius, J. (1984) Toxicity of an overproduced foreign gene product in Escherichia coli and its use in plasmid vectors for the selection of transcription terminators. *Gene* **27,** 161–172.

7. Muller-Hill, B., Crapo, L., and Gilbert, W. (1968) Mutants that make more lac repressor. *Proc. Natl. Acad. Sci. USA* **59,** 1259–1264.

8. Selzer, G., Som, T., Itoh. T., and Tomizawa, J. (1983) The origin of replication of plasmid p15A and comparative studies on the nucleotide sequences around the origin of related plasmids. *Cell* **32,** 119–129.

9. Movva, N. R., Nakamura, K., and Inouye, M. (1980) Amino acid sequence of the signal peptide of ompA protein, a major outer membrane protein of Escherichia coli. *J. Biol. Chem.* **255,** 27–29.

10. Chung, C. T., Niemela, S. L., and Millar, R. H. (1989) One-step preparation of competent Escherichia coli: transformation and storage of bacterial cells in the same solution. *Proc. Natl. Acad. Sci. USA* **86,** 2172–2175.

11. Weir, A. N. C. and Bailey, N. A. (1997) Process for obtaining antibodies utilising heat treatment. US Patent No. 5,665,866.

# 3

## Secretory Production of Therapeutic Proteins in *Escherichia coli*

### Sang Yup Lee, Jong Hyun Choi, and Sang Jun Lee

## 1. Introduction

*Escherichia coli* has been the workhorse for the production of recombinant proteins *(1,2)*. However, problems often occur in recovering substantial yields of correctly folded proteins. *E. coli* cannot produce some proteins containing complex disulfide bonds or mammalian proteins that require posttranslational modification for activity. Overexpressed proteins are often produced in the form of inclusion bodies, from which biologically active proteins can only be recovered by complicated and costly denaturation and refolding processes. Various techniques have been developed to solve these problems, including the use of different promoters to regulate the level of expression, using different host strains, coexpression of chaperones, reduction of culture temperature, and secretion of proteins into the periplasmic space (*see* **Fig. 1**).

Secretory production of recombinant proteins provides several advantages when compared to cytosolic production *(3)*. First, the N-terminal amino acid residue of the secreted product can be identical to the natural gene product, resulting from the cleavage of the signal peptide by a specific signal peptidase. Second, there are much less protease activities in the periplasmic space than in the cytoplasm *(4)*. Third, purification of recombinant proteins is much simpler because of the presence of less contaminating proteins in the periplasmic space. Fourth, the correct formation of disulfide bonds can be facilitated because periplasmic space provides more of an oxidative environment than the cytoplasm.

Generally, proteins found in the outer membrane or periplasmic space are synthesized in the cytoplasm as premature proteins *(5)*. These proteins contain a short (15–30) specific amino acid sequence (signal sequence) that allows proteins to be exported to the outside of the cytoplasm. Numerous signal sequences have been used for efficient secretory production of recombinant proteins in *E. coli*, including pectate lyase B (PelB), outer membrane protein A (OmpA), alkaline phosphatase (PhoA), endoxylanase, and heat-stable enterotoxin 2 (StII). Although sequence diversity exists

From: *Methods in Molecular Biology, vol. 308: Therapeutic Proteins: Methods and Protocols*
Edited by: C. M. Smales and D. C. James © Humana Press Inc., Totowa, NJ

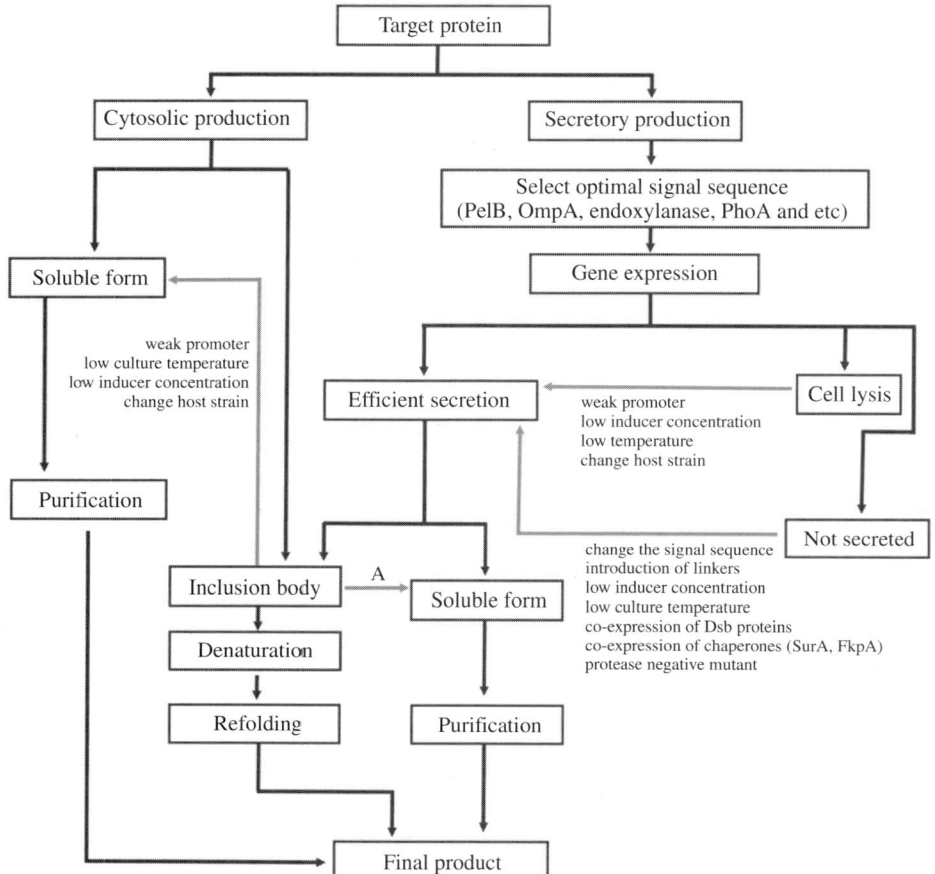

Fig. 1. Strategies for the production of recombinant proteins in *E. coli*. Arrow **A** can be achieved by one of the methods described for the conversion of "Not secreted" to "Efficient secretion."

among these signal sequences, some common structural features have been identified. **Table 1** shows the representative signal sequences that have been used for the secretory production of recombinant proteins in *E. coli*. The signal sequences are typically rich in hydrophobic amino acids, e.g., alanine, valine, leucine, which are essential for the secretion of proteins into the periplasmic space of *E. coli*. During the transport of proteins outside of the cytoplasm, the signal sequence is cleaved by signal peptidase to yield a mature protein product. Thus, the selection of an optimal signal sequence is important for effective secretory production of recombinant proteins.

This chapter discusses protocols for the secretory production of therapeutic proteins in *E. coli* with human leptin (*6,7*) and human granulocyte colony-stimulating factor (hG-CSF) (*8*) as model proteins.

**Table 1**
**Representative Signal Sequences**
**for the Secretory Production of Recombinant Proteins in *E. coli***

| Signal sequences | Amino acid sequences |
|---|---|
| Endoxylanase from *Bacillus* sp. | **MFKFKKK**FLVGLTAAFMSISMFS<u>ATASA</u> |
| PhoA | **MKQS**TIALALLPLLFT<u>PVTKA</u> |
| StII | **MKK**NIAFLLASMFVFSIA<u>TNAYA</u> |
| PelB from *Erwinia carotovora* | **MKY**LLPTAAAGLLLLAAQ<u>PAMA</u> |
| OmpA | **MKKT**AIAIAVALAG<u>FATVA</u>QA |
| OmpF | **MMKR**NILAVIVPALLVA<u>GTANA</u> |
| PhoE | **MKKS**TLALVVMGIVA<u>SASVQA</u> |

The signal sequence is composed of three domains: N, H, and C. The N domains of signal sequences are shown in bold, whereas the C domains are underlined.

## 2. Materials

1. *E. coli* XL1-Blue, BL21(DE3), HB101, and MC4100. Competent cells can be prepared using the standard calcium chloride method.
2. Recombinant plasmids: pUCOb *(6)*, pTrcSOb4 *(7)*, pTrcSObD *(7)*, pTHKCSFmII *(8)*, and pJS101ΔP *(9; see* **Fig. 2** and **Note 1**).
3. Chromosomal DNA of *E. coli* W3110.
4. Luria–Bertani (LB) medium and LB agar plates: sterilized by autoclaving.
5. Ampicillin: used for both liquid and agar cultivation at a concentration of 50 μg/mL (Sigma, St. Louis, MO).
6. TE buffer, pH 8.0.
7. Isopropyl-β-D-thiogalactopyranoside (IPTG) (Sigma).
8. 1 *M* Tris-HCl, pH 8.0.
9. Sucrose: 40% (w/v).
10. 0.5 *M* EDTA.
11. 5 m*M* MgSO$_4$.
12. H$_2$O: sterilize before use.
13. Polymerase chain reaction (PCR) purification kit (Qiagen, Chatsworth, CA).
14. Expand High Fidelity PCR system (Roche, Mannheim, Germany).
15. Agarose (FMC Bioproducts, Rockland, ME).
16. Sterile toothpick.
17. Restriction enzymes (New England Biolabs [NEB], Beverly, MA).
18. T$_4$ DNA ligase (Roche).
19. QIAquick gel extraction kit (Qiagen).
20. Ni-nitrilotriacetic acid (Ni-NTA) superflow column (Qiagen).
21. Factor Xa (NEB).
22. Ex-Taq DNA polymerase (TaKaRa, Tokyo, Japan).
23. Dialysis membrane (molecular weight cut-off [MWCO] of 3500, Spectrum Lab Inc., Laguna Hills, CA).

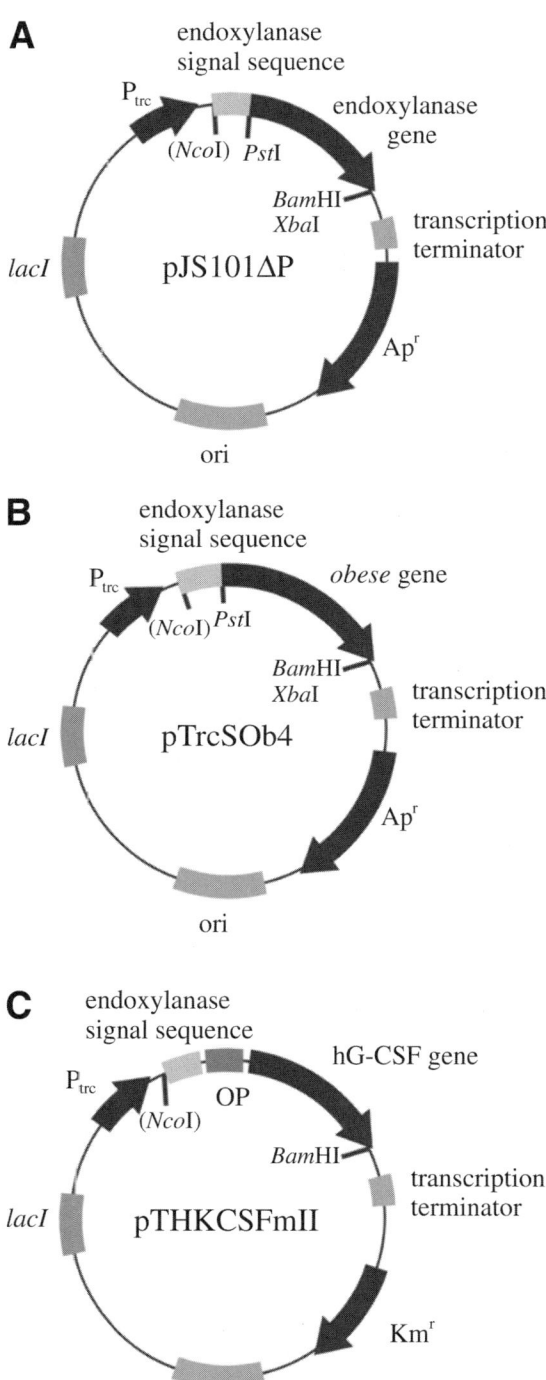

Fig. 2. Structures of plasmids: (A) pJS101ΔP, (B) pTrcSOb4, and (C) pTHKCSFmII.

## 3. Methods

### 3.1. Construction of Secretion Vector of Human Leptin

#### 3.1.1. Vector Preparation

1. Add the following to a sterile Eppendorf tube: 10 µL of a 100 ng/µL solution of pJS101ΔP, 2 µL 10X *Bam*HI NEB buffer, 8 µL $H_2O$, and 0.4 µL *Bam*HI and *Pst*I (NEB).
2. Mix the contents, and incubate the solution at 37°C for 2 h.
3. Separate the digested gene product by electrophoresis on a 0.7% (w/v) agarose gel.
4. Purify the DNA fragments using a gel extraction kit.
5. The DNA sample can be stored at –20°C.

#### 3.1.2. Preparation of Insert DNA

1. Amplify the human *obese* gene (*see* **Note 2**) using pUCOb (6) as a template for the PCR reaction with 50 µL that contains 5 µL 10X PCR reaction buffer, 2 µL each of 20 µ*M* primer (*see* **Note 3**), 1 µL of 0.2 ng/µL template, and 1.25 U Ex-Taq DNA polymerase (*see* **Note 4**).
2. Purify the PCR product with the PCR purification kit following the manufacturer's instructions.
3. Partially digest the purified PCR product with the *Pst*I restriction enzyme and fully digest with *Bam*HI restriction enzyme.
4. Separate the digested gene product by electrophoresis on a 0.7% agarose gel.
5. Purify the DNA fragments using a gel extraction kit.
6. The DNA sample can be stored at –20°C.

#### 3.1.3. Construction of Human Leptin Secretion Vector

1. Ligate digested DNA vector and PCR product (*obese* gene) by adding 100 ng vector and 300 ng PCR product into the 20 µL ligation mixture ($T_4$ DNA ligase and buffer).
2. Incubate for 6 h at 16°C.
3. Transform *E. coli* XL1-Blue competent cells (200 µL) with 5 µL ligation mixture.
4. Dilute with 800 µL LB medium without the antibiotic.
5. Incubate cells at 37°C for 1 h.
6. Spread the cells on LB plates with ampicillin (or appropriate antibiotics if different antibiotic markers are used).
7. Incubate the plates overnight at 37°C.
8. Individual colonies are screened for the correct clone by miniplasmid preparation, then by restriction enzyme digestion.
9. Verify the correct sequence of the *obese* gene in the constructed vector (pTrcSOb4; *see* **Fig. 2B**) by DNA sequencing (*see* **Note 5**; 7).

### 3.2. Protein Expression and Periplasmic Fractionation

#### 3.2.1. Protein Expression

1. Transformation of *E. coli* HB101 with the constructed vector pTrcSOb4 (*see* **Note 6**).
2. Inoculate *E. coli* HB101 (pTrcSOb4) into the LB medium that contains ampicillin.
3. Cultivate the cells to an optical density ($OD_{600}$) of approx 0.6 with shaking (150 rpm) at 37°C.
4. Add IPTG to a final concentration of 1 m*M* for gene expression (*see* **Note 7**).
5. Cultivate the cells for an additional 5 h.

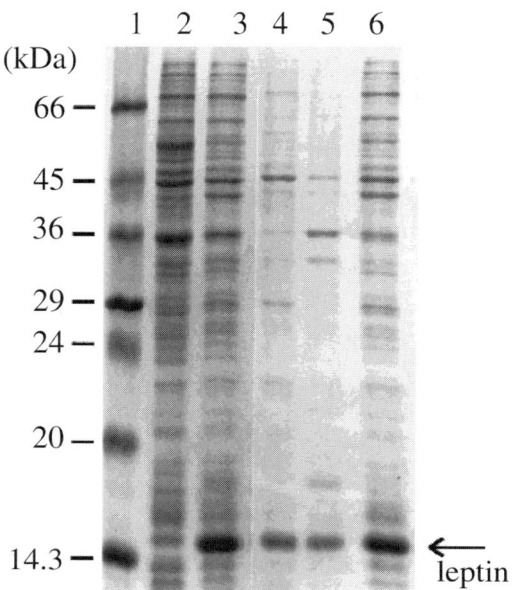

Fig. 3. SDS-PAGE analysis of periplasmic fractionation from *E. coli* BL21(DE3) harboring pTrcKObD. Lane 1, molecular mass standard; lane 2, *E. coli* BL21(DE3; no plasmid); lane 3, total proteins; lane 4, periplasmic proteins; lane 5, insoluble proteins; lane 6, soluble proteins. (Figure reproduced with permission from **ref. 7**.)

### 3.2.2. Periplasmic Fractionation

1. Centrifuge 1 mL induced culture sample in a microcentrifuge for 10 min at 7000$g$ and 4°C.
2. Discard the supernatant (*see* **Note 8**).
3. Resuspend the cell pellet thoroughly in 0.4 culture volumes of 30 m$M$ Tris-HCl, pH 8.0, plus 20% sucrose.
4. Add EDTA to a final concentration of 2 m$M$ (*see* **Note 9**).
5. Incubate for 10 min at room temperature with shaking.
6. Collect the cells by centrifugation for 10 min at 10,000$g$ and 4°C.
7. Remove all of the supernatant as far as possible.
8. Resuspend the pellet in the same volume of ice-cold 5 m$M$ MgSO$_4$ (*see* **Note 10**).
9. Stir the cell suspension slowly for 15 min on ice (*see* **Note 11**).
10. Centrifuge for 10 min at 10,000$g$ and 4°C.
11. Transfer sample from the supernatant to a microcentrifuge tube (periplasmic fraction).
12. Resuspend the pellet in the same volume of TE buffer for preparing the cytoplasmic fraction (*see* **Note 12**).
13. Store at –20°C until sodium dodecyl sulfate polyacrylamide gel electrophoresis (SDS-PAGE) analysis or activity assay (*see* **Note 13** and **Fig. 3**).

### 3.2.3. Soluble and Insoluble (Inclusion Body) Fractionation

1. Centrifuge 1 mL induced culture sample in a microcentrifuge for 10 min at 7000$g$ and 4°C.
2. Discard the supernatant.

3. Resuspend the cell pellet thoroughly in 0.2 mL TE buffer, pH 8.0.
4. Disrupt the cells by sonication using Sonics Vibra Cell (Branson Ultrasonic Co., Danbury, CT) for 1 min at 40% output.
5. Centrifugation for 10 min at 10,000$g$ and 4°C.
6. Transfer sample from the supernatant to a microcentrifuge tube (soluble fraction).
7. Resuspend the pellet in the same volume of TE buffer, pH 8.0, to prepare the insoluble fraction.
8. Store at –20°C until SDS-PAGE analysis.

### 3.3. Enhancing the Efficiency of Secretory Production of Human Leptin

### 3.3.1. Vector Construction for DsbA Coexpression

1. Amplify the *dsbA* gene *(7)* by PCR using *E. coli* W3110 chromosome as a template. The PCR reaction is carried out in 50 µL solution that contains 5 µL 10X PCR reaction buffer, 2 µL each of 20 µ*M* primer, 1 µL of 0.2 ng/µL template, and 1.25 U Ex-Taq DNA polymerase (*see* **Note 14**).
2. Purify the PCR product with the PCR purification kit following manufacturing instructions.
3. Digest pTrcSOb4 and the purified PCR product with *Nde*I restriction enzyme.
4. Separate the digested gene product by electrophoresis on 0.7% (w/v) agarose gel.
5. Purify the DNA fragments using a gel extraction kit.
6. Ligate digested DNA vector (100 ng) and PCR product (300 ng) by using T$_4$ DNA ligase.
7. Verify the correct sequence of the insert DNA in the constructed vector (pTrcSObD) by DNA sequencing (*7*; *see* **Note 15**).

### 3.3.2. Protein Expression and Fractionation

1. Transformation of *E. coli* BL21(DE3) with the constructed vector pTrcSObD.
2. Cultivate the cells to an OD$_{600}$ of approx 0.6 with shaking at 37°C.
3. Add IPTG at a final concentration of 1 m*M* for gene expression.
4. Cultivate for another 5 h.
5. Fractionation of induced cells for the SDS-PAGE analysis or purification as described earlier (*see* **Fig. 4**).

### 3.4. Expression of hG-CSF in E. coli

### 3.4.1. Expression of hG-CSF in E. coli

1. Transformation of *E. coli* MC4100 with the plasmid pTHKCSFmII (*8*; *see* **Note 16**).
2. Cultivate the cells to an OD$_{600}$ of approx 0.6 with shaking at 37°C.
3. Add IPTG at a final concentration of 1 m*M* for gene expression.
4. Cultivate for an additional 5 h.
5. Fractionation of induced cells for the SDS-PAGE analysis or purification as described earlier.

### `3.4.2. Purification of Secreted hG-CSF

1. Centrifuge 50 mL induced culture for 10 min at 7000$g$ and 4°C, and discard the supernatant.
2. Resuspend the cell pellet in 50 mL resuspension buffer: 50 m*M* NaH$_2$PO$_4$, 0.3 *M* NaCl, and 10 m*M* imidazole, pH 8.0.
3. Disruption by sonication with Sonics Vibra Cell for 20 min at 40% output.
4. Centrifugation for 30 min at 14,000$g$ and 4°C.

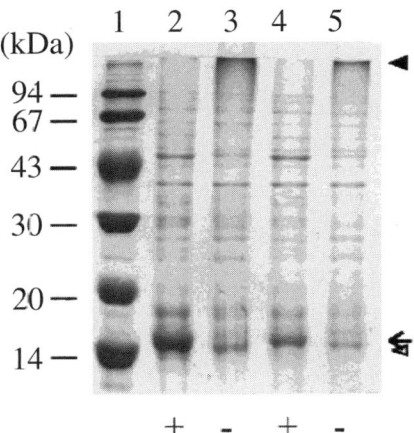

Fig. 4. SDS-PAGE analysis of reduced and nonreduced inclusion body of human leptin produced in *E. coli* BL21(DE3) harboring pTrcKOb4 and pTrcKObD. Lane 1, molecular mass standard; lanes 2 and 3, *E. coli* BL21(DE3) (pTrcKOb4); lanes 4 and 5, *E. coli* BL21(DE3) (pTrcKObD); (+) with dithiothreitol (DTT); (–) without DTT; (→) reduced leptin; (▷) oxidized leptin; (▶) intermolecular disulfide-linked inclusion bodies. (Figure reproduced with permission from **ref. 7**.)

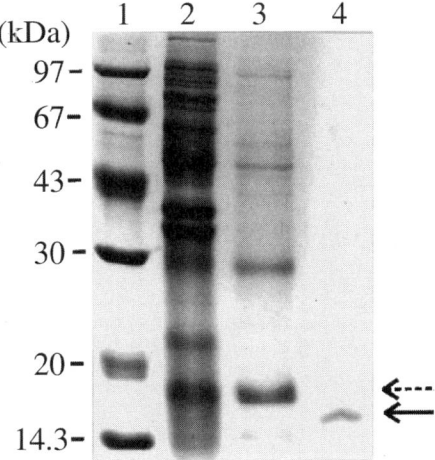

Fig. 5. SDS-PAGE analysis of the samples from each purification step. Lane 1, molecular mass standards; lane 2, total protein of *E. coli* MC4100 harboring pTHKCSFmII; lane 3, sample after Ni-NTA superflow column; lane 4, sample after anion-exchange chromatography. Dashed arrow, the oligopeptide-fused hG-CSF protein; solid arrow, the mature hG-CSF protein. (Figure reproduced with permission from **ref. 8**.)

5. Purification of hG-CSF protein (His-tagged) using a Ni-NTA superflow column as recommended by the manufacturer (*see* **Note 17**).
6. Dialysis of eluted fraction against 20 m*M* Tris-HCl buffer, pH 8.0, using dialysis membrane.
7. Incubate dialyzed hG-CSF with 200 μg Factor Xa at 23°C for 8 h.
8. Purification of hG-CSF by anion-exchange column chromatography (*see* **Note 18**).
9. Store at –20°C until SDS-PAGE analysis or activity assay (*see* **Fig. 5**).
10. If desired, fed-batch culture of this recombinant *E. coli* can be carried out for the scaled-up production of hG-CSF (*see* **Note 19**).

## 4. Notes

1. One advantage of secretory protein production is that the authentic N-terminal amino acid sequence without the fMet extension can be obtained after cleavage by the signal peptidase. This can be achieved only when the gene of interest is correctly fused to the cleavage site. This secretion vector containing the endoxylanase signal sequence has the *Pst*I site in the cleavage site (A-S-A) within the endoxylanase signal sequence, which can be directly connected to the mature protein sequence of interest. Using this endoxylanase signal sequence, human leptin, hG-CSF, and alkaline phosphatase could be efficiently secreted into the *E. coli* periplasm *(7–11)*.
2. The codons of therapeutic proteins of mammalian origin can be altered based on the preferred codon usage of *E. coli* for efficient expression *(12)*. Human proteins have been shown containing codons, which are rare in *E. coli*, e.g., Arg (AGG/AGA), Arg (CGA), Leu (CUA), and Pro (CCC) may be inefficiently expressed in *E. coli*. If it is difficult to alter the codons; a different expression host, such as BL21-CodonPlus (Stratagene, La Jolla, CA) or similarly constructed strains, can be used. BL21-CodonPlus strain contains extra copies of the *argU*, *ileY*, and *leuW* tRNA genes, and rescues heterologous expression of human genes containing unusual codons.
3. To ensure that the target gene is in frame with the endoxylanase signal sequence, the forward primer 5'-GCGAATT<u>CTGCAG</u>TGCCCATCCAAAAAGTCC-3' was designed to contain a *Pst*I site (underlined). The reverse primer 5'-GCC<u>GGATCC</u> TTATTAGCACCCAGGGCTGAGG-3' was designed to contain a *Bam*HI site (underlined).
4. As a general PCR protocol, we use the following condition: 1 min at 94°C; 30 cycles of 40 s at 94°C; 40 s at 56°C (annealing temperature); 1 min 20 s (extension time) at 72°C; and 10 min at 72°C. The annealing temperature and extension time should be optimized depending on the PCR product to be amplified and the primer length.
5. The choice of signal sequence is important for the efficient secretion of recombinant proteins, as some signal sequences do not always ensure secretion of recombinant proteins. This is because of cell lysis owing to a weakened outer membrane, incomplete processing by signal peptidase, characteristics of proteins to be secreted, and degradation of secreted proteins by proteases in the periplasmic space of *E. coli*. To date, no general rule applies in selecting a proper signal sequence for a given protein to guarantee its successful secretion. Several signal peptides, such as those listed in **Table 1**, are examined in a trial-and-error type approach.
6. *Bacillus subtilis* can be an appealing alternative to *E. coli* as a host for the secretory production of recombinant proteins because it has several advantages over *E. coli*. The greatest advantage is its ability to secrete proteins into the culture medium and accumulate them to a high level in a relatively pure state. Furthermore, *B. subtilis* is a GRAS strain *(13)*.

7. The IPTG concentration should be optimized for different proteins. The recombinant proteins to be overexpressed under the strong promoter may cause the formation of inclusion bodies in the periplasmic space or cell lysis by metabolic burden during translocation and cannot be fractionated by the osmotic shock method *(9)*. To possibly prevent the formation of inclusion bodies, cells can be induced with a lower IPTG concentration (e.g., 0.1 m*M*).

8. Cell pellets cannot be stored in the refrigerator after centrifugation, and cells refrigerated cannot be fractionated by the osmotic shock method because they are lysed during thawing.

9. Lysozyme treatment (500 μg/mL) can increase the yield of recombinant protein released from the periplasmic space. Some secreted recombinant proteins may be entrapped within the peptidoglycan matrix. Lysozyme cleaves the crosslinked peptidoglycan matrix.

10. EDTA binds divalent cations (e.g., magnesium) that are required for the stability of the bacterial outer membrane.

11. The shape of cells after osmotic shock can be monitored with a light microscope.

12. Avoid removing any loose pellets with the supernatant.

13. To confirm the correctly folded proteins, circular dichroism, analysis of redox state, or activity assay can be used (*see* **Fig. 4**).

14. To amplify the *dsbA* gene, two primers, 5'-GGCAAGCTT<u>CATATG</u>TTC GCACCGCTG AAATCG-3' and 5'-GCGGATC<u>CATATG</u>TTATTATTTTTTCTCGGA CAG-3', were designed to contain *Nde*I sites (underlined) for vector construction. The coexpression of foldases (Dsb proteins, Skp and SurA) or molecular chaperones (DnaKJ and GroELS), can increase secretion efficiency, folding, and solubility of recombinant proteins in the periplasmic space *(7,14–16)*.

15. Possible competition exists between β-lactamase (*bla*) and the recombinant protein to be secreted during translocation because β-lactamase is also secreted into the periplasmic space. Therefore, another antibiotic marker is desirable to increase the amount of secreted recombinant proteins.

16. The recombinant plasmid pTHKCSFmII allows secretory production of hG-CSF under the *trc* promoter in *E. coli* *(8)*. When the hG-CSF gene was directly fused to the endoxylanase signal sequence, all recombinant *E. coli* cells were completely lysed, and hG-CSF could not be detected by SDS-PAGE analysis. hG-CSF is known to be a very hydrophobic protein. Generally, the hydrophobic proteins are different to transport to the periplasmic space because hydrophobicity disturbs the membrane structure during translocation. To alter the hydrophobicity, small oligopeptide (13 amino acids) containing the histidine hexamer and the factor Xa cleavage sequence were inserted between the endoxylanase signal sequence and hG-CSF. The histidine hexamer allows convenient recovery of protein by metal affinity chromatograph and introduces positive charges to hydrophobic proteins. The Factor Xa cleavage site allows the production of hG-CSF with authentic N-terminal amino acid sequence without fMet extension after the cleavage by Factor Xa.

17. Secreted hG-CSF can be easily purified by the Ni-NTA column because histidine hexamers are translationally fused to hG-CSF.

18. The protein samples are loaded onto an anion-exchange column (Bio-Scale Q2 column, Bio-Rad) pre-equilibrated with 50 m*M* Tris-HCl, pH 7.0, and are then eluted by a linear gradient of NaCl (from 0 to 1.0 *M* in the same buffer) at 90 mL/h. After collecting the fractionation containing hG-CSF, NaCl was removed by dialysis against 1 L of 20 m*M* sodium phosphate buffer, pH 6.5, for 24 h with three buffer exchanges.

19. The Fed-batch culture of recombinant *E. coli* allowed production of 4.2 g/L of hG-CSF *(11)*.

## Acknowledgments

Work described in this chapter was supported by the Korean Systems Biology Research grant (M10309020000-C3B5002-00000) from the Ministry of Science and Technology, Center for Ultramicrochemical Process Systems, and by the BK21 project.

## References

1. Lee, S. Y. (1996) High cell density cultivation of *Escherichia coli. Trends Biotechnol.* **14,** 98–105.
2. Makrides, S. C. (1996) Strategies for achieving high-level expression of genes in *Escherichia coli. Microbiol. Rev.* **60,** 512–538.
3. Choi, J. H. and Lee, S. Y. (2004) Secretory and extracellular production of recombinant proteins using *Escherichia coli. Appl. Microbiol. Biotechnol.* **64,** 625–635.
4. Park, S. J., Georgiou, G., and Lee, S. Y. (1999) Secretory production of recombinant protein by a high cell density culture of a protease negative mutant *Escherichia coli. Biotechnol. Prog.* **15,** 164–167.
5. Pugsley, A. P. (1993) The complete general secretory pathway in gram-negative bacteria. *Microbiol. Rev.* **57,** 50–108.
6. Jeong, K. J. and Lee, S. Y. (1999) High-level production of human leptin by fed-batch cultivation of recombinant *Escherichia coli* and its purification. *Appl. Environ. Microbiol.* **65,** 3027–3032.
7. Jeong, K. J. and Lee, S. Y. (2000) Secretory production of human leptin in *Escherichia coli. Biotechnol. Bioeng.* **67,** 398–407.
8. Jeong, K. J. and Lee, S. Y. (2001) Secretory production of human granulocyte colony-stimulating factor in *Escherichia coli. Protein Expr. Purif.* **23,** 311–318.
9. Choi, J. H., Jeong, K. J., Kim, S. C., and Lee, S. Y. (2000) Efficient secretory production of alkaline phosphatase by high cell density culture of recombinant *Escherichia coli* using the *Bacillus* sp. Endoxylanase signal sequence. *Appl. Microbiol. Biotechnol.* **53,** 640–645.
10. Jeong, K. J., Lee, P. C., Park, I. Y., Kim, M. S., and Kim, S. C. (1998) Molecular cloning and characterization of an endoxylanase gene of *Bacillus* sp. in *Escherichia coli. Enzyme Microb. Technol.* **22,** 599–605.
11. Yim, S. C., Jeong, K. J., Chang, H. N., and Lee, S. Y. (2001) High level secretory production of human G-CSF by fed-batch culture of recombinant *Escherichia coli. Bioproc. Biosyst. Eng.* **24,** 249–254.
12. Kane, J. F. (1995) Effects of rare codon clusters on high-level expression of heterologous proteins in *Escherichia coli. Curr. Opin. Biotechnol.* **6,** 494–500.
13. Jeong, K. J. and Lee, S. Y. (2000) Secretory production of human leptin in *Bacillus subtilis. J. Microbiol. Biotechnol.* **10,** 753–758.
14. Kurokawa, Y., Yanagi, H., and Yura, T. (2001) Overproduction of bacterial protein disulfide isomerase (DsbC) and its modulator (DsbD) markedly enhances periplasmic production of human nerve growth factor in *Escherichia coli. J. Biol. Chem.* **276,** 14393–14399.
15. Wulfing, C. and Rappuoli, R. (1997) Efficient production of heat-labile enterotoxin mutant proteins by overexpression of *dsbA* in a *degP*-deficient *Escherichia coli* strain. *Arch. Microbiol.* **167,** 280–283.
16. Qiu, J., Swartz, J. R., and Georgiou, G. (1998) Expression of active human tissue-type plasminogen activator in *Escherichia coli. Appl. Environ. Microbiol.* **64,** 4891–4896.

# 4

## Expression of Recombinant LTB Protein in *Marine vibrio* VSP60

### Wang Yili and Si Lusheng

## 1. Introduction

The main toxin of pathogenic *Escherichia coli* responsible for infant diarrhea and traveler's diarrhea is *E. coli* heat-labile enterotoxin (LT), and it is the cause of death for several hundred thousand children under 5 yr old every year, particularly in underdeveloped areas *(1,2)*. LT is a bacterial adenosine phosphate (ADP)-ribosylating exotoxin composed of a single A subunit (27 kDa) and five monomeric B subunits (11.6 kDa). Subunit A is the toxin portion, causing diarrhea with ADP-ribosyltransferase activity. Subunit B (LTB) is the receptor-binding site, which can bind to the GM1 ganglioside present on the eukaryotic cell membrane and has no pathogenic activity *(3–5)*. LT holotoxin and its B subunit are highly immunogenic, invoking neutralizing antibody and blocking the receptor-binding activity and diarrhea-causing activity in immunized individuals *(6,7)*. Furthermore, LTB has been shown to be the most powerful mucosal immunogen and mucosal adjuvant, markedly strengthening the immunoresponse to coadministered antigens *(8,9)*. In addition, LTB has an important role in the activity and control of immunomodulation, whereby it will result in a Th1 reaction switching to a Th2 reaction *(10,11)*. Consequently, much attention has been focused on LTB preparation using genetic engineering techniques. However, recombinant LTB (rLTB) has been expressed mainly in *E. coli*, in which the recombinant protein is secreted into the periplasm (*see* **Note 1**). To obtain the purified target protein, it is necessary to destruct the cell wall, which makes purification more difficult. *Cholerae vibrio* can express and efficiently secrete rLTB into the culture medium, and a low-pathogenic mutant strain has been used to prepare rLTB *(12)*. *Marine vibrio*, an organism closely related to *C. vibrio*, is very weakly pathogenic to humans and is similar to *C. vibrio* in terms of its expression and secretion of rLTB, with the ability to express this foreign protein very efficiently and secrete the recombinant protein into the media in its natural conformation *(13–15)*. Therefore, the use of *M. vibrio* for LTB expression is much more efficient. The methods for LTB expression in *M. vibrio* and

From: *Methods in Molecular Biology, vol. 308: Therapeutic Proteins: Methods and Protocols*
Edited by: C. M. Smales and D. C. James © Humana Press Inc.. Totowa, NJ

purification of the secreted protein using Sephacryl S-100 chromatography are described.

## 2. Materials

1. *E. coli* standard strain.
2. Plasmids: pMM66EH and pGEM-T.
3. Bacteria DH5a (Invitrogen) and host bacteria *M. vibrio* (VSP60).
4. Restriction enzymes: T4 DNA ligase.
5. Rabbit polyclonal antibody against rLTB.
6. Murine anti-rabbit immunoglobulin G (IgG) conjugated with horseradish peroxidase.
7. Ampicillin.
8. Isopropyl-β-D-thio-galactopyranoside (IPTG).
9. Luria-Bertani (LB) media and high-salt LB media containing 2% NaCl.
10. Agarose.
11. 1.5% Agar LB media.
12. Solution A: 0.991 g glucose, 0.303 g Tris base, and 20.372 g EDTA-Na, dissolved in 85 mL double-distilled (dd)$H_2O$. Adjust to pH 8.0 with HCl and add dd$H_2O$ to 1 L.
13. Solution B: Mix 10 *M* NaOH (2 mL), 10% sodium dodecyl sulfate (SDS; 10 mL), and 88 mL sterilized dd$H_2O$.
14. Solution C: Dissolve 29.44 g potassium acetate in sterilized dd$H_2O$ (60 mL), add glacial acetic acid (11.5 mL), and then make up to 100 mL with dd$H_2O$, pH 4.8–5.2. Store at 4°C.
15. TE buffer, pH 8.0: 10 m*M* Tris-HCl, pH 7.5, TBE electrophoresis buffer, and 1 *M* $CaCl_2$.
16. SDS-polyacrylamide gel electrophoresis (SDS-PAGE) and electroblotting equipment.
17. LKB chromatography system.
18. Sephacryl S-100 resin (Pharmacia).
19. Protein-dissolving solution: 29 m*M* Tris-HCl buffer, pH 8.0, containing 0.9% NaCl.
20. Ammonium sulfate.

## 3. Methods

The methods described in this chapter include the (1) construction of the target protein expression plasmid, (2) expression of the recombinant protein, (3) purification of the expressed recombinant protein, and (4) characterization of the protein.

### 3.1. Construction of Expression Plasmid

The construction of rLTB expression plasmid is described in **Subheadings 3.1.1.– 3.1.3.** and includes *LTB* gene cloning, a description of the pMMB66EH expression plasmid, and a discussion of the construction of the pCMMB-LTB plasmid (which is derived from the pMMB66EH plasmid with the *LTB* gene inserted).

### 3.1.1. LTB Gene Cloning

1. The *LTB* gene (17–391 bp) was obtained from an *E. coli* standard strain by polymerase chain reaction (PCR) with the following specific primers: a forward primer (5-CC GAATTCGGGATGAATTATGAATAAAG 3') and a reverse primer (5'-CGGGGA TCCCTAGTTTTCCATACTGAT-3') with *Eco*R1 and *Bam*H1 sites in the 5'- and 3'-ends.
2. PCR conditions were set as follows: 94°C for 30 s; 60°C for 30 s; and 72°C for 30 s. This program was performed for 30 cycles.

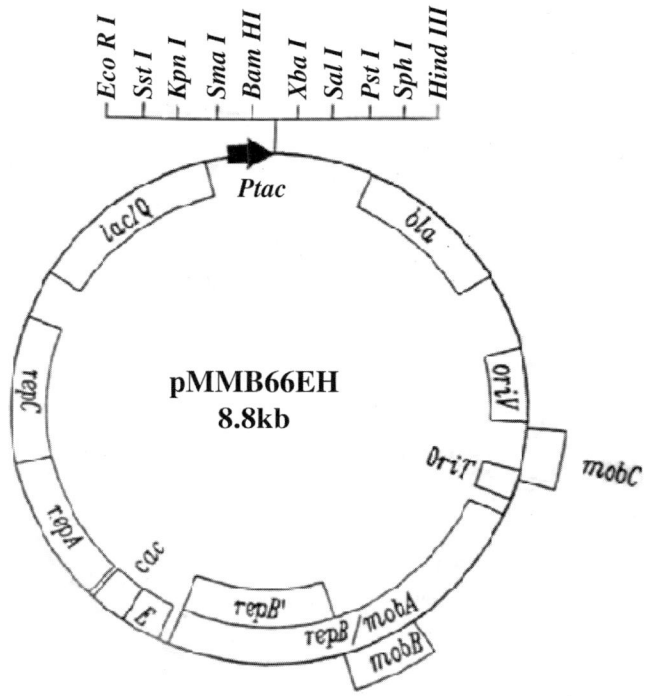

Fig. 1. A schematic circle map of the pMMB66EH expression plasmid.

3. After PCR, the generated fragment is cloned into the pGEM-T vector *(16)*. Transform competent DH5a cells with the recombinant pGEM-T/*LTB* plasmid. Plate the cells on LB media containing ampicillin.
4. Select single colonies culturing in LB media with ampicillin.
5. Isolate the plasmid for characterization with endonuclease digestion and sequencing.

### 3.1.2. pMMB66EH Plasmid

The plasmid pMMB66EH (**Fig. 1**) is a cyclic double-stranded DNA (8807 bp), derived from pBR322, containing the *Amp* and *LacZ* genes for antibiotic and color selection. The plasmid also contains an IPTG-inducible *Tac* promotor, as well as a polycloning site downstream from the promotor. The plasmid has a broad host range, including Gram-negative bacteria.

### 3.1.3. Construction of the pMMB-LTB Plasmid Containing the LTB Gene

1. Cut the pGEM-T/*LTB* plasmid with *Eco*R1 and *Bam*H1 to generate the *LTB* fragment.
2. Insert the fragment into the pMMB66EH expression plasmid after cutting with *Eco*R1 and *Bam*H1 using the standard molecular biology technique. Briefly, harvest the *LTB* gene fragment, digest the pMMB66EH plasmid, and then carry out the ligation. The resulting recombinant plasmid is pMMB-*LTB*.

### 3.2. Expression of the Recombinant Protein

The next steps (**Subheadings 3.2.1.–3.2.3.**) in the process involve the transformation of *M. vibrio* VSP60 with the pMMB-*LTB* expression plasmid, growth of the engineered host cells, followed by the IPTG induction to initiate protein expression, and finally, collection of the expressed protein.

### 3.2.1. Transformation of M. vibrio *VSP60*
### With pMMB-LTB Recombinant Plasmid

1. Transform competent VSP60 cells with the pMMB-LTB recombinant plasmid using standard molecular biology methods.
2. Plate the cells onto LB plates containing ampicillin, IPTG, and X-gal. Incubate overnight at 37°C.
3. Select single white colonies, and grow overnight at 37°C in LB media with ampicillin. Be careful to select single colonies from the plates, and avoid cross-contamination from other colonies.
4. Grow the host cells, isolate the plasmid, and verify the success of transformation by restriction digestion and sequencing.
5. Aliquot the host cells grown overnight (1.5 mL/vial), and freeze at –70°C in LB media containing 25% glycerol.

### 3.2.2. Recombinant LTB Expression

1. Plate the engineered VSP60 stock from a frozen aliquot onto an agar LB plate with ampicillin, and incubate it overnight at 37°C.
2. Select a single colony of engineered VSP60 from the plate, and grow it in approx 5 mL LB media containing ampicillin at 37°C overnight.
3. Transfer the bacteria into a flask containing 1 L LB media with ampicillin, and grow to an optical density ($OD_{600}$) of 0.2–0.3 at 30°C. Then, induce the cells with IPTG at a concentration of 0.5 m$M$ further for 20 h (*see* **Note 2**).

### 3.2.3. Harvesting the Recombinant Protein

1. Recombinant LTB expressed by the engineered VSP60 is designed to be secreted into the supernatant; therefore, harvest the supernatant of cultures by centrifugation.
2. Add $(NH_4)_2SO_4$ with gentle stirring to 30% saturation.
3. After precipitating at 4°C for 4 h, remove the precipitate by centrifugation (4°C for 30 min).
4. Add $(NH_4)_2SO_4$ to the remaining supernatant until 85% saturation, leave the solution at 4°C for 10 h, and then harvest the precipitate by centrifugation at 4°C for 30 min.
5. Dissolve the precipitate in protein-dissolving solution, and transfer into a dialysis bag (molecular weight cut-off 8000–10,000), and dialyze against 0.9% NaCl solution, pH 7.5, until no $(NH_4)_2SO_4$ is present.

### 3.3. Purification of the Protein

The purification of rLTB is described in **Subheadings 3.3.1.–3.3.4.** and includes sample preparation: preparation of the Sephacryl 100 gel column, sample loading, and the collection of the eluted target protein.

### 3.3.1. Sample Preparation

Centrifuge the dialyzed protein solution at 5600$g$ and 4°C for 20 min, and discard the precipitate. (If the volume is too large, concentrate the sample.) Aliquot the sample into convenient sized samples, then boil one aliquot for 5 min immediately before loading.

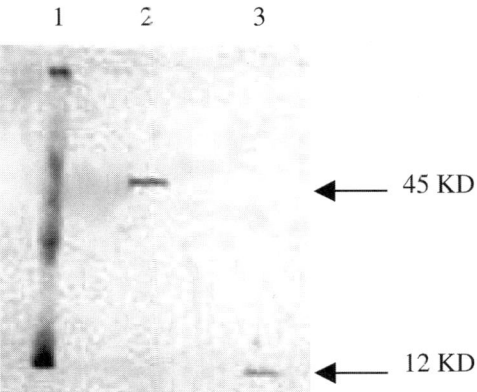

Fig. 2. SDS-PAGE electrophoresis analysis of purified rLTB protein. Lane 1, rainbow Marker; lane 2, purified samples without boiling prior to electrophoresis; lane 3, purified sample with boiling prior to electrophoresis.

### 3.3.2. Preparation of the Sephacryl 100 Gel Column

1. Prepare the Sephacryl 100 suspension, and fill up a $4 \times 100$-cm glass column with the resin. (Be careful that the packed column is distributed evenly and without lamella and bubbles.)
2. After pouring the column, connect the column to the LKB automate chromatography system, and flash the column with at least three column volumes of water to remove the remaining alcohol used for suspending and storing the Sephacryl 100.
3. Equilibrate the column with 100 m$M$ Tris-HCl buffer (approx three column volumes are required).

### 3.3.3. Sample Loading

1. Load the boiled sample (protein < 10 mg and vol < 2 mL) dissolved in 100 m$M$ Tris-HCl buffer onto the column.
2. Elute with the same buffer (1.8 mL/min current velocity), and monitor the elution by ultraviolet at 280 nm.
3. Collected the eluted protein peak.
4. Process the unboiled sample using the same procedure.

### 3.4. Characterization of the Purified Protein

1. Identify the fraction of interest using a Coomassie blue (G250)-stained 15% SDS-PAGE gel followed by Western blotting using standard methods. There will be an extra protein band (45 kDa) corresponding to the pentamer of LTB for the unboiled sample and two bands (12.5 and 45 kDa) corresponding to the monomer and pentamer for the boiled sample on an SDS-PAGE gel (**Fig. 2**). After blotting, the protein bands can be positively identified using an antibody against LTB.
2. Quantify the total protein and generated protein using the Biuret reaction, and calculate the yield rate. The yield rate should be approx 51.5–52%. Therefore, the yield from 1 L LB culture of engineered bacteria is 5.1–5.3 mg.

## 4. Notes

1. rLTB has been expressed in transgenic plants, such as tobacco and potato *(17–21)*, in which rLTB is present in a very stable pentamer form that is almost impossible to depolymerize, even after boiling. Compared with the conventional prokaryotic expression systems, transgenic plants offer many great advantages, mainly ease, simplicity, convenience, and cost. Compared with injection vaccines, edible transgenic plant LTBs are much easier to apply, cheap, and safe. Transgenic plant LTB has been shown to protect mice against *E. coli* LT, which causes diarrhea.

2. Multiple variables can affect the expression of LTB in the *M. vibrio* system. Based on our experience, the optimal conditions for generating LTB are as follows. Culture the host cells in LB medium at 30°C. When the $OD_{600}$ is 0.2–0.3, add IPTG to the medium at a concentration of 0.5 m$M$ and induce for 18–22 h.

## Acknowledgments

We thank Drs. Wang Jing and Zheng Jin for their help.

## References

1. Sack, B. R. (1975) Human diarrheal disease caused by enterotoxingenic *Escherichia coli. Annu. Rev. Microbiol.* **29,** 331–353.
2. Katouli, M., Jafari, A., Farhoudi-Moghadam, A. A., and Ketobi, G. R. (1990) Etiological studies of diarrheal disease in infants and children in Iran. *J. Trop. Med. Hyg.* **93,** 22–27.
3. Dallas, W. S. and Falkow, S. (1980) Amino acid sequence homology between cholera toxin and heat-labile toxin. *Nature* **228,** 491–501.
4. Moss, J. H. and Richardson, S. H. (1978) Activation of adenylate cyclase by heat-labile enterotoxin. Evidence of ADP-ribosyl transferase activity similar to that of choleragen. *J. Clin. Invest.* **62,** 281–285.
5. Spangler, B. D. (1992) Structure and function of cholera toxin and the related Escherichia coli heat-labile enterotoxin. *Microbiol. Rev.* **56,** 622–647.
6. Nashar, T. O., Webb, H. M., Eaglestone, S., Williams, N. A., and Hirst, T. R. (1996) Potent immunogenicity of the B subunit of the Escherichia coli heat-labile enterotoxin: receptor binding is essential and induces differential modulation of lymphocyte subsets. *Proc. Natl. Acad. Sci. USA* **93,** 226–230.
7. de Haan, L., Verweij, W. R., Feil, I. K., Holtrop, M., Hol, W. G., Agsteribbe, E., and Wilschut, J. (1998) Role of GM1 binding in the mucosal immunogenecity and adjuvant activity of the Escherichia coli heat-labile enterotoxin and its B subunit. *Immunology* **94,** 424–430.
8. Millar, D. G., Hirst, T. R., and Snider, D. P. (2001) Escherichia Coli heat-labile enterotoxin B subunit is a more potent mucosal adjuvant than its closely related homology, the B subunit of cholera toxin. *Infect. Immun.* **69,** 3476–3482.
9. Richards, C. M., Aman, A. T., Hirst, T. R., Hill, T. J., and Williams, N. A. (2001) Protective mucosal Immunity to ocular herpes simplex virus type-1 infection in mice by using Escherichia coli heat-labile enterotoxin B subunit as an adjuvant. *J. Virol.* **75,** 1664–1671.
10. Willams, N. A. (2000) Immune modulation by the cholera-like enterotoxin B subunit: from adjuvant to immunotherapeutic. *Int. J. Med. Microbiol.* **290,** 447–453.
11. Luross, J. A., Heaton, T., Hirst, T. R., Day, M. J., and Williams, N. A. (2002) *Escherichia coli* heat-labile enterotoxin B subunit prevents autoimmune arthritis through induction of regulatory CD4+ T cells. *Arthritis Rheum.* **46,** 1671–1678.

12. Ramesh, A., Panda, A. K., Maiti, B. R., and Mukhopadhyay, A. (1995) Studies on plasmid stability and LTB production by recombinant *Vibrio cholerae* in batch and chemostat cultures: A lesson for optimizing conditions for chemical induction. *J. Biotechnol.* **45,** 45–51.

13. Leece, R. and Hirst, T. R. (1992) Expression of the B subunit of Escherichia coli heat-labile enterotoxin in a marine Vibrio and in a mutant that is pleiotropically defective in the secretion of extracellular proteins. *J. Gen. Microbiol.* **138,** 719–724.

14. Marcello, A., Loregian, A., Palu, G., et al. (1994) Efficient extracellular production of hybrid E. coli heat-labile enterotoxin B subunits in a Marine vibrio. *FEMS Microbiol. Lett.* **117,** 47–51.

15. Amin, T. and Hirst, T. R. (1994) Purification of the B-subunit oligomer of Escherichia coli heat-labile enterotoxin by heterologous expression and secretion in a marine vibrio. *Protein Expr. Purif.* **5,** 198–204.

16. pGEM-T and pGEM-T Easy Vector Systems manual.

17. Haq, T. A., Mason, H. S., Clements, J. D., and Arntzen, C. J. (1995) Oral immunization with a recombinant bacterial antigen produced in transgenic plants. *Science* **268,** 714–716.

18. Lauterslager, T. G., Florack, D. E., van der Wal, T. J., Molthoff, J. W., Langeveid, J. P., Bosch, D., et al. (2001) Oral immunization of naïve and primed animals with transgenic potato tubers expressing LT-B. *Vaccine* **19,** 2749–2755.

19. Tacket, C. O., Mason, H. S., Losonsky, G., Clememts, J. D., Levine, M. M., and Arntzen, C. J. (1998) Immunogenecity in humans of a recombinant bacterial antigen delivered in a transgenic potato. *Nat. Med.* **4,** 607–609.

20. Arakawa, T., Chong, D. K., Merritt, J. L., and Langridge, W. H. (1997) Expression of cholera toxin B subunit oligomers in transgenic potato plants. *Transgenic Res.* **6,** 403–413.

21. Mason, H. S., Haq, T. A., Clements, J. D., and Arntzen, C. J. (1998) Edible vaccine protects mice against Escherichia coli heat-labile enterotoxin (LT) potatoes expressing a synthetic LT-B gene. *Vaccine* **16,** 1336–1343.

# 5

## Heterologous Gene Expression in Yeast

### Lee J. Byrne, Kenneth J. O'Callaghan, and Mick F. Tuite

## 1. Introduction

*Saccharomyces cerevisiae* and *Pichia pastoris* provide appealing alternatives to bacterial and mammalian protein expression systems for the production of recombinant proteins. The main advantages of using yeast include an extensive toolbox of genetic modification strategies, production of authentic eukaryotic products, and low culture costs when compared with mammalian systems. Target recombinant proteins can be produced intracellularly or engineered to enter the yeast eukaryotic secretory pathway, resulting in secretion into the culture medium. This is essential if the target protein requires additional folding steps (e.g., disulfide bond formation) or modification (e.g., glycosylation). However, glycosylation patterns produced by yeast cells may differ from the native protein and may be potentially immunogenic (*1*). Therefore, if native glycosylation patterns are an important consideration, then the use of a mammalian expression system may be preferred.

In recent years, there has been a trend to move from *S. cerevisiae* to *P. pastoris* for production of recombinant proteins. This trend is caused by the preference of *P. pastoris* for respiratory growth and the extremely high level of expression obtained when heterologous gene expression is driven by the alcohol oxidase gene (*AOX1*) promoter (*2*). Over 100 heterologous proteins from bacteria to humans have been successfully expressed in *P. pastoris*; examples include human insulin (*3*), human amyloid precursor protein (APP; *4*), hepatitis B surface antigen (*5*) and human immunodeficiency virus type 1 (HIV-1) envelope glycoprotein, gp120 (ENV; *6*). In considering expression strategies using either yeast species, we begin with a general consideration of heterologous gene expression in these organisms before moving on to the procedures for their introduction, maintenance, and expression. Finally, we suggest methods for efficient recovery of the recombinant protein and its concentration prior to subsequent purification to homogeneity.

From: *Methods in Molecular Biology, vol. 308: Therapeutic Proteins: Methods and Protocols*
Edited by: C. M. Smales and D C. James © Humana Press Inc., Totowa, NJ

**Table 1**
**Useful Yeast Strains**

| Strain designation | Genotype | Comment | ATCC* number |
|---|---|---|---|
| *S. cerevisiae* | | | |
| W303 | *MATα, ade2-1, ura3-1, leu2-3, trp1-1* | Standard laboratory strain | 200060 |
| BJ2168 | *MATa, leu2, trp1, ura3-52, prb1-1122, pep4-3, prc1-407, gal2* | Protease-deficient | 208277 |
| BJ3501 | *MATα, pep4::HIS3, prb1-Δ1.6R, his3-Δ200, ura3-52, can1. gal2* | Protease-deficient | 208280 |
| BJ35C5 | *MATα, pep4::HIS3, prb1-Δ1.6R, his3-Δ200, lys2-801, trp1-Δ101 (gal3), ura3-52 (gal2), can1* | Protease-deficient | 208281 |
| *P. pastoris* | | | |
| Y-11430 | Wild-type | Wild-type | 76273 |
| GS115 | *his4* | His⁻ | 20864 |
| KM71 | *Δaox1::SARG4 his4 arg4* | Protease-deficient | 201178 |

*American type culture collection (ATCC) reference number.

## 2. Materials

### 2.1. Yeast Strains and Media

1. Many laboratory strains of *S. cerevisiae* and *P. pastoris* are available for heterologous gene expression that have been genetically manipulated to enable expression vector selection and, in some strains, improved target gene expression and/or recombinant product recovery *(2,7)*. To optimize recombinant protein production, it is clearly necessary to choose a host strain compatible with the vector in which the target heterologous gene is cloned. Yeast strains are available from the Yeast Genetics Stock Centre, which has a searchable website (www.lgcpromochem-atcc.com/SearchCatalogs/Fungi_Yeasts.cfm) to aid the selection of a strain with a suitable genotype. Strains of *S. cerevisiae* and *P. pastoris* that have been used previously to express recombinant proteins are specified in **Table 1**.
2. YEPD medium: 1% (w/v) yeast extract, 2% (w/v) bactopeptone, and 2% (w/v) glucose. For solid agar medium, add 2% (w/v) granulated agar. Sterilize by autoclaving for 20 min at 15 lb/in.²
3. Synthetic complete (SC) medium: 0.67% (w/v) yeast nitrogen base with ammonium sulfate but without amino acids, 2% (w/v) glucose, and 0.1% (w/v) amino acid "drop-out"

mix (*see* **Note 1**). For solid agar medium, add 2% (w/v) granulated agar. Sterilize by autoclaving 20 min at 15 lb/in.$^2$.

4. YNBG medium: 1.34% (w/v) yeast nitrogen base with ammonium sulfate but without amino acids, 2% (w/v) glycerol, and $4 \times 10^{-5}$% (w/v) biotin. For solid agar media, add 2% (w/v) granulated agar. Sterilize by autoclaving 20 min at 15 lb/in.$^2$.

5. Phosphate-buffered saline (PBS): 140 m$M$ NaCl, 2.7 m$M$ KCl, 10 m$M$ Na$_2$HPO$_4$, and 0.2 m$M$ KH$_2$PO$_4$, pH 7.4. Sterilize by autoclaving 20 min at 15 lb/in.$^2$.

## *2.2.* S. cerevisiae *Transformation*

### *2.2.1. Protocol 1: Lithium Acetate and Heat-Shock Reagents*

1. YEPD medium: *see* **Subheading 2.1.1.**, **item 1**.
2. TE buffer: 10 m$M$ Tris-HCl, pH 7.6, and 0.1 m$M$ EDTA. Sterilize by autoclaving 20 min at 15 lb/in.$^2$.
3. Lithium acetate solution: 0.1 $M$ lithium acetate in TE buffer. Sterilize by autoclaving 20 min at 15 lb/in.$^2$.
4. Sheared and denatured salmon sperm DNA (5 mg/mL), prepared by passing through the fine point of a 23G 0.6-mm needle 50 times, incubating at 100°C for 5 min, and immediately cooling on ice.
5. Polyethylene glycol (PEG) 4000: 40% (w/v) in sterilized 0.1 $M$ lithium acetate solution in TE buffer (*see* **item 3**).

### *2.2.2. Protocol 2: Plate Method Reagents*

1. Sheared and denatured salmon sperm DNA (*see* **Subheading 2.2.1.**, **item 4**).
2. Plate buffer: 40% (w/v) PEG 4000, 0.1 $M$ lithium acetate, 10 m$M$ Tris-HCl, pH 7.5, and 1 m$M$ EDTA. Sterilize by autoclaving 20 min at 15 lb/in.$^2$.

## *2.3.* P. pastoris *Transformation*

1. YEPD medium (*see* **Subheading 2.1.1.**, **item 1**).
2. YEPD/HEPES: 1% (w/v) yeast extract, 2% (w/v) bactopeptone, 2% (w/v) glucose, and 0.167 $M$ HEPES, pH 8.0. Sterilize by autoclaving 20 min at 15 lb/in.$^2$.
3. 1 $M$ Sorbitol in H$_2$O. Sterilize by autoclaving 20 min at 15 lb/in.$^2$.
4. Sterile 0.2-cm electroporation cuvet (e.g., Bio-Rad).
5. Electroporator (e.g., Bio-Rad Gene Pulser Xcell™).

## *2.4. Heterologous Gene Expression in Yeast*

### *2.4.1. Induction of the* GAL *Promoter in* S. cerevisiae

1. Glucose SC medium: *see* **Subheading 2.1.1.**, **item 2** and **Note 1**.
2. Preinduction medium: 0.67% (w/v) yeast nitrogen base with ammonium sulfate but without amino acids, 2% (w/v) raffinose, and 0.1% (w/v) amino acid drop-out mix (*see* **Note 1**). Sterilize by autoclaving 20 min at 15 lb/in.$^2$.
3. 20% (w/v) galactose stock in H$_2$O (filtered through 0.2-μm disposable sterile filter).

### *2.4.2. Induction of the* AOX1 *Promoter in* P. pastoris

1. YNBG: 1.34% (w/v) yeast nitrogen base with ammonium sulfate without amino acids and 2% (w/v) glycerol. Sterilize by autoclaving 20 min at 15 lb/in.$^2$. Once cool, add filter-sterilized biotin to a final concentration of $4 \times 10^{-5}$% (w/v).

2. YNBM: 1.34% (w/v) yeast nitrogen base with ammonium sulfate without amino acids, 100 m$M$ sodium phosphate buffer, pH 6.0, and 1% (w/v) casamino acids (*see* **Note 2**). Sterilize by autoclaving 20 min at 15 lb/in.$^2$. When cool, add methanol to a final concentration of 1% (v/v) and also filter-sterilized biotin to a final concentration of $4 \times 10^{-5}$% (w/v).

## 2.5. Recovery of Expressed Protein

### 2.5.1. Extraction of Intracellular Protein
### With Glass Beads in Nondenaturing Buffer

1. Nondenaturing lysis buffer, pH 6.8: 25 m$M$ Tris-HCl, 150 m$M$ NaCl, 10% (v/v) glycerol, 2% (v/v) of 100 m$M$ phenylmethyl sulfonylfluoride (PMSF) stock solution (dissolved in isopropanol), and 1X Complete (Roche) protease inhibitors. The latter will result in a final concentration of 1 m$M$ EDTA (a metalloprotease inhibitor), but additional EDTA may be added (up to 5 m$M$ final) if high levels of metalloprotease activity are suspected. Complete protease inhibitors can also be used at 2X concentration.
2. Glass beads of approx 0.5-m$M$ diameter (BDH, United Kingdom).

### 2.5.2. Concentration of Extracellular Recombinant Proteins

1. Amicon stirred ultrafiltration cell model 8400 (Millipore).
2. Amicon YM-10 membrane ∅76 mm (Millipore).
3. A regulated N$_2$ gas cylinder and supply.

### 2.5.3. Concentration of Protein Samples

1. Centricon™ devices (Millipore) come with a range of nominal molecular weight cutoffs: YM-3, 3 kDa; YM-10, 10 kDa; YM-30, 30 kDa; YM-50, 50 kDa; YM-100, 100 kDa.
2. Centricon™ devices require the use of a fixed-angle rotor with inserts and adaptors capable of accepting 17 × 100-mm tubes, e.g., a JA-20 rotor with rubber inserts (Beckman).

## 3. Methods

### 3.1. Culturing S. cerevisiae and P. pastoris

Yeast growth medium is made from simple and cheap ingredients that provide both a carbon and nitrogen source. Although yeasts can be grown anaerobically, aerobic growth is preferred to maximize yields of recombinant protein. Ideally, culture volumes should not exceed 10–20% of the total flask volume and must be shaken vigorously (≥200 rpm) to ensure adequate aeration of the culture.

YEPD is a complex-rich growth medium ideal for growing any yeast strain when selective pressure does not need to be supplied. SC medium is a defined medium supplemented with amino acids and other nutrients required for growth, except that provided by the complementing vector (*see* **Note 3**). Culturing *P. pastoris* in a defined minimal medium (e.g., YNBG) before methanol induction of secreted recombinant protein results in fewer contaminants.

Typically, small cultures (up to 50 mL) are inoculated with individual yeast colonies from YEPD or SC drop-out agar plates using a sterile loop, whereas larger cultures (≥500 mL) require inoculation from a previously grown preculture. Cultures are incubated at 30°C with vigorous shaking (≥200 rpm). Yeast culture growth can be most readily measured spectrophotometrically by monitoring the optical density (OD) of the culture at 600 nm, with an $OD_{600}$ of 1.0 being approximately equal to $1 \times 10^7$ cells/mL (*see* **Note 4**). Alternatively, appropriate dilutions of the culture in sterile PBS can be counted on a hemocytometer.

### 3.2. Transformation of S. cerevisiae

Two protocols are used for the introduction of plasmid DNA into the host strain of *S. cerevisiae*. The heat-shock method usually has greater transformation efficiency over the Plate method (*8*), but the Plate method has the advantage that cells can be used directly from the stationary phase.

### 3.2.1. Protocol 1: Lithium Acetate and Heat Shock

1. Inoculate 50 mL YEPD with the desired *S. cerevisiae* strain, and grow overnight (12–16 h) at 30°C with shaking (180 rpm) to mid-log phase ($OD_{600}$ 0.4–0.5).
2. Harvest cells (3000 rpm for 5 min at 4°C) and wash twice in TE buffer.
3. Resuspend the cell pellet in 5 mL lithium acetate solution and centrifuge (3000 rpm for 5 min at 4°C). Discard the supernatant.
4. Resuspend the cell pellet in 0.5 mL lithium acetate solution (*see* **Note 5**).
5. Mix 1 µg plasmid DNA to be transformed with 5 µL sheared salmon sperm DNA. Add 60 µL resuspended cells.
6. Mix and add 400 µL PEG 4000 solution. Incubate at 30°C for 45 min.
7. Transfer to a 42°C water bath and incubate for 15 min.
8. Pellet cells gently (1000 rpm for 2 min at room temperature) and resuspend in 0.5 mL fresh YEPD. Incubate at 30°C for 1 h.
9. Plate 100 µL onto YNB selective agar. Colonies usually appear after 2–5 d of incubation at 30°C.

### 3.2.2. Protocol 2: Plate Method

1. Mix 1 µg plasmid DNA to be transformed with 10 µL sheared salmon sperm carrier DNA (5 mg/mL) in a 1.5-mL microfuge tube.
2. Add 500 µL Plate mix.
3. If a stationary culture is to be used, take 0.5 mL and centrifuge at 3000 rpm for 5 min at 4°C, and then resuspend the cell pellet in the DNA/Plate mixture. If yeast cells are used from a previously cultured agar plate, use a toothpick to scrape off $10^7$–$10^8$ cells and resuspend in the Plate/DNA mixture.
4. Incubate the mixture overnight (12–16 h) at room temperature, and then plate 50–200 µL transformation mix onto selective agar medium. Incubate at 30°C for 2–5 d or until colonies appear.

### 3.3. Transformation of P. pastoris by Electroporation

There are several different methods for *P. pastoris* transformation (*see* **Note 6**), but electroporation is the most convenient, allowing high-transformation frequency without the requirement for spheroplast generation (*9*). Most *P. pastoris* plasmid vectors are linearized prior to transformation to allow integration of the vector DNA into the host cell genome by homologous recombination (*1*). This ensures greater genetic stability of the transformed strains engineered to express a recombinant protein (*see* **Note 7**). Various methods for the preparation of electrocompetent cells exist, and the user may wish to follow the protocol suggested by the manufacturer of their preferred electroporator. The following protocol is based on a general protocol for preparation of competent cells for use with the Bio-Rad Gene Pulser Xcell electroporator.

1. Inoculate 50 mL YEPD, and grow it overnight at 30°C with vigorous shaking (200 rpm).
2. Use 0.5 mL overnight culture to inoculate 250 mL fresh YEPD, and grow this overnight again until the $OD_{600}$ is 1.5/mL.
3. Pellet the cells at 3000 rpm for 5 min at 4°C, discard the supernatant, and resuspend the cell pellet in 250 mL ice-cold sterile $H_2O$.
4. Pellet the cells again (3000 rpm at 5 min for 4°C), but now resuspend the cells in 50 mL sterile YEPD/HEPES.
5. Add 2.5 mL of 1 $M$ dithiothreitol (DTT), and swirl gently to mix. Incubate the cells at 30°C *without shaking* for 15 min.
6. Add 150 mL ice-cold 1 $M$ sorbitol. Centrifuge as before to pellet the cells.
7. Resuspend in 25 mL ice-cold 1 $M$ sorbitol, vortex if necessary, and add 1 $M$ ice-cold sorbitol to bring the volume to 250 mL.
8. Pellet the cells by centrifugation, and discard the supernatant. Resuspend the pellet in 12.5 mL sterile ice-cold 1 $M$ sorbitol, and transfer to a prechilled 30-mL Oakridge centrifuge tube.
9. Pellet the cells (3000$g$ for 5 min at 4°C), discard the supernatant, and resuspend in 0.5 mL ice-cold 1 $M$ sorbitol (*see* **Note 8**).
10. Take 50 μL resuspended cells, and mix with 5 μg linear DNA to be transformed (*see* **Note 9**), and incubate in a previously chilled 0.2-cm electroporation cuvet for 5 min on ice.
11. Place the cuvet in the electroporator, and pulse the cells according to the parameters recommended by the manufacturer (e.g., for the Bio-Rad Gene Pulser Xcell electroporator, these are $C = 25$ μF, $PC = 200$ Ω, $V = 2.0$ kV; *see* **Note 10**).
12. Remove the cuvet, and add 1 mL 1 $M$ ice-cold sorbitol directly into the cuvet, and transfer the transformation mix to a fresh 1.5-mL centrifuge tube.
13. Plate 200 μL onto selective agar medium, and incubate the plates at 30°C for 3–5 d until colonies appear.

### 3.4. Heterologous Gene Expression in Yeast

Few native yeast genes contain introns; therefore, any heterologous gene introduced should be as cDNA made from the full-length mRNA transcript. In yeast, there are several different ways in which the level of gene expression can be controlled. Primarily, this is achieved via a promoter placed upstream of the target gene, but the use of a multicopy vector over a single-copy vector can increase the level of transcript and resulting polypeptide. It is advisable to use a homologous yeast promoter rather than that of the heterologous gene. In practice, two types of promoter are useful for high-level expression: constitutive and regulated. Constitutive promoter–driven genes are continuously transcribed and drive high levels of expression throughout the growth phase of the cell (*see* **Note 11**). Practically, the host strain is grown in rich complex medium (YEPD) or SC medium (depending on whether vector selection is required) until the cell density has reached the stationary (nongrowth) phase. The expressed product is then simply recovered. (For examples of product recovery protocols, *see* **Subheading 3.5**.)

In contrast, regulated promoters are usually responsive to a nutrient or chemical trigger that is added directly to the culture medium (*see* **Note 12**), allowing a culture to reach a particular cell mass before initiating recombinant protein expression. Therefore, if the target product is suspected of having detrimental effects to cell growth, then the use of a regulated promoter may be preferred over a constitutive one. Sample regulated promoter induction procedures for *S. cerevisiae* and *P. pastoris* can be found in **Subheadings 3.4.1.** and **3.4.2.**

In addition, increased authenticity of the target protein, such as the formation of native disulfide bonds and N- and/or O-linked glycosylation, may also be provided when utilizing the secretory pathway of the yeast cell. By engineering a signal peptide onto the N-terminus of the target protein, it is possible to route the protein through the secretory pathway and out into the culture medium (*see* **Note 13**). This is advantageous, as yeast cells secrete few proteins; therefore, the recombinant protein can be relatively pure when compared to intracellular expression (*see* **Note 14**).

### 3.4.1. Galactose-Induced Gene Expression in S. cerevisiae

1. Inoculate a 50-mL culture of glucose SC medium (made with the appropriate drop-out mix) with previously transformed cells, and grow this overnight (16 h) with shaking (200 rpm at 30°C). Record the $OD_{600}$ of the culture.
2. Harvest the cells (3000 rpm at 25°C), and resuspend cells in 50 mL preinduction medium to give an $OD_{600}$ equal to 0.1 (*see* **Note 15**).
3. Grow the culture at 30°C (200 rpm) until $OD_{600}$ reaches 0.5/mL.
4. Add 0.5 mL 20% (w/v) galactose stock, and grow for 4–6 h at 30°C (200 rpm).
5. Cool the culture on ice and harvest cells (3000 rpm for 5 min at 25°C). Retain cells or the culture medium depending on target protein location.

### 3.4.2. Methanol-Induced Gene Expression in P. pastoris

1. Inoculate 5 mL YNBG medium and grow overnight at 30°C with vigorous shaking (250 rpm).
2. Use 0.5 mL overnight culture to inoculate 250 mL YNBG. Incubate overnight until the undiluted culture has an $OD_{600}$ between 20 and 40/mL.
3. Pellet the cells (3000 rpm for 5 min at room temperature), and resuspend in 1/5 of the original culture volume (i.e., 50 mL) with fresh YNBM, pH 6.0 (*see* **Note 16**).
4. Add methanol to a final concentration of 0.5% (v/v) to induce the *AOXI* promoter and incubate at 30°C with vigorous shaking (250 rpm).
5. Induce for 96 h, adding methanol to a final concentration of 0.5% (v/v) every 24 h to maintain the induction.
6. After 96 h, cool the culture on ice before pelleting the cells (3000*g* for 5 min at 4°C), retaining either the cell pellet or culture medium as appropriate.

## 3.5. Recovery of Expressed Protein

### 3.5.1. Extraction of Intracellular Proteins With Glass Beads in Nondenaturing Buffer

1. Cool the culture on ice, and pellet the cells by centrifugation at 3000 rpm for 10 min at 4°C.
2. Wash the cells in 5 mL sterile $H_2O$, centrifuge (3000 rpm at 4°C), discard the supernatant, and place the cell pellet on ice.
3. Resuspend the cell pellet in 500 µL cold nondenaturing lysis buffer, pH 6.8, and transfer to a 1.5-mL microfuge tube on ice.

4. Add cold acid-washed (**Note 17**) glass beads (~0.5-mm diameter) until their upper edge rests just below the sample meniscus.
5. Vortex at high speed for $5 \times 30$ s, returning tube to ice for 30 s between each vortexing.
6. Thoroughly clean the tapered tip of the tube with ethanol, rinse with sterile water, open lid (**Note 18**), and pierce the tube tapered tip with a hot narrow ($\leq 0.5$-mm diameter) needle.
7. Quickly place the pierced sample tube into a new sterile tube, close the lid of the sample tube, and briefly centrifuge (at 4°C) the two tubes together to obtain lysate without glass beads.
8. The protein sample is now ready for analysis by sodium dodecyl sulfate-polyacrylamide gel electrophoresis (SDS-PAGE) or further purification. Keep the lysates cold unless they are being prepared for electrophoresis. Do not store protein samples at room temperature. Lysates can be stored at 4°C (but they may degrade within hours or days), –20°C with 50% (v/v) glycerol (where samples are stable for months), or –80°C (where samples are stable for years, *see* **Note 19**).

### 3.5.2. Concentration of Extracellular Proteins

1. Cool the culture on ice, and centrifuge the cells at 3000 rpm for 10 min at 4°C to pellet the cells.
2. Prepare the ultrafiltration YM-10 Ø76 mm membrane by floating the glossy side down in a beaker of distilled water for 1 h, changing the water three times to remove glycerine and other preservatives used in membrane storage.
3. Assemble the Amicon 8400 (400 mL) stirred ultrafiltration cell components according to the manufacturers' instructions (*see* **Fig. 1**), ensuring that the glossy side of the YM10 Ø76 mm membrane faces up.
4. Affix the filtrate waste tubing, and insert the stirring assembly so that the stirrer arms rest on the ridge at the top of the cell body.
5. Decant the cleared media into the stirred cell and push on the cap, ensuring that the gas inlet port is opposite the filtrate waste tube.
6. Open the pressure-relief valve (horizontal position), and slide into the retaining stand.
7. Close the pressure-relief valve (vertical position), stand the unit on a magnetic stirrer, and attach the inert gas pressure line ($N_2$ recommended; *see* **Note 20**).
8. Open the gas valve, and allow the $N_2$ gas pressure to rise to an optimal 55 psi (3.7 kg/cm$^2$) or maximal 70 psi (4.7 kg/cm$^2$). The stirred cell assembly will move slightly within the retaining stand and lock into place.
9. Adjust the stirring rate until the vortex created is approximately one third of the depth of the medium volume.
10. Concentrate until the volume of medium has decreased to approx 20–50 mL. This may take several hours (*see* **Note 21**).
11. Turn off the gas supply and magnetic stirrer, and depressurize the stirred cell by slowly opening (horizontal) the pressure-relief valve (*see* **Note 22**).
12. Remove the stirred cell from the retaining stand, and remove the cap and stirring assembly.
13. Pour out the retained medium into a 50-mL Falcon tube. The retained medium is now ready for further concentration using the smaller 50-mL Amicon stirred ultrafiltration cell (Model 8050 following the above instructions) or by using Amicon Centricon Centrifugal Filter Devices as described in **Subheading 3.5.3**. Alternatively, the concentrated medium may be used directly for analysis and/or purification of the expressed recombinant protein.

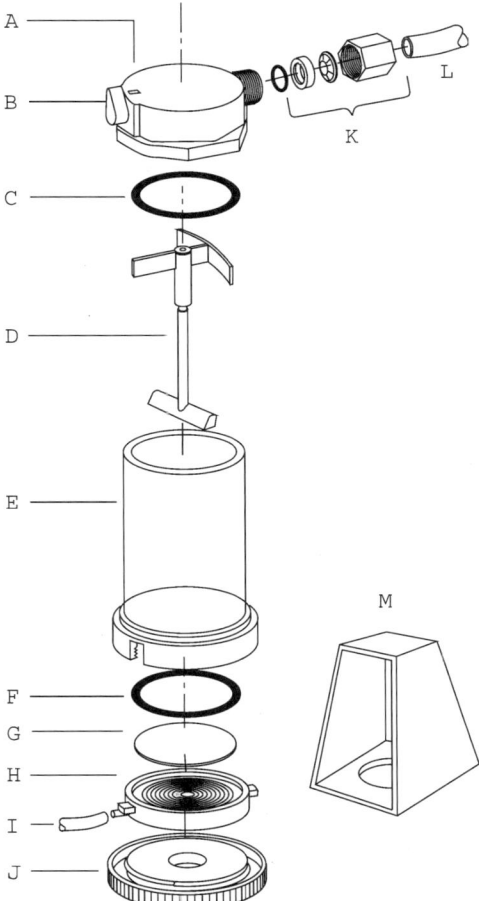

Fig. 1. Exploded diagram of the Amicon 8400 ultrafiltration stirred cell. A, cap assembly; B, pressure release valve; C, O-ring for cap assembly; D, stirrer assembly; E, body; F, O-ring; G, filter; H, base; I, filtrate waste tubing; J, retaining stand assembly; K, tubing fitting assembly; L, pressure tubing; and M, retaining stand. (Figure courtesy of Millipore Corporation.)

### 3.5.3. Concentration of Protein Samples

1. Remove contaminating membrane preservatives by adding 1 mL sterile $H_2O$ to the upper reservoir of the Centricon and centrifuging at 3600 rpm at 4°C in a Beckman JA20 rotor (or equivalent) until the $H_2O$ has passed through the device. Discard the $H_2O$ rinse, and refill with 2 mL fresh $H_2O$. Incubate at 4°C overnight (*see* **Note 23**).
2. Spin at 3600 rpm at 4°C until the $H_2O$ rinse has passed through from the upper to the lower reservoir (*see* **Fig. 2**). Discard the rinse.
3. Add 2 mL solution to be concentrated (e.g., concentrated medium) to the upper reservoir, and place the retentate vial on top of the upper reservoir to cover it.

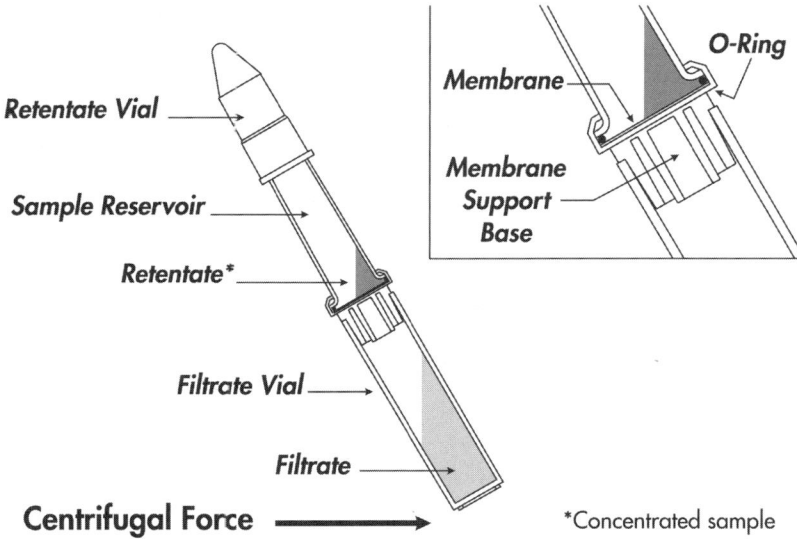

Fig. 2. Components of the Amicon Centricon centrifugal filter unit. (Figure courtesy of Millipore Corporation.)

4. Centrifuge at 3600 rpm at 4°C (*see* **Note 24**), stopping the centrifuge at regular intervals to check concentration progress (*see* **Note 25**). Typical final retentate volumes are approx 100–300 μL.

5. Remove the Centricon from the rotor, and separate the upper reservoir from the lower.

6. Ensure that the retentate vial is attached, and invert the upper reservoir so that the retentate vial is now at the bottom of the insert in the JA20 rotor. Counterbalance with an identical Centricon.

7. Centrifuge at 3600 rpm at 4°C for 2 min to transfer the concentrate into the retentate vial.

8. The retentate containing the recombinant protein is now ready for analysis by SDS-PAGE and any subsequent purification. Transfer the retentate from this vial to a 1.5-mL centrifuge tube and store at –80°C.

## 4. Notes

1. Amino acid drop-out mixture can be made as a dry powder stock. Any particular combination can be made using the amounts provided in **Table 2** or purchased commercially from Formedium Ltd (United Kingdom). Prepare the mixture by adding the amino acids/nutrients in the quantities given but omitting those components required for the selection of transformed cells. Grind to a fine powder using a pestle and mortar, and store this at room temperature in a suitable airtight container away from light.

2. The addition of casamino acids has sometimes been found to reduce the amount of degradation of the expressed recombinant product by secreted proteases (*10*).

3. The selectable markers used on *S. cerevisiae* plasmid vectors are usually genes encoding enzymes required for the synthesis of a particular amino acid or nucleotide. Therefore, host strains that contain corresponding deletions or mutations are complemented for auxotrophy by the introduced vector. *P. pastoris* plasmid vectors can be considerably smaller owing to their utilization of a single selectable marker (*Sh ble* gene), which confers resis-

**Table 2**
**Amino Acid Dry Mix**
**"Drop-Out" Components**

| Constituent | Amount |
| --- | --- |
| Adenine sulfate | 2 g |
| Uracil | 2 g |
| L-Tryptophan | 2 g |
| L-Histidine-HCl | 2 g |
| L-Arginine-HCl | 2 g |
| L-Methionine | 2 g |
| L-Tyrosine | 2 g |
| L-Leucine | 2 g |
| L-Isoleucine | 2 g |
| L-Lysine-HCl | 2 g |
| L-Phenylalanine | 2 g |
| L-Glutamatic acid | 2 g |
| L-Aspartic acid | 2 g |
| L-Valine | 6 g |
| L-Threonine | 8 g |
| L-Serine | 16 g |
| L-Proline | 2 g |
| L-Cysteine | 2 g |
| L-Glycine | 2 g |

tance to Zeocin (Invitrogen) in both bacteria and *P. pastoris* (e.g., pPICZ series *P. pastoris* plasmid vectors are 3.3–3.6 kb in size; *11*).

4. Cultures with an $OD_{600}$ over 0.6 are diluted to ensure measurement within the linear range of the spectrophotometer.

5. Unlike bacterial cells, competent *S. cerevisiae* cells cannot retain high-transformation efficiency indefinitely at –80°C. It is advisable to make fresh competent cells for each new transformation.

6. Other methods of transformation include using spheroplasts, PEG 1000, and lithium chloride and heat shock (*9*).

7. As in *S. cerevisiae*, linear DNA fragments are more recombinogenic in *P. pastoris* than circular DNA plasmids. Therefore, the circular plasmid is linearized using a suitable restriction enzyme before electroporating into *P. pastoris* (*12*).

8. It is advisable to make competent *P. pastoris* cells fresh for each transformation as freezing sorbitol-treated cells results in a significant decrease in transformation efficiency.

9. Most *P. pastoris* vectors utilize homologous regions, either within the *AOX1* gene or *HIS4* gene, to target integration into the host cell genome. Plasmids that confer resistance to Zeocin can be used to select transformants that have increased target gene expression levels: transformants hyperresistant to the drug (100–1000 µg/mL Zeocin™) have multiple integrated copies of the plasmid (*2,9,11*).

10. The specific parameters for electroporating *P. pastoris* may vary depending on the electroporator used; it is therefore advisable to use parameters for electroporation as recommended by the manufacturer.

11. In *S. cerevisiae*, constitutive promoters are usually derived from genes of the glycolytic pathway, e.g., promoters from the alcohol dehydrogenase (*ADH1*) or phosphoglycerate kinase (*PGK1*) gene (*13*). In *P. pastoris*, the promoter from the glyceraldehyde-3-phosphate dehydrogenase (*GAP*) gene is typically used (*9*).

12. Examples of regulated *S. cerevisiae* promoters include those of galactose metabolic enzymes (*GAL1, 10*), copper metallothionein (*CUP1*), alcohol dehydrogenase (*ADH2*), acid phosphatase (*PHO5*), and *O*-acetyl homoserine sulphydrylase (*MET25*) (*13*). In *P. pastoris*, the *AOX1* promoter is primarily used (*9*). The strength of induction can vary significantly between promoters: e.g., the addition of galactose to *S. cerevisiae* cells results in a 1000-fold increase in expression of genes under the control of the *GAL1, 10* promoter compared to only a 20-fold increase under *CUP1* promotion following induction by $Cu^{2+}$ (*14–16*). In *P. pastoris*, the transcription product of a target gene under the control of the *AOX1* promoter may represent a significant proportion of the total polyA$^+$ RNA for the cell (*9*).

13. The most common signal sequence used to translocate recombinant proteins into the endoplasmic reticulum (ER) of both *S. cerevisiae* and *P. pastoris* is that derived from the *S. cerevisiae* α-mating factor (*2,9,15*). This requires the DNA sequence-encoding α-factor to be spliced to the coding sequence of the target cDNA. On expression, the signal sequence is recognized, and the target protein is directed to the ER, where the signal sequence is cleaved upon entering the ER by signal peptidase (*17*).

14. The major disadvantage with secreting the target recombinant protein is that it may become quite dilute in the culture medium and consequently requires extensive concentration. Intracellular expression has the benefit that the product is already concentrated within the yeast cells themselves.

15. As glucose is a potent inhibitor of the *GAL1, 10* promoter, the culture is passaged and allowed to grow for 2–3 generations in 2% (w/v) raffinose to ensure maximum induction upon the addition of galacatose.

16. If the target recombinant protein is to be secreted, buffering the medium to pH 6.0 may aid stability of the target by inhibiting any secreted proteases. Buffering the medium may also aid in maintaining the biological activity of the expressed protein (*10*).

17. To acid-wash beads, place them into a glass flask, add concentrated hydrochloric or nitric acid, and stir several times during a 1-h incubation at room temperature in a fume hood. Pour off the acid, rinse beads exhaustively in tap water, rinse once in $ddH_2O$, autoclave, and dry beads in a drying oven.

18. Pushing the hot needle into a closed tube will cause a momentary pressure increase that discharges the sample lysate out of the aperture.

19. Protein can be easily removed for analysis many times because 50% (v/v) glycerol stops lysates from freezing at –20°C. Frozen protein is susceptible to damage; it cannot be freeze-thawed repeatedly. Therefore, although storage at –80°C is desirable, samples should be stored in small aliquots for once-only removal.

20. Use of compressed $N_2$ gas is recommended over compressed air, as the latter can cause large pH shifts owing to dissolution of carbon dioxide and result in oxidation of the target recombinant protein.

21. Monitor concentration. Do not allow all the retentate to run through.

22. Continuing to stir for a few minutes after depressurization can help to maximize recovery of retained substances.

23. Do not allow the membrane to run dry once wet. If the Centricon is not to be used immediately, leave some residual fluid on the membrane.

24. Each type of Centricon must be centrifuged according to its specification. The following is the centrifugal limit for each type of Centricon: YM-3, $7500g$; YM-10, YM-30, and YM-50, $5000g$; YM-100, $1000g$.
25. To aid in determining how concentration is progressing, Centricons can be marked for the desired retentate volume by transferring the equivalent volume of water into the upper reservoir body of the Centricon and marking the outside of the body with an indelible pen. The water is then centrifuged or decanted out of the device prior to use.

## Acknowledgments

The authors would like to thank Dr. Angela Dunn and Dr. Louise Emberson for their comments and advice on protocols for expression in these organisms and the Millipore Corporation for provision of figures for the Amicon 8400 and Centricon centrifugal filter units. This work in the author's laboratory is supported by the BBSRC.

## References

1. Higgins, D. R. and Cregg, J. M. in Pichia *Protocols* (Higgins, R. D. and Cregg, J. M., eds.), Humana, Totowa, NJ, 1998, 1–15.
2. Cereghino, J. L. and Cregg, J. M. (2000) Heterologous protein expression in the methylotrophic yeast *Pichia pastoris. FEMS Microbiol. Rev.* **24**, 45–66.
3. Kjeldsen, T., Pettersson, A. F., and Hatch, M. (1999) Secretory expression and characterization of insulin in *Pichia pastoris. Biotechnol. Appl. Biochem.* **29**, 79–86.
4. Henry, A., Masters, C. L., Beyreuther, K., and Cappai, R. (1997) Expression of human amyloid precursor protein ectodomains in *Pichia pastoris*: analysis of culture conditions, purification and characterization. *Protein Expr. Purif.* **10**, 283–291.
5. Cregg, J. M., Tschopp, J. F., Stillman, C., Siegel, R., Akong, M., Craig, W. R., et al. (1987) High level expression and efficient assembly of hepatitis B surface antigen in the methylotrophic yeast, *Pichia pastoris. Biotechnology* **5**, 479–485.
6. Scorer, C. A., Buckholz, R. G., Clare, J. J., and Romanos, M. A. (1993) The intracellular production and secretion of HIV-1 envelope protein in the methylotrophic yeast *Pichia pastoris. Gene* **136**, 111–119.
7. Jones, E. W. (1991) Tackling the protease problem in *Saccharomyces cerevisiae. Meth. Enzymol.* **194**, 428–453.
8. Elble, R. (1992) A simple and efficient procedure for transformation of yeasts. *BioTechniques* **13**, 18–20.
9. Cregg, J. M. and Russell, K. A. in Pichia *Protocols* (Higgins, R. D. and Cregg, J. M., eds.), Humana, Totowa, NJ, 1998, 27–40.
10. Clare, J. J., Romanos, M. A., Rayment, F. B., Rowedder, J. E., Smith, M. A., Payne, M. M., et al. (1991) Production of mouse epidermal growth factor in yeast: high-level secretion using *Pichia pastoris* strains containing multiple gene copies. *Gene* **105**, 205–212.
11. Higgins, D. R., Busser, K., Comiskey, J., Whittier, P. S., Purcell, T. J., and Hoeffler, J. P. in *Pichia Protocols* (Higgins, R. D. and Cregg, J. M., eds.), Humana, Totowa, NJ, 1998, 41–51.
12. Cregg, J. M., Barringer, K. J., Hessler A. Y., and Madden, K. R. (1985) *Pichia pastoris* as a host for transformations. *Mol. Cell. Biol.* **5**, 3376–3385.
13. Schena, M., Picard, D., and Yamamoto, K. R. (1991) Vectors for constitutive and inducible gene expression in yeast. *Meth. Enzymol.* **194**, 389–398.

14. Schneider, J. C. and Guarente, L. (1991) Vectors for expression of cloned genes in yeast: regulation, overproduction, and underproduction. *Meth. Enzymol.* **194,** 373–388.
15. Moir, D. T. and Davidow, L. S. (1991) Production of proteins by secretion from yeast. *Meth. Enzymol.* **194,** 491–507.
16. Mascorro-Gallardo, J. O., Covarrubias, A. A., and Gaxiola, R. (1996) Construction of a *CUP1* promoter-based vector to modulate gene expression in *Saccharomyces cerevisiae*. *Gene* **172,** 169–170.
17. Haguenauer-Tsapis, R. (1992) Protein-specific features of the general secretion pathway in yeast: the secretion of acid phosphatase. *Mol. Microbiol.* **6,** 573–579.

# 6

## Pharmaceutical Proteins From Methylotrophic Yeasts

### Eric C. de Bruin, Erwin H. Duitman, Arjo L. de Boer, Marten Veenhuis, Ineke G. A. Bos, and C. Erik Hack

## 1. Introduction

Because of their favorable properties, methylotrophic yeasts have become increasingly important as cell factories for the production of biomaterials, therapeutic proteins, and vaccines. As a eukaryote. yeast can perform most of the posttranslational modifications that are required to ensure the functionality and/or stability of recombinant human proteins, such as N- and O-linked glycosylation, phosphorylation, and formation of disulfide bonds. In contrast to other yeast systems, foreign genes can be expressed at high levels under control of strong inducible promoters derived from genes encoding proteins that are involved in methanol metabolism. Furthermore, heterologous proteins can be secreted at high levels into the culture medium, which, in combination with the fact that few endogenous proteins are secreted, significantly facilitates purification of the desired protein. Finally, as unicellular microorganisms, methylotrophic yeasts have major advantages in industrial fermentation.

Within the methylotrophic yeast species, *Pichia pastoris* and *Hansenula polymorpha* have been used most extensively as cell factories for the production of recombinant proteins. These yeasts have proven to be robust industrial fermentation organisms that can be grown to high cell densities in cheap, chemically defined media. At present, over 200 different proteins have been produced by the *P. pastoris* expression system *(1)*. Although fewer examples have been described, the thermotolerant yeast *H. polymorpha* has the same potential for high-level expression of heterologous genes *(2)*. The protocols described in this chapter focus on foreign gene expression in *P. pastoris* and *H. polymorpha*, two ascomycetous yeasts that belong to different genera.

One issue in heterologous protein production in yeast is protein degradation. Some secreted proteins are unstable in the culture medium, where they are rapidly degraded by endogenous yeast proteases. These degradation processes pose a problem particularly in fermentor cultures, as lysis of only a small percentage of cells in a high cell density culture results in significant amounts of proteases in the medium. The use of

From: *Methods in Molecular Biology, vol. 308: Therapeutic Proteins: Methods and Protocols*
Edited by: C. M. Smales and D. C. James © Humana Press Inc., Totowa, NJ

production strains deficient in major vacuolar proteases has proven to reduce the (partial) degradation of the protein of interest *(3)*.

For optimal stability in human sera, therapeutic proteins often require posttranslational addition of glycans to the side-chain amide nitrogen of asparagine residues in the consensus sequence Asn-X-Thr/Ser, allowing N-glycosylation. Although several mammalian cell lines are able to perform human-like N-glycosylation, low yields, long fermentation times, and risks of viral infections hamper the use of such cell lines as protein factories. Yeasts are also capable of N-glycosylation, and they do not have the disadvantages mentioned above. Unfortunately, several recombinant glycoproteins derived from yeast systems have been shown to contain N-linked high mannose oligosaccharides, which are immunogenic and rapidly cleared from the blood when introduced intravenously into mammals. Several research groups are currently engineering glycosylation pathways in both *P. pastoris* and *H. polymorpha* *(4,5)* to obtain human-like glycosylation in recombinant proteins. First results with these glycoengineered yeasts are very promising *(6)*, leading methylotrophic yeasts to be even more appealing in the future as a platform for the production of recombinant proteins for pharmaceutical use in humans. This chapter details some methods for the secreted production of heterologous proteins in *P. pastoris* and *H. polymorpha*.

## 2. Materials

### 2.1. Pichia pastoris *(see Note 1)*

1. Strain GS115 (*his4*) and plasmid pPIC9 (Invitrogen, Carlsbad, CA).
2. Yeast-extract peptone dextrose medium (YPD) media: dissolve 10 g yeast extract, 20 g peptone, and 20 g glucose in 1 L water and autoclave.
3. YPD plates: As YPD, but with 1.5% (w/v) agar.
4. Minimal glucose agar (MG) plates: Dissolve 11.5 g $KH_2PO_4$, 2.66 g $K_2HPO_4$, 6.7 g yeast nitrogen base (without amino acids), 20 g glucose, and 15 g agar in 1 L water, pH 6.0, and autoclave.
5. Buffered minimal glycerol (BMG) media: Dissolve 11.5 g $KH_2PO_4$, 2.66 g $K_2HPO_4$, and 6.7 g yeast nitrogen base (without amino acids) with 0.5% (v/v) glycerol in 1 L water, pH 6.0, and autoclave.
6. Buffered minimal methanol (BMM) media: Same as BMG, except that glycerol is replaced by 0.5% (v/v) methanol.
7. Buffered glycerol-rich (BMGY) media: Same as BMG, with the addition 10 g yeast extract and 20 g peptone.
8. Buffered methanol-rich (BMMY) media: Same as BMGY, except that glycerol is replaced by 0.5% (v/v) methanol.
9. Oligo nucleotide primers: 5'AOX1-primer: 5'-GACTGGTTCCAATTGACAAGC-3'; 3'AOX1-primer: 5'-GCAAATGGCATTCTGACATCC-3'; and primers designed for amplification of the heterologous gene (if desired, without the part encoding the native leader sequence, preferably *Xho*I [*see* **Note 2**] and *Not*I can be introduced for cloning purposes).
10. Competent *Escherichia coli* cells.
11. Ampicillin.
12. Restriction enzymes: *Taq* DNA polymerase and T4 DNA ligase.
13. Agarose gel equipment.

14. Electroporation device (e.g., GenePulser, Bio-Rad, Richmond, CA).
15. Sterile electroporation cuvets (0.2-cm gap).
16. PBD: 50 m$M$ potassium phosphate buffer, pH 7.5, containing 25 m$M$ dithiothreitol (DTT). Add DTT freshly from 1 $M$ frozen stock.
17. 1 $M$ Sorbitol.
18. Sodium dodecyl sulfate-polyacrylamide gel electrophoresis (SDS-PAGE)/Western blot equipment.
19. Fermentation equipment.
20. Chromatography equipment.

## 2.2. Hansenula polymorpha *(see Note 3)*

1. Strain NCYC495 (*leu1.1*) and plasmid pHIPX4 *(7,8)*.
2. YPD media (*see* **Subheading 2.1.**, **item 2**).
3. YPD plates: As YPD, but with 1.5% (w/v) of agar.
4. Yeast nitrogen base dextrose medium (YND) media: Dissolve 1.7 g yeast nitrogen base (without amino acids) and 5 g $(NH_4)_2SO_4$ in 1 L water and autoclave. After sterilization, add 1% (w/v) glucose.
5. YND plates: As YND, but with 1.5% (w/v) agar.
6. Mineral medium (MM): Dissolve 2.5 g $(NH_4)_2SO_4$, 0.2 g $MgSO_4$, 0.7 g $K_2HPO_4$, 3 g $NaH_2PO_4$, and 0.5 g yeast extract in 1 L water; add 1 mL Vishniac solution [per liter: 10 g EDTA, 4.4 g $ZnSO_4$, 1.01 g $MnCl_2$, 0.32 g $CoCl_2$, 0.315 g $CuSO_4$, 0.22 g ammonium heptamolybdate, 1.47 g $CaCl_2$, and 1 g $FeSO_4$] and autoclave. After sterilization, add 1 mL vitamin solution (per liter: 20 mg D-biotin and 600 mg thiamine) and 0.5% carbon source (glucose or methanol).
7. Competent *E. coli* cells.
8. Kanamycin.
9. Restriction enzymes: *Taq* DNA polymerase and T4 DNA ligase.
10. Agarose gel equipment.
11. Electroporation device (e.g., Gene-Pulser, Bio-Rad).
12. Sterile electroporation cuvets (0.2-cm gap).
13. TED: 100 m$M$ Tris-HCl, pH 8.0, 50 m$M$ EDTA, and 25 m$M$ DTT. Add DTT freshly from 1 $M$ frozen stock.
14. STM: 270 m$M$ Sucrose, 10 m$M$ Tris-HCl, pH 8.0, and 1 m$M$ $MgCl_2$.
15. Southern blot equipment.
16. DNA probe, preferably a DNA fragment of approx 600 bp containing the alcohol oxidase (AOX) promoter region.
17. SDS-PAGE/Western blot equipment.
18. Fermentation equipment.
19. Chromatography equipment.

## 3. Methods

The production of a heterologous gene product in *P. pastoris* and *H. polymorpha* requires three basic steps: (1) construction of a suitable expression plasmid containing the gene of interest; (2) integration of the expression vector into the yeast genome; and (3) examination of recombinant strains for production of the recombinant protein.

## 3.1. Construction of Expression Vectors

Plasmid vectors designed for the expression of recombinant proteins in *P. pastoris* and *H. polymorpha* share several common features. They contain an origin of replication and an antibiotic resistance marker for maintenance and selection in *E. coli*, a marker for selection in the yeast of interest, a promoter to drive expression of the heterologous gene and a transcription terminator. In addition, *H. polymorpha* vectors frequently contain a *Hansenula* autonomously replicating sequence (*HARS*). Selection in yeast may be achieved by using auxotrophic markers (e.g., *LEU2* or *HIS4*), or dominant markers (resistance to geneticin or zeocin). The choice of selection marker strongly influences the number of plasmid copies that are integrated into the host chromosome. Promoter elements are typically derived from genes, encoding proteins involved in methanol metabolism, such as the strong AOX promoters (*Paox1* and *Paox* for *P. pastoris* and *H. polymorpha*, respectively).

For secretion of the heterologously produced protein, the native signal sequence of this protein of interest can often be used. Also, the heterologous signal sequence can be used, e.g., for the signal sequence of the *Saccharomyces cerevisiae* α-mating factor pheromone (*see* **Note 4**). C-terminal fusions of the gene with a *myc*-epitope or polyhistidine affinity tag can be considered for detection and/or immobilized-nickel purification of the heterologous protein. Apart from their use as markers or promoters, the yeast-derived vector sequences provide homologs regions for stable integration of the plasmids into the host chromosome by homologous recombination. As examples, described is the use of pPIC9 and pHIPX4 in combination with the appropriate host strains for the secreted production of heterologous proteins by *P. pastoris* and *H. polymorpha*, respectively. Using these plasmids, heterologous genes will be expressed under control of the strong AOX promoters of the respective organisms.

## 3.2. Transformation of P. pastoris and H. polymorpha

When the plasmids for expression of a heterologous gene in *P. pastoris* or *H. polymorpha* have been constructed, they are introduced into the host of choice, where they become integrated into the genome. Integration can be targeted to a specific locus or can be random. Site-specific integration via a single crossover results in the integration of the entire vector, whereas a double crossover results in the replacement of a yeast gene by the heterologous gene.

*P. pastoris* GS115 has a nonfunctional histidinol dehydrogenase gene (*HIS4*). This deficiency is complemented by the *HIS4* gene on pPIC9. The plasmid contains an origin of replication and ampicillin resistance gene for maintenance and selection in *E. coli*. Prior to transformation to *P. pastoris*, the expression vectors (1–5 µg) are linearized by restriction enzymes at a unique site in the *Paox1* or *HIS4* region of pPIC9. Linearization of the vector enhances its integration at the homologous locus (at the *AOX1*– or *HIS4* region, respectively) in the genome of *P. pastoris*. Integration at the *AOX1* locus may result in two phenotypes of transformants—Mut[+] and Mut[S]—for wild-type or slow growth rates on methanol. These two phenotypes arise from the fact that *P. pastoris* has two *AOX* genes: *AOX1* and *AOX2*. When integration occurs via a single crossover event, the *AOX1* gene is retained (Mut[+]). In contrast, when integra-

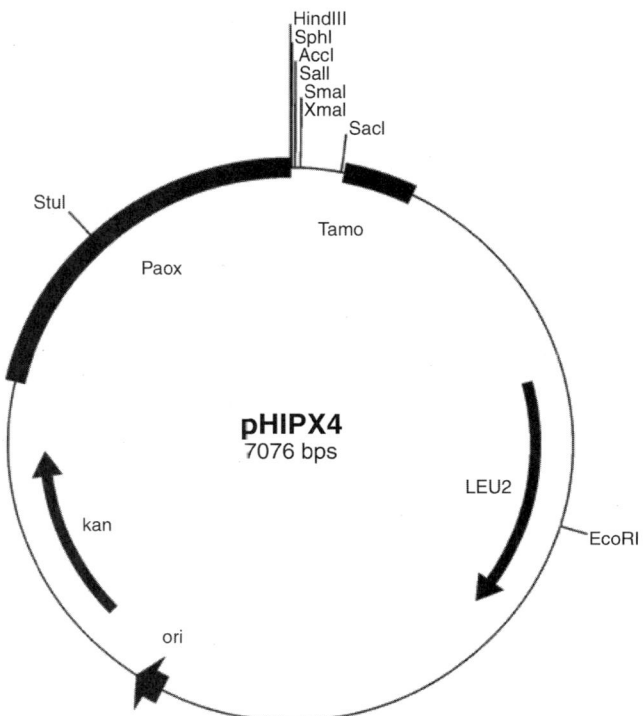

Fig. 1. Plasmid map of the *E. coli-H. polymorpha* shuttle vector pHIPX4, as an example of an expression vector that can be used for the production of heterologous proteins in *H. polymorpha* (**8**). Expression is mediated by the alcohol oxidase promoter (Paox), and transcriptional termination is obtained by the use of the amino oxidase transcriptional terminator (Tamo). The plasmid also contains a kanamycin resistance marker (kan) and an origin of replication (ori) for cloning purposes in *E. coli*. The *LEU2* gene, derived from *S. cerevisiae*, is used for selection in *H. polymorpha*.

tion is via a double crossover between the *AOX1* and 3'*AOX1* regions on the plasmid and corresponding regions on the host chromosome, the *AOX1* gene is lost. This loss leads to the reduction of AOX activity, resulting in reduced growth rate on methanol (Mut^S) compared to the wild-type strain. This is because methanol metabolism in Mut^S strains is dependent on the *AOX2* gene, which is transcribed less efficiently than *AOX1* (**9**).

In *H. polymorpha*, plasmids may be integrated in, e.g., the *AOX* promoter, *LEU2* gene, or *rDNA* cluster. More frequently, however, the *H. polymorpha* system uses plasmids that contain *HARS* sequences. Although these plasmids are initially present in an (unstable) episomal state, mitotically stable integrants can be obtained by culturing the initial transformants for 50–200 generations under selective/nonselective conditions or by using a sequence of selective and nonselective cultivations—a process called *stabilization* (**10**). As an example for integration in *H. polymorpha*, the use of plasmid pHIPX4 is described (**Fig. 1**). This plasmid contains an origin of replication and a kanamycin resistance gene for maintenance and selection in *E. coli*—the strong, methanol-inducible *AOX* promoter to drive heterologous gene expression in *H. polymorpha*, the *AMO* transcription terminator, and the *LEU2* gene from *S. cerevisiae*. The *LEU2* marker can complement *H. polymorpha* strains that are defective in the gene for β-isopropylmalate dehydrogenase, such as NCYC495 *leu1.1*. A single copy of the *S. cerevisiae LEU2* marker does not fully complement deficiency of the host strain, which allows selection of multicopy integrants. Colony size on selective plates corresponds to a large extent to the number of integrated copies (typically 1–8). The plasmid is linearized with a restriction enzyme that cuts in the *AOX* promoter to promote integration in the chromosomal *AOX* locus by homologous recombination (via single crossover); the resulting strain is Mut⁺.

Several methods to introduce DNA into *P. pastoris* and *H. polymorpha* have been described, of which electroporation is most efficient and convenient *(11,12)*. A recent study showed that pretreatment of *P. pastoris* with DTT further increased transformation efficiencies *(13)*. The various steps for the preparation of competent *P. pastoris* and *H. polymorpha* cells and subsequent transformation of the yeasts by electroporation are described in more detail.

### 3.2.1. Preparation of Competent P. pastoris Cells

1. Inoculate 10 mL YPD with a single fresh yeast colony of the strain to be transformed from a YPD plate. Grow overnight with shaking at 30°C.
2. Dilute the overnight culture by transferring 5, 10, and 15 µL into three baffled flasks of 1 L containing 100 mL YPD medium and incubate overnight with shaking.
3. Subsequently, select the culture that has an optical density ($OD_{600}$) of 1.3–1.5, harvest the cells by centrifugation at 2000$g$ for 15 min at 4°C, resuspend the cells in 40 mL PBD, and incubate for 15 min at 30°C.
4. Wash the cells twice with 100 and 50 mL ice-cold 1 *M* sorbitol, respectively.
5. Resuspend the cells in the smallest volume of 1 *M* sorbitol possible.
6. Divide the cell mixture in batches of 40 µL, and use immediately or freeze in liquid $N_2$ and store at –80°C for later use.

### 3.2.2. Preparation of Competent H. polymorpha Cells

1. Inoculate 10 mL YPD with a single fresh *H. polymorpha* colony of the strain to be transformed from a YPD plate. Grow overnight with shaking at 37°C.
2. Dilute 2 mL into 200 mL prewarmed YPD and grow at 37°C to an $OD_{600}$ of 1.2–1.5. (This will take approx 6 h.)
3. Pellet cells 10 min at 2000$g$ at room temperature (RT).
4. Resuspend the cells in 50 mL TED.
5. Incubate 15 min at 37°C with shaking (250 rpm).
6. Pellet cells for 10 min at 2000$g$ at RT.
7. Wash the cells with 200 mL ice-cold STM.
8. Pellet cells for 10 min at 2000$g$ at 4°C.
9. Wash the cells with 100 mL ice-cold STM.
10. Pellet cells for 10 min at 2000$g$ at 4°C.
11. Resuspend the cells in 1 mL ice-cold STM.
12. Divide the cell mixture in batches of 60 µL, and use immediately or freeze in liquid $N_2$ and store at –80°C.

### 3.2.3. Transformation of P. pastoris by Electroporation

1. Add between 0.5 and 1 μg (in no more than 10 μL) of linearized vector DNA to a tube containing 40 μL competent cells and transfer to a 2-mm gap electroporation cuvet held on ice (*see* **Note 5**).
2. Pulse cells with instrumental settings of 1.5 kV, 25 μF, and 200 Ω (*see* **Note 6**).
3. Immediately add 1 mL ice-cold sorbitol, and transfer the contents of the cuvet into a sterile 1.5-mL Eppendorf tube.
4. Spread selected aliquots onto MG plates, and incubate the plates for 2–4 d at 30°C.

### 3.2.4. Transformation of H. polymorpha by Electroporation

1. Add linearized vector DNA (0.5–1 μg in maximum 3–4 μL) to a tube containing 60 μL cells.
2. Transfer cell/DNA mixture to a 2-mm gap electroporation cuvet (*see* **Note 5**).
3. Pulse settings: 1.5 kV, 50 μF, and 129 Ω (*see* **Note 6**).
4. Add immediately 940 μL YPD (at RT) to the cell/DNA mixture and transfer to a 2-mL Eppendorf tube.
5. Incubate cells 1 h at 37°C with shaking (250 rpm).
6. Pellet cells for 2 min at 3000*g* in an Eppendorf centrifuge.
7. Wash the cells with 1 mL YND.
8. Pellet cells for 2 min at 3000*g* in an Eppendorf centrifuge.
9. Resuspend in 1 mL YND and place 1%, 10%, and 89%. Respin in an Eppendorf centrifuge and resuspend in small volume on YND plates.
10. Colonies appear after 2–3 d.

## 3.3. Selection of Transformants

### 3.3.1. P. pastoris

Colonies can be subjected to polymerase chain reaction (PCR) without any pretreatment to determine if the gene of interest has integrated into the yeast genome. For *Pichia* Mut[+] integrants, two different DNA fragments are expected using the above-mentioned *AOX1* primers. One fragment is 2.2 kb in size, which corresponds to the wild-type *AOX1* gene, and the other has an expected size of 492 bp (for pPIC9) plus the size of the inserted gene. In Mut[S] integrants, in which the *AOX1* gene is disrupted, only one band of 492 bp plus the size of the inserted gene is expected (*see* **Note 7**).

### 3.3.2. H. polymorpha

Integrant and 1, 2, and more than 2 copy number selection:

1. Pool transformants of 89% plate in 20 mL YPD.
2. Grow for approx 50 generations in YPD (may contain 50 mg/L ampicillin to prevent bacterial infections). Dilute a culture that has been grown to stationary phase 200-fold in fresh YPD, and allow the culture to reach the stationary phase again; repeat this step six times.
3. Make dilutions from the last culture in YND and plate on YND plates. Normally, 50 μL of 10³-, 10⁴-, and 10⁵-fold dilution gives rise to individual colonies.
4. After 2–4 days, small-, medium- and large-sized colonies are able to be discriminated. Small colonies routinely have one copy of the expression plasmid integrated, medium-sized have two copies, and large have three or more copies. Selecting these different sized colonies will give a range of expression levels of the heterologous gene.

To determine site-specific integration of the vector in the genome, transformants are analyzed by Southern blot analysis. This technique is used to verify integration at the *AOX* promoter (rather than at a random locus by nonhomologous recombination) and to estimate the number of copies of the expression vector that have integrated. A DNA fragment containing part of the *AOX* promoter is used as a probe. Chromosomal DNA of the transformants and of the original wild-type strain (as a control) is digested with a suitable restriction enzyme, such as *Eco*RI. For the wild-type strain, a single band of 3.0 kb (containing the *AOX* promoter) is detected. This band is no longer present if one or more copies have integrated. Instead, two new bands appear for a single-copy integrant, because a *Eco*RI site is also present in the plasmid. A third band corresponding to the size of the integrated plasmid is present if two or more copies of the vector are present, and the intensity of this band increases with increasing copy numbers.

## 3.4. Protein Induction

### 3.4.1. P. pastoris

1. Single colonies of Mut⁻ integrants (verified by PCR) are inoculated in a 250-mL baffled flask containing 25 mL BMG or BMGY (*see* **Notes 8** and **9**).
2. Grow at 30°C in a shaking incubator (250 rpm) until the culture has reached an $OD_{600}$ between 4 and 6 (usually after overnight culture).
3. To induce synthesis of the heterologous protein, cells grown in medium containing glycerol are harvested by centrifugation at 2000$g$ for 5 min at RT. The cell pellet is resuspended to an $OD_{600}$ of 1.0 in 100 mL medium containing 0.5% (v/v) methanol as the sole carbon source (BMM or BMMY, respectively).
4. After taking a 1-mL sample ($t = 0$), the medium is transferred to 1-L baffled flasks, which is returned to the incubator for further growth.
5. Take a 1-mL sample of the cultures every 24 h; cells are removed by centrifugation (20,000$g$ for 1 min), and supernatants are transferred into a 1.5-mL Eppendorf tube and stored at –20°C to analyze protein production later. Also, store the cell pellets at –20°C.
6. After each time taking a sample at 24 ($t = 1$), 48 ($t = 2$), and 72 ($t = 3$) h, respectively, of growth on methanol, 1 mL 50% (v/v) methanol is added to maintain induction (*see* **Note 10**).

### 3.4.2. H. polymorpha

1. Inoculate cells into MM plus glucose (0.5%) from a relatively fresh liquid culture or a fresh YND plate. Use a lot of material (turbid culture), not a single colony.
2. Grow until late exponential growth phase: $OD_{600} < 2.0$.
3. Dilute into fresh (prewarmed) media to an $OD_{600}$ of 0.1.
4. When culture has reached maximal growth rate in mid-exponential growth phase: Dilute two more times to MM plus glucose (0.5%).
5. Dilute mid-exponential cells ($OD_{600}$ 1–1.5) to 0.1 in MM plus methanol (0.5%).
6. Grow overnight (15 h) until $OD_{600}$ is 1.5.
7. After cell removal by centrifugation (2000$g$), the supernatant is stored at –20°C.

## 3.5. Analysis of Production by Enzyme-Linked Immunosorbent Assay (ELISA), SDS-PAGE, or Functional Assays

Thaw the supernatants of a production experiment, and analyze protein production by ELISA, SDS-PAGE, and/or Western blotting (**Fig. 2**). As a negative control,

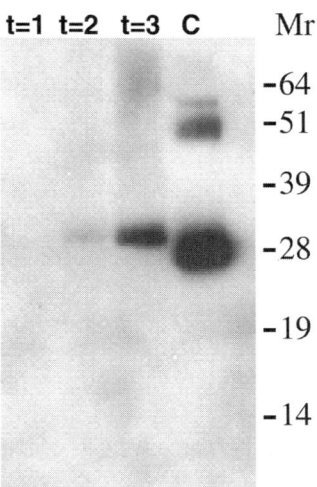

Fig. 2. Western blot analysis, using a chemiluminiscent method, of extracellular proteins of a *P. pastoris* transformant containing the cDNA of human granzyme B (kindly provided by M.C. Strik, Sanquin Research, Amsterdam, The Netherlands). Lanes *t* = 1 to *t* = 3: Supernatants of a production experiment in a shake flask after 24, 48, and 72 h of methanol induction. Lane C: Positive control; lysate of YT-Indy cells.

samples of culture supernatants of yeast transformed with the original plasmid without an insert can be tested. These controls will reveal background production of native extracellular yeast proteins. Often, the supernatants can be tested directly without prior purification of the recombinant protein. Functional assays can also be performed directly with the culture medium, but dialysis against a suitable assay buffer may be necessary (*see* **Note 11**).

### 3.6. Characterization of the Recombinant Product

N-terminal amino acid sequencing of secreted proteins is generally required to ensure that the secreted protein has been properly processed. After SDS-PAGE analysis under reducing conditions, proteins are blotted onto a polyvinylidene fluoride (PVDF) membrane. Subsequently, proteins on the membrane are stained with Coomassie Brilliant Blue, and selected protein bands are cut out. N-terminal sequencing is performed using Edman degradation.

For determining N-glycosylation on the recombinant protein, a gel-shift assay can be conducted. This relatively simple method can be used, because only high-mannose-type oligosaccharides are found on yeast glycoproteins. Two different enzymes—endoglycosidase H and peptide-N-glycosidase F—are frequently used for the respective removal of high-mannose glycans and Asn-linked N-acetyl glucosamine residues. The resulting decrease in the apparent molecular mass can be detected by SDS-PAGE as, e.g., has been shown for the recombinant human C1-inhibitor *(14)*.

Further characterization of the product can be performed by mass spectrometry (electrospray mass spectrometry or matrix-assisted laser desorption ionization-time-of-flight), which can eventually reveal other post-translational modifications and/or proteolytic degradation in the recombinant product.

## 3.7. High Cell Density Fermentation

The conditions for growth of *P. pastoris* and *H. polymorpha* are ideal for large-scale production of human pharmaceuticals. These yeast species can grow on inexpensive defined media that do not contain ingredients that may be sources of pyrogens, toxins, or infectious agents. Because of the controlled environment in a fermentor, cell densities in excess of 150 g dry cell weight per liter can be reached. The concentration of secreted protein in the medium is roughly proportional to the cell density; therefore, production yields in bioreactors are usually much higher than in shake flasks.

Essentially, fermentations of both *P. pastoris* and *H. polymorpha* can be performed using a three step-procedure *(15–17)*. The first step is a glycerol- or glucose-batch phase for *P. pastoris* or *H. polymorpha*, respectively, to generate cell mass. Upon depletion of glycerol or glucose, a transition phase is initiated in which a mixture of glycerol/glucose and methanol is added to the culture *(18)*. Finally, pure methanol is fed to the culture to further induce expression. Throughout the fermentation, culture samples are taken to monitor cell densities and the concentration of the recombinant protein is used to determine the harvest time (*see* **Note 12**).

## 4. Notes

1. For more reading on the *Pichia* expression system: *Pichia Protocols, Methods in Molecular Biology, vol. 103* (Higgins, D. R. and Gregg, J. M., eds.), Humana, Totowa, NJ and *Pichia* Expression Kit Instruction Manual, version M, Invitrogen, Carlsbad, CA.
2. For the native N-terminal amino acid sequence in the recombinant product, the *Xho*I site in pPIC9 has to be used for cloning. However, the sequence between the *Xho*I and *Sna*BI sites encoding the dibasic Kex-2 endopeptidase processing site (E-K-R-X, where X can be any amino acid residue except for proline, which inhibits Kex-2 cleavage) must be recreated for efficient secretion of the product. In some cases, the tertiary structure of the heterologous protein can protect the Kex-2 cleavage site, thus inhibiting efficient removal of the α-mating factor prepro sequence.
3. For further reading on the *Hansenula* expression system: *Hansenula polymorpha*: Biology and Applications (Gellissen, G., ed.), Wiley-VCH, Weinheim, Germany.
4. When using the α-mating factor signal for secreted production of the protein of interest, the gene must be cloned in-frame behind the α-mating factor prepro sequence (*see* **Note 2**).
5. Always use a linearized original plasmid (lacking the gene of interest) as a positive transformation control. Use water instead of vector as a negative control.
6. With instrumental settings for electroporation as mentioned, time constants must be between 4 and 5 ms. A lower time constant indicates the salt concentration is too high in the transformation mixture, which results in poor transformation efficiency.
7. If PCR problems are encountered using whole-yeast cells, genomic DNA can be isolated by using the Nucleon MiY DNA extraction kit (AP-Biotech, Buckinghamshire, England).
8. Different yields can be obtained in minimal medium (BMM) vs rich medium (BMMY). Evaluation of production level of the protein in both media is therefore recommended. In some cases, proteolytic degradation can be prevented by using BMMY.

9. For selection of high-level-producing clones, numerous single colonies can be inoculated in 50-mL tubes containing 5 mL BMG or BMGY, which are subsequently induced by adding methanol (0.5%) every 24 h for the next 3 d.

10. Too high local methanol concentrations should be avoided, as these can cause cell death and lysis, resulting in the release of intracellular proteases.

11. A failure to detect the desired protein in the culture medium can be because of the following causes: (1) low-production levels, (2) inefficient secretion, and (3) degradation of the protein product. Analysis of cell lysates is required to determine whether intracellular accumulation of the recombinant protein has occurred. *P. pastoris* and *H. polymorpha* can both be lysed easily with the Y per reagent of Pierce (Rockford, IL).

12. Lowering the pH of the culture media at the beginning of the methanol induction phase in fermentations may help prevent proteolytic degradation of the heterologously produced protein, because yeast proteases are inhibited at a lower pH. Alternatively, if the heterologous protein is sensitive to low pH, addition of amino acid–rich supplements (e.g., casamino acids) to the culture medium may be considered to reduce extracellular proteolysis *(19)*.

## References

1. Cereghino, J. L. and Cregg, J. M. (2000) Heterologous protein expression in the methylotrophic yeast *Pichia pastoris*. *FEMS Microbiol. Rev.* **24,** 45–66.

2. Gellissen, G. (2000) Heterologous protein production in methylotrophic yeasts. *Appl. Microbiol. Biotechnol.* **54,** 741–750.

3. Gleeson, M. A. G., White, C. E., Meininger, D. P., and Komives, E. A. (1998) Generation of protease-deficient strains and their use in heterologous protein expression, in *Methods in Molecular Biology, vol. 103: Pichia Protocols* (Higgins, D. R. and Gregg, J. M., eds.), Humana, Totowa, NJ, pp. 81–94.

4. Choi, B. K., Bobrowicz, P., Davidson, R. C., Hamilton, S. R., Kung, D. H., Li, H., et al. (2003) Use of combinatorial genetic libraries to humanize N-linked glycosylation in the yeast *Pichia pastoris*. *Proc. Natl. Acad. Sci. USA* **100,** 5022–5027.

5. Kim, M. W., Rhee, S. K., Kim, J. Y., Shimma, Y., Chiba,Y., Jigami, Y., and Kang H. A. (2004) Characterization of N-linked oligosaccharides assembled on secretory recombinant glucose oxidase and cell wall mannoproteins from the methylotrophic yeast *Hansenula polymorpha*. *Glycobiology* **14,** 243–251.

6. Hamilton, S. R., Bobrowicz, P., Bobrowicz, B., Davidson, R. C., Li, H., Mitchell, T., et al. (2003) Production of complex human glycoproteins in yeast. *Science* **301,** 1244–1246.

7. Gleeson, M. A. G. and Sudbery, P. E. (1988) Genetic analysis in the methylotrophic yeast *Hansenula polymorpha*. *Yeast* **4,** 293–303.

8. Gietl, C., Faber, K. N., van der Klei, I. J., and Veenhuis, M. (1994) Mutational analysis of the N-terminal topogenic signal of watermelon glyoxysomal malate dehydrogenase using the heterologous host *Hansenula polymorpha*. *Proc. Natl. Acad. Sci. USA* **91,** 3151–3155.

9. Cregg, J. M., Madden, K. R., Barringer, K. J., Thill, G. P., and Stillman, C. A. (1989) Functional characterization of the two alcohol oxidase genes from the yeast *Pichia pastoris*. *Mol. Cell. Biol.* **9,** 1316–1323.

10. Gatzke, R., Weydemann, U., Janowicz, Z. A., and Hollenberg, C. P. (1995) Stable multicopy integration of vector sequences in *Hansenula polymorpha*. *Appl. Microbiol. Biotechnol.* **43,** 844–849.

11. Becker, D. M. and Guarente L. (1991) High-efficiency transformation of yeast by electroporation. *Methods Enzymol.* **194,** 182–187.

12. Faber, K. N, Haima, P., Harder, W., Veenhuis, M., and AB, G. (1994) Highly-efficient electrotransformation of the yeast *Hansenula polymorpha. Curr. Genet.* **25,** 305–310.

13. Wu, S. and Letchworth, G. J. (2004) High efficiency transformation by electroporation of *Pichia pastoris* pretreated with lithium acetate and dithiothreitol. *Biotechniques* **36,** 152–154.

14. Bos, I. G., de Bruin, E. C., Karuntu, Y. A., Modderman, P. W., Eldering, E., and Hack, C. E. (2003) Recombinant human C1-inhibitor produced in *Pichia pastoris* has the same inhibitory capacity as plasma C1-inhibitor. *Biochim. Biophys. Acta* **1648,** 75–83.

15. *Pichia* Fermentation Process Guidelines, version B, Invitrogen, Carlsbad, CA.

16. Stratton, J., Chiruvolu, V., and Meagher, M. (1998) High cell-density fermentation, in *Methods in Molecular Biology, vol. 103: Pichia Protocols* (Higgins, D. R. and Gregg, J. M., eds.), Humana, Totowa, NJ, pp. 107–120.

17. de Bruin, E. C., de Wolf, F. A., and Laane, N. C. (2000) Expression and secretion of human alpha1(I) procollagen fragment by *Hansenula polymorpha* as compared to *Pichia pastoris. Enzyme Microb. Technol.* **26,** 640–644.

18. Potter, K. J., Zhang, W., Smith, L. A., and Meagher, M. M. (2000) Production and purification of the heavy chain fragment C of botulinum neurotoxin, serotype A, expressed in the methylotrophic yeast *Pichia pastoris. Protein Expr. Purif.* **19,** 393–402.

19. Clare, J. J., Romanos, M. A., Rayment, F. B., Rowedder, J. E., Smith, M. A., Payne, M. M., et al. (1991) Production of mouse epidermal growth factor in yeast: high-level secretion using *Pichia pastoris* strains containing multiple gene copies. *Gene* **105,** 205–212.

# 7

## Expression of Human Papillomavirus Type 16 L1 in Baculovirus Expression Systems

*A Case Study*

### Zheng Jin, Si Lusheng, and Wang Yili

## 1. Introduction

With the great achievements of recombinant DNA technology, a variety of expression systems, including yeast, baculovirus, adenovirus, and attenuated Salmonella expression systems, have been used for the overexpression of recombinant proteins. Recombinant baculovirus–insect cell systems have become widely used because: (1) baculoviruses have a restricted host range, which is limited to specific invertebrate species. It is much safer to work with these viruses than most mammalian viruses. (2) Baculovirus has a large genome that accommodates large exogenous genes. Recombinant baculovirus can still propagate in insect cell lines or larvae from different insect species. (3) Recombinant proteins expressed by baculovirus are processed, modified, and targeted to their appropriate cellular locations, where they are functionally similar to their authentic counterparts.

Many papillomavirus (PV) capsid proteins have successfully been produced and purified in the baculovirus expression system, such as bovine PV, cottontail rabbit PV, and many types of human PV. However, among the existing purification protocols of PVL1 proteins, CsCl ultracentrifugation is usually used. This procedure is labor-intensive and costly. To overcome this problem, we have improved the expression and purification system of PVL1 proteins using a baculovirus expression vector with a 6X histidine (His) tag, expressing PVL1 protein in *Spodoptera frugiperda* (Sf9) insect cells and purifying the product using a Ni-resin column. Containing an extra 6X His peptide, the generated L1 proteins have identical biological activity to their wild-type counterpart. The preparation of human papillomavirus type 16 L1 (HPV16L1) protein will be used to exemplify the methods utilized for expressing PV proteins.

From: *Methods in Molecular Biology, vol. 308: Therapeutic Proteins: Methods and Protocols*
Edited by: C. M. Smales and D. C. James © Humana Press Inc., Totowa, NJ

## 2. Materials

### 2.1. Bac-to-Bac Baculovirus Expression Systems (Invitrogen)

1. pFastBac HT donor plasmid.
2. HPV16L1 cDNA.
3. DH10Bac-competent cells.
4. *Escherichia coli*-competent cells.
5. Cellfectin reagent.
6. Restriction endonucleases: T4 DNA ligase.
7. Agarose and DNA sequencing gel equipment.
8. Ampicillin, gentamicin, kanamycin, and tetracycline.
9. Bluo-gal and isopropyl-β-D-thio-galactopyranoside.
10. NaOH, sodium dodecyl sulfate (SDS), KOAc, isopropanol, 70% ethanol.
11. SOC medium, Luria Bertani (LB) Medium, and LB agar plates.
12. Sf9 cells (ATCCCRL-1711).
13. Grace's insect cell culture medium, penicillin, streptomycin, and 10% heat- inactivated fetal bovine serum.
14. Six-well tissue culture-treated plates.
15. SDS-polyacrylamide gel electrophoresis (SDS-PAGE) equipment.
16. Anti-HPV16 L1 monoclonal antibody.
17. Diaminobenzidine (DAB).
18. Lysis buffer: 50 m$M$ Tris-HCl, pH 8.5, 5 m$M$ 2-mercaptoethanol, 100 m$M$ KCl, and 1 m$M$ phenylmethylsulfonyl fluoride (PMSF), 1% NP-400.
19. Blocking solution: 5% nonfat dry milk (w/v) in Tween-TBS buffer.

### 2.1. ProBond™ Purification System

1. Ni-resin columns.
2. Leupeptin.
3. PMSF.
4. Native binding buffer: 20 m$M$ sodium phosphate and 500 m$M$ sodium chloride, pH 7.2.
5. Native wash buffer: 20 m$M$ sodium phosphate and 500 m$M$ sodium chloride, pH 6.0.
6. Native imidazole elution buffers:

| 3 $M$ Imidazole, pH 6.0); wash buffer, pH 6.0 | | |
| --- | --- | --- |
| 50 m$M$ (10 mL) | 0.16 mL | 9.84 mL |
| 200 m$M$ (10 mL) | 0.66 mL | 9.34 mL |
| 350 m$M$ (10 mL) | 1.16 mL | 8.84 mL |
| 500 m$M$ (10 mL) | 1.66 mL | 8.34 mL |

## 3. Methods

The methods described below outline the (1) construction of the recombinant donor plasmid; (2) transposition and isolation of recombinant bacmid DNA; (3) transfection of Sf9 cells with recombinant bacmid DNA; (4) identification of the expressed protein; (5) purification of the protein; and (6) biological activity analysis of the purified protein.

### 3.1. Recombinant Donor Plasmid

The construction of the donor plasmid for HPV16L1 protein is described in **Sub-headings 3.1.1.–3.1.3.** This includes the description of the donor plasmid and the HPV16L1 cDNA and cloning.

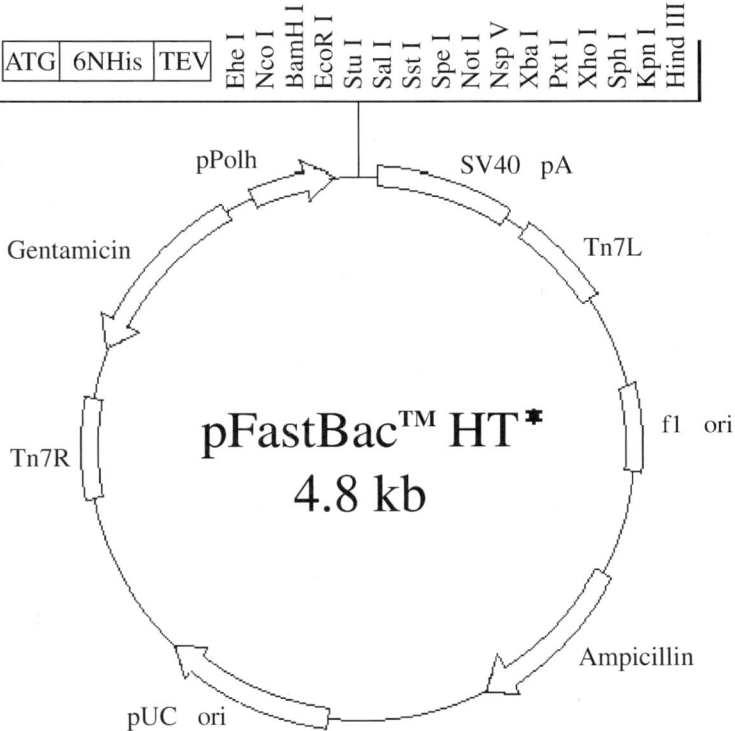

Fig. 1. A schematic circle map of the pFastBac-HT donor plasmid.

### 3.1.1. pFastBac HT Donor Plasmid

The pFastBac HT donor plasmid is used to express polyhistidine-tagged proteins that can be rapidly purified on metal affinity resins. The mini-Tn7 in a pFastBac HT donor plasmid contains an expression cassette consisting of a Gm$^r$ gene, a baculovirus-specific promoter (polyhedrin promoter), a multiple-cloning site, and a SV40 poly(A) signal inserted between the left and right arms of Tn7. Genes to be expressed are inserted into the multiple-cloning site of a pFastBac HT donor plasmid downstream from the baculovirus-specific promoter. The pFastBac HT donor plasmid also contains an ampicillin resistance gene for selectivity.

### 3.1.2. cDNA

The plasmid pGEMT-HPV16L1 containing the cDNA for HPV16L1 was constructed by Zheng Bin *(2)*. It contains the initiation methionine codon followed by 1545 nt coding for the 515 amino acid HPV16L1 protein.

### 3.1.3. Cloning

The HPV16L1 gene is obtained by digesting pGEM-T-HPV16 with *Xba*I plus *Hind*III and purified by agarose gel, then cloned into the prepared pFastBac-HTb donor plasmid *(see* **Fig. 1***)* baculovirus expression system. The DNA is transformed into

*E. coli* DH5α cells by standard methods. The *E. coli* DH5α cells were then plated on LB plates containing ampicillin (100 µg/mL) and incubated at 37°C for 16–20 h. Single colonies were selected and grown in LB medium with ampicillin for 10 h. The plasmid DNA was then isolated and checked for the presence of the insert and for the correct orientation using *Xba*I plus *Hin*dIII restriction cleavage and DNA sequencing.

### 3.2. Transposition and Isolation of Recombinant Bacmid DNA

The method to generate recombinant baculoviruses is based on site-specific transposition of an expression cassette into a baculovirus shuttle vector (bacmid) propagated in *E. coli* DH10Bac cells. The bacmid contains the low-copy-number mini-F replicon, a kanamycin resistance marker, and a segment of DNA encoding the *lacZα* peptide from a pUC-based cloning vector. Inserted into the N-terminus of the *lacZα* gene is a short segment containing the attachment site for the bacterial transposon Tn7 (mini-attTn7), which does not disrupt the reading frame of the *lacZα* peptide. Recombinant bacmids are constructed by transposing a mini-Tn7 element from a pFastBac HT donor plasmid to the mini-attTn7 attachment site on the bacmid when the Tn7 transposition functions are provided *in trans* by a helper plasmid. The next steps in this process involve the transposition and isolation of recombinant bacmid DNA.

The recombinant plasmid pFB-HPV16L1 was transformed into DH10Bac-competent cells, which contained the bacmid with a mini-attTn7 target site and the helper plasmid. The mini-Tn7 element on the pFastBac HT plasmid could transpose to the mini-attTn7 target site on the bacmid in the presence of transposition proteins provided by the helper plasmid. Colonies with recombinant bacmid were identified by disruption of the lacZ gene, then cultured at 37°C for at least 24 h (colonies are very small, and blue colonies may not be discernible prior to 24 h), after which recombinant bacmid DNA was isolated.

### 3.3. Transfection and Harvesting
### of the Transfected Cells and Recombinant Baculovirus

The following steps outline the procedure for transfection of Sf9 cells with recombinant bacmid, harvesting of recombinant baculovirus containing the HPV16L1 gene, and expression of the HPV16L1 protein from transfected cells (**Fig. 2**).

### 3.3.1. Transfection

1. Sf9 cells were cultured in Grace's insect cell culture media supplemented with 50 U/mL penicillin, 50 µg/mL streptomycin, and 10% heat-inactivated fetal bovine serum at 27°C. Cells $9 \times 10^5$ (in 2-mL media) were seeded in a 35-mm culture plate.
2. While waiting for the cells to settle, prepare the required transfection solutions. (Solution A: 5 µL bacmid DNA in 100 µL Grace's insect cell culture media without antibiotics; solution B: 6 µL CellFectin reagent in 100 µL of the same media. Mix the two solutions gently and incubate for 20 min at room temperature.) Wash the cells once with culture media, then overlay the transfection solution onto the cells. Replace the transfection solution with 2 mL culture media containing antibiotics after 5 h, then culture the transfected cells further for 48 h at 27°C. Meanwhile, the cells will appear rounded, swollen, more

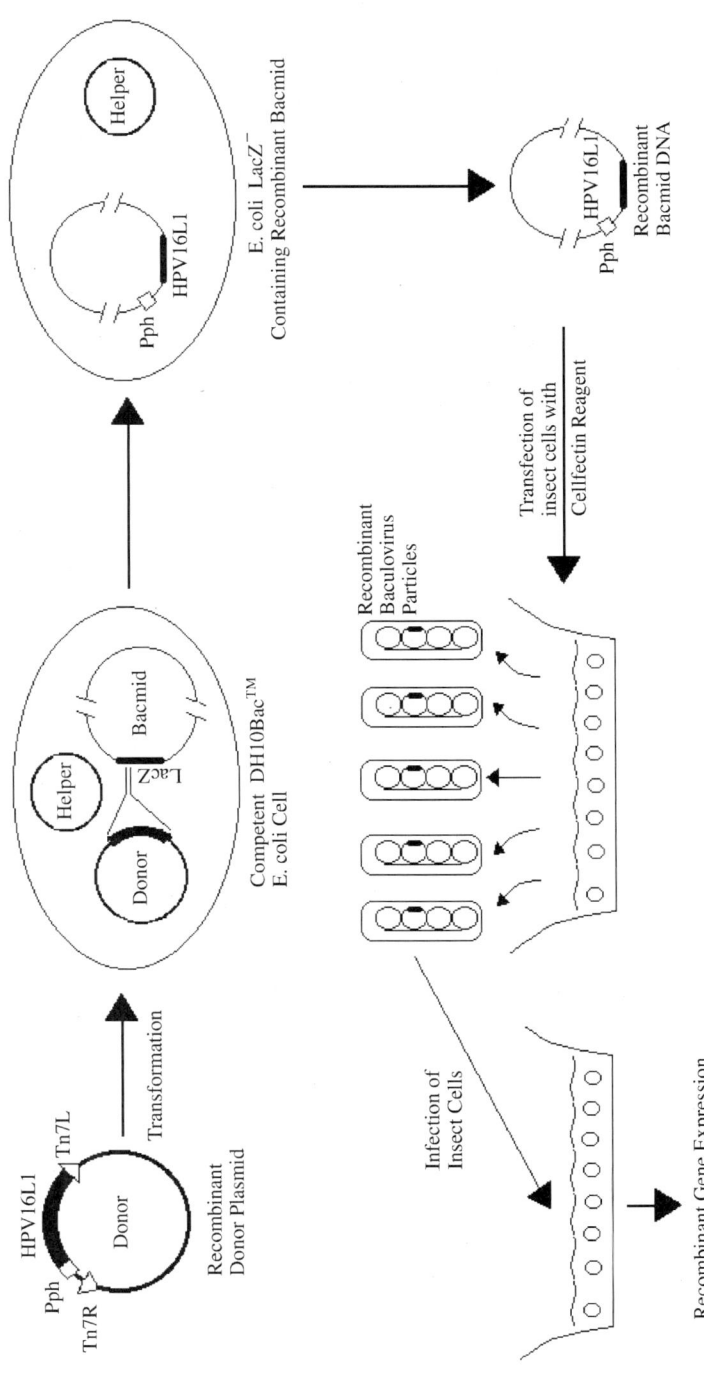

Fig. 2. A flowchart depicting the steps involved in the generation of recombinant baculoviruses and gene expression with the Bac-to-Bac Expression System.

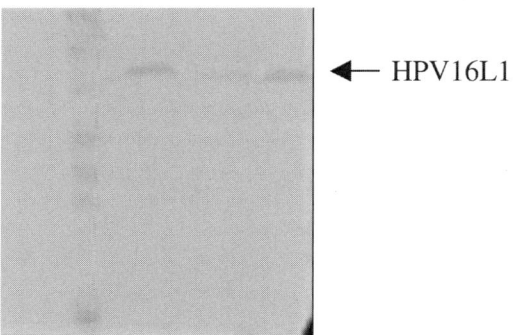

Fig. 3. Identification of expressed protein by Western blotting.

reflective, and will gradually shed. This difference is significant when compared to untransfected normal cells (*see* **Note 1**).

### 3.3.2. Harvesting Recombinant Baculovirus

1. Transfer the transfected cell culture supernatant (2 mL) to a sterile tube, then centrifuge at 500*g* for 5 min, and pipet the supernatant to another sterile capped tube. Virus titers of 2–4 × 10⁷ pfu/mL can be expected after the initial transfection.
2. Store the virus at 4°C in the dark. For long-term storage of the virus, add fetal bovine serum to a final concentration of 2% and keep at –70°C.

### 3.3.3. Harvesting Transfected Cells

1. Harvest the transfected cells by centrifugation (500*g* for 5 min).
2. Resuspend the cell pellet in lysis buffer.
3. Invert the tube end-over-end for 1 min, then remove the cell debris by centrifugation at 10,000*g* for 10 min.

## 3.4. Identification/Characterization of the Expressed Protein

To confirm HPV16L1 protein is expressed, the lysis supernatant is analyzed by SDS-PAGE and Western blotting *(1)*. The membrane blots were blocked by incubation in blocking solution at room temperature, then washed and probed with HPV16L1 McAb (Neomarkers) (at a dilution of 1:200) and anti-mouse immunoglobulin G (IgG) conjugated with horseradish peroxidase (DAKO). DAB was then used for color development (*see* **Fig. 3**).

## 3.5. Purification of Recombinant Protein

The steps described in **Subheadings 3.5.1.–3.5.3.** include preparation of the ProBond column *(2)*, preparation of the extract, and purification of recombinant protein.

### 3.5.1. Preparation of ProBond Columns (see **Note 2**)

1. Transfer the desired amount of Ni-resin to a centrifuge tube and centrifuge at 1500*g* for 2 min (*see* **Note 3**).

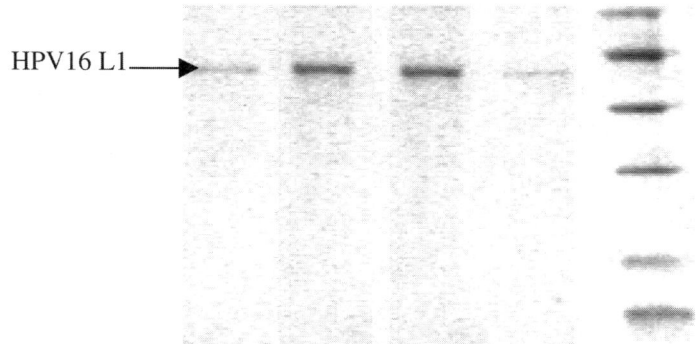

HPV16 L1 →

50 mM  200 mM  350 mM  500 mM  Marker

Fig. 4. HPV16L1 protein eluted from the Ni-resin column with increasing imidazole elution buffer concentrations as analyzed by 12% SDS-PAGE.

2. Remove the supernatant, and mix the resin with one equal volume of binding buffer, then centrifuge at 1500*g* for 2 min.
3. Remove the supernatant, and repeat **step 2** a total of 10 times to equilibrate the Ni-resin column.

### 3.5.2. Preparation of Extract

1. Take a 50-mL culture of Sf9 cells infected with high-titer viral stock at an multiplicity of infection of 5, which is an optimal size for initial purification of the protein.
2. Harvest the cells by centrifugation (500*g* for 5 min or use cells stored at −70°C).
3. Resuspend the cell pellet in native binding buffer containing leupeptin at a concentration of 0.5 µg/mL and 1 m*M* PMSF (*see* **Note 4**), then freeze–thaw twice (freeze in liquid nitrogen and thaw in a 37°C water bath).
4. Remove the cell debris by centrifugation at 10,000*g* for 10 min.
5. Transfer the supernatant to another tube (*see* **Note 5**).

### 3.5.3. Purification of Expressed Recombinant Protein

1. Mix the supernatant with the prepared Ni-resin column gently for 20 min at 4°C, then centrifuge at 1500*g* for 2 min.
2. Remove the supernatant carefully.
3. Wash the column three times with two column volumes of native binding buffer by resuspending the resin, rocking for 2 min, and then separating the resin from the supernatant by gravity or centrifugation (1500*g* for 2 min).
4. Sequentially resuspend the resin in increasing concentrations of imidazole elution buffer (i.e., 50, 200, 350, and 500 m*M*), rock gently for 5 min, then separate the resin form the supernatant by gravity or centrifugation (1500*g* for 2 min).
5. Save all the eluting supernatants, and assay for the presence of the desired protein (*see* **Fig. 4** and **Note 6**).

## 3.6. Biological Activity Analysis of the Purified Protein

The native HPV16L1 protein can self-assemble into virions in vitro and has the ability to agglutinate murine erythrocyte *(3)*. To identify the purified protein, the pro-

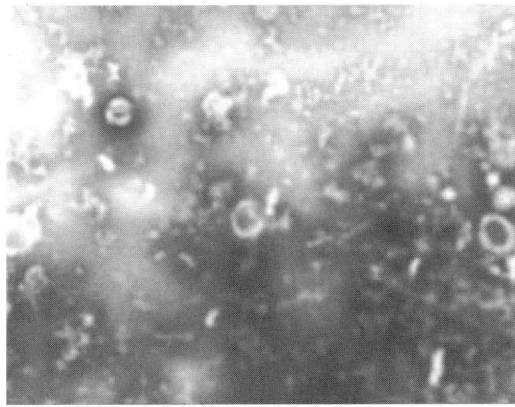

Fig. 5. Self-assembled HPV16 VLPs negatively stained with phosphotingstic acid analyzed under an electron microscope (×35,000).

tein was analyzed by a transmission electron microscope, hemagglutination assay, and hemagglutination inhibition assay (HAI).

### 3.6.1. Analysis of Virus-Like Particles (VLPs) by Transmission Electron Microscopy

1. Purified L1 protein is dropped onto carbon-coated copper grids and allowed to absorb for 3 min.
2. The grid is negatively stained with phosphotungstic acid for 2 min.
3. Specimens are then examined under an H600 transmission electron microscope.

When visualizing by transmission microscopy, the hollow spherical virions (50-nm diameter) are comparable with the HPVs virions of native L1 protein; i.e., the purified HPV16L1 protein possessed the ability to self-assemble into virions (*see* **Fig. 5**).

### 3.6.2. Hemagglutination Assay

1. Indicator erythrocyte suspensions were prepared from C57BL/6 fresh blood as described elsewhere *(6)*.
2. Purified HPV16L1 VLPs (1 mg/mL) are diluted in 50 µL phosphate-buffered saline (PBS), plated into 96 U-well plates, and then mixed with an equal volume of 1% (v/v) erythrocytes suspended in PBS.
3. Incubate for 3 h at 4°C.
4. The purified L1 protein should now have shown the ability to agglutinate murine erythrocytes (**Fig. 6A**).

### 3.6.3. Hemagglutination Inhibition Assay

The ability of HPV VLPs to agglutinate murine erythrocyte can be abolished by specific neutralizing antibodies *(3)*. The HAI procedure is basically the same as the murine hemagglutination procedure, with the exception of incubating the monoclonal antibody with VLPs before adding the erythrocyte suspension to the plate *(4)*. Sera from nonimmunized mice and PBS should be used as a negative control (*see* **Fig. 6B**).

**A**

A1

A2

A3

**B**

B1

B2

B3

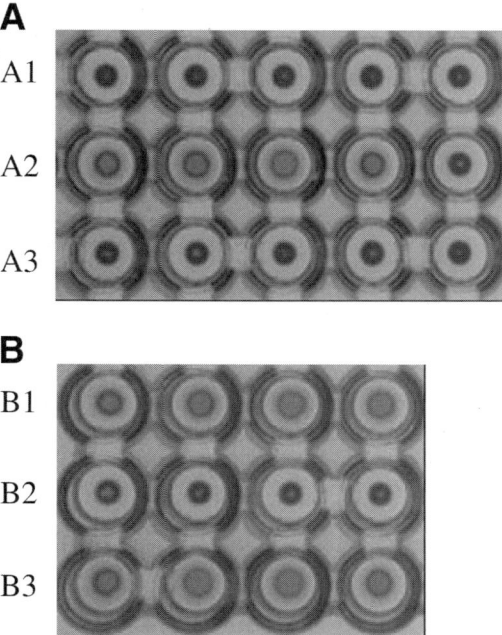

Fig. 6. Biologic activity analysis of purified HPV16L1 proteins. (**A**) Hemagglutination assay; (A1) PBS control; (A2) purified HPV 16L1 protein; (A3) and imidazole elution buffer control. (**B**) Hemagglutination inhibition assay (HAI); (B1) sera from nonimmunized mice; (B2) HPV 16L1 monoclonal antibody; and (B3) PBS control.

## 4. Notes

1. Cellfectin reagent is a lipid suspension that may settle with time. Mix thoroughly by inverting the tube 5–10 times before removing a sample for transfection to ensure that a homogeneous sample is taken.
2. Do not use dithiothreitol, EDTA, or other chelating agents in the buffers, which will reduce the $Ni^{2+}$ ion concentration.
3. 1 mL Resin is sufficient to purify approx 4 mg 6X his affinity-tagged protein.
4. Add fresh PMSF to the solution. PMSF loses efficacy within 30 min of dilution into an aqueous solution.
5. To avoid possible clogging of the column, spin down the debris, and load only the supernatant onto the column.
6. Fractions containing imidazole should be heated to 37°C for 10 min instead of boiling for SDS-PAGE analysis. This will avoid imidazole-mediated cleavage of labile peptide bonds.

## References

1. Bin, Z., Jianwei, W., Huiying, J., Jianguo, Q., Yili, W., Lusheng, S., and Xiaoping, D. (2001) Expression of human papillomavirus type16 L1 protein by insect-baculovirus expression system. *Chinese J. Exp. Clin. Virol. (China)* **15,** 314–317.
2. ProBond Purification System Manual, Invitrogen.

3. Roden, R. B., Hubbert, N. L., Kirnbauer, R., Breitburd, F., Lowy, D. R., and Schiller, J. T. (1995) Papillomavirus L1 capsids agglutinate mouse erythrocytes through a proteinaceous receptor. *J. Virol.* **69,** 5147–5151.
4. Zheng, L. S., Si, J. M., Song, J., Sun, X., Yu, J., and Wang, Y. (2003) Enhanced immune response to DNA-based HPV16L1 vaccination by costimulatory molecule B7-2. *Antiviral Res.* **59,** 61–65.

# 8

## Large-Scale Transient Expression
## of Therapeutic Proteins in Mammalian Cells

### Sabine Geisse, Martin Jordan, and Florian M. Wurm

### 1. Introduction

Research of the molecular and cellular biology of mammalian cells would be highly restricted if not for the development of methods to deliver exogenous DNA into cultivated cells. Transient transfection of mammalian cells became a routine research tool following a landmark publication by Graham and Van der Eb, who presented the calcium phosphate method as an assay to test the infectivity of purified viral DNA in 1973 (1). For the first time, this technique allowed the delivery of genes into animal cells without the help of a virus. Novel techniques for nonviral gene transfer have been developed and improved since then. Today, numerous commercial transfection agents for a wide range of cell lines are available (2). This chapter focuses on the two leading methods for large-scale applications of transient gene expression (TGE): calcium phosphate DNA coprecipitates (3,4) and polyethylenimine (PEI)–DNA polyplexes (5,6). Although calcium phosphate DNA coprecipitation is a well-established method that has been modified for high efficiency by many authors (7–12), PEI has only recently been developed as a transfection vehicle (13–15), but it has achieved rapid acceptance for large-scale suspension transfections because of its simplicity in handling and its efficacy of gene transfer to many different cell types.

#### 1.1. Large-Scale Transient Expression

Transient transfection protocols for suspension cells have been developed using rapidly growing and well-characterized host cells, such as human embryo kidney 293 (HEK-293) and Chinese hamster ovary (CHO) cells (16). This allows the transfection of 100–10,000 times more cells than usually employed for analytical approaches, but it requires uncomplicated transfection procedures and affordable reagents. The large-scale term should not be misunderstood—the largest culture vessel used for TGE thus far is a 100-L bioreactor (17). By comparison, classical recombinant protein expression

From: *Methods in Molecular Biology, vol. 308: Therapeutic Proteins: Methods and Protocols*
Edited by: C. M. Smales and D. C. James © Humana Press Inc., Totowa, NJ

from stable mammalian cell lines is routinely undertaken in reactors of 10,000 L or more.

Currently, large-scale TGE is principally used for the synthesis of secreted proteins of pharmaceutical interest, usually for the purpose of early structure–function analyses. In this context, large-scale TGE is particularly effective for the rapid synthesis of small quantities of many variants of the target protein for selecting the optimum candidate for clinical development *(18)*.

## 1.2. Polyethylenimine

Intensive studies on PEI as a potential transfection reagent for gene therapy have been undertaken at a small scale using multiple different cell lines, including primary cells *(13,19)*. For large-scale transient transfection, HEK-293 EBNA (293-EBNA) cells are the preferred host system *(5,6,20)*. Only recently, serum-free PEI transfections in CHO cells have been proposed as an alternative *(21)*.

The most important parameter for efficient PEI–DNA complex formation is the amount of PEI vs the amount of plasmid DNA. This ratio can be expressed as the number of amino nitrogens from PEI vs the number of phosphates from DNA (N/P ratio; *see* **Note 1** for details on calculation). Whereas an N/P ratio of 2–3 results in complete condensation of DNA into particles *(22)*, high-level expression rates are only obtained at much higher N/P ratios *(13)*.

PEI–DNA complexes with a weak positive charge can efficiently adsorb onto cells through electrostatic interactions. Once bound, they rapidly enter the cells via endocytosis. The large proton buffer capacity of the PEI molecule is a key factor to escape from the endosomal/lysosomal compartment. Subsequently, the complexes may be translocated to the perinuclear region and then into the nucleus *(15,19,23)*.

## 1.3. Calcium Phosphate

In contrast to methods based on liposomes or synthetic polymers, the calcium phosphate approach is based only on chemicals that can be found in all standard culture media. The advantage is that cells and the recombinant protein are never exposed to complex and frequently nonhomogeneous compounds that are difficult to define in molecular terms. In addition, these simple chemicals are available in kilogram amounts at cell culture quality. Furthermore, the solutions required for precipitate formation can be easily made. Ready-to-use kits including detailed protocols are commercially available as well.

Mixing the HEPES/phosphate solution with the calcium/DNA solution induces a spontaneous formation of a precipitate, which will also incorporate the plasmid DNA *(10)*. When added to adherently growing cells, the complexes slowly settle onto the cells and are taken up by phagocytosis/endocytosis *(24)*. In suspension cultures, the interaction between particles and cells is thought to be mediated by the agitation/mixing of the culture. Although the formation of the precipitate is probably the most difficult part of the method, carefully prepared buffers and strict control of physicochemical conditions provide efficient and reproducible transfection results for HEK-293 cells.

## 2. Materials

### 2.1. Plasmids and Cell Lines

1. Plasmids: pREP4 and pCEP4 (Invitrogen, Carlsbad, CA; *see also* **Note 2**).
2. 293-EBNA (American Type Culture Collection [ATCC], cat. no. CRL-10852, 293 c18, and until recently, also marketed by Invitrogen).
3. HEK-293 cell line (ATCC, cat. no. CRL-1573).
4. Dulbecco's modified Eagle medium (DMEM) or DMEM/F12 medium fortified with 10% fetal calf serum (FCS). For serum-free suspension cultivation, a selection of suitable media is shown in **Table 1**. If the cultivation medium does not support PEI-mediated transfection, RPMI 1640 medium can also be used during the transfection process (*see* **Note 3**).

### 2.2. Cell Culture Equipment and Instruments

1. Standard cell culture plasticware, centrifuge, $CO_2$ incubator, and incubators for roller bottles or spinner flasks.
2. Shaker with humidity and $CO_2$ control (e.g., ISF-1-W, Kühner, Switzerland).
3. The WAVE Bioreactor system, C20SPSS, and corresponding cell bag, CBN20L cat. no. 520.050, Wave Biotech, Bridgewater, NJ) or fully equipped laboratory bioreactors.

### 2.3. Commercial PEI Polymers

1. 25-kDa Branched PEI (Aldrich, Milwaukee, WI).
2. 25-kDa Linear PEI (Polysciences, Warrington, PA).
3. 25-kDa Linear PEI solution available as JetPEI™ (Polyplus Transfection, Illkirch, France).
4. 22-kDa Linear PEI solution available as ExGen 500 (Fermentas, St. Leon-Rot, Germany; *see* **Note 4**).

### 2.4. Calcium Phosphate Method

1. Calcium solution: 250 m$M$ $CaCl_2$ in water, sterilized by filtration. For storage, *see* **Note 5**.
2. HEPES/phosphate buffer: 140 m$M$ NaCl, 50 m$M$ HEPES, and 1.5 m$M$ $NaH_2PO_4$; adjust pH exactly to 7.05 with HCl or NaOH. For storage, *see* **Note 5**.
3. Transfection medium: DMEM/F12 supplemented with 2% FCS and 20 m$M$ HEPES, pH 7.4.
4. Phosphate stock solution: 150 m$M$ $NaH_2PO_4$, pH 7.05.

## 3. Methods

### 3.1. Plasmid Purification

Commercially available kits, such as NucleoBond™ (Macherey-Nagel, Oensingen, Switzerland), are used according to the manufacturer's instruction. Such kits can cover the needs of plasmid DNA for transient expression, which is approx 1–2 mg DNA per liter of transfected cells (*see* **Note 6**). The plasmid DNA is dissolved in sterile TE buffer (10 m$M$ Tris-HCl and 1 m$M$ EDTA, pH 7.6) at a concentration of 1 μg/μL for transfections.

### 3.2. Cell Cultivation

Seed cultures are maintained at 37°C in a humidified 5% $CO_2$ atmosphere in tissue culture flasks in serum-supplemented medium. The cultures are usually passaged twice per week at a split ratio of 1:10. Adaptation to serum-free culture medium is achieved

**Table 1**
**Overview of Cell Culture Media Available for Serum-Free Cultivation of 293-EBNA Cells and their Suitability for Transfection Using PEIs**

| Culture medium | Supplier | Specification | Price per L [CHF.-][a] | Growth characteristics in suspension | Transfection using PEI |
|---|---|---|---|---|---|
| Hektor S | Cell Culture Technologies GmbH, cat. no. HEK S | Medium for production after transient transfection | 30.- | Adaptation to serum-free suspension growth not achieved | Not assessed |
| CD 293 | Invitrogen, cat. no. 11913-019 | Protein-free, chemically defined medium for suspension culture | 102.- | Max. cell density: $8 \times 10^5$ c/mL | Possible |
| 293 SFM II | Invitrogen, cat. no. 11686-029 | Serum-free, low-protein medium for suspension culture | 132.- | Max. cell density: $5 \times 10^5$ c/mL | Not possible |
| FreeStyle 293 | Invitrogen, cat. no. 12338-018 | Serum and protein-free medium for suspension culture | 133.- | Max. cell density: $2 \times 10^6$ c/mL | Possible |
| Pro 293 s-CDM | BioWhittaker, cat. no. 12-765Q | Chemically defined medium for 293 suspension culture | 80.- | Max. cell density: $6 \times 10^5$ c/mL | Not possible |
| Ex-Cell VPRO | JRH Biosciences, cat. no. 14560 | Serum-free medium developed for cultivation of PER.C6 cells | 85.- | Max. cell density: $6 \times 10^6$ c/mL | Not possible |
| Ex-Cell 293 | JRH Biosciences, cat. no. 14570 | Serum-free medium developed for cultivation of HEK-293 cells | 85.- | Max. cell density: $6 \times 10^6$ c/mL | Not possible |
| Novartis M11 | Not commercial | In-house developments for serum-free suspension culture | 24.- | Max. cell density: $3.5 \times 10^6$ c/mL | Not possible |
| Novartis M11V3 | Not commercial | In-house developments for serum-free suspension culture | 50.- | Max. cell density: $2.6 \times 10^6$ c/mL | Possible |

[a]CHF, Swiss Francs.

by the reduction of serum-containing medium and replacement by commercially available serum-free media. Many cultures that are maintained in high-serum concentrations ($\geq 10\%$) can rapidly adapt to growth in some of the serum-free media listed in **Table 1**. If the direct transfer into a serum-free medium does not work, then gradual "weaning" of the culture to serum-free conditions is required, usually in a stepwise manner. As soon as the culture shows stable growth rates and high viability ($\geq 90\%$), a subsequent round of serum reduction should be initiated until a fully adapted culture is obtained.

Cells adapted to, e.g., serum-free Excell 293 medium (Cambrex BioWhittacker), can be maintained in roller bottles, shake flasks, or spinner cultures. Cells are seeded at $3 \times 10^5$ cells per milliliter and incubated in a $CO_2$ incubator with ventilated or open caps. To avoid any limitations of oxygen supply, the culture volume should be adapted to the high cell density ($6 \times 10^6$ cells/mL) that can be achieved with this medium. The cells can be diluted 10-fold by adding fresh medium every 3–4 d.

### 3.3. PEI Method

Detailed protocols for PEI-mediated transfection of suspension-adapted 293-EBNA cells at scales of 100 mL and 10 L are described.

### 3.3.1. Preparation of 1 mg/mL (w/v) PEI Solution

1. Dissolve the appropriate amount of PEI in distilled water.
2. Adjust pH to 7.0 by adding 1 *M* HCl.
3. Filter-sterilize (0.22-µm filter), prepare aliquots, and store frozen at –80°C until use.

### 3.3.2. PEI Transfection at 100-mL Culture Scale

1. In a sterile 50-mL cell culture tube, pellet $5 \times 10^7$ 293-EBNA cells from exponential growth phase by centrifugation (5–10 min at 500$g$).
2. Remove supernatant; gently resuspend cell pellet in 36 mL RPMI 1640 medium. Cell density is now $1.4 \times 10^6$ cells/mL (*see* **Notes 3** and **7**).
3. Transfer cell suspension into a small roller bottle (496 cm$^2$ surface area, Corning, cat. no. 430195). Aerate roller bottle with a 95% air/5% $CO_2$ mixture, close the caps, and incubate at 37°C with a turning speed of 6.5 rpm.
4. Preparation of DNA–PEI complexes: In a sterile 50-mL polystyrene tube, mix 100 µg plasmid DNA with 7 mL RPMI 1640 medium. Similarly, mix 200 µg PEI with 7 mL RPMI 1640 medium in a second tube (i.e., 200 µL stock solution, DNA:PEI ratio = 1:2 (µg:µg) or N/P ratio = 15.5; *see* **Note 1**). Gently mix the contents of both tubes by shaking and leave at room temperature for 15 min (*see* **Note 8**).
5. Add the PEI solution to the DNA solution while gently mixing, then briefly vortex for 1 s and incubate for 15 min at room temperature to allow complex formation.
6. Briefly mix again before adding the complexes to the cells. The resulting transfection volume is 50 mL at a density of $1.0 \times 10^6$ cells/mL. Aerate as above, and incubate the culture at 37°C with a turning speed of 6.5 rpm for 4 h.
7. Add 50 mL cultivation medium, resulting in a 100-mL culture at a density of $5 \times 10^5$ cells/mL, and continue incubation under the forementioned conditions.
8. Take daily samples for analysis until maximal production titers have been obtained. Productivity is highly protein-specific and dependent on intracellular vs secreted expression and thus needs to be determined empirically.

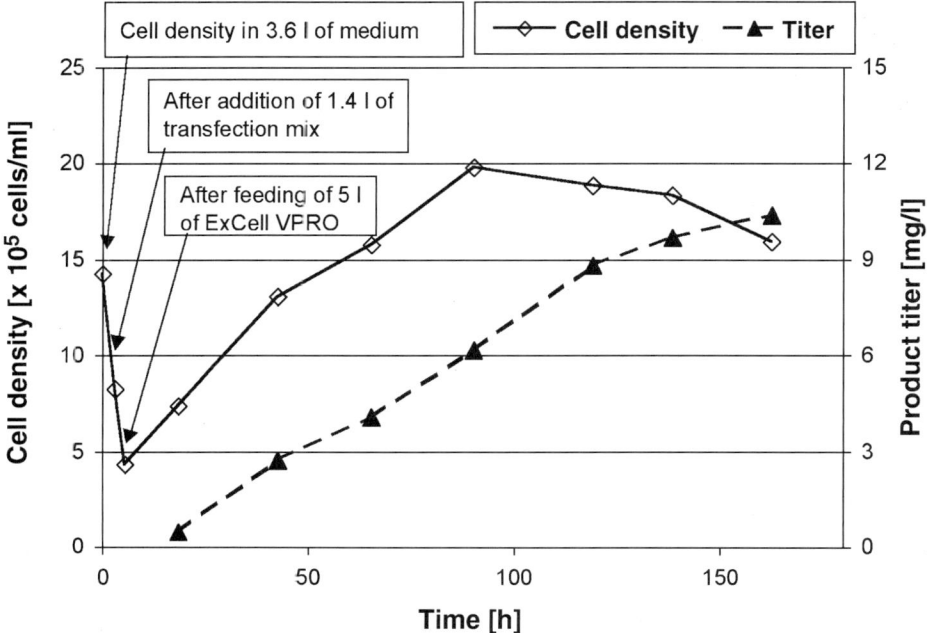

Fig. 1. The graph shows a typical fermentation run of 293-EBNA cells transfected and cultivated in the WAVE bioreactor in M11V3 (Novartis proprietary medium) and fed with 5 L VPRO medium for the production phase. Final volume at harvest: 10 L.

### 3.3.3. PEI Transfection at the 10-L Scale

This protocol can be adapted to 10 L from the previous using the Wave™ Bioreactor by multiplication of volumes and quantities by a factor of 100 (without changing the actual concentrations and ratios). A scale-down protocol to cell culture flasks and multiwell plates (e.g., for construct screening prior to large-scale transient transfection) is likewise possible by applying the same overall conditions.

1. Maintain a sufficiently large back-up culture of cells in roller bottles or shake flasks.
2. Transfer 3.6 L culture at a cell density of $1.4 \times 10^6$ cells/mL to the reactor medium (*see* **Notes 3** and **7**).
3. Prepare a transfection mix of 1.4 L in two roller bottles using 10 mg plasmid DNA and 20 mg PEI in 700 mL RPMI 1640 medium each. Following mixing and incubation, the 1.4 L of complexes are added to the reactor, yielding a transfection volume of 5 L.
4. After 4-h incubation, the final volume of 10 L is achieved with the addition of 5 L growth medium, reducing the cell density to approx $5 \times 10^5$ cells/mL at the beginning of the production phase (*see* **Note 9**). A typical production run using this protocol is depicted in **Fig. 1**.

### 3.4. Calcium Phosphate Technique

Detailed transfection protocols optimized for improved reproducibility are described for both adherent and suspension cultures of mammalian cells. A turbidity assay is suggested as a quality control measure that allows a quick check on the quality of the precipitate.

### 3.4.1. Turbidity Assay to Evaluate Precipitates

It is recommended to test new batches of $CaCl_2$ and HEPES/phosphate solutions. Until recently, the only reliable test was a transfection experiment with results available within 1 d. Jordan and coworkers established a faster turbidity assay that detects the formation of the precipitate by measuring the absorbance at 320 nm *(10)*. The assay works with or without DNA; however, the presence of DNA does affect the signal.

1. Add 0.5 mL HEPES/phosphate to 0.5 mL 250 m*M* $CaCl_2$ and mix quickly.
2. After 50 s transfer the mixture into a cuvet.
3. At exactly 1 min after mixing, measure the optical density (OD) at 320 nm.
4. The OD value should be approx 0.15 (or ~0.24 in the presence of 25 μg DNA/mL), or it should correspond to a value obtained by a set of solutions that have worked well in a previous transfection.
5. If the OD differs from the expected value, the final phosphate concentration in the mixture can be adjusted (*see* **Note 10**).

### 3.4.2. Transfection of Adherently Growing HEK-293 Cells

The transfection of adherent cells will remain an important tool when moderate quantities of several variants of a given target molecule are needed, requiring a high-throughput transfection approach *(18)*. Small-scale transfections can be executed in 12-well plates with adherent HEK-293 cells using a simplified calcium phosphate transfection method. The protocol for calcium phosphate DNA coprecipitation has been adapted to the HEK-293 cells, which easily detach from the culture flask. The glycerol shock treatment was omitted and replaced by a gentle medium exchange step after transfection (*see* **Note 11**).

1. Seed HEK-293 cells into a 12-well plate 12–20 h prior to transfection. For each well, prepare 1 mL transfection medium containing $1 \times 10^5$ cells (*see* **Note 12**).
2. On the next day, briefly check if the cells are attached and visually check the pH before preparing the transfection cocktail (*see* **Note 13**).
3. To transfect one well, add 2.5 μg DNA into 50 μL $CaCl_2$ solution.
4. Remove the 12-well plate from the $CO_2$ incubator.
5. Add 50 μL HEPES/phosphate to the calcium/DNA mixture and briefly mix with a pipet (the addition and mixing will not take more than 5 s), giving a solution in which precipitate formation will occur spontaneously.
6. After 1 min, add the 100 μL transfection cocktail to a single well. If you have prepared and timed everything properly, you can continue to add additional cocktails to the remaining wells of the plate. However, the entire transfer procedure should not take more than 5 min. Subsequently, return the plate to the incubator (5% $CO_2$).
7. After 4 h, carefully remove the transfection medium (avoid detaching the cells), and add your preferred medium.
8. Cells will express the protein during the next few days.

### 3.4.3. Calcium Phosphate Transfection
### of Suspension Cultivated HEK-293 Cells

Transfecting cells in suspension differs from the previous protocol for attached cells in several ways.

For cells grown in suspension (spinner flasks, bioreactors, or shaken systems) different serum-free media are available that prevent cell aggregation and provide good cell growth for HEK-293 cells (**Table 1**). Unfortunately, such culture media are not adapted for transfection with calcium phosphate. Thus, the transfection is undertaken in DMEM/F12 medium. Also, 1–2% FCS must be present during the transfection.

Upon addition of the precipitate, cells start to form aggregates, facilitated by the elevation of calcium ion concentration. The size of the aggregates remains acceptable only in the presence of serum. Transfection efficiency does not appear to be negatively affected by the formation of aggregates. Cells throughout these aggregates have been shown to express a green fluorescent protein reporter gene *(25)*.

The medium exchange after the transfection is replaced by a dilution step that eventually leads to dissolution of the precipitate.

1. Harvest the cells by centrifugation, and resuspend them in prewarmed transfection medium at $5 \times 10^5$ cells/mL.
2. Seed a spinner flask with this cell suspension at 50% of the regular working volume. Now the cells are ready to receive the transfection cocktail within the next 2 h.
3. Add 2.5 mg DNA into 50 mL $CaCl_2$ solution per liter of culture.
4. Add 50 mL HEPES/phosphate solution per liter of culture, quickly mix the two solutions, and incubate for 1 min (*see* **Note 14**).
5. Add 100 mL transfection cocktail per liter of cell suspension and incubate at 37°C for 4 h.
6. Add 1 vol of fresh medium (any medium that supports cell growth), and incubate cells until the product is harvested (*see* **Note 15**).

## 4. Notes

1. The calculation of the N/P ratio is based on the assumption that one repeating unit of PEI featuring one nitrogen corresponds to 43.1 g/mol, and one repeating unit of DNA featuring one phosphate corresponds to 330 g/mol. To calculate the N/P ratio for a 1 mg/mL solution of 25 kDa PEI, the equation is as follows:

$$N/P = \frac{(\mu L \text{ PEI stock solution}) \times 23.2 \text{ m}M \text{ nitrogen residues}}{(\mu g \text{ plasmid DNA}) \times 3 \text{ nmol phosphate}}$$

   In this example, for 100 µg plasmid DNA and 200 µg PEI (= 200 µL), the N/P ratio is 15.5.
2. Any recombinant vector for mammalian hosts prepared in sufficient quantity and quality can be used. If episomal replication of the plasmid in 293-EBNA is desired, then the expression vector must also feature an origin of replication (oriP) from the Epstein–Barr virus (EBV). Alternatively, for episomal replication in HEK-293, both oriP and EBNA-1 gene from EBV must be present on the vector. Episomal replication positively affects the levels of protein expression *(6,26,27)*.
3. Most of these media have been developed specifically for suspension cultures of HEK-293 cells. Unfortunately, the composition of these media is proprietary to the vendor.

Some media compositions appear to be incompatible with PEI-mediated DNA uptake. In particular, the Excell-VPRO (marketed by JRH Biosciences) as an excellent growth medium is not useful for PEI-mediated transfection. Here, a change of medium to DMEM/F12 or RPMI 1640 prior to transfection is required. Invitrogen's FreeStyle medium supports acceptable cell growth and PEI-mediated transfection, but it is not very cost-effective on a large scale. Recently, we have found that for most of these growth media, low-transfection efficiencies can be overcome by using up to fivefold higher PEI:DNA ratios.

4. PEI–DNA complexes formed with linear PEI appear to result in better transfection efficiencies when compared to branched PEI. However, the transfection efficiency with the branched 25 kDa PEI was improved when complemented with dioleoylmelittin *(5)*. This two-component mixture is now marketed as Xtreme GENE Ro-1539 by Roche Applied Science *(28)*.

5. Some authors suggest the frozen storage of calcium and phosphate solutions. However, the solutions can be left at room temperature without any loss of activity when stored as aliquots in tightly closed vessels (evaporation or uptake of $CO_2$ from the laboratory atmosphere might change the properties of the solution over time). If the solutions are not stored at room temperature, they should be allowed to warm to about 20°C before use, because precipitation is sensitive to temperature changes—complex formation will be slowed at lower temperatures!

6. A 2-L *E. coli* culture at a high cell density ($OD_{600}$ ~15) usually results in 30 g wet cell pellet and a plasmid yield of approx 10–15 mg. For these and larger amounts of plasmid DNA, conventional purification of DNA by anion-exchange and size-exclusion chromatography has been described *(29,30)*.

7. RPMI 1640 without serum does not support cell growth, but its use during the 4-h transfection does not harm cells either. In shaken cultures, the addition of 0.1% Pluronic F-68 is recommended to protect the cells from the damaging effects of the medium–air interface *(31)*. After transfection, a richer medium that supports cell growth should be added. Media may also contain peptones, particularly gelatine peptone N3 (OrganoTechnie S.A., La Courneuve, France) at a concentration of 5% (w/v). The addition of this peptone has been shown to lead to a fourfold enhancement in volumetric productivity *(20)*.

8. Frequently, the PEI mixture is prepared in 150 m$M$ NaCl solution; however, no adverse effect has been found when dissolving the PEI stock solution in culture medium *(28,32)*. We even observed an increase in recombinant protein titer under these conditions, presumably because the diluting effect of the NaCl solution on cell culture nutrients is eliminated. A second important aspect relates to the order of addition: the PEI solution must be added to the DNA solution. This has been shown to impact transfection rates by a factor of 10 when the solution was added dropwise *(13)*. By adding the solutions more rapidly, we could no longer see a significant difference in transfection efficiency.

9. For transfection at the 10-L scale, the cell density at the time of transfection is important. For 293-EBNA cells, a density of $\geq 4 \times 10^5$ cells/mL leads to undisturbed cell growth after transfection, whereas lower densities result in a lag phase in growth during the production phase.

10. Although the OD values do not directly correlate with the transfection efficiency, they allow a more objective description of the precipitate than visual judgments like "opalescence" or "cloudiness." If the OD is too low, insufficient amounts of particles are created to complex all the DNA molecules. Increasing the phosphate concentration by adding concentrated phosphate will solve the problem. If the OD is too high, the complexes get

coarser and are less efficient than those that are small. In this case, the phosphate concentration should be decreased (e.g., by reducing the volume of the HEPES/phosphate solution within the mixture, mixing at a ratio of 1:0.8).

Having solutions or mixing ratios that work well is a prerequisite for successful transfection. Another critical parameter in this context is the method of mixing. Adding the HEPES/phosphate rapidly in one step or adding it dropwise while constantly mixing does not lead to the same kind of precipitate. Mixing affects both the OD values and transfection results.

The turbidity assay serves as a tool to find a good range in which transfections work. Chowdhury (11) used it to modify the solutions for forming a precipitate at a pH of 7.6, which is significantly higher than the range of 6.95–7.1 that is reported in the literature, and obtained higher transfection efficiencies.

11. This protocol—addition of DNA precipitates by dropwise pipetting, 1-day incubation in serum-containing media, gentle medium exchange, and then incubation for 5–6 d for accumulation of secreted product—proved to be efficient and reproducible, resulting in 2–10 mg/L recombinant human tissue plasminogen activator (18).

12. Cells are seeded the day before the transfection to get them firmly attached. Because mitosis favors the uptake of DNA (33), cells should be in the exponential growth phase during transfection. Also, cells should be transfected when they are subconfluent to allow at least one more cell doubling. Depending on the amount of recombinant protein required the transfection can be undertaken in multiwell plates, 10-cm plates, or T-flasks of any size.

13. If cells do not attach firmly to the culture plates in the presence of only 2% FCS, higher concentrations may be used. The transfection works at serum concentrations of 1–10%; therefore, a medium exchange prior to transfection is not needed. If cells grow fast or are seeded at higher cell densities, a decline of the pH value below 7.2 might favor the dissolution of the precipitate. In such a case, the medium should be exchanged before transfection.

14. Mixing and forming the precipitate at a larger scale have not been studied systematically. By comparing results from small- and large-scale experiments, we frequently noticed that the solutions did not perform the same. Considering that the volumes used for the larger scale generally demand more time for addition/mixing, it is reasonable to assume that such differences are owing to mixing issues. Thus, for different scales, adding the HEPES/phosphate solution to the $CaCl_2$-DNA solution should be considered a critical parameter. After mixing, we always use a 1-min incubation time for practical reasons. If the addition of the phosphate is undertaken very slowly, longer standing times are possible. Seelos investigated the standing time and found that 1 min was the best for fast mixing, but 20 min gave better results for slow mixing (34).

The DNA concentration is another critical factor for the formation of the precipitate. It should not be modified once the protocol is established. We routinely use 25 µg DNA per milliliter of precipitate. When less DNA is available, an empty vector or a noninterfering plasmid or even eukaryotic chromosomal DNA is added as "carrier" DNA.

15. Despite the dilution step, elevated concentrations of calcium remain within the medium during the production phase. This fact has to be acknowledged when the product is harvested. Avoid adding the supernatant into slightly basic solutions that are rich in phosphate, because this would immediately provoke the formation of a precipitate! Another issue is serum: if the product can only be purified from a serum-free culture, the transfection medium can be exchanged 4 h after transfection by a centrifugation step and then replaced by serum-free medium.

## References

1. Graham, F. L. and van der Eb, A. J. (1973) A new technique for the assay of infectivity of human adenovirus 5 DNA. *Virology* **52,** 456–467.
2. Peel, A., ed. Transfection of mammalian cells, in *Methods*, vol. 33, No 1, 2004.
3. Jordan, M., Koehne, C., and Wurm, F. M. (1998) Calcium-phosphate mediated DNA transfer into HEK-293 cells in suspension: control of physicochemical parameters allows transfection in stirred media. *Cytotechnology* **26,** 39–47.
4. Meissner, P., Pick, H., Kulangara, A., Chatellard, P., Friedrich, K., and Wurm, F. M. (2001) Transient gene expression: recombinant protein production with suspension-adapted HEK 293-EBNA cells. *Biotechnol. Bioeng.* **75,** 197–203.
5. Schlaeger, E.-J. and Christensen, K. (1999) Transient gene expression in mammalian cells grown in serum-free suspension culture. *Cytotechnology* **30,** 71–83.
6. Durocher, Y., Perret, S., and Kamen, A. (2002) High-level and high-throughput recombinant protein production by transient transfection of suspension-growing human 293-EBNA1 cells. *Nucleic Acids Res.* **30,** 2–9.
7. Gorman, C., Padmanabhan, R., and Howard, B. H. (1983) High efficiency DNA-mediated transformation of primate cells. *Science* **221,** 551–553.
8. Chen, C. and Okayama, H. (1987) High-efficiency transformation of mammalian cells by plasmid DNA. *Mol. Cell. Biol.* **7,** 2745–2752.
9. O'Mahoney, J. V. and Adams, T. E. (1994) Optimization of experimental variables influencing reporter gene expression in hepatoma cells following calcium phosphate transfection. *DNA Cell Biol.* **13,** 1227–1232.
10. Jordan, M., Schallhorn, A., and Wurm, F. M. (1996) Transfecting mammalian cells: optimization of critical parameters affecting calcium-phosphate precipitate formation. *Nucleic Acids Res.* **24,** 596–601.
11. Chowdhury, E. H., Sasagawa, T., Nagaoka, M., Kundu, A. K., and Akaike, T. (2003) Transfecting mammalian cells by DNA/calcium phosphate precipitates: effect of temperature and pH on precipitation. *Anal. Biochem.* **314,** 316–318.
12. Girard, P., Porte, L., Berta, T., Jordan, M., and Wurm, F. M. (2001) Calcium phosphate transfection optimization for serum-free suspension culture. *Cytotechnology* **35,** 175–180.
13. Boussif, O., Lezoualc'h, F., Zanta, M. A., Mergny, M. D., Scherman, D., Demeneix, B., and Behr, J.-P. (1995) A versatile vector for gene and oligonucleotide transfer into cells in culture and *in vivo*: Polyethylenimine. *Proc. Natl. Acad. Sci.* **92,** 7297–7301.
14. Boussif, O., Zanta, M. A., and Behr, J.-P. (1996) Optimized galenics improve *in vitro* gene transfer with cationic molecules up to 1000 fold. *Gene Therapy* **3,** 1074–1080.
15. Kircheis, R., Wightman, L., and Wagner, E. (2001) Design and gene delivery activity of modified polyethylenimines. *Adv. Drug Deliv. Rev.* **53,** 341–358.
16. Wurm, F. M. and Bernard, A. (1999) Large-scale transient expression in mammalian cells for recombinant protein production. *Curr. Opin. Biotechnol.* **10,** 156–159.
17. Girard, P., Derouazi, M., Baumgartner, G., Bourgeois, M., Jordan, M., Jacko, B., and Wurm, F. M. (2002) 100 Liter-transient transfection. *Cytotechnology* **38,** 15–21.
18. Bennett, W. F., Paoni, N. F., Keyt, B. A., Botstein, D., Jones, A. J. S., Presta, L., et al. (1991) High resolution analysis of functional determinants on human tissue-type plasminogen-activator. *J Biol. Chem.* **226,** 5191–6201.
19. Merlin, J.-L., N'Doye, A., Bouriez, T., and Dolivet, G. (2002) Polyethylenimine derivatives as potent nonviral vectors for gene transfer. *Drug News Perspect.* **15,** 445–451.
20. Pham, P. L., Perret, S., Doan, H. C., Cass, B., St. Laurent, G., Kamen, A., and Durocher, Y. (2003) Large-scale transient transfection of serum-free suspension-growing HEK293-

EBNA1 cells: Peptone additives improve cell growth and transfection efficiency. *Biotechnol. Bioeng.* **84,** 333–342.

21. Derouazi, M., Girard, P., van Tilborgh, F., Iglesias, K., Muller, N., Bertschinger, M., and Wurm, F. M. (2004) Serum-free large scale transfection of CHO cells. *Biotechnol. Bioeng.* **87,** 537–545.

22. Kunath, K., von Harpe, A., Fischer, D., Petersen, H., Bickel, U., Voigt, K., and Kissel, T. (2003) Low-molecular-weight polyethylenimine as non-viral vector for DNA delivery: comparison of physicochemical properties, transfection efficiency and in vivo distribution with high-molecular weight polyethylenimine. *J. Control Release* **89,** 113–125.

23. Thomas, M. and Klibanov, A. M. (2003) Non-viral gene therapy: polycation-mediated DNA delivery. *Appl. Microbiol. Biotechnol.* **62,** 27–34.

24. Loyter, A., Scangos, G., Juricek, D., Keene, D., and Ruddle F. H. (1982) Mechanisms of DNA entry into mammalian cells. *Exp. Cell Res.* **139,** 223–234.

25. Pick, H. M., Meissner, P., Preuss, A. K., Tromba, P., Vogel, H., and Wurm, F. M. (2002) Balancing GFP reporter plasmid quantity in large-scale transient transfections for recombinant anti-human rhesus-D IgG1 synthesis. *Biotechnol. Bioeng.* **79,** 595–601.

26. Shen, E. S., Cooke, G. M., and Horlick, R. A. (1995) Improved expression cloning using reporter genes and Epstein-Barr virus ori-containing vectors. *Gene* **156,** 235–239.

27. Cho, M. S., Yee, H., and Chan, S. (2002) Establishment of a human somatic hybrid cell line for recombinant protein production. *J. Biomed. Sci.* **9,** 631–638.

28. Schlaeger, E.-J., Kitas, E. A., and Dorn, A. (2003) SEAP expression in transiently transfected mammalian cells grown in serum-free suspension culture. *Cytotechnology* **42,** 47–55.

29. Schmid, G., Schlaeger, E.-J., and Wipf, B. (2001) Non-GMP plasmid production for transient transfection in bioreactors. *Cytotechnology* **35,** 157–164.

30. Wright, J. L., Jordan, M., and Wurm, F. M. (2003) Transfection of partially purified plasmid DNA for high level transient protein expression in HEK293-EBNA cells. *J. Biotechnol.* **102,** 211–221.

31. Jordan, M., Sucker, H., Einsele, A., Widmer, F., and Eppenberger, H. M. (1993) Interactions between animal cells and gas bubbles: the influence of serum and pluronic F68 on the physical properties of the bubble surface. *Biotechnol. Bioeng.* **43,** 446–454.

32. Kichler, A., Zauner, W., Ogris, M., and Wagner, E. (1998) Influence of the DNA complexation medium on the transfection efficiency of lipospermine/DNA particles. *Gene Therapy* **5,** 855–860.

33. Grosjean, F., Batard, P., Jordan, M., and Wurm, F. M. (2002) S-phase synchronized CHO cells show elevated transfection efficiency and expression using CaPi. *Cytotechnology* **38,** 57–62.

34. Seelos, C. (1997) A critical parameter determining the aging of DNA-calcium-phosphate precipitates. *Anal. Biochem.* **245,** 109–111.

# 9

## Site-Specific Integration for High-Level Protein Production in Mammalian Cells

### Bhaskar Thyagarajan and Michele P. Calos

## 1. Introduction

Recombinant proteins have various applications in basic and applied research, as well as in therapy. Numerous systems have been developed to produce large protein amounts *(1–3)*. The most convenient systems are those that involve the overexpression of recombinant proteins in bacteria, mainly because they are fast and easy. However, there is a need for eukaryotic overexpression systems, because several proteins require either the proper folding environment or posttranslational modifications for physiological activity *(4)*. The ideal system for eukaryotic overexpression would have integration of the expression cassette in the genome of the target cell at a location that permits strong and long-term expression. However, most current methods for achieving integration of plasmid DNA into the genome involve random integration of DNA *(5–8)*. Expression of the recombinant protein in clones generated by such methods is dependent on the chromatin context at the site of integration and often leads to low expression levels *(9,10)*. Therefore, many clones should be screened to find those expressing the protein of interest at the desired level.

We have developed a mammalian expression system that minimizes the context effects seen with random integration and provides long-term expression of recombinant proteins. Our system uses the bacteriophage $\phi$C31 integrase—a site-specific recombinase that has been shown to function efficiently in mammalian cells *(11,12)*. The $\phi$C31 integrase mediates recombination between two 30–40-bp sequences termed *attB* and *attP*. If an *attB* site is placed on the incoming plasmid to be integrated, in the presence of $\phi$C31 integrase, unidirectional integration will occur at a limited number of sites in the genome that resemble *attP*. As shown in several cell types and tissues, the $\phi$C31 integrase directs integration of the *attB* plasmid at these pseudo *attP* sites in mammalian genomes and leads to robust expression levels in these cells *(12–15)*.

In addition to requiring sequence identity to *attP*, the $\phi$C31 integrase appears to choose its genomic target sites on the basis of chromatin features that allow the enzyme

From: *Methods in Molecular Biology, vol. 308: Therapeutic Proteins: Methods and Protocols*
Edited by: C. M. Smales and D. C. James © Humana Press Inc., Totowa, NJ

to access chromosomal DNA. These features may include an open-chromatin conformation and have also proven to be favorable for gene expression. As a result, transgenes integrated with the φC31 integrase system are characterized by robust and stable gene expression. Therefore, we have utilized this system for the overexpression of recombinant proteins in Chinese hamster ovary (CHO) cells, which are commonly used for the production of recombinant proteins. We show that the φC31 integrase targets integration at genomic pseudo-*attP* sites in CHO cells and allows for high levels of expression of a reporter gene. Expression levels are indicated to be higher when integration is mediated by φC31 integrase, rather than by random integration, whether or not the cells are subjected to selection.

Briefly, the generation of cell lines with φC31 integrase involves transfection of two plasmids. The donor plasmid contains the φC31 *attB* site and expression cassette carrying the gene of interest, and the second plasmid contains an expression cassette driving the expression of the φC31 integrase. When both plasmids are transfected together, the integrase protein is produced and directs the integration of the donor plasmid containing the *attB* site into a pseudo-*attP* site located in the chromosomes. The transfected cells can be cultured in the presence or absence of a drug if a selectable marker gene is present on the donor plasmid. If selection is performed, individual clones can be analyzed for high expression. If individual clones are not desired, the transfected cells can be pooled and treated as a population. This chapter describes general methods that can be used to transfect CHO cells with the φC31 integrase and a donor plasmid containing a luciferase reporter gene and the analysis of luciferase expression levels in these cells.

## 2. Materials

### 2.1. Cloning and Preparation of a Donor Plasmid

1. Plasmid containing the gene of interest in an appropriate expression cassette (*see* **Note 1**) and plasmid containing the *attB* sequence.
2. Appropriate restriction enzymes and buffers.
3. Agarose gel and loading buffer.
4. Qiaex gel extraction kit (Qiagen) or equivalent.
5. Ligase and appropriate buffer.
6. Competent DH10B cells.
7. Luria-Bertani (LB) agar plates containing the appropriate drug.
8. LB medium containing the appropriate medium.
9. Qiafilter plasmid maxi kit (Qiagen) or equivalent.
10. The control plasmid, pCMVLacZ, and the integrase-expressing plasmid, pCMVInt, have been described previously (*11*).

### 2.2. Culture and Transfection of CHO Cells

1. Appropriate tissue culture plates.
2. F12K medium (Invitrogen) or equivalent.
3. Fetal bovine serum.
4. Trypsin: A solution of 0.25% trypsin and 1 m*M* EDTA (Invitrogen or equivalent).
5. Fugene 6 transfection reagent (Roche or equivalent; *see* **Note 2**).

6. Opti-MEM I medium (Invitrogen or equivalent).
7. Geneticin sulfate or other appropriate drug for selection.

## 2.3. Preparation of Luciferase Extracts and Assay for Luciferase (see Note 3)

1. Phosphate-buffered saline (PBS), pH 7.4.
2. Lysis buffer: 25 m*M* Tris-HCl, pH 7.8, 2 m*M* EDTA, 0.5% Triton X-100, and 5% glycerol.
3. Luciferase assay reagent (Promega) or equivalent.
4. TD-20e luminometer (Turner Designs) or equivalent.

# 3. Methods

## 3.1. Cloning and Preparation of a Donor Plasmid

1. Clone the φC31 *attB* region into the plasmid containing the expression cassette using the appropriate restriction enzymes and buffers.
2. Sequence the *attB* region and gene of interest to ensure that there are no mutations. For our experiments, we used the luciferase reporter gene to assay expression levels.

## 3.2. Culture and Transfection of CHO Cells

1. Culture CHO cells in F12K medium containing 9% fetal bovine serum.
2. Passage cells and plate them to be ready for transfection the next day. With Fugene 6, the manufacturer typically recommends 50–80% confluency on the day of transfection. For our experiments, we transfected cells in wells of a 24-well plate (*see* **Note 4**).
3. Transfect the cells according to the manufacturer's instructions (*see* **Note 2**). We typically use 3 μL Fugene 6 for 1 μg DNA. We transfected 5 ng reporter plasmid (pNBL2) with 200 ng of either a control plasmid (pCMVlacZ) or a plasmid encoding the φC31 integrase (pCMVInt).
4. For experiments to be performed without selection (*see* **Note 5**), move the cells to a larger plate 24 h after transfection. For our experiments, we transfected cells in 24-well plates and transferred them to 35-mm plates.
5. After a further 48 h, harvest one third of the cells for a luciferase extract, and plate the remaining cells on a fresh 35-mm plate. This sample will be the first time point for protein assay. Continue to passage the cells every 3–4 d, except that at future time points, plate only one third of the cells, and reserve two thirds for an extract. Luciferase extracts can be made at each time point or once a week.
6. If a selection experiment is being performed, transfer the transfected cells to a larger plate 24 h after transfection. For selection experiments, we typically transfected wells on 24-well plates and transferred them to 100-mm plates after 24 h.
7. Remove nonselective medium from the plates, and apply medium containing the appropriate drug to start selection 48 h posttransfection. For our experiments, we used geneticin sulfate (G418 sulfate) at an effective concentration of 600 μg/mL medium (*see* **Note 6**). Apply fresh medium to the cells every 2–3 d to ensure that the medium on the plate is not too acidic.
8. If individual clones are to be analyzed, wait until well-isolated colonies are formed on the selected plates before carefully transferring them to a 24-well plate (*see* **Note 7**). It takes approx 12–14 d for the appearance of well-formed colonies.
9. If pooled populations of cells are to be analyzed with selection, follow the same procedure as for the unselected pool (**step 5**) until the first time point is harvested. After this point, start the selection process with the appropriate medium. Continue the selection process until colonies are visible on the plate. Then, dislodge the colonies from the plate,

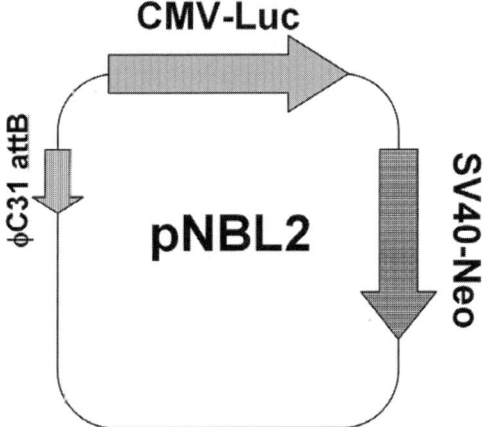

Fig. 1. Donor plasmid used in the experiment. The plasmid pNBL2 has the firefly luciferase gene driven by the cytomegalovirus (CMV) promoter (expression marker), a neomycin phosphotransferase gene driven by the SV40 promoter (selectable marker), and a φC31 *attB* site.

and transfer the cells to a fresh plate (*see* **Note 8**). Passage the cells every 3–4 d. Harvest two thirds of the cells for a luciferase extract at every passage.

### 3.3. Preparation of Luciferase Extracts and Assay for Luciferase

1. Once cells are harvested for preparation of a luciferase extract, wash them three times in ice-cold (4°C) PBS. From this point on, make sure that the cells and all the solutions are kept on ice.
2. Resuspend the cells in the appropriate amount of ice-cold lysis buffer. Use 50 µL lysis buffer for every million cells harvested.
3. Incubate the suspension for 5 min on ice.
4. Centrifuge the lysed cell suspension for 5 min in a microcentrifuge at 13,000g.
5. Carefully remove the supernatant, transfer into aliquots, and store at –80°C.
6. For the assay, thaw the extracts on ice and assay according to the protocol prescribed by the manufacturer of the assay reagent.

## 4. Notes

1. As shown in **Fig. 1**, the donor plasmid described in these experiments, pNBL2, has a selectable marker (neomycin phosphotransferase) and an expression marker (luciferase). The selectable marker provides resistance to geneticin in eukaryotic cells. Other genes commonly used for selection in mammalian cells are hygromycin phosphotransferase (hygromycin), *Sh ble* (zeocin), puromycin *N*-acetyl-transferase (puromycin), and *bsr* (blasticidin). Although any selectable marker can be used, one would have to perform the necessary experiments to determine the effective concentration of the drug required to kill cells that do not express the marker.

   The reporter gene on our plasmid was derived from firefly luciferase and allows the determination of expression from a population of cells. Any convenient expression marker may be used to establish the amount of protein being expressed by the cells. In our experience, luciferase and secreted alkaline phosphatase are generally better for determining average expression in a cell population. For expression in individual cells, green fluorescent protein is a convenient marker that can be assayed using a fluorescence-activated cell sorter.

2. We used Fugene 6 (Roche) as the transfection reagent in all our experiments. Because CHO cells are not particularly hard to transfect, any commercially available transfection reagent can be used. We typically obtain transfection efficiencies of 15–20% using Fugene. Lipofectamine Plus (Invitrogen) and electroporation are other transfection methods that have provided excellent transfection levels in mammalian cells in our experience. For all of our experiments, we typically used 3 µL Fugene for every microgram of DNA. If other transfection reagents are used, it may be necessary to optimize the amounts of reagent required in relation to the DNA.

3. We used the luciferase assay reagent from Promega Corporation (Madison, WI) and a Turner Designs TD-20e luminometer (Turner Designs, CA) for performing luciferase assays. There are several commercially available choices for assaying luciferase in extracts prepared from cells, including systems modified for use in high-throughput assays.

4. Because all our experiments involved transfection of cells in 24-well plates, we followed the conditions outlined in the **Methods** section. If transfection of larger plates is desired, the amount of each DNA and the transfection reagent should be optimized. We used low amounts of the reporter gene donor plasmid to ensure a low level of background, especially in the selection experiments. Using a small quantity of reporter DNA allowed us to efficiently pick single, well-isolated colonies. If the objective is to utilize a population of cells, larger amounts of donor DNA can be used.

5. Although higher expression levels are obtained when selection is performed for integration events, it may not be feasible to perform selection in some instances. In such cases, it is also possible to obtain elevated expression levels using the integrase system. As shown in **Fig. 2**, luciferase expression in cells that received the integrase along with the expression plasmid had approx 60-fold higher expression levels compared to cells that did not receive the integrase. As would be expected from the loss of unintegrated plasmid DNA, luciferase levels in both cases fell sharply in the first 2 wk after transfection before stabilizing. The levels of luciferase seen after day 17 are likely to indicate luciferase expression from integrated plasmid DNA.

6. Irrespective of the selection drug used, it is advisable to perform a kill curve to determine the optimal amount of drug required to kill cells not expressing the selectable marker. In our case, we obtained G418 from Invitrogen (Carlsbad, CA). The effective drug concentration supplied by other manufacturers may vary.

7. Isolation of individual clones may be performed using cloning cylinders. Alternatively, a 20-µL tip fixed to a pipet could be used to carefully dislodge and transfer the colony to a well of a 24-well plate. If the colonies are not well-separated, it may be necessary to plate the transfected cells into multiple dishes or to use lower amounts of DNA for the transfection. The results of a typical experiment are shown in **Fig. 3**. There were 44 G418-resistant colonies picked and analyzed for luciferase expression. As can be seen from **Fig. 3**, there are clones with high-expression levels (approx three to fourfold higher than the average expression of all the clones), as well as clones with no detectable expression. Analysis of genomic DNA from the high-expressing clones revealed integration into pseudo-*attP* sites mediated by φC31 integrase (data not shown). These data suggest that φC31 integrase can mediate integration into sites that enable high levels of transgene expression.

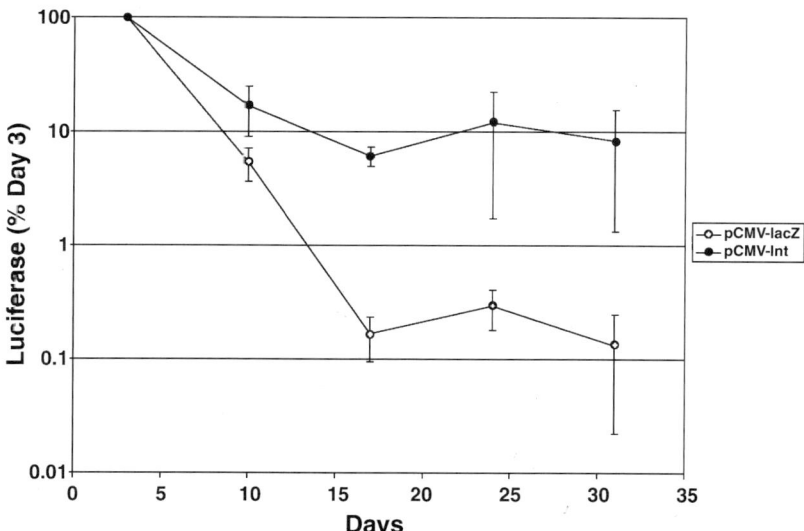

Fig. 2. Luciferase expression in the absence of selection. CHO cells were transfected as described, and protein extracts were prepared from the cells at various time points after transfection. The luciferase activity in the extracts is shown normalized to day-3 values (100%) and is the average of three independent experiments. Luciferase expression is approx 60-fold higher in the presence of φC31 integrase (filled circles) compared to a control plasmid lacking integrase (open circles). Error bars depict the standard error of the mean.

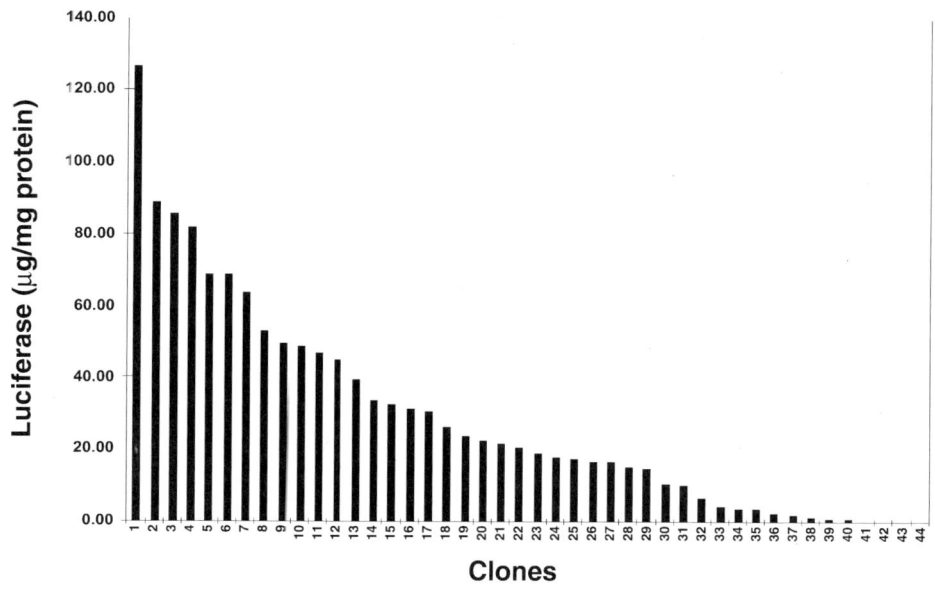

Fig. 3. Luciferase expression in individual clones. CHO cells were transfected with pNBL2 and pCMV-Int. After 14 d of selection, 44 G418-resistant colonies were isolated and analyzed for luciferase expression. In the highest expressing clones, integration occurred at pseudo-*attP* sites (data not shown).

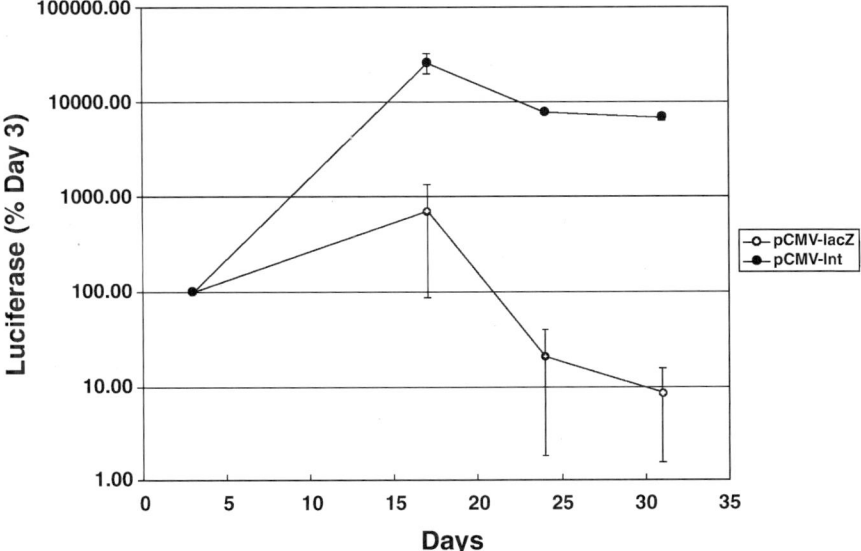

Fig. 4. Luciferase expression in the presence of selection. CHO cells were transfected as described and grown under selection. Protein extracts were prepared from these cells and assayed for luciferase activity. The luciferase activity in the extracts is shown normalized to day-3 values (100%) and is the average of three independent experiments. Luciferase expression in selected pools that received the φC31 integrase (filled circles) is approx 800-fold higher than the group that received a control plasmid lacking integrase (open circles). Error bars depict the standard error of the mean.

8.  In our experience, if the cells are disturbed during the early selection process, they tend not to stick well to the plate. Therefore, it is preferable to avoid harvesting cells for 2 wk after the start of selection. Expression levels in selected pools are much higher than those seen in unselected pools. Again, cells that received the integrase had much higher levels of luciferase expression (>800-fold) than cells that received the control plasmid, suggesting that the expression at pseudo-*attP* sites is much higher than expression in random integrants (**Fig. 4**). In this example, the expression 31 d posttransfection was almost 100-fold higher than the expression observed 3 d after transfection. By contrast, in the nonselected pools, expression at day 31 was 1/10 of the expression seen at d 3, indicating that for the highest expression levels, it would be advisable to select for φC31 integrase-mediated events.

## References

1.  Roy, P. and Jones, I. (1996) Assembly of macromolecular complexes in bacterial and baculovirus expression systems. *Curr. Opin. Struct. Biol.* **6,** 157–161.
2.  Makrides, S. C. (1996) Strategies for achieving high-level expression of genes in Escherichia coli. *Microbiol. Rev.* **60,** 512–538.
3.  Geisse, S., Gram, H., Kleuser, B., and Kocher, H. P. (1996) Eukaryotic expression systems: a comparison. *Protein Expr. Purif.* **8,** 271–282.

4. Weikert, S., Papac, D., Briggs, J., Cowfer, D., Tom, S., Gawlitzek, M., et al. (1999) Engineering Chinese hamster ovary cells to maximize sialic acid content of recombinant glycoproteins. *Nat. Biotechnol.* **17,** 1116–1121.

5. Kaufman, R. J., Davies, M. V., Wasley, L. C., and Michnick, D. (1991) Improved vectors for stable expression of foreign genes in mammalian cells by use of the untranslated leader sequence from EMC virus. *Nucleic Acids Res.* **19,** 4485–4490.

6. Kaufman, R. J. (2000) Overview of vector design for mammalian gene expression. *Mol. Biotechnol.* **16,** 151–160.

7. Lucas, B. K., Giere, L. M., DeMarco, R. A., Shen, A., Chisholm, V., and Crowley, C. W. (1996) High-level production of recombinant proteins in CHO cells using a dicistronic DHFR intron expression vector. *Nucleic Acids Res.* **24,** 1774–1779.

8. Cockett, M. I., Bebbington, C. R., and Yarranton, G. T. (1990) High level expression of tissue inhibitor of metalloproteinases in Chinese hamster ovary cells using glutamine synthetase gene amplification. *Biotechnology (NY)* **8,** 662–667.

9. Kalos, M. and Fournier, R. E. (1995) Position-independent transgene expression mediated by boundary elements from the apolipoprotein B chromatin domain. *Mol. Cell Biol.* **15,** 198–207.

10. Zahn-Zabal, M., Kobr, M., Girod, P. A., Imhof, M., Chatellard, P., de Jesus, M., et al. (2001) Development of stable cell lines for production or regulated expression using matrix attachment regions. *J. Biotechnol.* **87,** 29–42.

11. Groth, A. C., Olivares, E. C., Thyagarajan, B., and Calos, M. P. (2000) A phage integrase directs efficient site-specific integration in human cells. *Proc. Natl. Acad. Sci. USA* **97,** 5995–6000.

12. Thyagarajan, B., Olivares, E. C., Hollis, R. P., Ginsburg, D. S., and Calos, M. P. (2001) Site-specific genomic integration in mammalian cells mediated by phage phiC31 integrase. *Mol. Cell Biol.* **21,** 3926–3934.

13. Olivares, E. C., Hollis, R. P., Chalberg, T. W., Meuse, L., Kay, M. A., and Calos, M. P. (2002) Site-specific genomic integration produces therapeutic Factor IX levels in mice. *Nat. Biotechnol.* **20,** 1124–1128.

14. Ortiz-Urda, S., Thyagarajan, B., Keene, D. R., Lin, Q., Fang, M., Calos, M. P., and Khavari, P. A. (2002) Stable nonviral genetic correction of inherited human skin disease. *Nat. Med.* **8,** 1166–1170.

15. Ortiz-Urda, S., Thyagarajan, B., Keene, D. R., Lin, Q., Calos, M. P., and Khavari, P. A. (2003) PhiC31 integrase-mediated nonviral genetic correction of junctional epidermolysis bullosa. *Hum. Gene Ther.* **14,** 923–928.

# 10

## Production of Recombinant Therapeutic Proteins by Mammalian Cells in Suspension Culture

### Lily Chu, Ilse Blumentals, and Gargi Maheshwari

### 1. Introduction

Within the past 10 yr, 17 monoclonal antibodies (MAbs) have been approved in the United States. A survey of these approved MAbs reveals that the predominant platform is a serum-free, stirred-tank mammalian cell culture. In fact, all 17 products are grown in only four cell lines: Chinese hamster ovary (CHO) cells, mouse myeloma cells (NS0 and Sp2/0), or hybridoma cells. The trend toward platform consolidation enables the possibility to quicken process development timelines to establish a common method in the development of a large-scale production process. This chapter will outline the general format for producing a recombinant protein in stirred tank, serum-free cultures beginning from a recombinant plasmid and a mammalian host cell line. Although different cell lines are used in this chapter as examples for the various methods, these methods are applicable across most common production cell lines. The purification options following harvest of the bulk material are not discussed, nor are the details of regulatory requirements, but purification information is readily available in other chapters of this book, and regulatory guidance documents are available at the regulatory websites, such as www.fda.gov/cber and www.ich.org.

### 2. Materials

1. Biosafety hood.
2. Humidified $CO_2$ incubator.
3. Orbital shaker.
4. Pipettors.
5. Pipet tips.
6. Disposable graduated pipets.
7. Microscope.
8. Cell enumeration equipment and materials: The manual hemocytometer method or automated CEDEX instrument (Innovatis, Bielefeld, Germany) can be used.

From: *Methods in Molecular Biology, vol. 308: Therapeutic Proteins: Methods and Protocols*
Edited by: C. M. Smales and D. C. James © Humana Press Inc., Totowa, NJ

9. CsCl-purified (or equivalent) plasmid containing the gene of interest and expression vector system.
10. Mammalian host cell line (e.g., *dhfr⁻* CHO) and related cell culture medium.
11. LipofectAMINE 2000 (Invitrogen, Carlsbad, CA) or equivalent cationic lipid reagent.
12. Tissue culture plates (six-well, 24-well, and 96-well), tissue culture flasks (75 cm$^2$) and shake flasks (vented, 125 mL and 250 mL).
13. Trypsin.
14. Phosphate-buffered saline (PBS).
15. Culture media and media components (cell culture tested; *see* **Subheading 3.2.**)
16. Bioreactor system with 2-L culture vessels (B. Braun Biostat B, Allentown, PA or equivalent).
17. pH and dissolved oxygen (DO) probes (Mettler Toledo, Columbus, OH or equivalent).
18. Cylinders of compressed, medical grade, process gases $CO_2$, $O_2$, air, and $N_2$.
19. Sterile equipment for culture transfers and additions. The use of 1/8" ID C-Flex weldable tubing and tubing welder (Terumo SCD IIB, Somerset, NJ) is recommended.
20. Pluronic F-68.
21. Antifoam (Simethicone Emulsion, DowCorning, Midland, MI).
22. 7.5% $NaHCO_3$.
23. Analytical instruments to measure substrates and metabolic products (e.g., BioProfile, Nova Biomedical, Waltham, MA).

# 3. Methods

## 3.1. Cell Line Development

The generation of a high-producer clone of an adherent CHO cell line is described in **Subheadings 3.1.1.–3.1.4.** This process entails the (1) transfection of the parent cell line with the plasmid of interest; (2) selection and expansion of the transfectants; (3) measurement of specific productivities; and (4) evaluation of the genetic stability of high producers.

### 3.1.1. Transfection

The incorporation of a recombinant plasmid into mammalian cells can be accomplished through a variety of methods. Some more common methods include calcium phosphate coprecipitation *(1,2)*, electroporation *(3)*, and complex formation with cationic lipids *(4)*. Each has specific advantages and disadvantages; for simplicity, this section only describes a typical cationic lipid method (with LipofectAMINE™ 2000). The advantages of cationic lipids include convenience (commercially prepared reagents), simplicity, and scalability *(5)*. The following LipofectAMINE transfection conditions are based on vendor recommendations.

1. Passage CHO cells 1 d prior to transfection at a cell density that results in 90–95% confluency on the day of transfection.
2. Prepare LipofectAMINE and DNA solutions separately using serum-free medium, and combine these solutions prior to the application to CHO cells. A DNA:LipofectAMINE ratio of 1:2 or 1:3 (w:v) is usually sufficient for correct complexing (*see* **Note 1**; *6*).
3. Add LipofectAMINE/DNA solution to CHO cells.
4. Return cells to 37°C and incubate for 4–24 h. Replace medium with fresh growth medium. Growth medium is used for 48-h posttransfection to maximize cell recovery (*see* **Notes 2** and **3**).

5. Additionally, replace medium 24- and 48-h posttransfection with fresh growth medium.
6. On day-2 (48 h) posttransfection, trypsinize and count CHO cells. Inoculate 96-well plates with transfected CHO cells. The inoculation cell density per well should be based on two parameters: transfection efficiency and number of clones per well. The best practice is to target a majority of wells with only one colony per well, which will increase the probability of isolating a single clone for expansion. For example, a transfection efficiency of $10^{-4}$ would suggest a target inoculation cell density of $1 \times 10^4$ cells/well (*see* **Note 4**).
7. Incubate 96-well plates containing CHO transfectants at 37°C to select for successful transfectants.

## 3.1.2. Selection and Expansion

The selection and expansion phase will depend on the expression plasmid and cell line. For example, *dhfr⁻* CHO cells lack the *dihydrofolate reductase* (*dhfr*) gene, and only cells that successfully incorporate the expression plasmid containing a *dhfr* gene will survive the selection phase. Selection pressure for *dhfr⁻* CHO cells consists of medium lacking nucleosides, and this pressure is maintained throughout the expansion and production phases to ensure clone stability. Adherent CHO cells are expanded in well plates and T-flasks, then adapted to suspension culture if needed. The following protocol assumes *dhfr⁻* CHO cells were transfected in **Subheading 3.1.1.**

1. Screen 96-well plates for positive cell growth 3–4-wk posttransfection. Mark wells containing colonies and indicate wells containing single colonies (*see* **Note 5**).
2. Collect supernatant from each well and use enzyme-linked immunosorbent assay (ELISA, or equivalent assay) to determine MAb concentration per well. Rank positive growth wells based on MAb concentration, and expand as many of the top-ranked wells as feasible. To minimize population variability, only wells containing single colonies should be considered for expansion.
3. Transfer top-ranked transfectants into 24-well plates to expand (*see* **Note 6**).
4. After transfectants reach more than 70% confluency, expand cells into six-well plates. Collect supernatant samples from 24-well plates prior to transfer, and use ELISA (or equivalent assay) to determine MAb concentration per well. Rank transfectants again at the 24-well plate stage. The number of transfectants may be reduced again based on 24-well ranking, if desired.
5. Repeat the expansion process described above for expansion into T25 flasks. A benefit of measuring MAb concentration at each stage is to track any gross productivity losses due to genetic instability.
6. The expansion step from T25 flasks into T75 flasks allows the generation of enough cells for a token freeze. After T75 flasks are more than 90% confluent, prepare three vials of each transfectant for a token freeze. Typical cell culture cryopreservation protocols are sufficient (*7*).
7. The final expansion step consists of transferring cells from T75 flasks into 125-mL shake flasks. Inoculate shake flasks with at least $2 \times 10^5$ cells/mL and culture at a constant mixing rate (e.g., 100–150 rpm, 1 in. throw; *see* **Note 7**).
8. Continue expanding cells in shake flasks until cell growth is consistent for at least three passages. Suspension-adapted CHO transfectant cultures are ready for analysis at this point (*see* **Note 8**).

### 3.1.3. Growth Curves and Specific Productivity Evaluation

To determine which transfectants to carry into the medium and reactor optimization experiments, additional characterization of the transfectants is useful. These experiments will allow the determination of clone doubling times and specific productivity.

1. Inoculate 250-mL shake flasks with top-ranked transfectants at $2 \times 10^5$ cells/mL.
2. Sample cultures each day until day 7.
3. Determine cell concentration, and use ELISA (or equivalent) to determine MAb concentration for each sample.
4. Plot cell concentration (cells/mL) vs culture age (h) and use **Eqs. (1a)** and **(1b)** to determine the doubling time. **Equation (1a)** shows the relationship between cell concentration and culture age during exponential growth, and the growth rate ($\mu$) is calculated from an exponential curve fit. For **Eq. (1a)**, it is necessary to use data points in the exponential growth phase:

$$C_t = C_0 e^{\mu t} \tag{1a}$$

In **Eq. (1a)**, $C_t$ represents the cell concentration (cells/mL) at time $t$; $C_0$ is the initial cell concentration (at $t = 0$); $\mu$ is the growth rate; and $t$ is the culture age (h). The doubling time ($t_d$) is calculated via **Eq. (1b)**:

$$t_d = \frac{\ln 2}{\mu} \tag{1b}$$

5. Calculate the integral viable cell concentration (IVC) based on **Eq. 2**:

$$IVC_{i+1} = 0.5 \times (C_{i+1} + C_i) \times (t_{i+1} - t_i)/24 + IVC_i \tag{2}$$

In **Eq. 2**, IVC represents the integral viable cell concentration (cells-d/mL); $C$ is the cell concentration (cells/mL); and $t$ is the culture age (h). A simple trapezoidal rule is used to calculate the area under the curve for the IVC.
6. Plot MAb concentration (ng/mL) vs IVC (cells-d/mL), and use a linear curve fit to determine the specific productivity. The slope of the linear curve fit will represent the specific productivity (ng/cell/c). Specific productivity can be calculated during exponential cell growth or extended over the entire culture if desired.
7. Two to five transfectants are selected based on doubling times and specific productivity calculations (see **Note 9**).

### 3.1.4. Genetic Stability Evaluation

Another critical criterion is the genetic stability of the chosen transfectants. If this analysis is not performed before choosing a single transfectant for cell banking and subsequent medium/reactor optimization development, there is a risk of significant productivity loss that may occur upon extended expansion for scale-up. Performing genetic stability studies as early as possible will reduce this risk by incorporating genetic stability as a criterion for the selection of a transfectant.

1. Inoculate 125-mL shake flasks with top-ranked transfectants at $2 \times 10^5$ cells/mL.
2. Sample cultures for 2 d: day 3 or day 4 for serial passaging and day 7 for stability characterization.

3. On day 3 or day 4, determine cell concentration, and inoculate fresh 125-mL shake flasks at $2 \times 10^5$ cells/mL to continue cultures.
4. On day 7, determine cell concentration and use ELISA (or equivalent) to determine MAb concentration for each sample.
5. Repeat above protocol for at least 45 generations. If transfectant doubling time is 24 h, this will take 45 d. Generation time is calculated by **Eq. 3**, where $x_f$ is the final cell concentration (cells/mL); $x_o$ is the initial cell concentration (cells/mL); and $n$ is the generation number:

$$n = \frac{\ln(x_f / x_o)}{\ln 2} \tag{3}$$

6. Compare cell growth and MAb titers over 45 generations to evaluate genetic stability. Any transfectants that show minimal variation (specifically, the absence of decreasing MAb titers) should be chosen as potential production candidates.

## 3.2. Medium Development for Mammalian Serum-Free Cell Culture

A number of recent articles have focused on nutritional requirements of cells. The reader is referred to these references and their bibliographies for additional sources of information *(7–14)*. This section centers on certain practical considerations while developing a serum-free media for therapeutic protein production. With the rising cost of serum and existing issues with its sourcing, there is a significant emphasis on developing serum-free processes. Several serum-free media are now available for a variety of cell types, including CHO, NS0, and hybridomas. Most of these are proprietary in nature; hence, their detailed compositions are unavailable. However, they can serve as a good starting point for further enhancements for a specific cell clone.

Typical serum-free cell culture media contain amino acids, salts, trace elements, vitamins, hormones, carbon sources, growth factors, lipids and lipid precursors, and buffers. The amounts of these components vary with cell type and requirements of the process. For instance, media formulation may change depending on the mode of operation, such as batch, fed-batch, or perfusion. A general media development approach is described in the following section for batch operation. Modifications to the basal media and feed media for fed-batch and perfusion systems will be needed for per-cell requirements of specific nutrients.

### 3.2.1. Culture Systems and Analytical Equipment

The use of suspension cells introduces additional challenges when choosing a scale-down model system for media screens. Although a plate-based format (e.g., 96-well plate) may be used to derive preliminary information, shear and aeration among other differences between a plate and a stirred-tank bioreactor (STBR) are too significant to rely solely on a 96-well format for final media definition. Shake and spinner flasks permit the growth of suspension cells and are reliable and cost-effective screening systems for media development. Additionally, minibioreactor systems are now becoming available with pH and DO control capabilities *(15)*. Depending on their ease of use and cost, such miniaturized systems may prove to be useful for screening. It is important to verify the performance of intermediate milestone media in scale-down models of the expected production reactors prior to final medium definition. In addition, certain medium components may interact with the culture vessels and/or the medium-storage containers and impact the medium's performance. Risk of such interactions may be reduced by the verification of this performance in scale-down culture vessels using the same construction material as the expected production vessel. This may necessitate a modified delivery of certain components, such as lipids and essential proteins (if present), or changes in trace element concentrations in the medium.

Analytical tools to measure nutrient levels in the media during the course of cell culture are very valuable. Metabolite levels (e.g., glucose, lactate, ammonia, glutamate, and bicarbonate) are easily quantifiable using standard equipment. Amino acids and nucleosides can be quantified using well-established high-performance liquid chromatography (HPLC)-based methods *(16)*. Methods are available for the quantitation of vitamins and fatty acids, but their limits of quantitation may be close to the component levels typically found in the media. Trace element levels may be quantified using ion-coupled plasma mass spectrometry or other similar methods.

### 3.2.2. Medium Components

Glucose and glutamine are commonly used carbon sources. Other sources, such as pyruvate, fructose, galactose, and mannose, may also be used. Typical considerations when selecting a carbon source are the cost, sourcing of raw material, and the raw material's impact on cell metabolism. Glutamine is known to degrade in culture medium to produce ammonia, which may be toxic to the cells. Therefore, glutamine is typically added before the initiation of the culture batch. Glutamine modifications are commercially available that are reportedly more stable (*see* **Note 10**; *17*); however, their performance may vary between cell types. Amino acids serve as the primary nitrogen source for cells. It is beneficial to provide both essential and nonessential amino acids for more efficient metabolism. The levels of amino acids in the media may be determined based on the maximal cell density required and the mass-per-cell requirements of specific amino acids.

Typical cell culture media contain water-soluble vitamins, such as folic acid, riboflavin, choline chloride, niacinamide, pantothenic acid, biotin, pyridoxine, vitamin B12, and ascorbic acid. Vitamins A, D, E, and K may also influence cell metabolism at appropriate concentrations (*see* **Note 11**).

Typical buffers used to maintain the pH value of the cells are sodium bicarbonate, sodium phosphate, and HEPES. The levels of these buffers are best optimized in conjunction with the gassing strategy used in the culture vessels. Sodium chloride and potassium chloride are added to maintain the osmolality, typically in the 250–350-mOsm/kg range. Focus should be given to the $Na^+:K^+$ ratio to maintain the cell membrane potential. Phenol red may also be added as a pH indicator.

Trace elements (e.g., iron, copper, selenium, cobalt, vanadium, cobalt, molybdenum, zinc, nickel, magnesium, and calcium) have been shown to be beneficial for cell growth. When complex undefined components, such as serum, plant hydrolysates, and yeast extracts, are not added to the media, it then becomes essential to provide appropriate amounts of the trace elements via the addition of salts. When optimizing trace elements in the media, it is advantageous for them to be at levels that will mask the amount that may be introduced in the media adventitiously through the water supply or by any leaching in the culture vessels.

Most serum-free cell culture media contain growth factors, such as insulin, insulin growth factor 1, and epidermal growth factor, to stimulate cell growth. The amount and type of growth factors depend on the cell type of choice. Other proteins used are transferrin as an iron carrier and albumin (either bovine serum albumin or recombinant human albumin). Nonessential components (e.g., nucleosides) may also be added to reduce the cellular metabolic burden.

Pluronic F68 and dextran sulfate are protective agents that may also be added to reduce the shear sensitivity of the cells and to minimize cell clumping. Pluronic F68 may additionally be used to deliver hydrophobic components (e.g., fatty acids).

Antioxidants and reducing agents like vitamin E, ascorbic acid, glutathione, and β-mercaptoethanol may be included in the media to reduce the damage caused by free-radical generation in the presence of trace elements and lipids. These reducing agents can also change the valency of certain trace metals, which may, in turn, negatively impact cell growth. Hence, it is important to perform an overall optimization of the media using statistically designed experiments to avoid negative interactions.

Use of preformulated mixes of vitamins, amino acids, trace elements, and lipids may be useful. In addition, plant hydrolysates and/or yeast extracts may be used as rich sources of nutrients. However, use caution with such chemically undefined components because of the potential lot-to-lot variability of these components. Depending on the sensitivity of the cell line, such variability may significantly affect cell performance. When using these complex components, it is advisable to screen the lots for their impact on cell growth and productivity before addition to the basal media (*see* **Notes 12** and **13**). If significant enhancement in cell growth is obtained by adding a complex mixture like a plant hydrolysate, there is potential indicated for improving the basal media by adding specific chemically defined components to reduce the dependence on the hydrolysate. Further fortification of the basal media may also improve process robustness.

### 3.2.3. Media Development Strategy

Depending on the time and resources available for medium development, different strategies may be pursued to arrive at a final medium for the process. One empirical approach is to first identify various commercially available serum-free media developed for the cell line of choice. These media and their combinations in various ratios may then be screened for cell growth and productivity. Cells should be cultured in the media for multiple passages. Typical screening parameters include cell doubling times, maximal cell density, and volumetric productivity. Methods described in the previous section may be used for quantitation of these parameters. Based on the analytical tools available, spent media analysis can then be performed, and depleted nutrients can be supplemented into the media. Supplementation can also be undertaken with commercially complex mixtures, e.g., lipid mixtures, hydrolysates, and vitamin, amino acid, and trace element mixes. Although this approach is somewhat less resource-intensive, there is little fundamental understanding of the specific media components and the cell sensitivity to those components. This makes troubleshooting a challenge in the event of a failure of a particular media lot or unexplained process failures.

An alternative approach is to design media using a rational design approach that requires greater input of resources. This methodology is based on using a basal media of known composition as a starting point and making specific additions to the media, usually sequentially based on spent media analysis and what is known about the metabolism of the specific cell type. Minimum requirements and toxicity levels of specific media components can be determined using media deficient in that particular component. Based on this information, a range of acceptable levels can be established for that component. This range can then be further refined by experiments aimed at identifying interactions between specific components in the media. These refinements are best done using statistically designed experiments.

## 3.3. Bioreactor Process Development

The goal of bioreactor development is to generate a robust, reproducible production process with optimized protein expression levels. The development strategy is executed in stages that parallel the clinical progress of the therapeutic protein. As such, the first-generation process is often relatively simple, striving to employ standard/shelf technologies with minimal effort toward optimization. Later-generation processes integrate the use of high-producing cell lines, customized media formulations, and optimized production processes. Small-scale bioreactors (2 L) are used during the initial developmental stages to determine cell line stability, compare productivity of different cell lines, and establish baseline process conditions. At this stage, small-scale bioreactor data are also used to determine if benchmarks for productivity and quality have been achieved. In later-stage development, these bioreactors are used to develop process intensification strategies, define acceptable ranges for process variables, troubleshoot during process scale-up, and support process changes. This section outlines the general procedures and considerations for developing a cell culture process for therapeutic proteins at the 2-L bioreactor scale (**Note 14**).

### 3.3.1. Bioreactor Assembly and Set-up

All materials and bioreactor parts that come into contact with the cell culture must be sterile. First, the bioreactor is assembled following the manufacturer's recommendations and general bioreactor preparation considerations *(20)*. The appropriate connection lines are prepared, including gas inlet and outlet, inoculation port, sample port, feed port (optional), base addition, and harvest port. The pH and DO probes are calibrated according to vendor instructions and inserted in the bioreactor through the headplate. The bioreactor is filled with 1.5 L Milli-Q water (or water for injection or PBS) and autoclaved. After sterilization, the bioreactor is attached to the control unit via the appropriate utility and instrumentation connections. The contents are then pumped out of the bioreactor, and the bioreactor is filled with 2 L of growth medium using sterile connections. The bioreactor control is turned ON, the process variable set points (temperature, aeration, pH, and agitation) are specified, and the bioreactor is allowed to reach all set points. Once the bioreactor has stabilized, the DO probe read-out is calibrated to 100% air saturation, and the pH probe readout is adjusted based on offline pH measurements (*see* **Note 15**).

### 3.3.2. Bioreactor Inoculation

The inoculum culture is prepared from a frozen vial through a series of expansion steps in shake or spinner flasks following established cell culture procedures (*7*). Optimal inoculum concentration and dilution for cell expansion depends on the particular cell line and culture medium and must therefore be determined experimentally. Generally, inoculum cultures are transferred in the exponential phase to avoid lags and ensure consistent process performance. For this, a relatively conservative split ratio (1:5) is recommended. Bioreactor cultures are typically inoculated at $1 \times 10^5$–$3 \times 10^5$ cells/mL and usually achieve $1 \times 10^6$–$3 \times 10^6$ cells/mL. Most batch culture processes are 4–7 d long with culture, doubling times between 12 and 36 h.

### 3.3.3. Bioreactor Process Control

One important aspect in developing a robust cell culture process is to establish bioreactor operating conditions and control strategies. Although the optimal conditions will vary for different cell lines and culture media, general recommendations can be made.

DO is typically controlled between 30% and 60% air saturation. This can be accomplished by using one or a combination of the following methods: adjusting the $O_2$ concentration in the inlet gas mixture ($O_2/CO_2/N_2$ or $O_2/CO_2/air$), increasing the volumetric flow rate of the inlet gas mixture, and increasing the agitation speed rate. The choice of method depends on the oxygen demand of the culture and scale-up considerations. Because culture oxygen demand is a function of cell density and cell-specific oxygen uptake rate (OUR), oxygenation requirements will increase exponentially during logarithmic growth, then decrease as the culture enters the stationary phase. The amount of oxygen consumed by the culture at any time (OUR) is a very sensitive measurement and a good indicator of the cellular energy metabolism. For scale-up of cell culture processes, one major consideration is the accumulation of $CO_2$, which can result in suppression of growth and productivity (*21–23*). A minimum amount of $CO_2$ is required for cell growth (approx 5 mmHg), but an excess amount (150–180 mmHg) can have negative effects on cellular metabolism and protein production or glycosylation (*24*; *see* **Note 16**).

For sparged bioreactors, the inlet air gas flow rate is generally maintained between 0.05 and 0.25 vol of gas per volume of liquid per minute. For a 2-L culture, the volumetric gas flow range is 0.1–0.5 L/min; higher flow rates can generate excessive foaming. Cell damage caused by gas sparging is prevented by the addition of 1 g/L Pluronic F-68 (*25,26*) and foam is controlled by addition of chemical antifoam (Simethicone emulsion).

Efficient mixing in the bioreactor is required to achieve high oxygen-transfer rates and homogeneous conditions. For cell culture processes, low-shear impellers, such as marine and pitched blade designs, are widely used. For small-scale 2-L bioreactors, twin-pitched blade impellers and agitation rates between 100 and 150 rpm are recommended.

Culture pH is an important parameter for optimal growth and product expression; it has been shown that antibody productivity and glycosylation can be influenced by pH *(27)*. Culture pH is maintained constant by supplying $CO_2$ in the inlet gas mixture during the initial growth phase and by addition of a base, such as 7.5% $NaHCO_3$, later in growth (*see* **Note 17**). Culture pH is controlled within a relatively narrow range and might be different for cell growth and protein production steps. The control of pH, $DO_2$, and $CO_2$ are combined to avoid the accumulation or excessive stripping of $CO_2$ while maintaining good oxygen supply.

Temperature in the bioreactors is controlled by adjusting the temperature of the liquid that circulates within the jacket of the vessel. Although the optimal growth temperature for most cell lines is around 37°C, product optimization strategies might require growth and product expression steps at different temperatures. Lower temperatures (<37°C) have been shown to prolong cell viability and, in some cases, increase cell-specific productivity *(28)*.

### 3.3.4. Bioreactor Process Monitoring

Culture pH, DO, and temperature are monitored using the probes installed in the bioreactor. For cell density and metabolic activity, culture samples are withdrawn from the bioreactor using sterile connections. Viable culture cell density is measured using the trypan blue exclusion method *(7)*, and either manual hemacytometer counts or automated CEDEX measurements can be used. Consumption of key substrates in the growth medium, such as glutamine and glucose and production of lactate and ammonia, can be determined through BioProfile measurements. Concentration of other amino acids and media components must be determined following appropriate analytical techniques (*see* **Subheading 2.**).

### 3.3.5. Fed-Batch Cultures

Three main variables determine the productivity (titer) of a cell culture process: viable cell density, culture duration, and cell-specific productivity. Cell densities achieved in batch culture are relatively low, and specific productivities can vary significantly over the course of the batch. Fed-batch strategies have been developed and widely adopted to increase cell density, optimize specific productivity, and extend culture duration *(29)*. In fed-batch cultures, small volumes of concentrated nutrients are added over time to avoid nutrient depletion, minimize production of inhibitory byproducts (e.g., lactate and ammonia), and maintain a balanced nutrient concentration. Successful feeding strategies are based on the development of concentrated feed solutions that typically contain a mix of components, including glucose, amino acids, trace elements, vitamins, lipid precursors, metal chelators, and reducing compounds. Further supplementation with antiapoptotic chemicals or peptides for serum-free media has also been explored *(30)*.

Developing an optimal feeding strategy often requires significant effort and multiple process iterations *(31,32)*. The composition of the nutrient feed should reflect the stoichiometric ratio of nutrient consumption. Using the basal medium formulation, the specific consumption rates of nutrients and production of metabolites in the culture are determined. The stoichiometry of nutrient utilization is calculated based on a reference compound (e.g., glucose, lactate, and oxygen) by dividing the moles of nutrient consumed by the moles of reference compound consumed (or produced). The feed solution is prepared by mixing the depleted nutrients according to the stoichiometric ratios. Optimized compositions for the basal medium and feed solution are obtained after several iterations of testing and analysis, including evaluation of medium changes on product quality. For early-stage process development, the use of complete medium concentrates provides a quick and simple alternative for increasing culture productivity. A general approach is to prepare a 10X concentrated solution of the basal medium components without the salts *(33)*.

The control of nutrient addition is an important component of the feeding strategy. The nutrient solution can be fed continuously or intermittently following a predetermined plan or in response to culture performance (feedback control; *32–34*). Most feed controls rely on one or more metabolic measurements to infer the feeding rate. For example, glucose can be maintained at a certain concentration range by adjusting the rate of nutrient addition according to glucose concentrations in the culture *(35)*. If intermittent feeding is used, the amount of nutrients fed must be sufficient to sustain growth until the next feeding time. For cell-line evaluation and rapid production of small amounts of product, offline measurements of glucose, glutamine, and lactate should provide the necessary metabolic information to develop a baseline fed-batch process.

## 4. Notes

1. The researcher will need to optimize DNA levels, LipofectAMINE levels, relative ratios of DNA and LipofectAMINE, and cell concentration. Nonoptimized levels may result in excessive cell death. Vendor recommendations are good starting points.
2. If using serum-free medium immediately following transfection (recovery phase) and during selection, be sure to test medium compatibility with LipofectAMINE prior to transfections. Do not use antibiotics during LipofectAMINE transfections because antibiotic use may result in excessive cell death.
3. Incubation time in LipofectAMINE may need to be optimized, as cell health may be sensitive to cationic lipids.
4. To increase the probability of expanding a single colony, the inoculation cell density into 96-well plates should be optimized for a high probability of one colony per well postselection. This will be system-specific, because results are dependent on transfection efficiencies and cell survival.
5. The selection phase can be minimized by moving cells into full-selection pressure as quickly as possible. This is easily accomplished when working with adherent CHO cells, because a full medium exchange can be performed, but residual carryover of recovery medium can be a problem when working with suspension cultures.
6. The total number of transfectants expanded will be based on transfection efficiency, expected productivity, genetic stability, and timeline requirements. If the desired clone only has to meet minimum productivity requirements for product delivery, it is not necessary to expand an excessive number of transfectants. In fact, 20 transfectants should be sufficient. If there are large delivery requirements, more transfectants should be expanded through the stages to increase the probability of finding a higher producer.
7. This step and subsequent passaging in shake flasks are necessary for suspension adaptation of the transfectants prior to further analysis. Growth curves and specific productivity

are dependent on culture conditions, such as static vs suspension cultures; therefore, it is beneficial to adapt cells as close to final process conditions before performing an in-depth analysis.

8. CHO cells adapt easily to suspension culture. Initial CHO cultures in shake flasks may contain significant aggregation, but this should decrease as CHO cells adapt to suspension culture. Another criterion to reduce transfectant numbers is to exclude cultures that show excessive aggregation during suspension adaptation.

9. Highly productive clones with long doubling times are more desirable than less productive clones with shorter doubling times. Cultures will continue to adapt to process conditions, and doubling times will often shorten until they reach a typical range of 24–48 h.

10. Care should be taken to determine the stability of the media over time. The media may be formulated by a vendor either in a powder or liquid form. The powder forms have a longer shelf-life than the liquids. However, certain components (growth factors and fatty acids) may not be stable or delivered uniformly through a powder. Such components are best added at the time of hydration.

11. Certain components in the media (e.g., riboflavin, tyrosine, and trypotophan) are known to contribute to its light sensitivity. Appropriate precautions should be taken to avoid exposure of cell culture media to high-intensity light during storage or cell culture.

12. It is recommended to evaluate the multiple lots of raw materials that go into the media for media robustness. Chances of variability are higher with large molecular-weight components, which are not completely defined. This type of analysis will help identify components with higher sensitivities to process performance. Appropriate screening procedures can then be instituted to minimize the lot-to-lot variability. Additionally, to minimize the risk of adventitious biological contaminants, components of animal origin should be avoided *(18)*.

13. Special attention should be given to the addition of hydrophobic components. They tend to bind to medium storage containers and culture vessels. Certain hydrophobic components like fatty acids are available in water-soluble forms by their conjugation with cyclodextrin that may provide more uniform and stable delivery *(19)*.

14. Disposable bag bioreactors (Wave Biotech, Bridgewater, NJ) have recently been adopted by many companies and laboratories as a cost-effective strategy to generate therapeutic proteins for preclinical and early clinical development phases. These bioreactors consist of disposable sterilized bags that are filled with media and inflated to form a rigid gas-impermeable chamber; a rocking motion is used to ensure mixing and gas transfer. These bioreactors have been successfully used to grow a variety of mammalian cells, including CHO, NS0, hybridomas, HEK293, and PER.C6. For a description of the system and its applications, *see* **ref. *36*** and www.wavebiotech.com.

15. Media and culture samples for pH determination must be kept closed and with minimal exposure to air; thus, the use of syringes or sampling bags is recommended. If the media is exposed to air, the pH quickly increases because of the loss of dissolved $CO_2$. Offline pH measurements are made with the BioProfile unit or a blood-gas analyzer.

16. Small-scale bioreactors do not typically exhibit oxygen supply, $CO_2$ accumulation, or mixing problems even under high-density culture conditions. Conversely, culture heterogeneity and gas supply/stripping problems increase with scale. As such, normal operating conditions for small-scale reactors could fail to accurately predict culture performance and product quality under high and heterogeneous $CO_2$ concentrations. Specially designed experiments at the 2-L scale are required to test the sensitivity of the process to this variable. For a comprehensive review of scale-down considerations, *see* **ref. *37***.

17. Sodium hydroxide (NaOH) can be used instead of sodium bicarbonate for pH control to minimize $CO_2$ accumulation. When using NaOH for pH control, the additional port must be properly located, and the bioreactor must be well-mixed to prevent localized areas of high pH that would cause cell damage.

# References

1. Jordan, M., Schallhorn, A., and Wurm, F. M. (1996) Transfecting mammalian cells: optimization of critical parameters affecting calcium-phosphate precipitate formation. *Nucleic Acids Res.* **24,** 59–601.
2. Batard, P., Jordan, M., and Wurm, F. (2001) Transfer of high copy number plasmid into mammalian cells by calcium phosphate transfection. *Gene* **270,** 61–68.
3. Bebbington, C. R., Renner, G., Thomson, S., King, D., Abrams, D., and Yarranton, G. T. (1992) High-level expression of a recombinant antibody from myeloma cells using a glutamine synthetase gene as an amplifiable selectable marker. *Biotechnology* **10,** 169–175.
4. Felgner, P. L., Gadek, T. R., Holm, M., Roman, R., Chan, H. W., Wenz, M., et al. (1987) Lipofection: a highly efficient, lipid-mediated DNA-transfection procedure. *PNAS* **84,** 7413–7417.
5. Norton, P. A. and Pachuk, C. J. Methods for DNA introduction into Mammalian Cells, in *Gene Transfer and Expression in Mammalian Cells* (Makrides, S. C., ed.), Elsevier Science B. V., New York, NY, 2003, pp. 265–277.
6. Ciccarone, V., Chu, Y., Schifferli, K., Pichet, J.-P., Hawley-Nelson, P., Evans, K., et al. (1999) LipofectAMINE(tm) 2000 Reagent for rapid, efficient transfection of eukaryotic cells. *Focus* **21,** 54–55.
7. Freshney, I. R., ed. *Culture of Animal Cells: A Manual of Basic Technique.* John Wiley & Sons, Hoboken, NJ, 2002.
8. Barnes, D. and Sato, G. (1980) Serum-free cell culture: a unifying approach. *Cell* **22,** 649–655.
9. Sinacore, M. S., Drapeau, D., and Adamson, S. R. (2000) Adaptation of mammalian cells to growth in serum-free media. *Mol Biotechnol.* **15,** 249–257.
10. Ertola, R. J., Giulietti, A. M., and Castillo, F. J. (1995) Design, formulation, and optimization of media. *Bioprocess Technol.* **21,** 89–137.
11. Hu, W. S. and Piret, J. M. (1992) Mammalian cell culture processes. *Curr. Opin. Biotechnol.* **3,** 110–114.
12. Jayme, D. W. and Blackman, K. E. (1985) Culture media for propagation of mammalian cells, viruses, and other biologicals. *Adv. Biotechnol. Processes* **5,** 1–30.
13. Mather, J. P. and Barnes, D., eds. *Animal Cell Culture Methods*, vol 57. Academic Press, London, UK, 1998.
14. Jayme, D. W. (1991) Nutrient optimization for high density biological production applications. *Cytotechnology* **5,** 15–30.
15. www.xcellerex.com; www.bioprocessors.com; www.dasgip.de; www.fluorometrix.com.
16. Molnar-Perl, I. (2003) Quantitation of amino acids and amines in the same matrix by high-performance liquid chromatography, either simultaneously or separately. *J. Chromatogr. A.* **987,** 291–309.
17. Christie, A. and Butler, M. (1999) The adaptation of BHK cells to a non-ammoniagenic glutamate-based culture medium. *Biotechnol. Bioeng.* **64,** 298–309.
18. Jayme, D. W. (1999) An animal origin perspective of common constituents of serum-free medium formulations. *Dev. Biol. Stand.* **99,** 181–187.

19. Walowitz, J. L., Fike, R. M., and Jayme, D. W. (2003) Efficient lipid delivery to hybridoma culture by use of cyclodextrin in a novel granulated dry-form medium technology. *Biotechnol. Prog.* **19,** 64–68.

20. Gray, D. R. (2000) Bioreactor operations—preparation, sterilization, charging, culture initiation and harvesting, in *Encyclopedia of Cell Technology* (Spier, R. E., ed.), John Wiley & Sons, New York, NY, 2000, pp. 138–174.

21. Mostafa, S. S. and Gu, X. (2003) Strategies for improved $dCO_2$ removal in large-scale fed-batch cultures. *Biotechnol. Prog.* **19,** 45–51.

22. deZegontita, V. M., Kimura, R., and Miller, W. M. (1998) Effects of $CO_2$ and osmolarity on hybridoma cells: Growth, metabolism, and monoclonal antibody production. *Cytotechnology* **28,** 213–227.

23. Gray, D. R., Chen, S., Horwath, W., Inlow, D., and Maiorella, B. L. (1996) $CO_2$ in large-scale and high-density CHO perfusion culture. *Cytotechnology* **22,** 65–78.

24. deZegontita, V. M., Schmelzer, A. E., and Miller, W. M. (2002) Characterization of hybridoma cell response to elevated $pCO_2$ and osmolarity: intracellular pH, cell size, apoptosis, and medium metabolism. *Biotechnol. Bioeng.* **77,** 369–380.

25. Michaels, J. D., Nowalk, J. E., Mallik, A. K., Koczo, K., Wasan, D. T., and Papoutsakis, E. T. (1995) Analysis of cell-to-bubble attachment in sparged bioreactors in the presence of cell-protecting additives. *Biotechnol. Bioeng.* **47,** 407–419.

26. Michaels, J. D., Nowalk, J. E., Mallik, A. K., Koczo, K., Wasan, D. T., and Papoutsakis, E. T. (1995) Interfacial properties of cell culture media with cell-protecting additives. *Biotechnol. Bioeng.* **47,** 420–430.

27. Muthing, J., Kemminer, S. E., Conrad, H. S., Sagi, D., Nimtz, M., Karst, U., and Peter-Katalinic, J. (2003) Effect of buffering conditions and culture pH on producing rates and glycosylation of clinical phase I anti-melanoma mouse IgG3 monoclonal antibody R24. *Biotechnol. Bioeng.* **83,** 321–334.

28. Kaufmann, H., Mazur, X., Marone, R., Bailey, J. E., and Fussenegger, M. (2001) Comparative analysis of two controlled proliferation strategies regarding product quality, influence on tetracycline-regulated gene expression, and productivity. *Biotechnol. Bioeng.* **72,** 592–602.

29. Chu, L. and Robinson, D. K. (2001) Industrial choices for protein production by large-scale cell culture. *Curr. Opin. Biotechnol.* **12,** 180–187.

30. Arden, N. and Betenbaugh, M. J. (2004) Life and death in mammalian cell culture: strategies for apoptosis inhibition. *Trends Biotechnol.* (e-publication, doi 10.1016/j.tibtech. 2004.02.004).

31. Dempsey, J., Ruddock, S., Osborne, M., Ridley, A., Sturt, S., and Field, R. (2003) Improved fermentation processes for NSO cell line expressing human antibodies and glutamine synthetase. *Biotechnol. Prog.* **19,** 175–178.

32. Zhang, L., Shen, H., and Zhang, Y. (2004) Fed-batch culture of hybridoma cells in serum-free medium using an optimized feeding strategy. *J. Chem. Technol. Biotechnol.* **79,** 171–181.

33. Frahm, B., Lane, P., Märkl, H., and Pörtner, R. (2003) Improvement of a mammalian cell culture process by adaptive, model-based dialysis fed-batch cultivation and suppression of apoptosis. *Bioprocess Biosyst. Eng.* **26,** 1–10.

34. Portner, R., Schwabe, J. O., and Frahm, B. (2004) Evaluation of selected control strategies for fed-batch cultures of a hybridoma cell line. *Biotechnol. Appl. Biochem.* March 3 (e-publication, doi 10.1042/BA20030168).

35. Sauer, P. W., Burky, J. E., Wesson, M. C., Sternard, H. D., and Qu, L. (2000) A high-yielding generic fed-batch cell culture process for production of recombinant antibodies. *Biotechnol. Bioeng.* **67,** 585–597.
36. Singh, V. (1999) Disposable bioreactor for cell culture using wave-induced agitation. *Cytotechnol.* **30,** 149–158.
37. Palomares, L. A. and Ramirez, O. T. Bioreactor scale-down, in *Encyclopedia of Cell Technology* (Spier, R. E., ed.), John Wiley & Sons, New York, NY, 2000, pp. 174–183.

# 11

# Transgene Control Engineering in Mammalian Cells

## Beat P. Kramer and Martin Fussenegger

## 1. Introduction

Capitalizing on the generic design principle of the tetracycline-responsive expression systems *(1,2)*, many various transgene control modalities responsive to streptogramins *(3)*, macrolides *(4)*, and butyrolactones *(5)* have been constructed. All of these gene-regulation systems could be arranged in small-molecule-repressible off-type and small-molecule-inducible on-type configurations. Off-type systems consist of a transactivator and transactivator-dependent promoter. The transactivator is a fusion protein between a bacterial small-molecule-responsive transcription regulator and mammalian transactivation domain, e.g., *Herpes simplex*–derived VP16 *(6)*. The transactivator-dependent promoter consists of transactivator-specific (tandem) operator modules placed upstream of a minimal eukaryotic promoter, e.g., the minimal version of the human cytomegalovirus immediate early promoter ($P_{hCVMmin}$). Transactivator binding to cognate operators in the absence of regulating small molecules initiates promoter-dependent transcription, whereas promoter–transactivator interaction is abolished in the presence of adjusting molecules, as is transgene expression (**Fig. 1A**). On-type systems harbor the same prokaryotic response regulator that operates as a transrepressor on cognate chimeric promoters in its native form or optionally fused to a eukaryotic repression domain, such as the Kruppel-associated box (KRAB) domain of the human *kox-1* gene *(7)*. Transrepressor-dependent promoters consist of transrepressor-specific (tandem) operators cloned downstream of desired constitutive promoters. In the absence of regulating molecules, transrepressor–operator binding blocks/silences promoter-driven transgene expression, whereas small molecules interfering with transrepressor–operator interaction induce promoter-specific transcription (**Fig. 1B**).

Heterologous on- and off-type gene regulation systems are expected to comply with the following criteria at a high standard: (1) high-level transgene expression under induced conditions; (2) minute basal transcription under repressed conditions; (3) bioavailability; (4) high pharmacokinetic turnover of regulating small molecules

From: *Methods in Molecular Biology, vol. 308: Therapeutic Proteins: Methods and Protocols*
Edited by: C. M. Smales and D. C. James © Humana Press Inc., Totowa, NJ

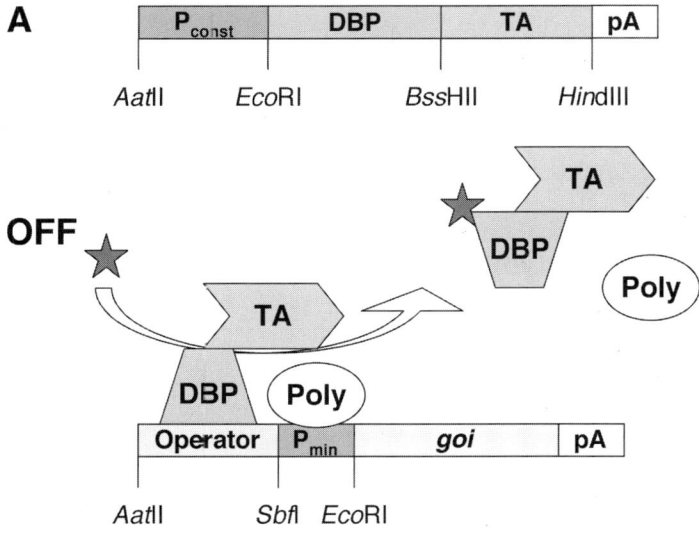

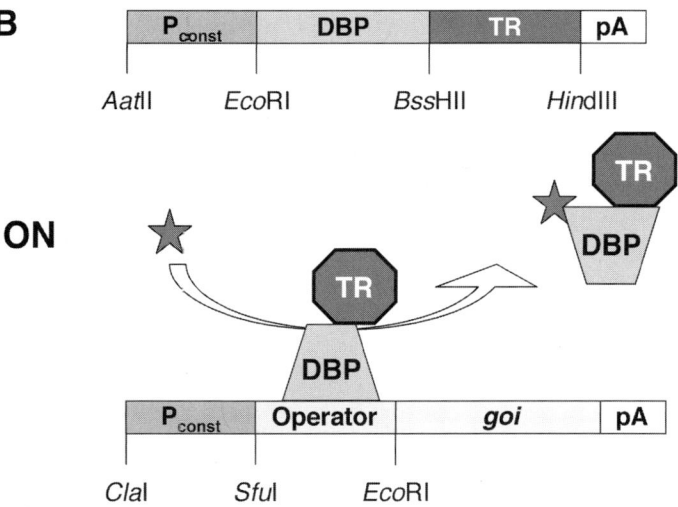

Fig. 1. Molecular configuration of off-type (**A**) and on-type (**B**) gene regulatory systems. (**A**) off-type systems. A DNA-binding protein (DBP), typically a bacterial response regulator, fused to a mammalian transactivation domain (TA) binds to a specific operator module and induces polymerase (Poly)-mediated transcription of the gene of interest (*goi*) from a minimal promoter ($P_{min}$). DBP-TA only binds to its operator in the absence of the regulating molecule (five-point star). Strategically important restriction sites are indicated. (**B**) on-type systems. The DBP, fused to a mammalian transcription repressor domain (TR), binds to the specific operator sequence located downstream of a constitutive promoter and thus represses its transcription. Upon addition of the regulating molecule, repression is abolished, and the gene of interest (*goi*) is expressed.

to mediate rapid expression switches in all cell types and tissues; (5) neglectable immunogenicity of the gene-control components; (6) absence of pleiotropic effects; and (7) compact one-vector design to limit the number of interventions on the host chromosome. Following configuration of a prototype gene regulation system in a highly modular manner (all key modules [repressor, transactivator/repressor domains, operators, and minimal promoters] are flanked by unique restriction sites), key performance characteristics could be fine-tuned in a desired manner and/or tailored to a particular cell phenotype by straightforward module swapping. For example, the gene regulation system's maximum and basal transgene expression profiles can be adjusted by the type of (minimal) promoter, spacing between the operator module and (minimal) promoter, type of transactivation and transrepression domains, and number of tandem operator repeats. Autoregulated expression configurations enabling simultaneous small-molecule-responsive expression of the transactivator and desired transgene in a one-vector format are among the most compact regulation designs. (1) Bidirectional autoregulated arrangements consist of a central operator module flanked by two divergent minimal promoters that drive simultaneous expression of the transactivator and the gene of interest in a small-molecule-adjustable manner *(8)*. (2) Small-molecule-dependent promoter-driven multicistronic expression units encoding the transgene(s) of interest along with the promoter-specific transactivator *(9)*. Autoregulated expression configurations rely on leaky transcripts resulting in sufficient transactivator to initiate small molecule-adjustable transgene expression.

Arrangement of several compatible transgene control systems may result in higher-order control modalities and/or achieve novel signal-integration circuits. Examples in mammalian cells include configurations for independent control of several transgenes *(10–12)*, combinatorial transcription logic *(13)*, serial interconnection of control systems to regulatory cascades for discrete expression management *(13)*, and engineering of epigenetic gene networks *(14)*.

This chapter provides comprehensive guidelines for the construction and management of heterologous mammalian gene regulation systems, along with strategies for their improvement and suggestions for artificial gene network design.

## 2. Materials

### 2.1. Reagents/Solutions

1. FMX-8 complete medium: FMX-8 medium (Cell Culture Technologies GmbH, Zurich, Switzerland) containing 10% fetal calf serum (FCS; PAN Biotech GmbH, Aidenbach, Germany, cat. no. 3302-P231902) and 1% penicillin–streptomycin solution (Sigma, St. Louis, MO, cat. no. P4458-100).
2. Trypsin solution: Trypsin-EDTA in phosphate-buffered saline (PBS) without calcium and magnesium (Teco Medical, Suhr, Switzerland, cat. no. P10-023100).
3. Sterile 1 $M$ $CaCl_2$ solution.
4. Phosphate solution: 50 m$M$ HEPES, 280 m$M$ NaCl, and 1.5 m$M$ $Na_2HPO_4$, pH 7.05, filter-sterilized and stored at 4°C for up to 6 mo.
5. Sterile glycerol.
6. Sterile cell-culture licensed dimethylsulfoxide (DMSO).

7. Erythromycin (Fluka, Buchs, Switzerland, cat. no. 4573) stock solution of 1 mg/mL in 96% ethanol. Store at –20°C.

8. Pristinamycin (Pyostacin; Aventis, Paris, France; lot no. RP27404) as stock solution in 500 µg/mL in DMSO. Store at –20°C and then at 4°C once thawed up.

9. Neomycin (G418 sulfate, Calbiochem, La Jolla, CA, cat. no. 345810), stock solution of 100 mg/mL in water. Store at –20°C.

10. Puromycin (Puromycin dihydrochloride, Alexis Biochemicals, Lausen, Switzerland, cat. no. 380-028-M500) as a stock solution of 10 mg/mL in water. Store at –20°C.

11. 2X SEAP buffer: 20 m*M* homoarginine, 1 m*M* MgCl$_2$, and 21% (v/v) diethanolamine/HCl, pH 9.8. Store in the dark at 4°C for up to 6 mo.

12. pNPP solution (120 m*M* para-nitrophenylphosphate [pNPP hexahydrate, Chemie Brunschwig, Basel, Switzerland, cat. no. 12886-0100]) in 2X SEAP buffer.

13. Dulbecco's PBS solution without magnesium and calcium: 130 m*M* NaCl, 2.7 m*M* KCl, 10.1 m*M* Na$_2$HPO$_4$, and 1.7 m*M* K$_2$HPO$_4$, pH 7.4 (DPBS, Teco Medical, Suhr, Switzerland, cat. no. P04-36010P).

14. Freezing medium: 90% FCS and 10% DMSO.

## 2.2. Hardware and Miscellaneous Material

1. CHO-K1 cells (American Type Culture Collection, cat. no. CCL 61).

2. Plasmid pPUR (Clontech, Palo Alto, CA).

3. Silica-based anion-exchange DNA purification kit (Genomed Jetstar 2.0 Midiprep, Genomed AG, Bad Oyenhausen, Germany).

4. Cell counting device (Casy1® counter, Schärfe System, Reutlingen, Germany).

5. SpectraMAX plus spectrophotometer (Molecular Devices, Sunnyvale, CA) for SEAP quantification.

6. Fluorescence-activated cell sorter (FACStar[Plus] and Cell Quest™ software, Becton-Dickinson, San Jose, CA) for single-cell sorting.

7. 50-µm filters (DAKO Diagnostics AG, Zug, Switzerland, cat. no. 150475).

8. Incubator (e.g., Nuaire Autoflow, NU-4500 E, Nuaire, Plymouth, MN) for the cultivation of mammalian cells at 37°C in a humidified atmosphere containing 5% CO$_2$.

9. StrataCooler cryopreservation module (Stratagene, La Jolla, CA, cat. no. 400005).

10. 2-mL cryotubes (Greiner, Kremsmuenster, Austria, cat. no. 122263).

11. Cell culture-certified disposable plasticware: six-, 24-, and 96-well plates; Petri dishes (TPP, Trasadingen, Switzerland) and pipets (Cellstar, Greiner bio-one, Kremsmuenster, Austria).

# 3. Methods

## 3.1. Molecular Setup of Heterologous Gene Regulation Systems

Molecular ingredients required for the design of a new gene regulation system include a protein that reversibly binds to a specific DNA fragment in response to a small molecule, preferably an antibiotic, any other clinically licensed compound, or a nontoxic side effect–free substance. After identification of these key elements, the design of a mammalian cell–compatible gene regulation system can be initiated. The following section provides detailed information on the modular design (facilitating quick exchange of single components) of new gene-regulation systems and sketches various strategies for their improvement and adaptation to specific expression environments. For amplification and cloning of DNA fragments and engineering of unique sites for restriction endonucleases, we suggest prior consultation of basic molecular biology protocols *(15,16)*.

### 3.1.1. Off-Type Systems

The generic design principle of an Off-type promoter is shown in **Fig. 1A**. For maximum flexibility and modularity, unique restriction sites will have to be introduced at strategic points. For example, we follow the convention of placing *Aat*II 5' of the operator module, *Sbf*I between operator module and minimal promoter, and *Eco*RI immediately 3' of the minimal promoter. This configuration enables straight-forward swapping of (1) the entire regulatable promoter by *Aat*II/*Eco*RI, (2) the operator module by *Aat*II/*Sbf*I and (3) the minimal promoter by *Sbf*I/*Eco*RI. The molecular setup of transactivators is outlined in **Subheading 3.2.4.**

### 3.1.2. On-Type Systems

**Figure 1B** shows the basic configuration of an on-type promoter. As a nonlimited example, we typically engineer a *Cla*I site 5' of the constitutive promoter, *Sfu*I in between promoter and operator site, and *Eco*RI 3' of the operator module. This setup enables the exchange of the entire on-type promoter by *Cla*I/*Eco*RI, swapping of the constitutive promoter subunit by *Cla*I/*Sfu*I, and replacement of the operator module by *Sfu*I/*Eco*RI. Information on the design of transrepressors is provided in **Subheading 3.2.4.**

## 3.2. Optimization of Off-Type Gene Regulation Systems

In principle, binary transactivator/promoter-based gene-regulation systems can be optimized by changing the (1) type of minimal promoter, (2) number of operator modules, (3) spacing between the operator and minimal promoter, and (4) type of transactivator and transrepressor.

### 3.2.1. Swapping of Minimal Promoters

Selection of the minimal promoter strongly influences maximum expression levels and tightness of any off-type system in a cell type-specific manner *(17,18)*. Minimal versions of the thymidine kinase, mouse mammary tumor virus, insect heat shock, and human cytomegalovirus immediate-early promoters are among well-established representatives in several off-type gene regulation systems. Alternative minimal promoters may be engineered into existing systems by applying the following steps.

1. Design oligonucleotides annealing to the desired minimal promoter containing *Sbf*I (5' primer) and *Eco*RI (3' primer) sites in their 5' overhangs.
2. Polymerase chain reaction (PCR) amplify the minimal promoter from a plasmid or genomic DNA.
3. Restrict the PCR product by *Sbf*I/*Eco*RI and ligate the resulting fragment into the corresponding sites of the target off-type promoter, thereby replacing the original minimal promoter.
4. Optionally, the minimal promoter may be entirely contained in the nonannealing part of the 5' primer and fused to the desired transgene by amplification with an appropriate transgene-specific 3' promoter. The off-type promoter will be reconstituted by joining the minimal promoter-transgene cassette with the operator module via *Sbf*I.

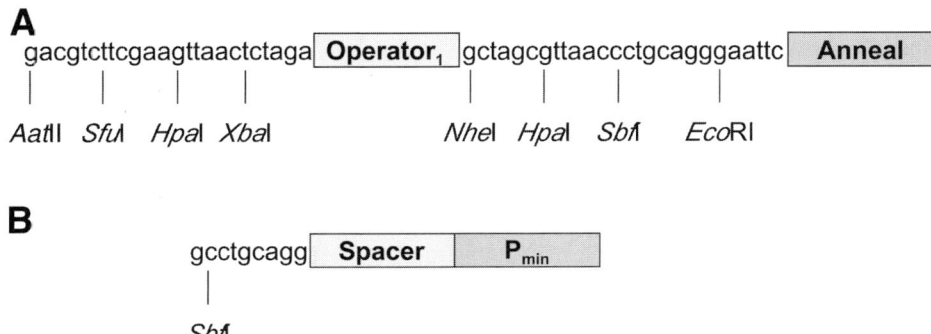

Fig. 2. Design of 5' primers for construction of tandem operator modules (**A**) and introduction of spacers between operator modules and minimal promoters (**B**). Both primers need a corresponding 3' counterpart for the amplification of the chosen sequence. (**A**) The indicated restriction sites have the following function: (1) *Xba*I/*Nhe*I enable multiplication of the operator site(s); (2) *Aat*II/*Sbf*I facilitate cloning of the operator sequence 5' of off-type minimal promoters; (3) operator sequences for on-type systems can be introduced by *Sfu*I/*Eco*RI; and (4) the two *Hpa*I sites produce a blunt-ended fragment for straightforward cloning. As the primers are merely used to multiply the number of operator sites, it is of no importance which gene is amplified. Therefore, this primer can be designed to anneal to any desired sequence (Anneal; preferably a minimal promoter). (**B**) In an off-type gene regulation system, the *Sbf*I restriction site enables cloning of the amplified minimal promoter 3' of the operator sequence.

### 3.2.2. Increasing the Number of Operator Modules

Extension of tandem operator modules within an off-type promoter may increase its maximum expression level *(18,19)*. There are two strategies for amplification of operator modules: custom synthesis (e.g., synthesis of triple-operator site PIR$_3$ for the streptogramin-responsive gene regulation system *[3]*; Operon Technologies, Alameda, CA) and sequential integration of operator sequences by tandem cloning steps using specific primers, along with the following procedure.

1. Design a 5' oligonucleotide with the operator site flanked upstream by *Aat*II-*Xba*I and downstream by *Nhe*I-*Sbf*I-*Eco*RI sites in the extension of the 5'primer (*see* **Fig. 2A** for more details). The 3' primer should include a unique restriction site, such as *Hin*dIII. The optimal choice is to use the minimal promoter (P$_{min}$) as a template and produce an operator fragment ready to replace existing operator sites via *Aat*II/*Sbf*I or an operator-P$_{min}$ fragment for control of the desired transgene(s).
2. PCR-amplify the chosen sequence with primers designed in the previous step.
3. Restrict the PCR product with *Aat*II and *Hin*dIII, and ligate it into any vector containing unique *Aat*II and *Hin*dIII sites to result in plasmid p1, which harbors a single-operator site upstream of P$_{min}$ (operator$_1$-P$_{min}$).
4. Restrict p1 with *Xba*I and *Hin*dIII to excise operator-P$_{min}$, and ligate it into plasmid p1 digested with *Nhe*I and *Hin*dIII (exclusively eliminating P$_{min}$ but not the operator site). Restriction with *Xba*I and *Nhe*I creates compatible cohesive ends, which can be ligated but can no longer be cleaved by either enzyme. The resulting plasmid p2 therefore con-

tains two operator sites in front of the minimal promoter (operator$_2$-P$_{min}$), which are flanked by one *Xba*I and one *Nhe*I site.

5. Digest plasmid p2 with *Xba*I and *Hin*dIII to excise operator$_2$-P$_{min}$, and ligate it into plasmid p2 digested with *Nhe*I and *Hin*dIII (exclusively eliminating P$_{min}$ but not the two operator sites). The resulting plasmid p3 contains four operator sites in front of the minimal promoter (operator$_4$-P$_{min}$).

6. Digest plasmid p3 with *Xba*I and *Hin*dIII to excise operator$_4$-P$_{min}$, and ligate it into plasmid p3 digested with *Nhe*I and *Hin*dIII (exclusively eliminating P$_{min}$ but not the four operator sites). The resulting plasmid p4 contains eight operator sites in front of the minimal promoter (operator$_8$-P$_{min}$).

Steps 4–6 can, in principle, be repeated to create tandem operator modules with more than eight repeats. Operator sites multiplied by this procedure can also be used for on-type systems. In that case, the multiple-operator sites need to be placed downstream of a constitutive promoter.

### 3.2.3. Spacing Between Operator and Minimal Promoter

The operator-minimal promoter spacing impacts both the distance and torsion angle between the operator-bound transactivator and transcription–initiation complex. These steric constraints have been reported to influence the overall regulation performance in a variety of different gene regulation systems *(17)*. The operator-minimal promoter distance can be conveniently modified by including the desired spacer sequence in the 5' primer tailored for PCR-mediated amplification of the minimal promoter. If 5' and 3' oligonucleotides harbor *Sbf*I and *Eco*RI sites, the spacer-minimal promoter can be conveniently swapped via *Sbf*I/*Eco*RI in between the operator module and transgene of interest (*see* above and **Fig. 2B**).

### 3.2.4. Transactivation and Transrepression Domains

Different human (p65 and E2F4) and viral transactivation domains (VP16 and its derivatives) tested in different expression configurations and cell lines revealed performance differences of up to three orders of magnitude *(17,20)*. To enable straightforward exchange of transactivation domains, we routinely place sites for *Eco*RI 5' of the transactivator, in frame between the DNA-binding protein and transactivation domain for *Bss*HII or *San*DI, and 3' of the transactivator for *Hin*dIII. In this configuration, PCR products encoding the DNA-binding protein may be exchanged via *Eco*RI-*Bss*HII/*San*DI, or the transactivation domain could be swapped using *Bss*HII/*San*DI-*Hin*dIII (**Fig. 1A**). It is key to note that *Bss*HII/*San*DI be introduced exactly in frame to ensure translation of the transactivator fusion protein. For most applications the transactivators are best driven by a strong constitutive promoter, including P$_{SV40}$, P$_{CMV}$, or P$_{EF-1\alpha}$. Yet, expression of transactivators from tissue-specific promoters will enable tissue-specific gene regulation *(21)*. As a convention, we recommend flanking transactivator-driving promoters with *Aat*II and *Eco*RI to enable straightforward exchange of tissue-specific promoters (*see also* **Note 1**).

In ON-type systems, transrepressors—fusion proteins between DNA-binding proteins and transcription repressor domains—are required to silence transcription in a small molecule-responsive manner (**Fig. 1B**). In a recent study, various transcription repressor domains have been evaluated for their capacity to silence basal or activated transcription in mammalian cells. The most potent, e.g., TRHBS and KOX derivatives, were functional in most cell types and could be engineered into artificial gene networks *(11,22)*. Construction of different transrepressors follows exactly the same design principle as outlined for transactivators.

### 3.3. Assessment of Modified Gene Regulation Systems in Cell Culture

Following assembly of a new mammalian gene regulation system, its performance needs to be evaluated in mammalian cells. The following protocol is based on a binary two-vector setting that consists of a constitutive transactivator- and transrepressor-expressing vector and a plasmid harboring the SEAP reporter gene driven by the transactivator- and transrepressor-dependent promoter. Transient transfection into Chinese hamster ovary (CHO)-K1 cells is performed using the following protocol.

1. Purify the transactivator- and adjustable reporter gene-encoding plasmids using a silica-based anion-exchange DNA purification kit according to the manufacturer's protocol (Genomed Jetstar 2.0 midiprep kit).
2. Seed 360,000 CHO-K1 cells into six wells of a 24-well cell culture plate 15 h prior to transfection.
3. Aspirate medium from a Petri dish containing CHO-K1 cells at 90% confluence.
4. Add 1.5 mL trypsin solution and incubate for 5 min.
5. Tap Petri dish gently to detach all cells.
6. Transfer cells to a 15-mL Falcon tube containing 3.5 mL FMX-8 complete medium.
7. Centrifuge the cell suspension for 3 min at 275 $g$.
8. Aspirate the supernatant.
9. Resuspend the cells in 5 mL FMX-8 complete medium.
10. Measure cell density using a Casy1 counter.
11. Dilute cells to a concentration of 150,000 cells/mL, and seed 0.4 mL cell suspension per well of a 24-well plate.

### 3.3.1. Cotransfection of Plasmids Into CHO-K1 Cells

1. Prepare 84 µL of a solution containing 3.6 µg of each plasmid and 500 m$M$ CaCl$_2$ in a 50-mL Falcon tube.
2. Add 84 µL phosphate solution and vortex immediately for 5 s to allow calcium phosphate–DNA complex formation.
3. Add 2.4 mL FMX-8 medium supplemented with 2% FCS after exactly 25 s to stop CaPO$_4$–DNA complex formation.
4. Aspirate medium of CHO-K1 cultures.
5. Add 0.4 mL of the calcium phosphate-DNA complex solution produced in step 3 to each of the six wells while rocking gently.
6. Put the cells in an incubator for 5 h.
7. Aspirate the medium and add 0.4 mL FMX-8 medium supplemented with 2% FCS and 15% glycerol to each well.
8. Aspirate the glycerol-containing medium after exactly 30 s, and replace it with 1 mL FMX-8 complete medium. Rock gently for optimal washing.
9. Replace the washing medium with 0.4 mL FMX-8 complete medium.
10. Add the regulating molecule at appropriate subtoxic concentrations (determined in separate growth assessment; *see* **Note 2**) to three of the six wells.
11. Cultivate the cells in an incubator for 48 h.

## 3.3.2. Quantification of SEAP
## in the Culture Supernatants 48-h Posttransfection

1. Remove 200 µL supernatant from each well. Do not forget to get a blank sample from untransfected cells.
2. Heat-inactivate supernatants for 10 min at 65°C.
3. Centrifuge heat-inactivated supernatants for 2 min at maximum speed.
4. Add 100 µL 2X assay buffer (prewarmed to 37°C) to one well per sample in a flat-bottom 96-well plate.
5. Transfer 80 µL culture supernatant to wells containing 2X assay buffer.
6. Add 20 µL substrate solution to each well.
7. Quantify absorbance at 405 nm for up to 1 h. Subtract values of the blank from sample values.
8. Calculate enzymatic activity.

$$EA \text{ [U/L]} = \Delta Abs/min \times v/(\varepsilon \times d) \times 10^6$$

where
Dilution factor $v$ = volume measurement/volume sample = 200/80
Absorption factor of $\varepsilon = 18,600 \text{ M}^{-1}\text{cm}^{-1}$
Lightpath d = 0.5 cm
Calculate regulation factor = $EA_{induced}/EA_{uninduced}$

If the regulation performance remains inconclusive, consider the following points for troubleshooting.

1. Vary the concentration of regulating molecule.
2. Vary the relative plasmid concentrations.
3. Confirm transactivator production by VP16-directed Western blot analysis.
4. Modify the number of tandem operator modules.
5. Test regulation system in other cell lines.

## 3.4. Modular Expression Vectors
## for Multicistronic and Multiregulated Gene Expression

Many vectors for compact and/or simultaneous regulation of several transgenes are already available *(10,23–25)*. If the design advice provided in **Subheadings 3.1.** and **3.2.4.** was followed, the novel gene regulation system will be compatible with an impressive expression vector portfolio using one-step swapping of expression units and/or modules. The next section provides a concise overview on different generic expression vector platforms accessible for gene regulation systems.

### 3.4.1. Multicistronic Expression

The pTRIDENT vector family enables expression of up to four transgenes under control of a single (regulated) promoter *(9,23–26)*. Although the first cistron is translated in a cap-dependent manner, subsequent citrons rely on internal ribosome entry sites (IRES) for cap-independent translation-initiation. The standardized molecular configuration of pTRIDENT vectors containing several multiple-cloning sites, unique restriction sites flanking strategic modules (promoter, IRES, and pA), and a large variety of (adjustable) promoters provide a highly flexible regulated multicistronic expression vector platform (**Fig. 3A**). After following the aforementioned design guidelines, regulated promoters can now be placed in any pTRIDENT via *Aat*II/*Eco*RI and thus drive regulated expression of multicistronic transcripts (**Fig. 3A**). The latest update of the pTRIDENT vector family includes size-optimized IRES elements and sites specific for rare-cutting homing endonucleases placed at strategic positions to enhance module swapping *(26)*.

### 3.4.2. Autoregulation

Owing to possible interference between the enhancer of the promoter driving transactivator expression and the molecule-responsive promoter, which may result in deregulated gene expression, the classical off-type gene regulation is best arranged in a two-vector format. This entails a transactivator-encoding plasmid and one harboring the small molecule-responsive transgene expression unit. To prevent the aforementioned promoter interference following cointegration into the same chromosomal locus, the best regulation performance of any off-type gene regulation system is achieved by sequential stable integration of the two plasmids. This procedure is time-consuming and multiplies the risk for integration-associated side effects.

Autoregulated expression configurations represent an alternative to the classical two-vector setting. In an autoregulated expression unit, the transactivator is driven by its own target promoter (**Fig. 3B,C**). Leaky transcripts originating from the small-molecule-responsive promoter produce initial transactivators, which will activate their own, as well as the gene of interest's, expression in the absence of regulating molecules. Although autoregulated expression units may, in principle, be multiplasmid-based, this configuration provides an opportunity for compact one-vector design, including autoregulated multicistronic or bidirectional arrangements. For construction

---

Fig. 3. *(opposite page)* Modular setup of generic pTRIDENT (**A**), autoregulated bidirectional expression (pBiRex5, **B**) and dual-autoregulated (pAutoRex4, **C**) vectors. (**A**) The pTRIDENT vector family consists of a constitutive or regulated promoter that enables transcription of a multicistronic mRNA. Three different transgenes can be inserted into the three MCS. The IRES enable cap-independent transcription-initiation of 3'-encoded cistrons. All pTRIDENT vectors contain a high-copy number pBluescript-based backbone, including the *ori* and β-lactamase-encoding gene for amplification in *E. coli*. The identical modular setup of all pTRIDENT vectors enables swapping of promoters and expression cassettes. In this context, the rare-cutting homing endonucleases (I-*Ceu*I, I-*Sce*I, I-*Ppo*I, and PI-*Psp*I) are of particular interest for module swapping. (**B**) pBiRex5 harbors a bidirectional asymmetric streptogramin-responsive promoter-driving expression of the streptogramin-dependent transactivator (PIT) via a minimal version of the *Drosophila* heat-shock protein 70 promoter ($P_{hsp70min}$) and the reporter gene SEAP by a cytomegalovirus immediate early promoter derivative. The reporter gene SEAP can be replaced by any gene of interest via indicated restriction sites. (**C**) pAutoRex4 contains two independent streptogramin- and tetracycline-responsive off-type expression units for one-step construction of dual-regulated expression systems. Both autoregulated expression units are dicistronic and contain the transactivators tTA and PIT in their first cistrons. Translation of genes of interest, which could be cloned into MCS II and III of **Fig. 3A**, are initiated by IRES of polioviral origin.

## A pTRIDENT

| Promoter | IRES I | IRES II | pA |
|---|---|---|---|

| | MCS I | MCS II | MCS III | |
|---|---|---|---|---|
| Aatll | I-Scel | Notl | Xhol | PI-Pspl |
| I-Ceul | EcoRI | Nael | Spel | Sall |
| | Xbal | Srfl | Pacl | Hincll |
| | Sall | Smal | Swal | Xbal |
| | Pstl | Ascl | Mlul | |
| | HindIII | BssHII | Pmel | |
| | | Bcll | Bglll | |
| | | Clal | I-Ppol | |
| | | Pmel | | |

## B pBiRex5

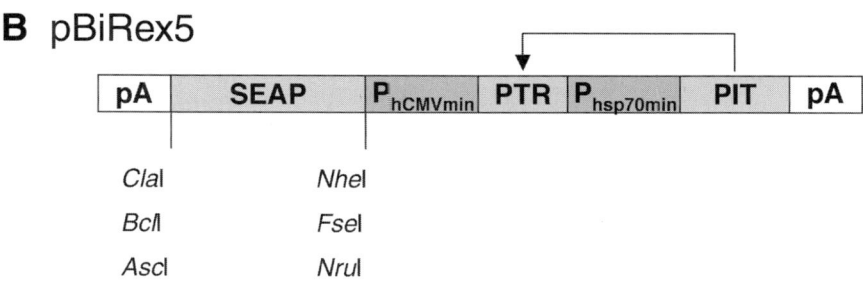

| pA | SEAP | P_hCMVmin | PTR | P_hsp70min | PIT | pA |

| Clal | Nhel |
|---|---|
| Bcll | Fsel |
| Ascl | Nrul |

## C pAutoRex4

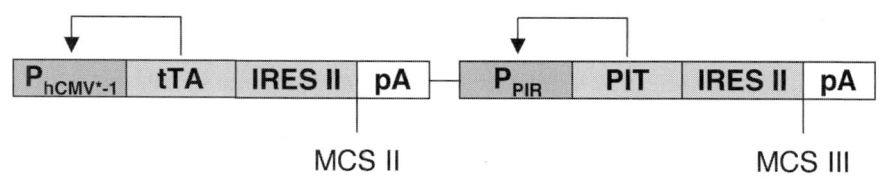

| P_hCMV*-1 | tTA | IRES II | pA | P_PIR | PIT | IRES II | pA |

MCS II      MCS III

Fig. 3.

of autoregulated multicistronic expression units, the transactivator is placed within a multicistronic message (preferably in the terminal cistron to enable self-selective autoregulation of preceding cistrons) transcribed from a transactivator-dependent promoter (**Fig. 3B**; *9,10*). Autoregulated bidirectional expression units contain a central bidirectional promoter with an operator module flanked by two (different) minimal promoters, one that drives transactivator expression and the other that initiates transcription of the gene of interest (**Fig. 3B**; *8,27,28*). Construction of bidirectional expression vectors requires several cloning steps: PCR-mediated amplification of an operator-minimal promoter and another minimal promoter module with 5′ primers containing sites for head-to-head ligation and 3′ primer harboring multiple-cloning sites (MCS and MCSI-$P_{minI}$-operator-$P_{minII}$-MCSII) compatible with those of the pTRIDENT vector family. Fux and coworkers have covered this topic in great detail *(8,28)*.

### 3.4.3. Dual Regulation

Many applications, including clinical, will require independent regulation of two different transgenes *(10)*. Owing to their compatibility, most off-type expression systems can be combined to control different gene activities. Arrangement of several transactivators in a multicistronic expression format is straightforward, whereas the interference-free arrangement of two expression units adjustable by two different molecules is challenging. Rigorous quantitative analysis of divergent, convergent, and consecutive expression units revealed that convergent orientation is optimal for mutually exclusive gene expression, but consecutive configuration enables greatest regulation flexibility (on/on, off/on, on/off, off/off) at near-optimal regulation performance *(29)*. The most advanced dual-regulated expression configuration harbors two consecutive autoregulated multicistronic expression units on a single-expression vector (pAutoRex4; **Fig. 3C**).

### 3.4.4. Lentiviral Expression Vectors

In recent years, lentiviral vectors have become significant among current gene therapy initiatives owing to their high-capacity stable transduction of a wide variety of mitotically active and inert cell phenotypes. The latest-generation HIV-I-based transduction systems consist of (1) a helper plasmid encoding viral proteins *gag*, *pol*, and *rev*, (2) a *vsv-g* expression vector (required for pseudotyping), and (3) the transgene(s)-encoding lentivector, which remains the only genetic material transferred to target cells. The transgene expression cassettes are flanked by *cis*-acting elements required for encapsulation, reverse transcription, and integration. Although lentivectors have traditionally been used for the constitutive expression of a single transgene, recent advances have exemplified multicistronic expression of up to three transgenes *(30)*, regulated transcription fine-tuning of heterologous genes, and bidirectional autoregulated transgene expression *(31)*. Because of their compatibility with the pTRIDENT expression format, the same design concepts outlined in **Subheadings 3.4.1.–3.4.3.** apply for lentiviral expression vectors.

## 3.5. Case Studies in Gene Network Engineering

Next-generation gene-therapy approaches for the treatment of complex disease phenotypes will have to deal with precise titration of key (regulatory) proteins and complementation of entire gene regulatory networks. Moreover, the design of artificial gene regulatory networks can result in completely new control modalities. Two examples of artificial mammalian gene networks will be presented: the regulatory cascade that enables titration of a transgene to four distinct levels and an epigenetic mammalian regulation circuit.

### 3.5.1. Artificial Regulatory Cascades for Discrete Multilevel Transgene Control

Mathematical models predicted that serial linkage of different off-type regulation systems will result in an artificial regulatory cascade that is able to adjust transgene expression to several distinct levels depending on the type of regulatory molecule added *(32)*. To be compatible with artificial regulatory cascade design, mammalian gene regulation systems need to comply with the following criteria.

- Availability in the repressible OFF-type format.
- Clinically licensed inducers.
- Expression abolished by a wide range of inducer concentrations.

Considering these parameters we assembled the three antibiotic-responsive gene regulation systems: $E_{OFF}$ *(4)*, $PIP_{OFF}$ *(3)*, and $Tet_{OFF}$ *(1)*. All three systems were assembled in a compact two-vector format. pBP167 harbors an autoregulated dicistronic expression cassette encoding the tetracycline- and macrolide-dependent transactivator (tTA and ET) under control of the tetracycline-responsive promoter ($P_{hCMV*-1}$). pBP138 contains expression units for macrolide-responsive ($P_{ETR2}$) expression of the streptogramin-dependent transactivator (PIT) and streptogramin-responsive ($P_{PIR8}$) transcription of a model product gene like SEAP (**Fig. 4A**; *32*). The following transcription cascade is triggered in the absence of any antibiotic: (1) the tetracycline-dependent transactivator (tTA) activates $P_{hCMV*-1}$ and drives tTA and ET1 expression in an autoregulated manner; (2) ET1 initiates $P_{ETR2}$-driven PIT expression; and (3) PIT initiates maximum SEAP transcription from $P_{PIR8}$ (no antibiotic: 100% SEAP expression). In the presence of tetracycline, the pBP167-encoded $P_{hCMV*-1}$-driven expression unit is shut down. Yet, leaky transcripts resulting in minute intracellular ET1 concentrations are expected to induce low-level PIT expression, which, in turn, results in decreased SEAP production (+tetracycline: 70% SEAP expression). Following addition of macrolide antibiotics, only basal PIT expression is available for SEAP expression (+erythromycin: 40% SEAP expression). In the presence of streptogramins, PIT binding to $P_{PIR8}$ is abolished, which results in complete SEAP repression (+pristinamycin I: undetectable SEAP expression; **Fig. 4**).

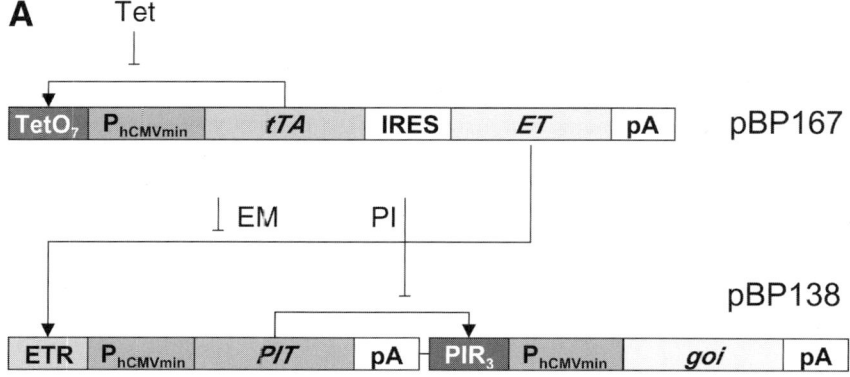

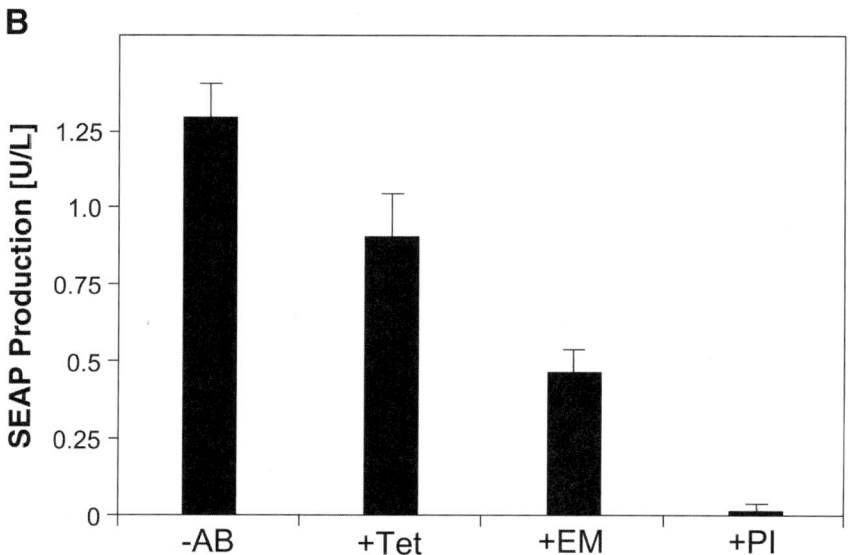

Fig. 4. Design and regulation performance of the regulatory cascade in CHO-K1 cells. (A) The regulatory cascade consists of a linear assembly of the tetracycline (TET$_{OFF}$)-, macrolide (E$_{OFF}$)-, and streptogramin (PIP$_{OFF}$)-responsive gene regulation systems. The network is constructed in a two-vector format (pBP138, P$_{hCMV*-1}$-tTA-IRES-ET1-pA; pBP167, P$_{ETR2}$-PIT-pA-P$_{PIR8}$-SEAP-pA). The tetracycline-responsive promoter (P$_{hCMV*-1}$, [TetO$_7$-P$_{hCMVmin}$]) drives in an autoregulatory fashion a dicistronic expression unit encoding the tTA, as well as ET1, which is under translation control of a polioviral IRES. Although tTA will autotransactivate P$_{hCMV*-1}$ in the absence of tetracycline (Tet), ET1 binds and activates the macrolide-dependent promoter (P$_{ETR2}$, [ETR-P$_{hCMVmin}$]) in an EM-responsive manner. Induction of P$_{ETR2}$ results in the expression of the streptogramin-dependent transactivator (PIT), which targets and transactivates the streptogramin-responsive promoter (P$_{PIR8}$, [PIR$_3$-P$_{hCMVmin}$]), thereby driving expression of the reporter gene SEAP in the absence of PI. (B) Regulation characteristics of the gene network were assessed by SEAP quantification 48-h post-cotransfection of pBP138 and pBP167 into CHO-K1 cells and cultivation in the absence of any antibiotic or in the presence of Tet, EM, or PI.

### 3.5.2. Design of an Epigenetic Mammalian Gene Switch

To adjust classic gene regulation systems to desired expression levels, the sustained presence of regulating molecules is required. Because long-term administration of regulating molecules in a clinical setting may result in side effects, the design of gene networks enabling stable and reversible on/off expression switches after integration of a transient signal are of high clinical priority. Capitalizing on a recent design principle established in *E. coli (33)*, we constructed a mammalian epigenetic expression circuit in a two-vector format by linking two on-type expression systems that mutually repress each other (*see also* **Note 2**). pBP62 contains a single cistron expressing E-KRAB under control of the streptogramin-inducible promoter ($P_{PIR}ON$), whereas pBP181 contains a multicistronic PIP-KRAB- and SEAP-encoding expression unit driven by the $P_{ETR}ON$ promoter. To facilitate construction of stable cell lines, pBP181 was also equipped with a neomycin resistance-conferring expression cassette (**Fig. 5A**; *14*).

### 3.5.3. Transient Assessment of Bistability

To assess bistable epigenetic expression readout, plasmids pBP62 and pBP181 were cotransfected into CHO-K1 according to the protocol outlined in **Subheading 3.3.1.**, up to item 9) and resulting cultures handled as follows:

1. Add erythromycin (EM) to wells 1–6 at a final concentration of 2 μg/mL.
2. Add pristinamyin (PI) to wells 7–12 at a final concentration of 2 μg/mL.
3. Incubate cells for 24 h.
4. At 24-h posttransfection/antibiotic addition, collect all supernatants for SEAP expression profiling (*see* **Subheading 3.3.2.**).
5. Wash cells twice with PBS (*see* **Note 3**).
6. Add 0.4 mL fresh FMX-8 complete medium to the cells.
7. Add EM to wells 1–3 but not to 4–6.
8. Add PI to wells 7–9 but not to 10–12.
9. Incubate cells for another 24 h.
10. At 24-h postmedium exchange, collect all supernatants for SEAP expression profiling (*see also* **Note 4**).

### 3.5.4. First Steps of Stable Engineering of Epigenetic Gene Expression in Mammalian Cells

Stable engineering of epigenetic expression control requires sequential transfection and selection of pBP181 and pBP62. To enable straightforward selection of first-order transgenic cells, we suggest first generating a cell line stable for pBP181. CHO-K1 cells were stably transfected with pBP181 following **Subheading 3.3.** Instead of 60,000 cells in a 24-well plate, use 300,000 CHO-K1 cells in a six-well plate. Adapt this protocol by multiplying all volumes and weights fivefold. Handle mock-transfected control cells the same. For selection of transfected cells, continue as follows.

1. At 48-h posttransfection, aspirate the medium, add 0.5 mL trypsin solution, and place the culture in an incubator for 5 min. Tap the culture plates to completely detach the cells.
2. Transfer the detached cells into a Petri dish containing 8 mL selection medium (FMX-8 complete medium supplemented with 400 μg/mL neomycin).

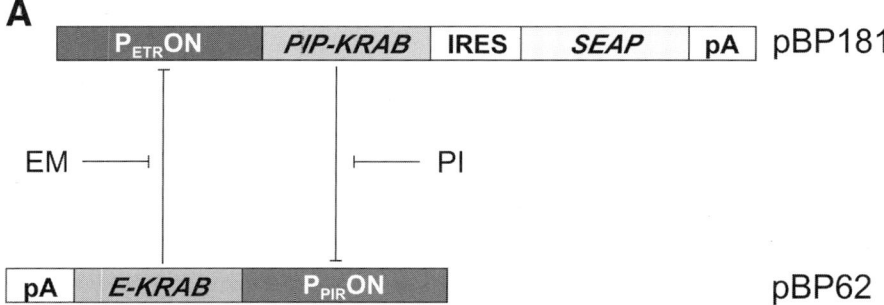

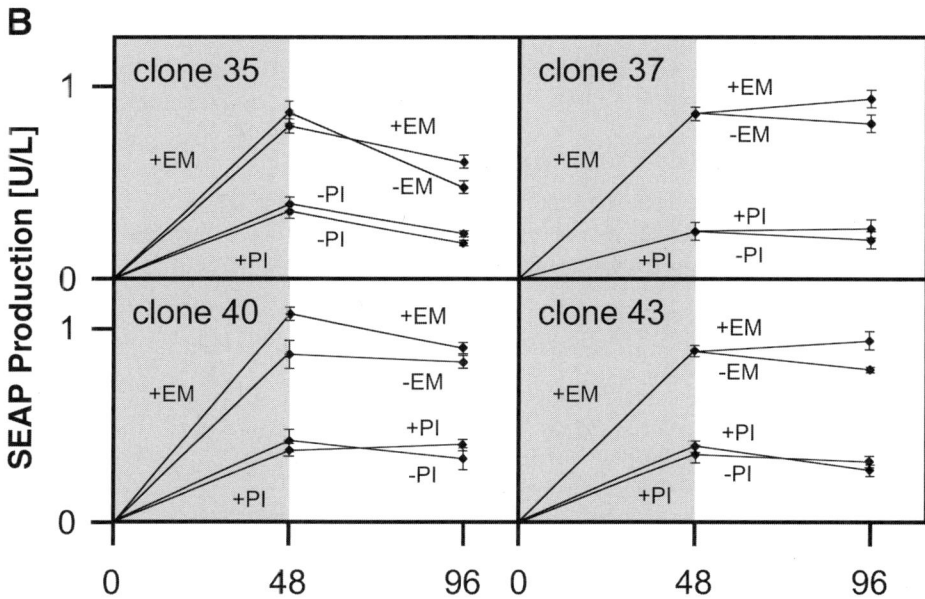

Fig. 5. Molecular setup and assessment of a generic epigenetic gene circuitry. (**A**) The two on-type gene regulation systems $E_{ON}$ and $PIP_{ON}$ have been assembled on the two plasmids, pBP62 and pBP181, such that the transcriptional repressors E-KRAB and PIP-KRAB inhibit each other's expression. Induced and repressed SEAP expression states fixed by transient administration of regulating antibiotics remain stable even in the absence of the inducers EM and PI. EM inhibits the repressive effect of E-KRAB, thereby inducing PIP-KRAB and (high-level) SEAP expression. Following production of sufficient PIP-KRAB, E-KRAB expression is inhibited; hence, EM is no longer required to prevent E-KRAB action. PI inhibits PIP-KRAB, thus inducing E-KRAB expression. As a consequence, SEAP expression drops to a low level. High E-KRAB levels keep the expression of SEAP and PIP-KRAB low; no PI is required to inhibit PIP-KRAB activity. (**B**) High and low SEAP expression states were induced for 48 h in clones of the double-transgenic $_{toggle}$CHO cell line. This induction phase was followed by another 48-h cultivation period in the presence/absence of EM/PI. Although the expression states of clone 35 seem to be unstable, three clones (37, 40, and 43), whose expression only marginally changed upon removal of the inducers, exhibited bistable behavior.

3. Exchange medium every other day until the neomycin-sensitive control population is eliminated (may take 8–10 d) and can be discarded. Continue cultivating transfected cell populations in selection medium until colonies of at least 2-mm diameter emerge.
4. Quantify SEAP expression in the culture supernatant. Proceed to fluorescence-activated cell sorting (FACS)-mediated single-cell sorting if cultures produce more SEAP compared to control cultures.

### 3.5.5. FACS-Mediated Single-Cell Sorting and Selection of pBP181-Transgenic CHO Clones

Transgenic monoclonal cell lines are generated from mixed stable populations by FACS-mediated single-cell sorting. To screen for pBP181-transgenic cell clones, follow these methods.

1. Prepare five 96-well plates containing 200 µL selective medium per well. Prewarm the plates to 37°C.
   a. Aspirate the medium from the Petri dish and add 1.5 mL trypsin solution, incubate for 5 min in an incubator, and tap the dish gently to detach all cells.
   b. Transfer cells into a 15-mL Falcon tube containing 3.5 mL FMX-8 complete medium.
   c. Centrifuge at 1200 rpm for 3 min.
   d. Aspirate medium and resuspend the cells in 5 mL sterile PBS (calcium- and magnesium-free; *see* **Note 5**).
   e. Centrifuge cell suspension at 1200 rpm for 3 min, aspirate supernatant, and resuspend cell in 1 mL PBS. Filter the cell suspension using a 50-µL filter to remove cell aggregates.
2. FACS-mediated single-cell sorting into nonperipheral wells of the 96-well plate (1 cell/well; 60 wells/plate; *see also* **Notes 6–8**).
3. Incubate the 96-well plates for up to 10 d to allow for clonal expansion.
4. Mark wells containing single, one-cell-derived colonies for subsequent SEAP profiling.
5. For SEAP expression analysis of single clones, remove 100 µL of each marked well, and continue with **Subheading 3.3.2., steps 1–7**.
6. Select 20 clones showing different SEAP expression levels.
7. Remove the remaining medium (not the cells) from the 96-wells containing the chosen clones.
8. Add 50 µL trypsin solution to each well and incubate for 5 min.
9. Transfer detached cells into individual wells of six-well plates containing 2 mL selection medium, and cultivate them until 90% confluence.
10. Seed six wells of a 24-well plate with 60,000 cells of each clone, and cultivate the rest in a Petri dish for backup and storage (*see* **Note 9** for freezing cells).
11. Transfect the clones in the 24-well plates with pBP62 following **Subheading 3.3.1.** Use 1.2 µg DNA per well. Add EM (1 µg/mL) to wells 1–3 and PI (1 µg/mL) to wells 4–6.
12. At 48-h posttransfection, analyze the supernatants for regulation of SEAP expression (**Subheading 3.3.2.**). Expression in the presence of EM should be increased (*see* **Note 7**). Select a cell clone with optimal SEAP regulation to continue with stable integration of pBP62 (*see* **Note 8**). For the sake of simplicity, we will refer to pBP181-transgenic CHO cells as CHO-BP181.

### 3.5.6. Final Steps of Stable Engineering
### of Epigenetic Gene Expression in Mammalian Cells

To stably transfect CHO-BP181 with pBP62, use **Subheading 3.5.4.** with the following modifications. Because pBP62 harbors no resistance marker, use 1.2 µg pBP62 and 0.08 µg pPUR (plasmid conferring puromycin resistance; Clontech, Palo Alto, CA) for transfection. Perform a control transfection exclusively with pBP62. Stable mixed populations are selected in medium containing puromycin and neomycin. FACS-mediated single-cell sorting is done according to **Subheading 3.5.5.** However, screening of positive clones is slightly different. Proceed as follows once wells containing individual clones have been identified.

1. For each clone, prepare two wells in a 24-well plate. Add 0.8 mL selection medium (containing 10 µg/mL puromycin and 400 µg/mL neomycin) to one well, while leaving the other one temporarily empty.
2. Aspirate the medium from wells containing single clones.
3. Add 50 µL trypsin solution to each clone and incubate for 5 min.
4. Transfer clones to the corresponding wells containing 0.8 mL selection medium.
5. Mix the cells and medium by pipetting the cell suspension up and down. Subsequently, transfer 0.425 mL cell suspension to the corresponding empty wells. Add EM (1 µg/mL) to one and PI (1 µg/mL) to the other half of the wells.
6. Incubate for 48 h.
7. Profile SEAP production. Expand clones exhibiting regulated SEAP expression (increased SEAP level in EM- compared to PI-containing wells) in six-well plates containing selection medium.
8. Per clone, seed 12 wells of a 24-well plate (*see* **Subheading 3.3.2.**) with 60,000 cells per well in 0.4 mL selection medium. Wells 1–6 were supplemented with EM (1 µg/mL) for maximum SEAP production and wells 7–12 with PI (1 µg/mL) to repress SEAP expression. Incubate for 48 h.
9. Remove 200 µL culture supernatants for SEAP quantification (*see* **Subheading 3.3.2.**); aspirate the rest of the medium. Wash each well twice with 1 mL sterile PBS.
10. Supply fresh selection medium to the clones. Add EM (1 µg/mL) to wells 1–3 and PI (1 µg/mL) to wells 7–9. As the expression states should be stable even in the absence of the inducers, wells 4–6 and 10–12 are cultivated in selection medium only. Incubate for 48 h.
11. Remove 200 µL supernatants for analysis of SEAP expression. Clones showing SEAP expression between 48 and 96 h, independent of the presence of the inducer (EM) or repressor (PI), are selected as bistable clones (**Fig. 5B**).
12. Expand and freeze (**Note 9**) selected clones for further analysis.

## 4. Notes

1. Theoretically, the genetic system could have been placed on a single construct. To circumvent the possible cross-silencing effect of the KRAB domains, which have alleged silencing range of several kilobases, two constructs were made. Using this strategy, we ensured that the two plasmids would integrate at different sites of the genome during the establishment of the double-transgenic cell line.
2. To determine appropriate concentrations of the regulating small molecule, growth and reporter gene expression of a transgenic production cell line (e.g., producing SEAP) are

quantified following cultivation in media supplemented with different inducer concentrations. Concentrations that do not influence growth and reporter gene expression are considered appropriate.

3. These washing steps are necessary to eliminate all inducer antibiotics, because even small amounts could lead to experimental artifacts.

4. When cells are transfected transiently, expression levels decrease over time. Hence, it is not advisable to follow expression for more than 48 h.

5. Cells can aggregate in the presence of calcium and magnesium that is undesired for FACS-mediated single-cell sorting.

6. No cells are placed in the peripheral wells, which are solely used as an evaporation barrier.

7. The clones with stably integrated pBP181 constitutively express SEAP, provided that there is no repressor-inhibiting expression from the $P_{ETR}ON$-promoter, which controls SEAP expression. pBP62 expresses this repressor that inhibits SEAP expression in the transfected cells. Because the transfection efficiency is not even close to 100%, many cells will still express SEAP constitutively after pBP62 transfection. The observed regulation factor is therefore much lower than it will be in a double-transgenic cell line. An effect pointing in the right direction is good enough to proceed.

8. High maximum-expression levels typically correlate with increased basal expression, whereas the best-regulated clones often exhibit poor maximum expression profiles. Choose the clones that best fit your experimental needs.

9. Freezing of cells.
   a. Detach cells grown to 90% confluence by following **Subheading 3.3.2., steps 1–6**.
   b. Resuspend the cells in 6 mL freezing medium, and aliquot cell suspensions in 1-mL cryotubes.
   c. Place the cryotubes in a cryopreservation module and store at –80°C for 2–3 d.
   d. Transfer the cryotubes to a liquid nitrogen tank.

# References

1. Gossen, M. and Bujard, H. (1992) Tight control of gene expression in mammalian cells by tetracycline-responsive promoters. *Proc. Natl. Acad. Sci. USA* **89**, 5547–5551.

2. Yao, F., Svensjo, T., Winkler, T., Lu, M., Eriksson, C., and Eriksson, E. (1998) Tetracycline repressor, tetR, rather than the tetR-mammalian cell transcription factor fusion derivatives, regulates inducible gene expression in mammalian cells. *Hum. Gene Ther.* **9**, 1939–1950.

3. Fussenegger, M., Morris, R. P., Fux, C., Rimann, M., von Stockar, B., Thompson, C. J., and Bailey, J. E. (2000) Streptogramin-based gene regulation systems for mammalian cells. *Nat. Biotechnol.* **18**, 1203–1208.

4. Weber, W., Fux, C., Daoud-El Baba, M., Keller, B., Weber, C. C., Kramer, B. P., et al. (2002) Macrolide-based transgene control in mammalian cells and mice. *Nat. Biotechnol.* **20**, 901–907.

5. Weber, W., Schoenmakers, R., Spielmann, M., Daoud-El Baba, M., Folcher, M., Keller, B., et al. (2003) *Streptomyces*-derived quorum-sensing systems engineered for adjustable transgene expression in mammalian cells and mice. *Nucleic Acids Res.* **31**, e71.

6. Triezenberg, S. J., Kingsbury, R. C., and McKnight, S. L. (1988) Functional dissection of VP16, the trans-activator of herpes simplex virus immediate early gene expression. *Genes Dev.* **2**, 718–729.

7. Moosmann, P., Georgiev, O., Thiesen, H. J., Hagmann, M., and Schaffner, W. (1997) Silencing of RNA polymerases II and III-dependent transcription by the KRAB protein domain of KOX1, a Kruppel-type zinc finger factor. *Biol. Chem.* **378**, 669–677.

8. Fux, C. and Fussenegger, M. (2003) Bi-directional expression units enable streptogramin-adjustable gene expression in mammalian cells. *Biotechnol. Bioeng.* **83,** 618–625.

9. Fussenegger, M., Moser, S., Mazur, X., and Bailey, J. E. (1997) Autoregulated multicistronic expression vectors provide one-step cloning of regulated product gene expression in mammalian cells. *Biotechnol. Prog.* **13,** 733–740.

10. Moser, S., Rimann, M., Fux, C., Schlatter, S., Bailey, J. E., and Fussenegger, M. (2001) Dual-regulated expression technology: a new era in the adjustment of heterologous gene expression in mammalian cells. *J. Gene Med.* **3,** 529–549.

11. Freundlieb, S., Schirra-Muller, C., and Bujard, H. (1999) A tetracycline controlled activation/repression system with increased potential for gene transfer into mammalian cells. *J. Gene Med.* **1,** 4–12.

12. Rossi, F. M., Kringstein, A. M., Spicher, A., Guicherit, O. M., and Blau, H. M. (2000) Transcriptional control: rheostat converted to on/off switch. *Mol. Cell.* **6,** 723–728.

13. Kramer, B. P., Fischer, C., and Fussenegger, M. (2004) BioLogic Gates enable logic transcription control in mammalian cells. *Biotechnol. Bioeng.* **87,** 478–484.

14. Kramer, B. P., Viretta, A. U., Daoud-El-Baba, M., Aubel, D., Weber, W., and Fussenegger, M. (2004) An engineered epigenetic transgene switch in mammalian cells. *Nat. Biotechnol.* **22,** 867–870.

15. Sambrook, J. and Russell, D. W. *Molecular Cloning, A Laboratory Manual.* Cold Spring Harbor Laboratory Press, Cold Spring Harbor, NY, 2001.

16. Ausubel, F. M., Brent, R., Kingston, R. E., Moore, D. D., Seidman, J. G., Smith, J. A., and Struhl, K. *Short Protocols in Molecular Biology.* John Wiley & Sons, New York, NY, 1999.

17. Weber, W., Kramer, B. P., Fux, C., Keller, B., and Fussenegger, M. (2002) Novel promoter/transactivator configurations for macrolide- and streptogramin-responsive transgene expression in mammalian cells. *J. Gene Med.* **4,** 676–686.

18. Hoffmann, A., Villalba, M., Journot, L., and Spengler, D. (1997) A novel tetracycline-dependent expression vector with low basal expression and potent regulatory properties in various mammalian cell lines. *Nucleic Acids Res.* **25,** 1078–1079.

19. Malphettes, L. and Fussenegger, M. (2004) Macrolide- and tetracycline-adjustable siRNA-mediated gene silencing in mammalian cells using polymerase II-dependent promoter derivatives. *Biotechnol. Bioeng.* **20,** 417–425.

20. Baron, U., Gossen, M., and Bujard, H. (1997) Tetracycline-controlled transcription in eukaryotes: novel transactivators with graded transactivation potential. *Nucleic Acids Res.* **25,** 2723–2729.

21. Grill, M. A., Bales, M. A., Fought, A. N., Rosburg, K. C., Munger, S. J., and Antin, P. B. (2003) Tetracycline-inducible system for regulation of skeletal muscle-specific gene expression in transgenic mice. *Transgenic Res.* **12,** 33–43.

22. Imhof, M. O., Chatellard, P., and Mermod, N. (2002) Comparative study and identification of potent eukaryotic transcriptional repressors in gene switch systems. *J. Biotechnol.* **97,** 275–285.

23. Fussenegger, M., Mazur, X., and Bailey, J. E. (1998) pTRIDENT, a novel vector family for tricistronic gene expression in mammalian cells. *Biotechnol. Bioeng.* **57,** 1–10.

24. Moser, S., Schlatter, S., Fux, C., Rimann, M., Bailey, J. E., and Fussenegger, M. (2000) An update of pTRIDENT multicistronic expression vectors: pTRIDENTs containing novel streptogramin-responsive promoters. *Biotechnol. Prog.* **16,** 724–735.

25. Weber, W., Marty, R. R., Keller, B., Rimann, M., Kramer, B. P., and Fussenegger, M. (2002) Versatile macrolide-responsive mammalian expression vectors for multiregulated multigene metabolic engineering. *Biotechnol. Bioeng.* **80,** 691–705.

26. Fux, C., Langer, D., Kelm, J. M., Weber, W., and Fussenegger, M. (2004) New-generation multicistronic expression platform: pTRIDENT vectors containing size-optimized IRES elements enable homing endonuclease-based cistron swapping into lentiviral expression vectors. *Biotechnol. Bioeng.* **86,** 174–187.

27. Baron, U., Freundlieb, S., Gossen, M., and Bujard, H. (1995) Co-regulation of two gene activities by tetracycline via a bidirectional promoter. *Nucleic Acids Res.* **23,** 3605–3606.

28. Fux, C., Weber, W., Daoud-El Baba, M., Heinzen, C., Aubel, D., and Fussenegger, M. (2003) Novel macrolide-adjustable bidirectional expression modules for coordinated expression of two different transgenes in mice. *J. Gene Med.* **5,** 1067–1079.

29. Fux, C. and Fussenegger, M. (2003) Toward higher order control modalities in Mammalian cells-independent adjustment of two different gene activities. *Biotechnol. Prog.* **19,** 109–120.

30. Mitta, B., Rimann, M., Ehrengruber, M. U., Ehrbar, M., Djonov, V., Kelm, J., and Fussenegger, M. (2002) Advanced modular self-inactivating lentiviral expression vectors for multigene interventions in mammalian cells and in vivo transduction. *Nucleic Acids Res.* **30,** e113.

31. Mitta, B., Weber, C. C., Rimann, M., and Fussenegger, M. (2004) Design and in vivo characterization of self-inactivating human and non-human lentiviral expression vectors engineered for streptogramin-adjustable transgene expression. *Nucleic Acids Res.* **32,** e106.

32. Kramer, B. P., Weber, W., and Fussenegger, M. (2003) Artificial regulatory networks and cascades for discrete multilevel transgene control in mammalian cells. *Biotechnol. Bioeng.* **83,** 810–820.

33. Gardner, T. S., Cantor, C. R., and Collins, J. J. (2000) Construction of a genetic toggle switch in Escherichia coli. *Nature* **403,** 339–342.

# 12

## Fusion to Albumin as a Means to Slow the Clearance of Small Therapeutic Proteins Using the *Pichia pastoris* Expression System

### *A Case Study*

**William P. Sheffield, Teresa R. McCurdy, and Varsha Bhakta**

## 1. Introduction

Of the numerous strategies that have been tested to slow the clearance of injected protein drugs from the body and circulation *(1)*, fusion to albumin offers several advantages. Albumin is the most abundant protein in mammalian plasma and one of the longest lived. It lacks posttranslational modifications, with the exception of extensive disulfide bonding *(2)*. If albumin can be fused in-frame with a therapeutic protein as a single-chain polypeptide, the novel protein may acquire the slow clearance profile of albumin, while retaining the activity important for clinical use. This acquisition derives primarily from an increase in the molecular volume of the therapeutic protein, such that it is no longer subject to loss via the kidneys. This approach has the potential to provide a more consistent and less heterogeneous product than one obtained, for instance, by chemical modification with polyethylene glycol. Whereas others have used *Kluveromyces (3,4)* and *Saccharomyces (5)* yeast species to produce human serum albumin (HSA) fusion proteins, we have used the methylotropic yeast, *Pichia pastoris*, to produce rabbit serum albumin (RSA) fusion proteins. This system is particularly well suited for albumin production *(6)*. This chapter summarizes our experience gained in expressing hirudin *(7)*, barbourin *(8)*, and reiterated RSA fusion proteins *(9)* in this system.

## 2. Materials

1. 10X Cloned *Pfu* polymerase buffer: 200 m$M$ Tris-HCl, pH 8.8, 20 m$M$ MgSO$_4$, 100 m$M$ KCl, 100 m$M$ (NH$_4$)$_2$SO$_4$, 1% (v/v) Triton X-100, and 1 mg/mL nuclease-free bovine serum albumin.
2. Cloned *Pfu* polymerase (Stratagene, La Jolla, CA, cat. no. 600153).

From: *Methods in Molecular Biology, vol. 308: Therapeutic Proteins: Methods and Protocols*
Edited by: C. M. Smales and D. C. James © Humana Press Inc., Totowa, NJ

3. Glycogen, MB grade (Roche Diagnostics, Laval QC, cat. no. 901 393).
4. TE: 10 m*M* Tris-HCl, pH 8.0 and 1 m*M* EDTA, pH 8.0.
5. 6X Agarose gel sample buffer: 30% (v/v) glycerol, 0.25% (w/v) bromophenol blue, and 0.25% xylene cyanol.
6. QIAquick Gel Extraction Kit (Qiagen, Chatsworth, CA, cat. no. 28704).
7. 10X DNA ligase buffer: 400 m*M* Tris-HCl, pH 7.8, 100 m*M* MgCl$_2$, 100 m*M* dithiothreitol (DTT), 5 m*M* adenotriphosphate (MBI Fermentas, Hamilton, Ontario, cat. no. EL0012).
8. T$_4$ DNA ligase (MBI Fermentas, cat. no. EL0011).
9. Plasmid Maxi Kit (Qiagen, cat. no. 12163).
10. Luria broth: 1% (w/v) bacteriological tryptone (BioShop Canada, Burlington, Ontario, cat. no. TRP402), 0.5 % (w/v) yeast extract (BioShop, cat. no. YEX401), and 1% (w/v) NaCl. Adjust pH to 7.0 with NaOH, autoclave, and store at room temperature protected from light.
11. Borosilicate glass tubes (13 × 100-mm; Kimble/Kontes, Vineland, NJ, cat. no. 73505-13100).
12. YPD medium: 1% (w/v) yeast extract, 2% bacteriological peptone (BioShop, cat. no. PEP403), and 2% dextrose (D-glucose). Sterilize by autoclaving. YPD plates are made by adding 20 g bacteriological agar per liter of YPD.
13. *P. pastoris* strain X-33 (Invitrogen, San Diego, CA, cat. no. C180-00).
14. Sorvall 250-mL Dry-Spin Bottle Assemblies (Mandel Scientific, Guelph, ON, Kendro, cat. no. 03431).
15. Zeocin (Invitrogen, cat. no. R250-01). Stock solutions obtained from the manufacturer are stored frozen at –20°C. Plates and media containing Zeocin should be stored at 4°C.
16. YPDS medium: YPD supplemented with 1.0 *M* sorbitol. YPDS/Zeocin plates are made by adding 20 g bacteriological agar per liter of YPDS and Zeocin to 0.1 mg/mL.
17. Gene Pulser II electroporator (Bio-Rad Laboratories, Hercules, CA, cat. no. 165-2661).
18. 13.4% YNB: This is a key stock solution for the preparation of yeast minimal media. Combine 34 g Yeast Nitrogen Base (Bioshop Canada, cat. no. YNB404; without ammonium sulfate) with 100 g ammonium sulfate in a total volume of 1 L double-distilled (dd) H$_2$O with moderate heat. Filter-sterilize and store protected from light at 4°C.
19. 1.0 *M* K$_2$HPO$_4$/KH$_2$PO$_4$ buffer, pH 6.0: Combine 1.0 *M* stock solutions of each form of phosphate in a 33:217 ratio (v/v), confirm pH as 6.0 (or adjust with concentrated phosphoric acid or KOH), autoclave, and store at room temperature.
20. 0.02% Biotin: Dissolve in ddH$_2$O, filter-sterilize, protect from light, and store at 4°C.
21. Buffered minimal methanol (BMM) medium: 100 m*M* potassium phosphate, pH 6.0, 1.34% (w/v) YNB, $4 \times 10^{-5}$% (w/v) biotin, and 0.5% (v/v) methanol. Filter-sterilize and store at 4°C protected from light. If made up ahead of time, omit methanol until just before use.
22. Buffered minimal glycerol (BMG) medium: Same as BMM, except with 1.0% (v/v) glycerol substituted for methanol.
23. 4X sodium dodecyl sulfate (SDS) sample buffer: 0.125 *M* Tris-HCl, pH 6.8, 5% (w/v) SDS, 25% (v/v) glycerol, and 400 m*M* DTT.
24. Phenylmethanesulfonyl fluoride (PMSF; Sigma-Aldrich Canada, Oakville, Ontario, cat. no. P7626). Prepare fresh as a 1 m*M* stock solution in dimethylsulfoxide. Toxic; handle with care using gloves and avoid any circumstances where powder might be inhaled. Consult Material Safety Data Sheet.
25. Centrifugal filter devices (Millipore, Nepean, Ontario, cat. no. UFC900508).
26. Bio-Rad Protein Assay Reagent (Bio-Rad Laboratories, Hercules, CA, cat. no. 500-0006).
27. Millipore Peristaltic Pump Drive (115 V/60 Hz; Millipore, cat. no. XX82 001 15).

28. Millipore Peristaltic Pump Head (480 mL; Millipore, cat. no. XX80 000 03).
29. Ni-NTA agarose (Qiagen, cat. no. 30210).
30. Ni-NTA equilibration buffer (NEB): 50 m$M$ sodium phosphate, pH 7.4, and 150 m$M$ NaCl.
31. Ni-NTA wash buffer (NWB): 50 mM sodium phosphate, pH 7.4, 150 m$M$ NaCl, and 20 m$M$ imidazole.
32. Ni-NTA elution buffer (NELB): 50 m$M$ sodium phosphate, pH 7.4, 150 mM NaCl, and 100 m$M$ imidazole.
33. Solutions and media used for the growth of *P. pastoris* are kept for 1 yr unless signs of contamination or suboptimal expression appear. Otherwise, they are replaced on the anniversary of their labeling date.

## 3. Methods (see Note 1)

### 3.1. Generation and Screening of Zeocin-Resistant P. pastoris Cell Lines

Initial attempts in our laboratory to use *P. pastoris* employing a first-generation commercial expression system were unsuccessful. This system required manipulation of the histidine utilization status of the yeast *(10)*. However, when second-generation vectors involving the use of a dominant selectable marker encoding resistance to Zeocin™ (a proprietary bleomycin/pleomycin-type antibiotic) became available, we found that the system could then easily be exploited to produce albumin fusion proteins. Our current approach involves a hybrid vector produced in our laboratory called pPICZ9ssamp *(7)* that combines the yeast prepro-α secretory signal of pPIC9, the *Sh ble* Zeocin-resistance gene of pPICZA, and elements common to both parental vectors, such as the methanol-inducible alcohol oxidase (AOX1) promoter. Ampicillin resistance was also introduced so that plasmid DNA manipulation in bacteria could be done using a less expensive antibiotic than Zeocin for at least part of the process. For the sake of brevity, we describe here specifically how to express a therapeutic protein fused to the amino-terminus of albumin, but the reverse geometry is also possible.

#### 3.1.1. Design of Expression Constructs

1. Obtain or construct plasmids containing the therapeutic protein and albumin cDNAs, respectively. All of our albumin work uses the RSA cDNA, but the same approach can be employed with any mammalian albumin cDNA.
2. Design oligonucleotides A and B for amplification of the therapeutic protein cDNA. Sense oligonucleotide A should commence with the sequence 5'-NNN NNN CTC GAG AAA AGA, immediately followed by the first seven codons of the therapeutic protein, as long as the first codon does not encode Pro (*see* **Note 2**). Antisense oligonucleotide B should similarly contain at least 20 nt of sequence that perfectly matches the strand opposite the terminal-coding region of the therapeutic protein cDNA at the 3' end, in addition to a sequence corresponding to six glycine codons and restriction site X. This site is not present in either the therapeutic protein cDNA or in albumin cDNA (*see* **Note 3**).
3. Design oligonucleotides C and D for the amplification of albumin cDNA. Sense oligonucleotide C should contain flanking nucleotides as in **Subheading 3.1.1., step 2**, in addition to nucleotides encoding restriction site X and the first seven codons of the mature albumin sequence of choice (*see* **Note 4**). Antisense oligonucleotide D should contain at least 20 nt of sequence perfectly matching the strand opposite the terminal-

coding region of the albumin protein cDNA at the 3' end, along with a sequence corresponding to six Histidine codons, a termination codon, an *Eco*RI site, and flanking nucleotides analogous to those in **Subheading 3.1.1.**, **step 2** (*see* **Note 5**).

## 3.1.2. DNA Manipulations

### 3.1.2.1. POLYMERASE CHAIN REACTION (PCR)

1. Combine 10 ng therapeutic protein-encoding plasmid DNA, 5 µL 10X cloned *Pfu* PCR buffer, 1 µL each of 50 µ*M* oligonucleotide A and 50 µ*M* oligonucleotide B, and sterile ddH$_2$O in a total volume of 48 µL.
2. Repeat **Subheading 3.1.2.1.**, **step 1**, substituting 10 ng albumin-encoding cDNA as the template and oligonucleotides C and D for A and B.
3. Heat both reactions to 95°C for 5 min, cool to 80°C, and add 2 µL *Pfu* polymerase (5.0 U) to each.
4. Perform 40 cycles at 95°C for 1 min, 55°C for 1.5 min, 72°C for 2.5 min, followed by 10 min at 72°C on each reaction in a thermocycler.

### 3.1.2.2. LIGATION OF RESTRICTED DNA TO FORM *P. PASTORIS* EXPRESSION PLASMID

1. Add 20 µg molecular biology–grade glycogen to the PCR reactions, 10 µL of 7.0 *M* ammonium acetate, and 130 µL absolute ethanol. Microcentrifuge for 10 min at 14,000*g*. Wash the pellet with 70% ethanol, repeat the microcentrifugation, and remove residual liquid. Resuspend in TE.
2. Restrict the PCR product containing the adapted therapeutic protein–encoding cDNA with *Xho*I and restriction enzyme X and the PCR product containing the adapted albumin cDNA with restriction Enzyme X and *Eco*RI. Restrict pPICZ9ssamp DNA with *Xho*I and *Eco*RI. Add 1/6 vol agarose gel sample buffer to each of the three restriction digests.
3. Electrophorese the PCR products on a 1% (w/v) agarose gel. Transfer the gel to an ultraviolet transilluminator, excise the bands, and purify using the QIAquick Gel Extraction Kit.
4. Combine the gel-purified DNA products in a 1:3:3 molar ratio (pPICZ9ssamp:PCR product:PCR product) with 2 µL of 10X DNA ligase buffer and 0.6 Weiss units of T$_4$ DNA ligase in a total volume of 20 µL. Control with a parallel ligation reaction lacking the two PCR products.

### 3.1.2.3. TRANSFORMATION OF *E. COLI* AND SELECTION OF EXPRESSION PLASMID CANDIDATES

Transformation of competent *E. coli* DH5α with a portion of the three-part ligation mixture from **Subheading 3.1.2.2.**, **step 4**, plating of transformants, miniplasmid DNA preparations, and restriction endonuclease screening of those DNAs are performed according to standard procedures (*10,11*; however, *see* **Note 6**).

## 3.1.3. Transformation of P. pastoris

1. Obtain a high-purity preparation of the therapeutic protein–albumin fusion construct in pPICZ9ssamp or an equivalent vector by employing a Plasmid Maxi Kit, using as starting material a 100-mL culture of *E. coli* DH5α transformed with the expression plasmid and grown to stationary phase in LB supplemented with 0.05 mg/mL sodium ampicillin.
2. Streak out a plate of *P. pastoris* X-33 cells on a YPD agar plate and incubate at 30°C overnight.

3. Using a sterile pipet tip on a P200 Pipetman or equivalent automatic pipetter, transfer a clearly isolated colony of apprcx 1-mm diameter to 5 mL YPD in a sterile, capped, boro-silicate glass tube. Incubate at 30°C with rotary shaking (225 rpm) overnight.

4. Inoculate a fresh 100-mL culture in a 500-mL flask with 0.5 mL overnight culture. Grow overnight at 30°C with rotary shaking to an optical density (OD) at 600 nm ($OD_{600}$) of 1.3–1.5.

5. Using aseptic technique, transfer the yeast culture into a sterile Sorvall Dry-Spin bottle assembly, including cap and O-ring, and centrifuge for 5 min at 1500$g$ at 4°C. Resuspend the pellet in 100 mL ice-cold sterile ddH$_2$O.

6. Repeat **Subheading 3.1.3.**, **step 5** two additional times, resuspending first in 50 mL, then 5 mL ice-cold sterile ddH$_2$O.

7. Centrifuge for 5 min at 1500$g$ at 4°C. Resuspend the pellet in 1.0 mL 1.0 $M$ ice-cold sorbitol.

8. Mix 80 µL *P. pastoris* cells suspended in sorbitol with 5 µg linearized expression plasmid DNA in 5 µL sterile ddH$_2$O, and transfer the mixture to a 0.2-cm electroporation cuvet (*see* **Note 7**).

8. Incubate the cuvet with the cells on ice for 5 min.

9. Insert the cuvet into the electroporation unit, and apply a pulse using settings of 1.5-kV charging voltage and 25-µF capacitance.

10. Add 1.0 mL ice-cold 1.0 $M$ sorbitol to the cuvet and incubate at 30°C for 1 h.

11. Plate 50 and 100 µL transformation reaction on separate YPDS/Zeocin plates and incubate at 30°C until colonies appear (3–4 d). Plated yeast colonies can then be stored at 4°C for up to 1 mo without a loss of viability.

## 3.1.4. Screening of Expression Candidate Cell Lines Using Milliliter-Scale Cultures

Individual colonies can now be picked from the YPDS/Zeocin plates and used to inoculate small cultures for the purposes of screening for cells expressing the albumin fusion protein of interest.

1. Inoculate a colony of Zeocin-resistant *P. pastoris* into 2.5 mL YPD/Zeocin and grow overnight as in **Subheading 3.1.3.**, **step 3.**

2. Centrifuge the cell suspension for 5 min at 1500$g$, decant the supernatant, and resuspend in BMM to an $OD_{600}$ of 1.0. Save a sample of the suspension and process as described in **Subheading 3.1.4.**, **step 4**.

3. Incubate the cells for 72 h at 30°C with rotary shaking (225 rpm), adding 100% methanol to 0.5% (v/v) final concentration every 24 h and removing a 0.1-mL sample of the cell suspension.

4. Microcentrifuge each sample of the cell suspension for 5 min at 14,000$g$, remove 60 µL supernatant, combine with 20 µL 4X SDS sample buffer, mix, and store at –20°C until all samples have been collected (*see* **Note 8**).

5. Analyze the samples by SDS-polyacrylamide gel electrophoresis (PAGE) and Coomassie Brilliant Blue staining of the completed gel using standard protocols (www.ruf.rice.edu/%7Ebioslabs/studies/sds-page/gellab2.html#gelprep; *see* **Note 9**).

6. Save each culture yielding a Coomassie-visible band of the appropriate size as a glycerin-ated culture at –70°C (*see* **Note 10**). Take 1.0 mL stationary-phase yeast culture grown in YPD/Zeocin and microcentrifuge for 5 min at 14,000$g$. Decant all but 0.3 mL supernatant, resuspend the cells, and combine with 0.1 mL sterile 80% (v/v) glycerol. Mix and store at –70°C.

### 3.2. Larger-Scale Production
### and Purification of Albumin Fusion Proteins From Liter-Scale Cultures

#### 3.2.1. Growth of Transformed Yeast
#### and Induction of Albumin Fusion Proteins on a Liter Scale

1. Inoculate 5.0 mL YPD/Zeocin in a sterile-capped borosilicate tube with a single colony or a frozen chunk of a glycerinated culture of approx 0.1-mL volume. Grow at 30°C overnight with shaking (225 rpm).
2. Centrifuge the cell suspension for 5 min at 1500$g$, decant the supernatant, and resuspend in 5 mL BMG. Transfer the suspension using aseptic technique to a 2-L Erlenmeyer flask containing 195 mL BMG. Grow at 30°C overnight with shaking (225 rpm).
3. Remove a 1.0-mL sample of the cell suspension and dilute with the addition of 9 mL BMG. If the OD$_{600}$ is less than 0.4, then continue the growth; otherwise, the culture is ready for induction.
4. Transfer the cell suspension to a Sorvall Dry-Spin bottle assembly and centrifuge for 5 min at 3000$g$. Decant the supernatant, and resuspend the cell pellet in BMM to an OD$_{600}$ of 1.0. Divide into culture flasks so that the culture volume does not exceed 25% of the volume of the flask. Typically, this corresponds to 5 × 2-L flasks for a total induction culture volume of 2.5 L.
5. Every 24 h, add 100% methanol to 1/200 culture volume (0.5% v/v).
6. Save aliquots of the culture as described in **Subheading 3.1.4.**, **steps 3** and **4**.
7. After 48 h, ensure that the expression induction of the fusion protein is proceeding as expected based on the results of the small-scale expression trials **(Subheading 3.1.4.)** by performing SDS-PAGE analysis as described in **Subheading 3.1.4.**, **step 5**; *see* **Note 11**. Typical results for the expression of a C-terminally His-tagged recombinant RSA expressed using the procedures detailed in this chapter are shown in **Fig. 1**.
8. After 72 h, transfer the cells to Sorvall Dry-Spin bottle assemblies, and centrifuge at 6000$g$ for 10 min at 4°C. Decant through bleached cheesecloth into clean, sterile plasticware (*see* **Note 12**).
9. Add PMSF to 0.1 m$M$ final concentration and sodium azide to 0.02%. At this point, the conditioned media can be stored at 4°C if processed within 48 h; otherwise, it should be frozen at −70°C until processed.

#### 3.2.2. Nickel-Chelate Affinity Chromatography Purification of Fusion Proteins

1. Centrifuge the conditioned media at 6000$g$ for 15 min at 4°C. Decant the supernatant into a clean beaker in preparation for adjusting the pH.
2. Using a pH meter or litmus paper, add 5.0 $M$ sodium hydroxide to adjust the pH of the conditioned media up to approx 7.5.
3. Repeat the centrifugation step as in **Subheading 3.2.2.**, **step 1** to remove precipitated material (*see* **Note 13**).
4. Decant the clarified, neutralized conditioned media, and concentrate it approx 10-fold using a 30-kDa molecular weight cut-off ultrafiltration tangential flow cartridge connected to a high-capacity peristaltic pump to a total volume of approx 200 mL.
5. Add PMSF and sodium azide as in **Subheading 3.2.1.**, **step 9**.
6. Pour a 20-mL (11 × 1.5 cm) column of Ni-NTA-agarose equilibrated with NEB, and apply the neutralized conditioned media to the column at a flow rate not exceeding 1 mL/min.
7. Wash the column with at least 10 column volumes of NWB, ensuring that no protein can be detected in the final column volume by Bio-Rad assay (*see* **Note 14**).

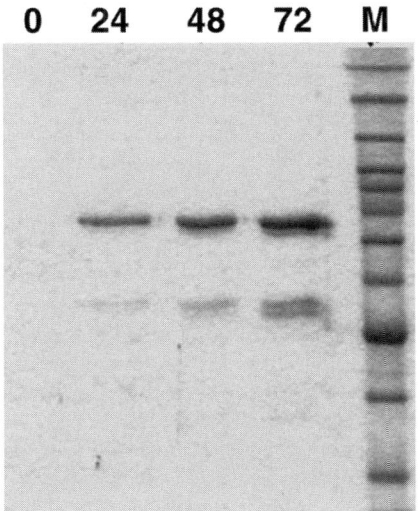

Fig. 1. Time-course of recombinant albumin production in transformed *P. pastoris*. A 10% SDS-polyacrylamide gel, electrophoresed under reducing conditions and stained with Coomassie blue, is shown. A Zeocin-resistant *P. pastoris* cell line expressing a form of RSA with an N-terminal hexahistidine tag and the C34A substitution *(9)* was observed over 72 h of methanol induction. Media samples conditioned by the cells were taken at intervals in hours specified above each lane. *M* denotes molecular mass markers corresponding to 220, 160, 120, 100, 90, 80, 70, 60, 50, 40, and 30 kDa.

8. Elute the bound fusion protein with NELB, collecting 20 fractions of 2–3-mL volume using a fraction collector (*see* **Note 15**).
9. Determine which fractions contain the eluted protein using either SDS-PAGE or protein assays. Pool appropriate fractions and concentrate using a centrifugal ultrafiltration device to approx 5 mL.
10. Dialyze the concentrated eluate vs 4 L NEB using a 12-kDa molecular weight cut-off dialysis tubing. Determine the protein concentration using a Coomassie blue dye-binding assay.
11. Aliquot and freeze the preparation at –70°C (*see* **Note 16**).

## 4. Notes

1. Many of our procedures are excerpts and/or modifications regarding specific locally available chemicals of the excellent protocols and procedures supplied by Invitrogen—the commercial supplier of a variety of *P. pastoris* expression strains, plasmids, and accessory products. Although originally provided at the time of purchase as hard-copy manuals, as the product line expanded over the last decade, they are now mostly available online (www.invitrogen.com).
2. The additional nucleotides at the 5' end are present to ensure efficient *Xho*I cleavage of the ensuing PCR product. The prohibition against Pro at position 1 of the therapeutic protein domain is owing to the properties of the propeptidase that cleaves the yeast prepro-α factor propeptide. We have employed Ile at position 1 of our hirudin–albumin fusion

protein *(7)*, Glu at position 1 of our barbourin–albumin fusion protein *(8)* and His at position 1 of our reiterated albumin *(9)*.

3. In the hirudin-albumin and barbourin-albumin proteins, X = *Nco*I (CCATGG recognition site).

4. Ensure that the alignment is to the mature albumin sequence, *not* to preproalbumin; e.g., in RSA, the first seven codons correspond to heptapeptide EAHKSEI. Oligonucleotides can be purchased from any reputable synthesis facility, such as the McMaster University Molecular Biology Institute Central Facility—the source of all oligonucleotides used in our laboratory.

5. The reader is directed to the exact sequences of the oligonucleotides used in this manner in expression of fusion protein $HLAH_6$ *(7)* as examples of how to create their own gene-specific oligonucleotides for uses described in these protocols.

6. The DNA sequence of the entire open-reading frame corresponding to the fusion protein in the final expression construct should be determined. The use of a high-fidelity polymerase (e.g., *Pfu* polymerase) in PCR lessens the chances of incorporation of an unintended mutation, but it is still high enough to make this exercise worth the effort.

7. The expression plasmid must be linearized to facilitate insertion of the expression construct into the *Pichia* genome by homologous recombination. We have used *Sac*I for this purpose; other restriction sites must be used if a *Sac*I site lies within either the therapeutic protein or albumin cDNAs being employed.

8. Our standard procedure is to perform SDS-PAGE under reducing conditions. For nonreducing electrophoresis, simply omit the DTT from the 4X SDS sample buffer. In theory, this would allow visualization of any disulfide-linked albumin dimers that form from disulfide bonding through Cys34, the free thiol conserved in most mammalian albumins. In practice, we have only seen this with purified samples of fusion protein $HLAH_6$ on overloaded gels *(7)*.

9. Any apparatus used for SDS-PAGE should suffice for this purpose. We use a Bio-Rad Mini-Protean II system (no longer sold by Bio-Rad, but very similar to the Mini-Protein III [Bio-Rad, cat. no. 165-3301]), typically employing 10% gels of 0.75-mm thickness. With care, 40 μL sample can be applied to each lane of the gels (30 μL conditioned media and 10 μL 4X SDS sample buffer).

10. In our experience, at least 50% of the Zeocin-resistant colonies will express the albumin fusion protein at a level sufficient that a Coomassie-visible band corresponding to the fusion protein (under the conditions specified) will be the single most abundant protein in the gel profile. If this is not the case, then immunoblotting should be undertaken using albumin-specific antibodies.

11. If the induction is proceeding as observed during the small-scale expression trials, proceed with the final 24 h of induction; if it is not, bleach and discard the culture, saving wasted effort on a failed induction. We have only observed problems in the methanol induction of expression if the Zeocin is omitted from the initial culturing of the yeast or if it has lost its activity.

12. Although sterile glassware can be used, the use of sterile plasticware minimizes adsorptive losses of protein.

13. Failure to raise the pH of the solution will result in the failure of the nickel-chelate affinity chromatography step owing to inability of the albumin fusion protein to bind to the column. Precipitates will form as the pH is raised, necessitating a centrifugation to clarify the neutralized conditioned media prior to its application to the column. If this is not done, the column will clog. Additionally, we have noted a sharp reduction in fusion protein recovery if the concentration step is performed prior to neutralization.

14. Using a low concentration of imidazole in the wash buffer helps to remove breakdown products from the final preparation of the albumin fusion protein. Most mammalian albumins contain a His at position 3 that forms part of a low-affinity metal-binding site, so that a breakdown product containing this structure, but lacking the C-terminal hexahistidine tag, will still bind to the column. Washing in a low-imidazole level and then increasing the imidazole concentration in the elution step provides optimal purity. However, the specific binding characteristics of each novel fusion protein should be determined, as some fusion proteins have eluted at lower imidazole concentrations than expected, despite the presence of an intact hexahistidine tag, as judged by immunoblotting with an antihexahistidine antibody.

15. The fusion protein should be completely eluted from the column by fraction 10.

16. A typical yield from this procedure should be 20–40 mg fusion protein. We have found this level of expression sufficient for in vitro testing of the characteristics of the fusion protein, clearance experiments using iodinated fusion protein, and testing the in vivo activity of the fusion proteins in intermediate-sized experimental animals (rabbits). Our procedure does not involve extensive screening of candidate *Pichia* cell lines for high expressers, but readers should be aware that rare multicopy integrations of the fusion protein constructs into the *Pichia* genome have been reported to manifest greatly increased levels of expression, and that this strategy can be employed to maximize expression levels. These issues are dealt with in more depth in another volume of the *Methods in Molecular Biology* series *(11)*.

## Acknowledgments

We wish to thank Ian Smith, a former member of the laboratory who constructed pPICZ9ssamp and who helped in our acquisition of experience with *P. pastoris*. This work was supported by Grant-In-Aid T4612 from the Heart and Stroke Foundation of Ontario and by infrastructure support from Canadian Blood Services. TRM holds a Canadian Blood Services Graduate Fellowship.

## References

1. Sheffield, W. P. (2001) Modification of clearance of therapeutic and potentially therapeutic proteins. *Curr. Drug Targets Cardiovasc. Haematol. Disord.* **1**, 1–22.
2. Peters, T. (1985) Serum albumin. *Adv. Prot. Chem.* **37**, 161–245.
3. Yeh, P., Landais, D., Lemaitre, M., Maury, I., Crenne, J. Y., Becquart, J., et al. (1992) Design of yeast-secreted albumin derivatives for human therapy: biological and antiviral properties of a serum albumin-CD4 genetic conjugate. *Proc. Natl. Acad. Sci. USA* **89**, 1904–1908.
4. Lu, H., Yeh, P., Guitton, J. D., Mabilat, C., Desanlis, F., Maury, I., et al. (1994) Blockage of the urokinase receptor on the cell surface: construction and characterization of a hybrid protein consisting of the N-terminal fragment of human urokinase and human albumin. *FEBS Lett.* **356**, 56–59.
5. Nomura, N., Matsubara, N., Horinouchi, S., and Beppu, T. (1995) Secretion by *Saccharomyces cerevisiae* of human apolipoprotein E as a fusion to serum albumin. *Biosci. Biotechnol. Biochem.* **59**, 532–534.
6. Kobayashi, K., Nakamura, N., Sumi, A., Ohmura, T., and Yokoyama, K. (1998) The development of recombinant human serum albumin. *Ther. Apher.* **2**, 257–262.
7. Sheffield, W. P., Smith, I. J., Syed, S. S., and Bhakta, V. (2001) Prolonged *in vivo* anticoagulant activity of a hirudin-albumin fusion protein secreted from *Pichia pastoris*. *Blood Coagul. Fibrinolysis* **12**, 433–443.

8. Marques, J. A., George, J. K., Smith, I. J., Bhakta, V., and Sheffield, W. P. (2001) A barbourin albumin fusion protein that is slowly cleared *in vivo* retains the ability to inhibit platelet aggregation *in vitro. Thromb. Haemost.* **86,** 902–908.

9. McCurdy, T. R., Gataiance, S., Eltringham-Smith, L. J., and Sheffield, W. P. (2004) A covalently linked recombinant albumin dimer is more rapidly cleared *in vivo* than wild-type and mutant C34A albumin. *J. Lab. Clin. Med.* **143,** 115–124.

10. Cregg, J. M., Vedvick, T. S., and Raschke, W. C. (1993) Recent advances in the expression of foreign genes in *Pichia pastoris. Biotechnology (NY)* **11,** 905–910.

11. Higgins, D. R. and Cregg, J. M., eds. *Pichia Protocols.* Humana, Totowa, NJ, 1998.

# 13

## High-Throughput Recovery of Therapeutic Proteins from the Inclusion Bodies of *Escherichia coli*

### An Overview

**Amulya K. Panda**

## 1. Introduction

*Escherichia coli* is widely used for the expression of recombinant proteins that do not require glycosylation for their bioactivity *(1,2)*. Many commercially available recombinant hormones and the majority of interleukins and interferons are all expressed and produced in *E. coli* systems *(3)*. The advantages of using *E. coli* as an expression system include the enormous volume of data on its cell biology, its fermentation process development, and its ability to produce relatively large and inexpensive quantities of recombinant protein *(4)*. However, the high-level expression of recombinant proteins in *E. coli* often results in the accumulation of the product as insoluble aggregates in vivo as inclusion bodies *(5)*. Inclusion body proteins are devoid of biological activity and require elaborate solubilization and refolding procedures to recover functional activity *(6,7)*. The renaturation of inclusion body proteins into a bioactive form is cumbersome, results in low recovery of the final product, and also accounts for the major cost in overall production of recombinant proteins using *E. coli (7,8)*. However, whereas high-yielding processes are developed for the refolding of the aggregated recombinant proteins, high-level expression of proteins as inclusion bodies provides a straightforward strategy for producing therapeutic proteins. The initial high level of expression compensates for loss during recovery of the protein of interest from inclusion bodies. Despite many successful refolding procedures for inclusion body proteins (even at industrial scale), the renaturation process for each protein is quite different. Most often, this process is carried out in an empirical way and causes poor recovery of the bioactive therapeutic protein. Thus, there is a need to develop high-throughput purification processes for the efficient recovery of bioactive therapeutic proteins from the inclusion bodies of *E. coli*.

From: *Methods in Molecular Biology, vol. 308: Therapeutic Proteins: Methods and Protocols*
Edited by: C. M. Smales and D. C. James © Humana Press Inc., Totowa, NJ

## 2. Problems With Protein Recovery From Inclusion Bodies

The accumulation of protein aggregates as inclusion bodies in *E. coli* is either caused by intermolecular interactions between the partially folded intermediates of nascent polypeptide chains or due to the inability of the cellular environment to correctly fold the protein of interest. Generally, inclusion bodies are solubilized by the use of high concentrations of denaturants, such as urea or guanidine hydrochloride, along with a reducing agent, e.g., β-mercaptoethanol *(9,10)*. Solubilized proteins are then refolded by slow removal of the denaturant in the presence of the oxidizing agent *(10–12)*. However, protein solubilization from inclusion bodies using high concentrations of chaotropic reagents results in the loss of secondary structure, leading to a random coil formation of protein structure and exposure of hydrophobic surfaces *(13)*. Loss of secondary structure during solubilization and the interaction among the denatured protein molecules and their consequent aggregation are considered to be the main reasons for poor recovery of bioactive proteins from inclusion bodies. For proteins with multiple disulfide bonds and a high percentage of hydrophobic stretches of amino acids, the problem of aggregation is severe and results in very poor recovery of bioactive protein. Most often, the overall yield of bioactive protein from inclusion bodies is approx 15–25% of the total expressed protein. Thus, the overall process yield and economic viability of recombinant *E. coil* fermentation processes mostly depends on the efficient recovery of bioactive protein from inclusion bodies *(8)*. Refolding at low-protein concentrations, which requires large volumes of extra-pure water and using urea for solubilization, have been reported as the most costly factors in the production of recombinant insulin from *E. coli* *(14)*. Although protein expression in the form of an inclusion body is often considered undesirable, their formation can be advantageous, because their isolation from cell homogenate is a convenient and effective way of purifying the protein of interest. Converting this inactive inclusion body protein into a soluble active form can cause high recovery of pure and active therapeutic recombinant protein.

## 3. Purification of Therapeutic Proteins From Inclusion Bodies

Recovery of bioactive therapeutic protein from inclusion bodies involves four steps: isolation of the inclusion bodies from *E. coli* cells; solubilization of the protein aggregates; refolding of the protein; and, finally, purification of the solubilized protein *(6,7)*. Among these steps, solubilization and refolding are the most crucial for the high recovery of bioactive protein. Protein aggregation leading to low recovery of bioactive protein mostly occurs from the use of suboptimal refolding procedures. Recombinant proteins expressed as inclusion bodies are solubilized using high concentrations (6–8 *M*) of chaotropic solvents. Chaotropic agents, e.g., urea, guanidine hydrochloride, and thiocyanate salts *(9,11)*, detergents, e.g., sodium dodecyl sulfate *(15)*, *N*-acetyl trimethyl ammonium chloride *(16)*, and sarkosyl (sodium *N*-lauroyl sarcosine; *17*), along with reducing agents, e.g., β-mercaptoethanol, dithiothreitol, or cysteine, have been extensively used for the solubilization of inclusion body proteins. Solubilized proteins are then refolded into their native state during the removal of chaotropic agents. Using additives during refolding often helps in improving the yield of bioactive protein *(18)*. Refolding followed by purification is generally preferable as some high-molecular-weight aggregates and contaminants can be separated in a single-step procedure.

One major problem associated with the low recovery of refolded protein from the solubilization mixture is aggregation *(12)*. Aggregation is a higher-order reaction, whereas refolding is a first-order reaction. Therefore, the aggregation rate is greater than the rate of refolding at high initial protein concentration. Because of this kinetic competition, yields of correctly folded protein decrease at increasing initial protein concentration. Protein concentrations in the range of 10–50 µg/mL are typically used during refolding. Dilution of the solubilized protein directly into the renaturation buffer is the most commonly used method for small-scale refolding of recombinant proteins. Refolding large amounts of recombinant protein using a dilution method requires a large refolding vessel, huge volumes of buffer, and additional concentration steps after protein renaturation, consequently adding to the cost of protein production. Pulse rena-turation, involving the addition of small amounts of solubilized protein to the renatur-ation buffer at successive time intervals, reduces the buffer volume and improves the overall performances of the protein-refolding process *(12,19)*. This success is based on the fact that once a small amount of denatured protein is refolded into the native form, it does not form aggregates with the unfolded protein. By carefully choosing the protein concentration and time of successive additions of the solubilized protein, large quantities of proteins can be refolded in the same buffer tank. Recently, alterna-tive methods for high-throughput refolding of inclusion body protein have been developed *(12,20,21)*.

## 4. Characteristics of Protein Aggregates in Inclusion Bodies

Inclusion bodies are dense particles of aggregated protein found in both the cyto-plasmic and periplasmic space of *E. coli* during high-level expression of foreign pro-teins *(22)*. These protein aggregates form electron-refracting particles in the cell that can be distinguished from other cell components and are therefore also referred to as refractile bodies. The exact reason and mechanism for protein aggregation in *E. coli* to form inclusion bodies is not yet known *(23)*. It is generally assumed that high-level expression of non-native protein (>2% of cellular protein) is more prone to accumu-late as inclusion bodies in *E. coli*. Recombinant protein deposition in inclusion bodies is commonly observed with hydrophobic proteins, as hydrophobic interaction among the partially folded protein molecules has been found to be responsible for aggrega-tion into inclusion bodies. Protein aggregates are also common to proteins containing disulfide bonds, because the reducing environment of the *E. coli* cytosol inhibits the formation of disulfide bonds. The formation of inclusion bodies is suggested to occur as a result of the intracellular accumulation of partially folded proteins, which aggre-gate through noncovalent hydrophobic, ionic interactions, or a combination of both.

The size of inclusion bodies in *E. coli* varies from 0.5 to 1.3 µm, and the aggregates can have either an amorphous or paracrystaline nature depending on their localization *(22,24)*. Inclusion bodies have a higher density (approx 1.3 mg/mL) than many cellu-lar fragments and contaminants; thus, they can be easily separated by high-speed centrifugation after cell disruption *(25)*. Protein aggregation and subsequent inclusion body formation has also been reported to be caused by specific intermolecular interactions between a single type of protein molecule *(26,27)*. Often, these inclusion bodies almost exclusively contain the overexpressed protein *(28)*. Other significant features of protein aggregates contained within inclusion bodies are the existence of native-like secondary structure of the expressed protein and their resistance to proteolysis *(29,30)*. Analysis of β-lactamase inclusion bodies from *E. coli* by Raman spectroscopy indicates the existence of native-like protein structure in inclusion bodies. Therefore, the expression of recombinant proteins as inclusion bodies does provide many advantages, such as the easy separation of large quantities of highly enriched protein, less protein degradation by cell proteases, and reduced toxicity to the host cell.

## 5. A Novel Method for the Improved Recovery of Bioactive Protein From Inclusion Bodies of *E. coli*

Based on new information on the structure and function of inclusion body proteins and the physicochemical properties of inclusion bodies, we have developed a novel solubilization and refolding process for the large-scale recovery of recombinant therapeutic proteins. This is described in detail in Chapter 14 of this volume. The intent was to develop an improved inclusion body protein solubilization and refolding process, which could be undertaken at high-protein concentrations with low-protein aggregation. If protein aggregation and inclusion body formation is highly specific, intact inclusion bodies, once formed, can be isolated from *E. coli* cells either using ultracentrifugation or detergent washing to more than 95% purity *(22,31)*. This helps to reduce the total number of protein purification steps.

Growing evidence exists indicating that proteins expressed as inclusion bodies in *E. coli* have extensive native-like secondary structure *(30)*, and that the formation of inclusion bodies is the result of specific aggregation between folding intermediates of protein molecules *(26)*. Several structural estimates of proteins expressed as inclusion bodies localized in different compartments of *E. coli* have yielded similar results, suggesting that inclusion body formation occurs at a late stage of the protein-folding pathway. Thus, proteins retain most of their secondary structure during aggregation into inclusion bodies *(22)*. During solubilization, if this existing native-like structure is preserved, the partially folded intermediates will improve the renaturation yield of bioactive protein. Solubilization profile of inclusion bodies in different buffers will yield information on the dominant forces involved in protein aggregation during the expression of recombinant proteins as inclusion bodies *(32)*. Such information can then be exploited to develop mild solubilizing buffers, which will protect the native-like secondary structure of the protein during subsequent solubilization. This can be achieved by manipulating the experimental conditions, such as pH, and by using different solubilizing agents in the presence of low concentrations of denaturants.

Protein aggregation occurs because of intermolecular interactions between partially folded intermediates before the formation of either stable intermediate or fully folded protein *(26)*. Furthermore, partially folded intermediates of protein molecules also do

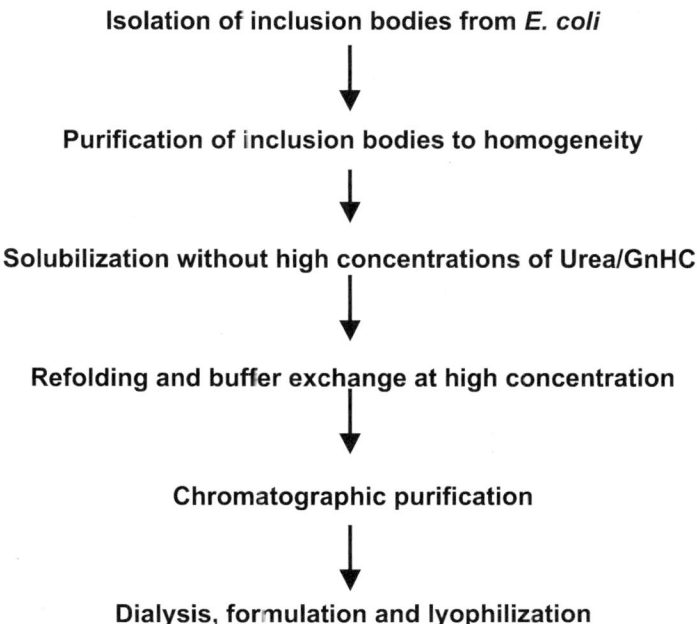

Fig. 1. A novel purification strategy for the improved recovery of bioactive proteins from inclusion bodies.

not interact with fully folded protein, as reported for the P22 tailspike protein *(33,34)*. This supports the concept of pulse renaturation processes for high-throughput refolding of denatured protein. This process exercises the theory that the overall yield of purified bioactive proteins from inclusion bodies can be improved if the existing secondary structure of the protein is protected during solubilization, and that the refolding is carried out so that the interactions between partially folded polypeptide intermediates is minimized. A schematic representation of how the isolation, solubilization, and refolding process of inclusion body proteins may be undertaken in this way is presented in **Fig. 1**. By applying such a novel concept, very high recoveries of bioactive proteins have been achieved for numerous proteins from the inclusion bodies of *E. coli* *(10)*. In the next chapter, details of the experimental protocols for the purification of bioactive human growth hormone from inclusion bodies of *E. coli* using such an approach are described.

## Acknowledgment

The author is thankful to Dr. Sandip K. Basu (Director, National Institute of Immunology, New Delhi) for his support and encouragement. This work was compiled during the author's visit to the College of Pharmacy, Nebraska Medical Center, Omaha, NE, under a Government of India, Dept. of Biotechnology, Overseas Associateship award for 2002–2003.

# References

1. Baneyx, F. (1999) Recombinant protein expression in *Escherichia coli. Curr. Opin. Biotechnol.* **10**, 411–421.
2. Swartz, J. R. (2001) Advances in *Escherichia coli* production of therapeutic proteins. *Curr. Opin. Biotechnol.* **12**, 195–201.
3. Walsh, G. (2003) Biopharmaceutical benchmarks—2003. *Nat. Biotechnol.* **21**, 865–870.
4. Makrides, S. C. (1996) Strategies for achieving high-level expression of genes in *Escherichia coli. Microbiol. Rev.* **60**, 512–538.
5. Kane, J. F. and Hartley, D. L. (1988) Formation of recombinant protein inclusion bodies in *Escherichia coli. Trends Biotechnol.* **6**, 95–101.
6. Rudolph, R. and Lilie, H. (1996) *In vitro* folding of inclusion body proteins. *FASEB J.* **10**, 49–56.
7. Clark, E. D. (2001) Protein refolding for industrial processes. *Curr. Opin. Biotechnol.* **12**, 202–207.
8. Datar, R. V., Cartwright, T., and Rosen, C. G. (1993) Process economics of animal cell and bacterial fermentations: a case study analysis of tissue plasminogen activator. *Biotechnology* **11**, 349–357.
9. Fischer, B., Sumner, I., and Goodenough, P. (1993) Isolation, renaturation and formation of disulfide bonds of eukaryotic proteins expressed in *E. coli* as inclusion bodies. *Biotechnol. Bioeng.* **41**, 3–13.
10. Panda, A. K. (2003) Bioprocessing of therapeutic proteins from the inclusion bodies of *Escherichia coli. Adv. Biochem. Eng. Biotechnol.* **85**, 43–93.
11. Rudolph, R., Böhm, G., Lilie, H., and Jaenicke, R. Folding proteins, in *Protein Function, a Practical Approach* (Creighton, T. E., ed.), IRL-Press, Oxford University Press, Oxford, 1997, pp. 57–99.
12. De Bernardez Clark, E., Schwarz, E., and Rudolph, R. (1999) Inhibition of aggregation side reactions during *in vitro* protein folding. *Methods Enzymol.* **309**, 217–236.
13. Dill, K. A. and Shortle, D. (1991) Denatured states of proteins. *Annu. Rev. Biochem.* **60**, 795–825.
14. Petrides, D., Sapidou, E., and Calandranis, J. (1995) Computer-aided process analysis and economic evaluation for biosynthetic human insulin production-a case study. *Biotechnol. Bioeng.* **48**, 529–541.
15. Stockel, J., Doring, K., Malotka, J., Jahnig, F., and Dornmair, K. (1997) Pathway of detergent-mediated and peptide ligand-mediated refolding of heterodimeric class II major histocompatibility complex (MHC) molecules. *Eur. J. Biochem.* **248**, 684–691.
16. Cardamone, M., Puri, N. K., and Brandon, M. R. (1995) Comparing the refolding and reoxidation of recombinant porcine growth hormone from a urea denatured state and from *Escherichia coli* inclusion bodies. *Biochemistry* **34**, 5773–5794.
17. Burgess, R. R. (1996) Purification of overproduced *Escherichia coli* RNA polymerase sigma factors by solubilizing inclusion bodies and refolding from Sarkosyl. *Methods Enzymol.* **273**, 145–149.
18. Yasuda, M., Murakami, Y., Sowa, A., Ogino, H., and Ishikawa, H. (1998) Effect of additives on refolding of a denatured protein. *Biotechnol. Prog.* **14**, 601–606.
19. Mark Buswell, A., Ettinger, M., Vertes, A. A., and Middelberg, A. P. (2002) Effect of operating variables on the yield of recombinant trypsinogen for a pulse-fed dilution-refolding reactor. *Biotechnol. Bioeng.* **77**, 435–444.
20. Tsumoto, K., Ejima, D., Kumagai, I., and Arakawa, T. (2003) Practical considerations in refolding proteins from inclusion bodies. *Protein Expr. Purif.* **28**, 1–8.

21. Lilie, H., Schwarz, E., and Rudolph, R. (1998) Advances in refolding of proteins produced in *E. coli. Curr. Opin. Biotechnol.* **9,** 497–501.
22. Bowden, G. A., Paredes, A. M., and Georgiou, G. (1991) Structure and morphology of protein inclusion bodies in *Escherichia coli. Biotechnology* **9,** 725–730.
23. Mitraki, A., Fane, B., Haase-Pettingell, C., Sturtevant, J., and King, J. (1991) Global suppression of protein folding defects and inclusion body formation. *Science* **253,** 54–58.
24. Taylor, G., Hoare, M., Gray, D. R., and Marston, F. A. O. (1986) Size and density of protein inclusion bodies. *Biotechnology* **4,** 553–557.
25. Georgiou, G. and Valax, P. (1999) Isolating inclusion bodies from bacteria. *Methods Enzymol.* **309,** 48–58.
26. Speed, M. A., Wang, D. I., and King, J. (1996) Specific aggregation of partially folded polypeptide chains: the molecular basis of inclusion body composition. *Nat. Biotechnol.* **14,** 1283–1287.
27. Rajan, R. S., Illing, M. E., Bence, N. F., and Kopito, R. R. (2001) Specificity in intracellular protein aggregation and inclusion body formation. *Proc. Natl. Acad. Sci. USA* **98,** 13060–13065.
28. Carrio, M. M. and Villaverde, A. (2001) Protein aggregation as bacterial inclusion bodies is reversible. *FEBS Lett.* **489,** 29–33.
29. Przybycien, T. M., Dunn, J. P., Valax, P., and Georgiou, G. (1994) Secondary structure characterization of β-lactamase inclusion bodies. *Protein Eng.* **7,** 131–136.
30. Oberg, K., Chrunyk, B. A., Wetzel, R., and Fink, A. L. (1994) Native-like secondary structure in interleukin-1 β inclusion bodies by attenuated total reflectance FTIR. *Biochemistry* **33,** 2628–2634.
31. Khan, R. H., AppaRao, K. B. C., Eshwari, A. N. S., Totey, S. M., and Panda, A. K. (1998) Solubilization of recombinant ovine growth hormone with retention of native-like secondary structure and its refolding from the inclusion bodies of *Escherichia coli. Biotechnol. Prog.* **14,** 722–728.
32. Patra, A. K., Mukhopadhyay, R., Mukhija, R., Krishnan, A., Garg, L. C., and Panda, A. K. (2000) Optimization of inclusion body solubilization and renaturation of recombinant human growth hormone from *Escherichia coli. Protein Expr. Purif.* **18,** 182–192.
33. Betts, S. and King, J. (1999) There's a right way and a wrong way: *in vivo* and *in vitro* folding, misfolding and subunit assembly of the P22 tailspike. *Structure Fold. Des.* **7,** R131–R139.
34. Kreisberg, J. F., Betts, S. D., Haase-Pettingell, C., and King, J. (2002) The interdigitated beta-helix domain of the P22 tailspike protein acts as a molecular clamp in trimer stabilization. *Protein Sci.* **11,** 820–830.

# 14

## Isolation, Solubilization, Refolding, and Chromatographic Purification of Human Growth Hormone from Inclusion Bodies of *Escherichia coli* Cells

### *A Case Study*

**Surinder M. Singh, A. N. S. Eshwari, Lalit C. Garg, and Amulya K. Panda**

## 1. Introduction

### 1.1. Isolation of Human Growth Hormone Inclusion Bodies From E. coli *Cells*

Inclusion bodies produced in *Escherichia coli* are composed of densely packed denatured protein molecules in the form of particles *(1,2)*. In addition to the recombinant protein of interest, inclusion bodies contain small amounts of host protein, ribosomal components, and DNA/RNA fragments *(3)*. It is advisable to purify the inclusion bodies from the cells to a high-degree purity before carrying out solubilization and purification. This will reduce the number of purification steps after solubilization and refolding, minimize the interference of other contaminating proteins during refolding, and result in a therapeutic protein free from other cellular contaminants, such as lipids, carbohydrate, and endotoxin *(4)*. Isolation of inclusion bodies from *E. coli* occurs by cell lysis with high-pressure disruption using a French press or sonication step followed by centrifugation *(5)*. Further purification can be achieved by washing with detergents and a low concentration of salt and/or urea *(5,6)*. The presence of contaminants, along with the protein of interest, is mainly because of incomplete purification of the inclusion bodies following cell lysis. With proper centrifugation and washing processes, more than 95% pure inclusion bodies of recombinant proteins can be isolated from *E. coli* cells *(7)*. As the inclusion bodies have a high density (approx 1.3 mg mL$^{-1}$), these are easily separated by high-speed centrifugation after cell disruption *(8)*. Centrifugal isolation, particularly sucrose gradient centrifugation, has been found to be the best method for isolating very pure inclusion bodies from *E. coli* cell lysate *(9)*. Expression of recombinant human growth hormone (r-hGH) in *E. coli* will illustrate

From: *Methods in Molecular Biology, vol. 308: Therapeutic Proteins: Methods and Protocols*
Edited by: C. M. Smales and D. C. James © Humana Press Inc., Totowa, NJ

the methods used for isolation and purification of intact inclusion bodies. The purified inclusion bodies will then be used for solubilization and refolding to obtain bioactive protein.

## 1.2. Solubilization and Refolding of Human Growth Hormone From Inclusion Bodies of E. coli

Inclusion bodies from *E. coli* are generally solubilized in high concentrations of denaturants, such as 8 *M* urea or 6 *M* guanidine hydrochloride with or without reducing agents (*10–12*). High concentrations of denaturants unfolds the protein completely, thus increasing the propensity of aggregation during refolding. Suboptimal buffer exchange during protein refolding also contributes to protein aggregation, which results in a low recovery of refolded protein from inclusion bodies. Reducing the protein aggregation improves the recovery of solubilized bioactive protein from the inclusion bodies. As the inclusion body proteins retain native-like secondary structure, it is advisable to protect this during solubilization. Partly folded protein conformations help in improved recovery of bioactive protein (*13*). Therefore, the use of a mild inclusion body solubilization process is the key step in successful recovery of high yields of therapeutic protein from inclusion bodies.

Recently, many new methods for solubilization of inclusion bodies without using high concentrations of denaturants have been reported, which involve using detergents (*14*), pressure (*15*), pH, or a combination of these (*7*). Furthermore, different solubilization buffers can be used to evaluate the importance of various protein interactions that lead to the accumulation of inclusion bodies (*16*). Such an approach is taken to further understand the nature of the protein aggregation leading to inclusion body formation. With this information, suitable buffers can be employed to solubilize the inclusion body proteins without unfolding the protein of interest into a random coil configuration. Recombinant ovine growth hormone (*7*) and *zona pellucida* glycoprotein (*17*) have both been successfully solubilized and refolded using such mild solubilization approaches. Developing an efficient and low-denaturant solubilization step is thus necessary for achieving high-throughput purification of recombinant protein. This chapter discusses the solubilization and refolding of r-hGH from inclusion bodies. Inclusion bodies are solubilized using mild concentrations of urea while "giving" a pH shock. Solubilized hGH is subsequently refolded into a soluble bioactive form using a pulse renaturation process.

## 1.3. Chromatographic Purification of Recombinant Human Growth Hormone

Ion-exchange chromatography followed by gel-filtration chromatography is the most commonly used procedure for the purification of recombinant proteins (*18,19*). The advantages of using ion-exchange chromatography include the fact that it can be operated at different salt, pH, and buffer compositions, and, perhaps most important, diluted and refolded proteins can be concentrated during this purification step. Thus, ion-exchange chromatography is the most suitable process for protein purification from the large volumes of dilute solution that are generated during inclusion body refolding processes. r-hGH has previously been purified using various ion-exchange chromatographic techniques *(20)*. As the isoelectric point of human growth hormone is 4.9, operating anion-exchange chromatography processes at a pH far from the isoelectric point results in the best separation *(21,22)*. The initial purification and concentration is undertaken using a Q-Sepharose (Amersham Pharmacia) column followed by gel-filtration chromatography using a Sephacryl S-200 (Amersham Pharmacia) matrix *(16)*.

During the refolding, high-molecular-weight aggregates are always associated with the pure monomeric protein. These aggregates constitute 2–10% of the total protein; therefore, it is necessary to remove them using gel filtration chromatography. As the hGH used as a case study in this chapter was refolded and purified in 2 *M* urea, the refolded protein is extensively dialyzed to remove the urea before gel filtration. During dialysis, the excipients and buffers required to promote protein stability are introduced. This chapter details ion-exchange and gel-filtration processes for the purification of hGH.

## 2. Materials

### 2.1. Isolation of Human Growth Hormone Inclusion Bodies From E. coli Cells

1. Transformed *E. coli* cells expressing hGH.
2. Luria Bertani (LB) medium: 10 g Bacto-tryptone, 5 g yeast extract, and 5 g NaCl per liter of Milli-Q water. Add 5 g/L glucose in the medium for enhanced cell growth.
3. Isopropyl-β-D-thio-galactopyranoside (IPTG) filter-sterilize.
4. French press.
5. High-speed centrifuge and microfuge.
6. Ultracentrifuge.
7. Probe sonicator.
8. Phenylmethylsulfonyl fluoride (PMSF).
9. Tris-EDTA buffer: 50 m*M* Tris-HCl, 5 m*M* EDTA, and 1 m*M* PMSF, pH 8.5.
10. Tris-HCl buffer: 10 m*M* Tris-HCl, pH 8.5.
11. Sucrose solution: Prepare in phosphate buffer.
12. Sodium deoxycholate: Dissolve 1% (w/v) in Tris-HCl buffer, pH 8.
13. Sodium dodecyl sulfate (SDS) sample buffer: 0.25 *M* Tris-HCl, pH 6.8, 40% glycerol, 8% SDS, 160 m*M* dithiothreitol, and 0.04 mg/mL bromophenol blue.
14. SDS-polyacrylamide gel electrophoresis (SDS-PAGE) equipment and buffers for running gel.
15. Molecular weight marker: 14-, 21-, 30-, 44-, 66-, and 90-kDa protein mixture markers (Amersham Pharmacia). Dissolve one vial of the marker in 1 mL SDS sample buffer and store at –20°C.
16. Coomassie blue stain: 0.1% (w/v) Phastgel Blue R (Amersham Pharmacia) in destaining solution.
17. Destaining solution: 40% (v/v) methanol and 10% (v/v) acetic acid.
18. Ampicillin and kanamycin.
19. Sodium phosphate buffer: 100 m*M*, pH 8.
20. Spectrophotometer (ultraviolet [UV]-visible range).
21. Micro-BCA protein assay kit (Pierce).

## 2.2. Solubilization and Refolding

1. Pure r-hGH inclusion bodies from *E. coli* cells.
2. High-speed centrifuge.
3. Spectrophotometer (UV-visible).
4. pH meter.
5. PMSF.
6. Magnetic stirrer.
7. Peristaltic pump.
8. Homogenizer.
9. Solubilization buffer: 100 m$M$ Tris-HCl, 1 m$M$ EDTA, and 2 $M$ urea, pH 12.5.
10. Refolding buffer: 50 m$M$ Tris-HCl, 1 m$M$ EDTA, 2 $M$ urea, 10% (w/v) sucrose, and 1 m$M$ PMSF, pH 8.
11. Membrane filter (0.45 μm).
12. Micro-BCA assay kit (Pierce).

## 2.3. Chromatographic Purification

1. Refolded r-hGH solution.
2. AKTA purifier (Amersham Pharmacia).
3. Liquid chromatography system (Amersham Pharmacia).
4. Glass jacket column (26 × 40 cm) for ion exchange.
5. Glass jacket column (16 × 100 cm) for gel filtration.
6. Q-Sepharose matrix (Amersham-Pharmacia).
7. Sephacryl S-200 matrix (Amersham-Pharmacia) for gel filtration.
8. Dialysis bag.
9. Spectrophotometer.
10. Lyophilizer.
11. Equilibration buffer: 50 m$M$ Tris-HCl, 1 m$M$ EDTA, 2 $M$ urea, 5% sucrose (w/v), and 1 m$M$ PMSF, pH 8.5.
12. Washing buffer: 10 m$M$ NaCl in equilibration buffer.
13. Elution buffer: 500 m$M$ NaCl in equilibration buffer.
14. PMSF.
15. Dialysis buffers.
    a. First change: 50 m$M$ Tris-HCl, 1 m$M$ EDTA, 1 $M$ urea, 5% (w/v) sucrose, and 1 m$M$ PMSF, pH 8.5.
    b. Second change: 50 m$M$ Tris-HCl, 0.5 m$M$ EDTA, 0.5 $M$ urea, 5% (w/v) sucrose, and 1 m$M$ PMSF, pH 8.5.
    c. Third change: 10 m$M$ Tris-HCl, 0.5 m$M$ EDTA, 0.25 $M$ urea, 1% (w/v) sucrose, and 1 m$M$ PMSF, pH 8.5.
    d. Fourth change: 10 m$M$ Tris-HCl and 1% (w/v) sucrose, pH 8.5.
16. Gel filtration buffer: 10 m$M$ Tris-HCl and 1% (w/v) sucrose, pH 8.5.
17. Micro-BCA protein assay (Pierce).

# 3. Methods

## 3.1. Isolation

hGH, an important therapeutic protein having a molecular weight of approx 22 kDa and expressed in *E. coli*, is used in this chapter as a model system for purification of inclusion bodies *(16)*. Two methods are described with the aim of obtaining pure

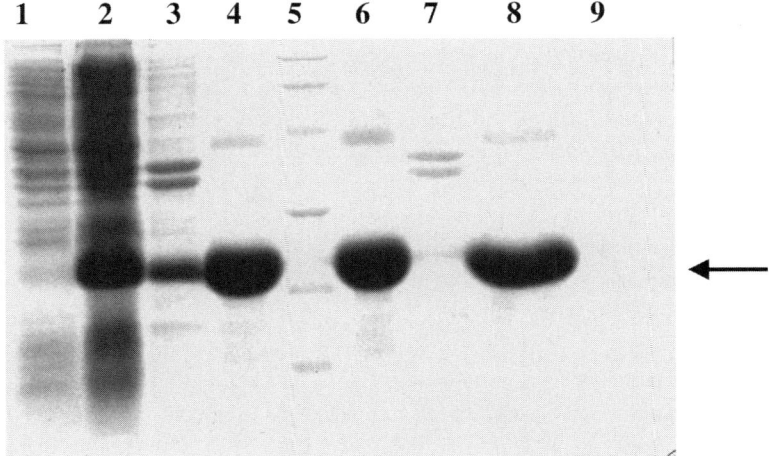

Fig. 1. SDS-PAGE analysis of r-hGH from inclusion bodies of *E. coli* cells. The arrow indicates the r-hGH band. Lane 1, uninduced cells; lane 2, induced cells; lane 3, inclusion bodies after cell lysis; lanes 4 and 6, pure inclusion body from sucrose gradient centrifugation; lane 5, molecular weight marker (14, 21, 30, 44, 66, and 90 kDa, from the bottom to top of the gel); lane 7, contaminating proteins; lane 8, pure inclusion bodies using deoxycholate treatment.

inclusion bodies of hGH from *E. coli* cells. These methods are sucrose gradient centrifugation and a sodium deoxycholate washing process. Therapeutic proteins expressed as inclusion bodies in *E. coli* can be isolated with the help of either of these methods with slight modification. Finally, an SDS-PAGE method has been described for the analysis of r-hGH expression and pure inclusion bodies. Growth and expression of r-hGH is also described, as it constitutes the starting point of the method.

### 3.1.1. Expression of hGH in E. coli

1. Inoculate *E. coli* strain expressing r-hGH from glycerol stock in to 10 mL sterilized LB media containing 100 µg/mL ampicillin and 25 µg/mL kanamycin. Grow the *E. coli* cells overnight at 37°C and at a shaker speed of 200 rpm in an orbital shaker.
2. Inoculate 10 mL overnight grown *E. coli* cells in to 1 L LB medium with 100 µg/mL ampicillin and 25 µg/mL kanamycin. Allow the cells to grow until the optical density (OD) of the culture at 600 nm reaches 0.5.
3. Induce *E. coli* cells with IPTG (1 m*M* final concentration) at an OD of 0.5 and grow for another 3 h (*see* **Note 1**).
4. After 3-h induction, harvest the *E. coli* cells by centrifugation at 10,000*g* for 30 min. Resuspend the *E. coli* cells in Tris-HCl buffer and centrifuge again at 10,000*g* for 30 min. Collect the cell pellet for inclusion body isolation. These cells can be directly used for inclusion body extraction or can be stored at –20°C for future processing.
5. Check *E. coli* cells for intracellular expression of r-hGH by SDS-PAGE analysis. Use both uninduced and induced cells to check the presence of r-hGH after induction with IPTG (**Fig. 1**).

### 3.1.2. Inclusion Body Purification Using Sucrose Gradient Centrifugation

This method is useful for the small-scale preparation of inclusion body proteins. Protein concentrations of up to 100 mg in the form of inclusion bodies can be processed using this method. As the inclusion bodies have higher densities than most membrane components, sucrose gradient centrifugation helps in separating these particles from contaminating cellular fragments/proteins. This method is useful and results in good yields of pure inclusion bodies, which can be further used for solubilization and refolding to obtain bioactive protein. This method does not require detergent treatment and can be carried out with the normal equipment available in a standard molecular biology laboratory.

1. Resuspend the *E. coli* cells expressing r-hGH from a 1-L culture (4 g wet cell pellet) in 10 mL 50 m$M$ Tris- EDTA buffer containing 5 m$M$ EDTA and 1 m$M$ PMSF, pH 8.5.
2. Pass the cells for three cycles through a French press at 15,000–18,000 psi for cell lysis. In the absence of a French press, use sonication (10 cycles of 1 min each with a 1-min gap between cycles, 50% duty, power 28 W, on ice) for cell lysis.
3. Centrifuge the cell lysate at 20,000$g$ for 20 min at 4°C. Discard the supernatant, and use the pellet for further purification of inclusion bodies.
4. Wash the pellet with Tris-HCl buffer and centrifuge once again as described in **step 3**.
5. Resuspend the inclusion body pellet in 2 mL sodium phosphate buffer (100 m$M$, pH 8).
6. Prepare sucrose step gradient in ultracentrifuge rotor tubes by dropwise addition of sucrose solution. Add 1 mL each of 72%, followed by 70%, 68%, 66%, 62%, and 60% (w/v) of sucrose solution from the bottom to the top of the tube to prepare the sucrose gradient.
7. Add 1 mL inclusion body suspension from **step 5** on top of the 60% sucrose layer in the tube, and centrifuge at 150,000$g$ for 6 h in a swinging rotor at 4°C.
8. Inclusion bodies will be seen in the sucrose gradient at densities between 65% and 70% as a dense layer. Carefully remove the dense layer by pipet without disturbing the other layers, and collect the inclusion body particles in an Eppendorf tube (*see* **Note 2**).
9. Add 0.5 mL Milli-Q water to the Eppendorf tube; vortex and centrifuge in a microcentrifuge for 20 min at 10,000$g$ at 4°C. Repeat **steps 6–8** to further purify inclusion bodies, or prepare gradients in two tubes to ultracentrifuge all of the inclusion bodies from **step 5**.
10. Wash the inclusion body pellet with Milli-Q water, then centrifuge at 10,000$g$ for 20 min at 4°C. Discard the supernatant. The remaining pellet consists of ultra-pure inclusion bodies of r-hGH. Run an SDS-PAGE gel, and check the homogeneity of the inclusion bodies.
11. Pure inclusion bodies run at approx 22 kDa on a SDS-PAGE gel with minor contaminants. High molecular aggregates of hGH may be seen along with the pure 22-kDa r-hGH (**Fig. 1**). Inclusion bodies have regular shapes with average particles sizes about 0.5 µm as observed by a scanning electron micrograph (**Fig. 2**).
12. Dissolve a small amount of r-hGH inclusion bodies in 1% SDS solution in 10 m$M$ Tris-HCl buffer (no EDTA), and prepare serial dilutions to measure the protein concentration in the inclusion bodies using a detergent-compatible Micro-BCA protein kit (*see* **Note 3**).

### 3.1.3. Isolation of r-hGH Inclusion Bodies by Deoxycholate Treatment

1. Suspend *E. coli* cells (4 g wet cell pellet) expressing r-hGH from a 1-L culture in 10 mL 50 m$M$ Tris-EDTA buffer containing 5 m$M$ EDTA and 1 m$M$ PMSF, pH 8.5. Lyse the cells as described in **Subheading 3.2.** using either a French press or sonicator.

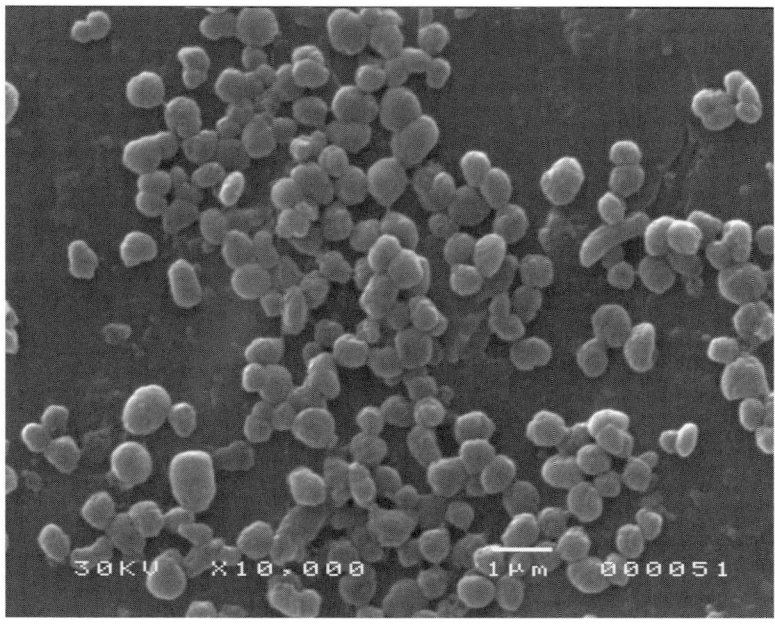

Fig. 2. Scanning electron microscopy of pure hGH inclusion bodies. The average size is approx 0.5 μm.

2. Centrifuge the cell lysate at 20,000g for 20 min at 4°C. Use the pellet for inclusion body purification.
3. Suspend the inclusion body pellet in 10 mL Tris-EDTA buffer and sonicate (10 cycles of 1 min each with a 2-min gap between cycles, 40–50% duty, power 28 W, on ice).
4. Centrifuge the inclusion body suspension from **step 3** at 20,000g for 20 min at 4°C. Decant the supernatant, and use the pellet for further processing and purification of inclusion bodies. Suspend the pellet in 10 mL 1% deoxycholate (sodium salt) solution.
5. Mix the solution thoroughly and sonicate (six cycles of 1 min each with 2-min gap, 40–50% duty, power 28 W, on ice). Centrifuge the suspension at 20,000g for 20 min at 4°C. Use the pellet for further processing to obtain ultra-pure inclusion bodies.
6. Dissolve the pellet again in 10 mL 1% deoxycholate solution and sonicate (six cycles of 1 min each with 2-min gap, 40–50% duty, power 28 W, on ice). Keep the inclusion body suspension in deoxycholate solution at room temperature overnight. This step helps in removing contaminating membrane proteins (*see* **Note 4**).
7. Vortex the inclusion body suspension thoroughly and centrifuge at 20,000g for 20 min at 4°C. Resuspend the inclusion body pellet after the second deoxycholate wash in 20 mL Tris-HCl buffer, pH 8.5. Mix thoroughly and centrifuge at 20,000g for 20 min at 4°C. Collect and process the pellet for further removal of detergent.
8. Wash the inclusion body pellet with 20 mL Milli-Q water to remove residual detergent and salts; centrifuge at 20,000g for 20 min at 4°C. The pellet obtained after centrifugation contains the pure inclusion bodies of r-hGH and looks similar when analyzed by SDS-PAGE to that described for sucrose gradient purification (**Fig. 1**).

### 3.1.4. SDS-PAGE Procedure for Protein Analysis

#### 3.1.4.1. CASTING OF THE GELS

Prepare 12% resolving/separating gel and 5% stacking gel. The following describes the composition of the gels.

| Preparation of a 12% resolving/separating gel (for one gel) | |
| --- | --- |
| 29.2% Acrylamide + 0.8% bisacrylamide | 4 mL |
| 1.5 *M* Tris-HCl, pH 8.8 | 2.5 mL |
| Distilled water | 3.3 mL |
| 10% SDS | 100 µL |
| 10% Ammonium persulfate (APS) | 100 µL |
| TEMED | 4 µL |

| Preparation of a 5% stacking gel (for one gel) | |
| --- | --- |
| 29.2% Acrylamide + 0.3% bisacrylamide | 0.5 mL |
| 1 *M* Tris-HCl, pH 6.8 | 0.38 mL |
| 10% SDS | 30 µL |
| Distilled water | 2.1 mL |
| 10% APS | 30 µL |
| TEMED | 3 µL |

#### 3.1.4.2. PREPARATION OF SAMPLES

Use *E. coli* cells expressing r-hGH for SDS-PAGE analysis. One milliliter of both uninduced and induced *E. coli* cells should be used and centrifuged at 10,000*g* for 10 min for cell isolation. Dissolve the cell pellet in 50 µL SDS sample buffer, and boil for 5 min. Vortex the solution, centrifuge at 10,000*g*, and load 20 µL supernatant onto a SDS-PAGE gel. Dissolve pure inclusion bodies in SDS sample buffer, boil, centrifuge as described previously, and load the clear soup onto the SDS-PAGE gel (*see* **Note 5**). Load 10 µL low-molecular-weight marker in a separate well in the SDS-PAGE gel along with samples.

#### 3.1.4.3. RUNNING SDS-PAGE GEL

1. Load the SDS-PAGE gel well with 20 µL *E. coli* cell extract and pure inclusion body solubilized and processed in sample dye and 10 µL molecular weight marker as described previously. Run the gel at a constant current of 30 mA. Stop electrophoresis when the tracker dye is approx 1 cm from the bottom of the glass plate.
2. Remove the gel gently, and place it in the container containing Coomassie blue stain. Stain the gel for 1 h by slowly mixing on a rocker. Place the gel in destaining solution (40% methanol and 10% acetic acid) and leave until the protein bands are visualized (*see* **Fig. 1**).

## 3.2. Solubilization and Refolding

### 3.2.1. Solubilization of r-hGH Inclusion Bodies at High pH

For hGH, ionic and hydrophobic interactions are the main forces causing protein aggregation in inclusion bodies *(16)*. High-alkaline pH (12.5) in the presence of 2 *M* urea promotes the solubilization of hGH from inclusion bodies. Use of 2 $M$ urea does not completely unfold the protein and preserves the existing native-like secondary structure. Using high pH also helps in better solubilization, as the buffer pH is far from the isoelectric point of hGH, which is 4.9. Therefore, a combination of alkaline pH and 2 $M$ urea helps in destabilizing both the ionic and hydrophobic interactions—the major cause of protein aggregation in inclusion bodies of r-hGH.

1. Suspend the pure inclusion body pellet of r-hGH (50 mg) in 10 mL 100 m$M$ Tris-EDTA buffer containing 1 m$M$ EDTA and 2 $M$ urea, pH 12.5 (*see* **Note 6**).
2. Homogenize the suspension thoroughly at 5000 rpm using a homogenizer for 5 min on ice, then leave for 15 min at room temperature. Following this, centrifuge the solution at 20,000$g$ for 20 min at 4°C (*see* **Note 7**).
3. Carefully aspirate off the supernatant obtained after centrifugation, and check the protein concentration in the supernatant by reading the OD at 280 nm. This will provide an indication of the protein solubility in the solubilizing buffer.
4. Check for the presence of hGH both in the pellet and supernatant by running a SDS-PAGE gel.
5. Repeat **steps 1–4** if a large amount of unsolubilized pellet remains.
6. Pool the solubilized hGH solution (total of ~40 mg r-hGH solubilized in 10 mL of solubilization buffer), and use immediately for refolding by the pulsatile renaturation process (*see* **Subheading 3.2.2.** and **Note 8**). Carry out the solubilization and refolding on the same day.

## 3.2.2. Refolding by Pulsatile Renaturation

Protein aggregation occurs due to the intermolecular interactions between partially folded intermediates well before the formation of either stable intermediate or fully folded protein *(23)*. However, it has been reported that partially folded intermediates of protein molecules do not interact with fully folded protein, as described for the P22 tail spike protein *(23)*. This supports using a pulse renaturation process for high-throughput recovery of denatured protein *(24,25)*. This process essentially involves the addition of small amounts of solubilized inclusion body proteins in batches to the same volume of refolding buffer. After a few minutes, further unfolded protein is added, assuming that the first batch of protein has been refolded into its native-like structure. As the unfolded protein does not interact with the folded protein, pulse addition keeps on refolding protein in the same buffer tank. This helps in reducing the volume of buffer required for refolding and results in high concentrations of refolded protein. Solubilized hGH is refolded using a pulse renaturation process as described here.

1. Take 90 mL refolding buffer (50 m$M$ Tris-HCl, 0.5 m$M$ EDTA, 2 $M$ urea, 10% glycerol, 5% sucrose, and 1 m$M$ PMSF, pH 8), and maintain it at 4–6°C with constant stirring (*see* **Note 9**).
2. Add solubilized r-hGH solution (~4 mg/mL) dropwise either using a peristaltic pump (keeping the flow rate at the 0.5 mL/min minimum), or use a micropipet to add small protein amounts at regular intervals.
3. Continue adding. It will take approx 1 h for the pulsatile renaturation of 10 mL of solubilized r-hGH into 100 mL refolding buffer. Refolding will also result in reducing the pH of the solution from a high-alkaline value to a pH of 8–8.5 (*see* **Note 10**).

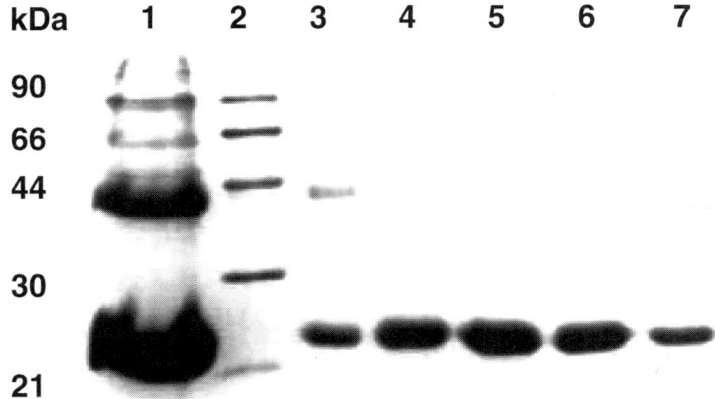

Fig. 3. SDS-PAGE analysis of the described r-hGH purification process using Q-Sepharose ion-exchange chromatography. Lane 1, solubilized inclusion body proteins; lane 2, molecular weight marker; lane 3, load on to Q-Sepharose column; lanes 4–7, purified fractions of hGH.

4. 10 mL r-hGH (solubilized in a buffer at pH 12.5 containing 2 *M* urea) is refolded in a pulsatile manner to give a final concentration of 400 µg/mL r-hGH.
5. After refolding, filter the protein solution using a 0.45-µm membrane filter to remove any aggregates. Obtain an estimate of the protein concentration using a Micro-BCA protein assay. The protein solution is now ready for chromatographic separation to obtain ultra-pure hGH. Check for the presence of r-hGH in the refolding buffer by SDS-PAGE.

## 3.3. Chromatographic Purification

### 3.3.1. Purification by Ion-Exchange Chromatography

1. Pack 25 mL Q-Sepharose into the glass column (~5-cm bed height). Equilibrate the column at flow rate of 2 mL/min with equilibration buffer. Use the AKTA purifier for ion-exchange chromatography.
2. Filter the refolded protein through a 0.45-µm filter, and load onto the pre-equilibrated Q-Sepharose ion-exchange column at a flow rate of 2 mL/min. Wash the column using three-column volumes of equilibration buffer containing 10 m*M* NaCl to remove nonspecific contaminants.
3. Elute the bound r-hGH using a linear continuous gradient from 10 to 500 m*M* NaCl in equilibration buffer.
4. Homogeneous r-hGH in the form of monomer elute at concentrations of between 150 and 250 m*M* NaCl, whereas protein aggregates elute at higher ionic strength (*see* **Note 11**).
5. The samples containing homogeneous r-hGH bands on SDS-PAGE (**Fig. 3**) should be pooled (approx 20 mL) and dialyzed against a lower urea gradient in each dialysis change, changing after every 8 h as described in the following buffers.
   a. First change: 250 mL 50 m*M* Tris-HCl, 1 m*M* EDTA, 1 *M* urea, 5% (w/v) sucrose, and 1 m*M* PMSF, pH 8.5.
   b. Second change: 250 mL 50 m*M* Tris-HCl, 0.5 m*M* EDTA, 0.5 *M* urea, and 2.5% (w/v) sucrose, pH 8.5.

   c. Third change: 250 mL 10 m*M* Tris-HCl, 0.5 m*M* EDTA, 0.25 *M* urea, and 1% (w/v) sucrose, pH 8.5.

   d. Fourth change: 250 mL 10 m*M* Tris-HCl, and 1% (w/v) sucrose, pH 8.5.

6. When the dialysis step is complete, lyophilize the r-hGH and use for the subsequent gel filtration step (*see* **Note 12**).

## 3.3.2. Purification by Gel Filtration

1. Pack the gel filtration column with Sephacryl S-200 resin up to 96 cm. Use the liquid chromatography system for gel filtration.
2. Dissolve the lyophilized hGH (approx 14 mg) in 3–4 mL 10 m*M* Tris-HCl buffer containing 1% (w/v) sucrose. Filter the solution through a 0.45-μm filter to remove any aggregates (*see* **Note 13**).
3. Load the protein solution in the form of a layer on top of the gel filtration matrix.
4. Run the column at a flow rate of 20 mL/h using a peristaltic pump, and collect fractions.
5. Check each fraction by SDS-PAGE for the presence of hGH, which will elute after approximately half a column volume of buffer.
6. Pool all the protein fractions containing pure hGH, and dialyze against 10 m*M* Tris-HCl buffer containing 1% sucrose. Check by SDS-PAGE for purity.
7. Lyophilize the protein and store for physicochemical and biological assay. Determine the protein concentration using the Micro-BCA assay procedure (*see* **Notes 14** and **15**).

# 4. Notes

1. The postinduction time required after IPTG addition for maximum gene expression is different for varying proteins and depends on the host vector relationship. Most often, optimal induction time is 3–6-hours postinduction for inclusion body accumulation. Do not grow *E. coli* cells overnight after induction, as cell lysis may expose the inclusion bodies to the culture medium and create problems during purification. Use freshly grown, induced cells for the isolation of inclusion bodies and subsequent protein refolding.
2. Remove the dense inclusion body layer very carefully without disturbing the other layers. If the homogeneity of the inclusion bodies is not sufficient, repeat the ultracentrifugation sucrose gradient again using the pellet of the first ultracentrifugation step. Depending on the density of the inclusion bodies, they will be deposited at different places in the sucrose gradient after ultracentrifugation.
3. Do not freeze the purified inclusion bodies to be used at a later date for solubilization. It is always advisable to isolate, purify, and refold the protein into a soluble form without freezing the protein. For quantification of the amount of protein in inclusion bodies, completely solubilize them in detergent solution (1% SDS in the case of r-hGH), and undertake a protein estimation using the detergent-compatible protein assay kit as described. Add detergent to the protein standard solution to minimize error. The presence of EDTA interferes with the Micro-BCA protein assay.
4. There is some loss of r-hGH in the supernatant during deoxycholate treatment. However, this step helps in eliminating contaminating membrane proteins. Depending on the nature of the therapeutic protein, deoxycholate concentrations may need to be optimized. Some proteins will be solubilized in 1% deoxycholate solution, whereas for others, this will not be the case. Additional use of 0.5–1 *M* NaCl and/or 2 *M* urea sometimes helps in solubilization of contaminating cellular proteins and may help improve the purity of inclusion bodies.

5. Use gloves while performing SDS-PAGE (acrylamide is a neurotoxin). Do not pipet the pellet at the bottom of the microfuge tube during sample loading for SDS-PAGE. Rinse the syringe a few times with distilled water after loading each individual well. If *E. coli* cells or the inclusion body pellet does not dissolve in sample dye, use 10–20 µL 10% SDS solution to dissolve them, then add SDS sample buffer for processing before loading onto the SDS-PAGE gel.

6. To determine the best buffer for solubilization of inclusion bodies of a particular protein, it is necessary to know the dominant forces that cause protein aggregation that lead to inclusion body formation. This can be determined by solubilization of aliquots of pure inclusion bodies in different buffers and monitoring the percent solubility either by protein assay or reduction in solution turbidity *(16)*. Alternatively, a sparse matrix approach can be used to design the solubilization protocol for a particular protein *(26)*. If disulfide bonds have an important role in the protein structure and stability, use β-mercaptoethanol in the solubilization buffer. If the protein of interest degrades at pH 12.5, adjust the pH of the buffer to pH 12.

7. Never use frozen inclusion body pellets for solubilization or keep the leftover pellet for solubilization at a later date. It is advisable to undertake the inclusion body purification, solubilization, and refolding in one attempt without storing the protein between any step.

8. Do not keep the solubilized protein at a high pH for any longer than absolutely necessary. Thiolate ion formation may result, leading to protein amidation and ultimately resulting in poor-quality bioactive protein. Use freshly prepared buffers, particularly adding the urea to the buffer just before the experiments to reduce cyanate ion formation. Carry out refolding at a low temperature to reduce the extent of protein aggregation during refolding.

9. The use of glycerol, sucrose, and 2 *M* urea helps to prevent protein aggregation during refolding and, more importantly, during gel filtration and lyophilization of the refolded protein. It is advisable to add these excipients to improve the protein stability during the different stages of processing.

10. Check for turbidity during the refolding process; otherwise, carry out the pulsatile-refolding process in large volumes, or dilute the solubilized protein to a concentration of 1 mg/mL before refolding. This helps to lower the aggregation of proteins. A different refolding process may be used instead of pulse dilution if this approach does not give a good recovery of soluble protein *(24)*.

11. The recombinant protein aggregates will usually elute at a higher ionic strength than the monomer during ion-exchange chromatography. However, sometimes the aggregates will coelute with the monomer. If significant amounts of protein coelute with the monomer, collect the mixture, dialyze, and run the gel filtration step as described to recover the monomeric protein.

12. Carry out dialysis with decreasing concentrations of urea at each step. The risk of protein aggregation is high during the removal of urea. Depending on the nature of a protein, the dialysis time and stepwise decreases of urea can be optimized to reduce the extent of any protein aggregation.

13. Refolded protein should be soluble in aqueous buffer. If the lyophilized r-hGH does not dissolve in aqueous buffer, this indicates that the protein is not refolded correctly. Try different excipients or concentrations of sucrose during lyophilization to achieve a better recovery. Filter the sample, and only use the soluble protein for further purification by gel filtration.

14. Calculate the total amount of r-hGH recovered at the end of the gel filtration process. The recovery should be around 40–50% (considering that the starting amount of r-hGH was 50 mg). The total recovery after gel filtration should be approx 20 mg of pure hGH.

15. If the recovery of protein is very low using the gel filtration process, it may be necessary to use different methods of refolding *(25)* as described in previous chapters of this book. Refolding by column chromatography *(27)*, reverse micelles *(28)*, or microfiltration *(29)* can be used to improve the recovery of protein.

## Acknowledgments

The authors thank Dr. Sandip K. Basu (Director, National Institute of Immunology, New Delhi, India) for providing core facilities for recombinant protein research. The work is also supported by grants from the Department of Biotechnology, Government of India to AKP and LCG.

## References

1. Mitraki, A., Fane, B., Haase-Pettingell, C., Sturtevant, J., and King, J. (1991) Global suppression of protein folding defects and inclusion body formation. *Science* **253**, 54–58.
2. Carrio, M. M., Cubarsi, R., and Villaverde, A. (2000) Fine architecture of bacterial inclusion bodies. *FEBS Lett.* **471**, 7–11.
3. Valax, P., and Georgiou, G. (1993) Molecular characterization of beta-lactamase inclusion bodies produced in *Escherichia coli*. 1. Composition. *Biotechnol. Prog.* **9**, 539–547.
4. Clark, E. D. B. (1998) Refolding of recombinant proteins. *Curr. Opin. Biotechnol.* **9**, 157–163.
5. Georgiou, G. and Valax, P. (1999) Isolating inclusion bodies from bacteria. *Methods Enzymol.* **309**, 48–58.
6. Lilie, H., Schwarz, E., and Rudolph, R. (1998) Advances in refolding of proteins produced in *E. coli*. *Curr. Opin. Biotechnol.* **9**, 497–501.
7. Khan, R. H., AppaRao, K. B. C., Eshwari, A. N. S., Totey, S. M., and Panda, A. K. (1998) Solubilization of recombinant ovine growth hormone with retention of native-like secondary structure and its refolding from the inclusion bodies of *Escherichia coli*. *Biotechnol. Prog.* **14**, 722–728.
8. Taylor, G., Hoare, M., Gray, D. R., and Marston, F. A. O. (1986) Size and density of protein inclusion bodies. *Biotechnology* **4**, 553–557.
9. Bowden, G. A., Paredes, A. M., and Georgiou, G. (1991) Structure and morphology of protein inclusion bodies in *Escherichia coli*. *Biotechnology* **9**, 725–730.
10. Rudolph, R. and Lilie, H. (1996) *In vitro* folding of inclusion body proteins. *FASEB J.* **10**, 49–56.
11. Misawa, S. and Kumagai, I. (1999) Refolding of therapeutic proteins produced in *Escherichia coli* as inclusion bodies. *Biopolymers* **51**, 297–307.
12. Clark, E. D. (2001) Protein refolding for industrial processes. *Curr. Opin. Biotechnol.* **12**, 202–207.
13. Creighton, T. E., Darby, N. J., and Kemmink, J. (1996) The roles of partly folded intermediates in protein folding. *FASEB J.* **10**, 110–118.
14. Cardamone, M., Puri, N. K., and Brandon, M. R. (1995) Comparing the refolding and reoxidation of recombinant porcine growth hormone from a urea denatured state and from *Escherichia coli* inclusion bodies. *Biochemistry* **34**, 5773–5794.

15. St John, R. J., Carpenter, J. F., Balny, C., and Randolph, T. W. (2001) High pressure refolding of recombinant human growth hormone from insoluble aggregates. Structural transformations, kinetic barriers, and energetics. *J. Biol. Chem.* **276,** 46856–46863.

16. Patra, A. K., Mukhopadhyay, R., Mukhija, R., Krishnan, A., Garg, L. C., and Panda, A. K. (2000) Optimization of inclusion body solubilization and renaturation of recombinant human growth hormone from *Escherichia coli. Protein Expr. Purif.* **18,** 182–192.

17. Patra, A. K., Gahlay, G. K., Reddy, B. V., Gupta, S. K., and Panda, A. K. (2000) Refolding, structural transition and spermatozoa-binding of recombinant bonnet monkey (*Macaca radiata*) zona pellucida glycoprotein-C expressed in *Escherichia coli. Eur. J. Biochem.* **267,** 7075–7081.

18. Evangelista Dyr, J. and Suttnar, J. (1997) Separations used for the purification of recombinant proteins. *J. Chromatogr. B, Biomed. Sci. Appl.* **699,** 383–401.

19. Amersham Pharmacia Biotech. *Ion Exchange Chromatography Principles and Methods.* Amersham Pharmacia Biotech, Uppsala, Sweden, 1999.

20. Ribela, M. T., Gout, P. W., and Bartolini, P. (2003) Synthesis and chromatographic purification of recombinant human pituitary hormones. *J. Chromatogr. B. Analyt. Technol. Biomed. Life Sci.* **790,** 285–316.

21. Becker, G. W. and Hsiung, H. M. (1986) Expression, secretion and folding of human growth hormone in *Escherichia coli.* Purification and characterization. *FEBS Lett.* **204,** 145–150.

22. Igout, A., Van Beeumen, J., Frankenne, F., Scippo, M. L., Devreese, B., and Hennen, G. (1993) Purification and biochemical characterization of recombinant human placental growth hormone produced in *Escherichia coli. Biochem. J.* **295,** 719–724.

23. Speed, M. A., Wang, D. I., and King, J. (1996) Specific aggregation of partially folded polypeptide chains: the molecular basis of inclusion body composition. *Nat. Biotechnol.* **14,** 1283–1287.

24. De Bernardez Clark, E., Schwarz, E., and Rudolph, R. (1999) Inhibition of aggregation side reactions during *in vitro* protein folding. *Methods Enzymol.* **309,** 217–236.

25. Tsumoto, K., Ejima, D., Kumagai, I., and Arakawa, T. (2003) Practical considerations in refolding proteins from inclusion bodies. *Protein Expr. Purif.* **28,** 1–8.

26. Lindwall, G., Chau, M., Gardner, S. R., and Kohlstaedt, L. A. (2000) A sparse matrix approach to the solubilization of overexpressed proteins. *Protein Eng.* **13,** 67–71.

27. Altamirano, M. M., Golbik, R., Zahn, R., Buckle, A. M., and Fersht, A. R. (1997) Refolding chromatography with immobilized mini-chaperones. *Proc. Natl. Acad. Sci. USA* **94,** 3576–3578.

28. Vinogradov, A. A., Kudryashova, E. V., Levashov, A. V., and van Dongen, W. M. (2003) Solubilization and refolding of inclusion body proteins in reverse micelles. *Anal. Biochem.* **320,** 234–238.

29. Batas, B., Schiraldi, C., and Chaudhuri, J. B. (1999) Inclusion body purification and protein refolding using microfiltration and size exclusion chromatography. *J. Biotechnol.* **68,** 149–158.

# 15

# Large-Scale Preparation of Factor VIIa from Human Plasma

## A Case Study

**Teruhisa Nakashima and Kazuhiko Tomokiyo**

## 1. Introduction

Human factor VII (FVII) is a glycoprotein with a molecular mass of 50 kDa, is synthesized in the liver as a single-chain precursor of activated FVII (FVIIa), and it circulates in the blood at a plasma concentration of 0.5 µg/mL. FVIIa is a serine protease, generated by limited proteolysis of zymogen FVII by activated factors Xa (FXa) or IXa (FIXa) in vivo. Upon binding to its receptor, cofactor tissue factor (TF), FVIIa gains full catalytic activity and triggers the extrinsic blood coagulation pathway, activating its substrates, factors X (FX) and IX (FIX).

Hemophilia A and B are congenital diseases in which patients lack factor VIII (FVIII) and FIX, respectively. Replacement therapies of FVIII and FIX have been applied for hemophiliacs using FVIII and FIX concentrates. When antibodies (inhibitors) against FVIII and FIX develop in hemophilia patients, however, those inhibitors make replacement therapies difficult. Recently, recombinant FVIIa (rFVIIa) was developed as a bypassing agent for hemophiliacs with inhibitors. High doses of rFVIIa (60–120 µg/kg body weight) are used to control bleeding episodes in these patients *(1)*.

Plasma-derived FVIIa (pdFVIIa) concentrate, which is prepared on an industrial scale from human plasma, is not available for clinical use for hemophiliacs with inhibitors against FVIII and FIX, but pdFVIIa should be more cost-effective than rFVIIa. To provide FVIIa concentrate from human plasma, it is necessary to solve the following problems:

1. Establishment of a large-scale manufacturing process with a high yield from plasma.
2. Optimization of the conversion process of zymogen FVII to the active form of FVIIa on an industrial scale.
3. Introduction of a production process to effectively eliminate and inactivate viruses.

The conversion of FVII to FVIIa is a key process in the production of FVIIa concentrate. In 1986, Bjoern reported that rFVII is autocatalytically activated to rFVIIa

From: *Methods in Molecular Biology, vol. 308: Therapeutic Proteins: Methods and Protocols*
Edited by: C. M. Smales and D. C. James © Humana Press Inc., Totowa, NJ

on an anion exchange chromatography resin *(2)*. Pedersen also reported that rFVII from serum-free cell culture is autoactivated in a solution containing poly-D-lysine and $Ca^{2+}$, and explained this activation on the basis of the autoactivation theory *(3)*. On an industrial scale, the activation of FVII on an anion exchange resin is a useful technique, because this reaction does not require other proteases, such as FXa, FIXa, or activated factor XII (FXIIa), which convert FVII to FVIIa. Thus, we established a production-scale method for the activation of FVII using autoactivation involving the following steps: (1) activation on an anion exchange resin; and (2) further activation in solution after elution from the resin. This method is suitable for the large-scale production of FVIIa with a high yield and quality. Furthermore, virus-spiking tests showed that immunoaffinity chromatography, nanofiltration, and dry heating effectively remove and inactivate the spiked viruses in this product, indicating that the FVIIa concentrate should be safe from the risk of virus transmission *(4)*.

## 2. Materials

### 2.1. Preparation of FVIIa Concentrate From Human Cryoprecipitate-Poor Plasma

#### 2.1.1. Preparation of $Ca^{2+}$-Dependent Anti-FVII Monoclonal Antibody (MAb)-Immobilized Gel

1. $Ca^{2+}$-dependent anti-FVII MAb (HFVII-1-20-25-133-19, IgG1, κ subtype, supplied by Kaketsuken).
2. Q-Sepharose Fast Flow (Q-FF; Amersham Pharmacia Biotech, Uppsala, Sweden).
3. Protein A Sepharose 4 Fast Flow (Amersham Pharmacia Biotech).
4. Virus removal filter: Planova® 15 N (Asahi Chemical, Tokyo, Japan).
5. Cyanogen bromide.
6. Sepharose Fast Flow (Amersham Pharmacia Biotech).

#### 2.1.2. Preparation of FVII

1. Cryoprecipitate poor plasma.
2. Q-FF (Amersham Pharmacia Biotech).
3. Equilibration buffer for Q-FF: 20 m$M$ citrate and 0.1 $M$ NaCl, pH 8.0.
4. Elution buffer for Q-FF: 20 m$M$ citrate and 0.5 $M$ NaCl, pH 8.0.
5. $Ca^{2+}$-dependent anti-FVII MAb-immobilized gel (anti-FVII MAb gel).
6. Equilibration buffer for anti-FVII MAb gel: 50 m$M$ Tris-HCl, 50 m$M$ NaCl, and 2.0 m$M$ $CaCl_2$, pH 7.0.
7. Washing buffer for anti-FVII MAb gel: 50 m$M$ Tris-HCl, 50 m$M$ NaCl, and 10 m$M$ EDTA, pH 7.2.
8. Regeneration buffer (I) for anti-FVII MAb gel: 20 m$M$ Tris-HCl and 2 $M$ NaCl, pH 7.3.
9. Regeneration buffer (II) for anti-FVII MAb gel: 6 $M$ urea, 1 $M$ NaCl, and 1% Tween-80, pH 7.2.

#### 2.1.3. Nanofiltration of FVII

1. Eluted FVII fraction from anti-FVII MAb gel.
2. Virus removal filter: Planova® 15 N (Asahi Chemical).

### 2.1.4. Partial Conversion of FVII to FVIIa on an Anion Exchanger

1. Nanofiltrated FVII.
2. DEAE-Sepharose Fast Flow (DEAE-FF; Amersham Pharmacia Biotech).
3. Equilibration buffer for DEAE-FF: 50 m$M$ Tris-HCl and 30 m$M$ NaCl, pH 7.8.
4. Elution buffer for DEAE-FF: 50 m$M$ Tris-HCl, 30 m$M$ NaCl, and 1.75 m$M$ CaCl$_2$, pH 7.8.

### 2.1.5. Full Activation of FVII in the Liquid Phase and Preparation of FVIIa

1. Eluted mixture of FVII and FVIIa fractions from DEAE-FF.
2. Dialysis buffer: 20 m$M$ citrate, 240 m$M$ NaCl, and 13.3 m$M$ glycine, pH 6.9.
3. Dialysis module: Artificial kidney (AM-30, Asahi Chemical).

## 2.2. Assay of Coagulation Factors

### 2.2.1. FVII and FVIIa Activity

1. CA 5000 automatic coagulation analyzer (Sysmex, Kobe, Japan).
2. FVII-depleted plasma (Dade Behring, Marburg, Germany).
3. Pooled normal human plasma (George King, St. Overland Park, KS).
4. Prothrombin time (PT) reagent (Dade Behring).
5. Soluble tissue factor (TF) (Kaketsuken).

### 2.2.2. FVII, FIX, and FX Antigens

1. Enzyme linked immunosorbent assay (ELISA) kits (Asserachrom® VII:Ag and FIX:Ag, Diagnostica Stago, Asnieres-Sur-Seine, France).
2. Ca$^{2+}$-dependent anti-FX MAb (FXM2-12-109, IgG1, κ subtype, supplied by Kaketsuken).
3. Peroxide-conjugated rabbit anti-FX polyclonal antibodies (DAKO, Glostrup, Denmark).

### 2.2.3. Sodium Dodecyl Sulfate-Polyacrylamide Gel Electrophoresis (SDS-PAGE) Analysis

1. Polyacrylamide gel (12.5% acrylamide concentration).
2. Dye: Coomassie Brilliant Blue R250 (CBB).
3. Densitometer (CS-9000, Shimadzu, Kyoto, Japan).

### 2.2.4. Gel Permeation Chromatography Analysis

1. High-performance liquid chromatography system: LC-10AS (Shimadzu).
2. Column: TSK-GEL G3000SW$_{XL}$ column (∅ 7.8 mm × 30 cm, Tosoh, Tokyo, Japan).
3. Loading buffer: 50 m$M$ phosphate and 300 m$M$ NaCl, pH 5.5.

## 3. Methods

## 3.1. Preparation of FVIIa From Human Cryoprecipitate-Poor Plasma

### 3.1.1. Preparation of Anti-FVII MAb Gel

1. Hybridoma cells (HFVII-1-20-25-133-19) were grown in bioreactors in serum-free medium.
2. After the cultured cells were removed by filtration, MAb was purified from the filtrate using a Q-FF column, followed by a protein A Sepharose 4 Fast Flow column.
3. The eluted MAb fraction was nanofiltrated by Planova® 15 N.
4. Sepharose Fast Flow was activated with cyanogen bromide, and MAb was coupled to the activated gel at a mean density of 5 mg MAb/mL of gel as previously described *(5)*.

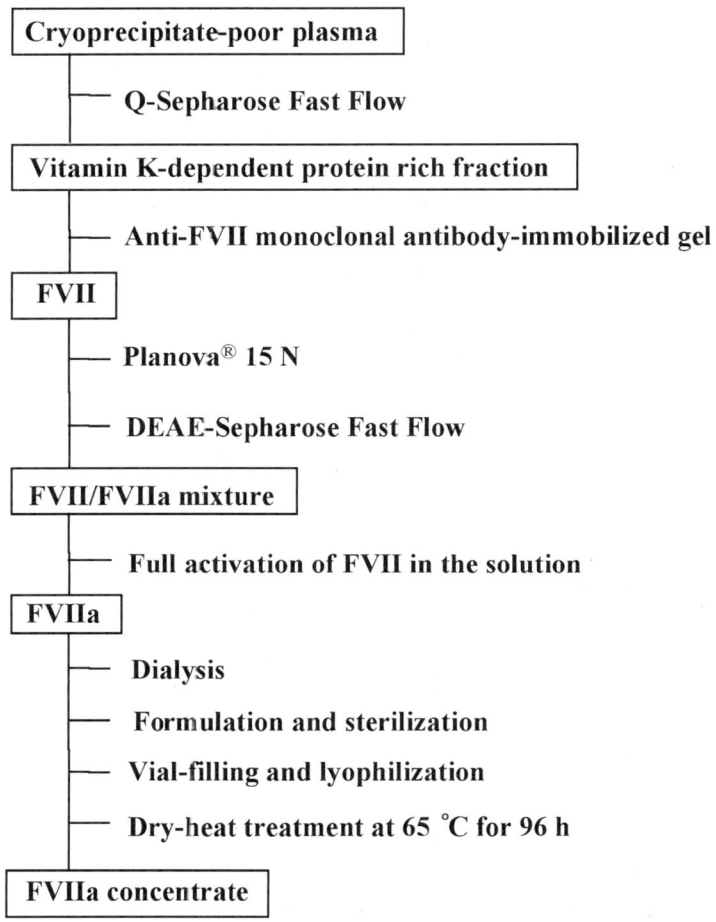

Fig. 1. Flow diagram of the FVIIa-purification process. *See* text for details.

### 3.1.2. Purification of FVII (see **Notes 1–4**)

Human pdFVIIa concentrate was prepared using the following procedures in which the temperature was kept at 4°C unless otherwise stated (*see* **Fig. 1**).

1. Human cryoprecipitate-poor plasma (3000 L) was applied to a Q-FF column (200-L gel volume) equilibrated with 20 m$M$ citrate, pH 8.0, containing 0.1 $M$ NaCl, and the bound fraction was eluted with the same buffer containing 0.5 $M$ NaCl (*see* **Note 1**).
2. The eluted fraction (vitamin K-dependent protein-rich fraction) containing 20 m$M$ CaCl$_2$ was applied to an anti-FVII MAb gel column ($\varnothing$ 25 cm × 2.0 cm) equilibrated with 50 m$M$ Tris-HCl, pH 7.0, containing 50 m$M$ NaCl and 2.0 m$M$ CaCl$_2$, at 400 mL/min.
3. The column was then extensively washed with the same buffer, and the bound FVII fraction was eluted with 50 m$M$ Tris-HCl and 50 m$M$ NaCl, pH 7.2, containing 10 m$M$ EDTA.

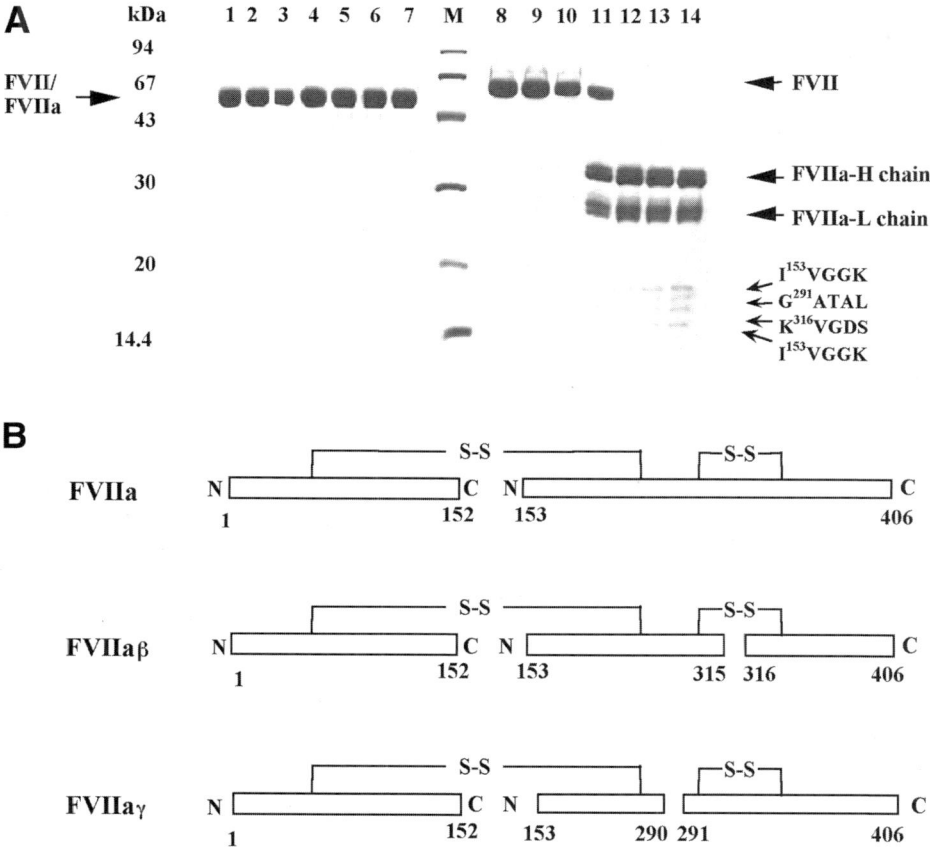

Fig. 2. SDS-PAGE analyses of FVIIa preparation on a large scale. (**A**) The obtained FVIIa was subjected to SDS-PAGE (12.5% acrylamide concentration) under nonreducing (lanes 1–7) and reducing (lanes 8–14) conditions. Lanes 1 and 8, before filtration using Planova® 15 N; lanes 2 and 9, after filtration using Planova®-15 nm; lanes 3 and 10, before application on DEAE-FF; lanes 4 and 11, the fraction eluted from DEAE-FF column; lanes 5 and 12, after incubation for 3 h; lanes 6 and 13, after incubation for 18 h; lanes 7 and 14, after dialysis; lane M, molecular mass markers. Four minor bands with molecular weights of 18, 17, 16, and 15 kDa were observed in lanes 13 and 14, and NH$_2$-terminal sequence analysis indicated that these bands had the sequence I$^{153}$VGGK, G$^{291}$ATAL, K$^{316}$VGDS, and I$^{153}$VGGK. (**B**) Schematic representation of intact FVIIa, FVIIaβ, and FVIIaγ.

4. The anti-FVII MAb gel was regenerated by successive washing with the following solutions: (1) 20 m*M* Tris-HCl, pH 7.3, containing 2 *M* NaCl and (2) 6 *M* urea, pH 7.2, containing 1 *M* NaCl and 1% Tween-80.

The purified FVII was shown as a single band under nonreducing conditions and was more than 90% pure as judged from the SDS-PAGE analyses (*see* **Fig. 2A**, lanes 1 and 8). The specific activity of the purified FVII was 1.8 U/μg and the content of the activated form (FVIIa) in the FVII preparation measured using soluble TF was less than 1.5%. FX and FIX antigens were not detected in the FVII solution by ELISA (<0.1 ng/mL), and prothrombin was also undetectable by a one-stage clotting assay using prothrombin-depleted plasma.

### 3.1.3. Nanofiltration of FVII (see **Notes 3** and **5**)

The eluted FVII fraction was nanofiltered for virus elimination using Planova® 15 N.

### 3.1.4. Partial Conversion of FVII to FVIIa (see **Note 6**)

1. Nanofiltered FVII fraction (22 L) yielding 4.5 g FVII from 15,000 L plasma was applied to a DEAE-FF column (∅ 9.0 cm × 5.0 cm) equilibrated with 50 m$M$ Tris-HCl, pH 7.8, containing 30 m$M$ NaCl, at 10°C at 320 mL/min.
2. After washing with the same buffer (10 column volumes) and at the same flow rate, the partially activated FVII was eluted with 50 m$M$ Tris-HCl, pH 7.8, containing 30 m$M$ NaCl and 1.75 m$M$ CaCl$_2$, at 250 mL/min.

### 3.1.5. Full Activation of FVII and Preparation of FVIIa (see **Note 7**)

1. For full activation of FVII in the FVII and FVIIa mixture, the eluate was incubated for 18 h at 10°C at a protein concentration of 1.5 mg/mL.
2. After incubation, the FVIIa preparation (2 L FVIIa solution, 1.5 mg/mL) was dialyzed against a dialysis buffer of 20 m$M$ citrate, pH 6.9, containing 240 m$M$ NaCl and 13.3 m$M$ glycine in an artificial kidney module of hollow fiber dialyzer, which was used as a membrane with a molecular weight cut-off of 10,000. The FVIIa solution (inner side of hollow fiber) and dialysis buffer (outer side of hollow fiber) flowed in antiparallel directions across the dialysis membrane to make a sequential diffusion in the module.
3. Dialyzed FVIIa (the eluted fraction from the outlet of the inside of the hollow fiber) was frozen and stored at –60°C.

As shown in **Table 1**, the mean recovery of FVII antigen after the activation of FVII and the specific activity of the collected FVIIa were 38% and 40 U/µg, respectively. Leaked anti-FVII MAb from the immunoaffinity gel was not detected in the FVIIa preparation (<0.25 ng/mL). Gel permeation chromatography of FVIIa showed that FVIIa was 97% pure, which was consistent with the SDS-PAGE analysis under nonreducing conditions (*see* **Fig. 2A**, lane 7). SDS-PAGE analyses of FVIIa under reducing conditions showed four minor bands presumably derived from degradation products as well as intact heavy and light chains of FVIIa (*see* **Fig. 2A**, lanes 13 and 14). NH$_2$-terminal amino acid sequence analyses of these bands produced the sequences for I$^{153}$VGGK (18 kDa), G$^{291}$ATAL (17 kDa), K$^{316}$VGDS (16 kDa), and I$^{153}$VGGK (15 kDa), indicating that traces of FVIIaβ and FVIIaγ were autocatalytically generated in the FVIIa preparation (*see* **Fig. 2B**).

### 3.1.6. Preparation of FVIIa Concentrate

1. The frozen FVIIa preparation was thawed in a water bath at 37°C, and human albumin was added to the solution to give a final concentration of 2.5%. The albumin was pasteurized in advance at 60°C for 10 h.
2. About 12,000 U/mL FVIIa solution was sterilized with a 0.22-µm filter, and 2.5 mL FVIIa was placed into glass vials.

**Table 1**
**Recovery of FVIIa From Large-Scale Production**

| Step | Specific activity (U/mg)[a] | FVII antigen recovery (%)[b] |
|---|---|---|
| Cryoprecipitate-poor plasma | $0.022 \pm 0.002$ ($n = 50$) | 100 |
| Q-FF | $0.85 \pm 0.05$ ($n = 50$) | $70 \pm 3.2$ ($n = 50$) |
| MAb-immobilized gel | $1830 \pm 130$ ($n = 50$) | $60 \pm 2.5$ ($n = 50$) |
| BMM-15 | $2330 \pm 210$ ($n = 5$) | $58 \pm 3.1$ ($n = 5$) |
| DEAE-FF/Full activation[c] | $40300 \pm 3500$ ($n = 5$) | $39 \pm 2.1$ ($n = 5$) |
| Dialysis | $42000 \pm 2140$ ($n = 5$) | $38 \pm 1.2$ ($n = 5$) |
| Postheat treatment | $470 \pm 49^d$ ($n = 5$) | $36 \pm 2.1$ ($n = 5$) |

Values are denoted as the mean value ± standard deviation, with the number of samples in parentheses.

[a]FVII and FVIIa activities were measured using the PT method with FVII-depleted plasma, and protein content was calculated as the average extinction coefficient of 1.0 (mg/mL)-1 cm at $A_{280}$ through the Q-FF fraction. In the subsequent steps, an extinction coefficient of 1.4 (mg/mL)-1 cm at $A_{280}$ was used (**9**).

[b]The recovery of FVIIa or FVII was calculated as FVII antigen recovery measured by ELISA.

[c]After partial activation of FVII using DEAE-FF, the FVII/FVIIa mixture was fully activated in the solution.

[d]Heat treatment (65°C for 96 h) was performed after lyophilization of the FVIIa solution (approx 12,000 U/mL), including 2.5% human serum albumin. FVII activity was determined after reconstitution by water for injection (*see* **Methods**).

3. Lyophilization was performed under the following conditions: Freezing temperature of less than –35°C; vacuum level of 30 mTorr; shelf temperatures for primary and secondary drying of 5°C and 30°C, respectively.

4. After lyophilization, the vials were stoppered under vacuum with rubber stoppers and sealed with flip caps, and they were further heated in a water bath at 65°C for 96 h. Even when FVIIa concentrate was dry-heated, residual FVIIa activity was retained at more than 95%, indicating that the dry-heating treatment had no significant effect on FVIIa activity (*see* **Note 8**).

## 3.2. Properties of the FVIIa Concentrate

The components and properties of the FVIIa concentrate are described in **Table 2**.

1. The mean moisture content of the FVIIa concentrate was $1.91 \pm 0.17\%$ ($n = 5$) and ranged from 1.7% to 2.1% (*see* **Table 2**).

2. The FVIIa concentrate was reconstituted with 5 mL water for injection and quality assurance of the FVIIa final preparation as listed in **Table 2**.

3. FVIIa activity of the final FVIIa preparation was $5900 \pm 600$ U/mL ($n = 5$), and its pH was $7.0 \pm 0.1$ ($n = 5$).

## 3.3. Stability of the FVIIa Concentrate

1. A stability test on the FVIIa concentrate was performed at 10°C for 2 yr. The pH, solubility, moisture content, and FVIIa activity did not change during this period.

2. FVIIa activity of the reconstituted FVIIa solution in 5 mL water for injection did not change at 25°C for 24 h.

**Table 2**
**Components and Properties**
**of FVIIa Concentrate From Large-Scale Production**

| Component/vial | Value |
| --- | --- |
| Glycine | 2.5 mg |
| Sodium chloride | 35 mg |
| Sodium citrate | 30 mg |
| Human serum albumin | 62.5 mg |

| Property | Value |
| --- | --- |
| Reconstitution volume | 5 mL |
| Reconstitution time | <1 min ($n = 5$) |
| pH | $7.0 \pm 0.1$ ($n = 5$) |
| FVII activity | $5900 \pm 610$ U/mL ($n = 5$) |
| Moisture content | $1.91 \pm 0.17\%$ ($n = 5$) |
| Murine antibody[a] | Not detected ($n = 5$) |
| HCV antibody[b] | Not detected ($n = 5$) |
| HIV antibody[c] | Not detected ($n = 5$) |
| HBs antigen[d] | Not detected ($n = 5$) |
| Safety | Sterile and nonpyrogenic ($n = 5$) |

Values are denoted as the mean value ± standard deviation, with the number of samples in parentheses

[a]Murine antibody was measured by ELISA.

[b]Hepatitis C virus (HCV) antibody was measured using a particle-agglutination test kit (Ortho HCV Ab PA test II, Ortho-Clinical Diagnostic, Tokyo).

[c]HIV antibody was measured using an Enzyme Immunoassay (EIA) kit (AXSYM, Dinabot, Tokyo).

[d]Hepatitis B surface (HBs) antigen was measured using a radioimmunoassay kit (AUSRIA II, Dinabot).

## 3.4. Coagulation Factors Assay Methods

### 3.4.1. FVII and FVIIa Activities

1. FVII activity was measured by the PT method using FVII-depleted plasma and pooled normal human plasma as a reference (1 U/mL).
2. FVIIa activity was also measured by the modified PT method using soluble TF to discriminate against FVII activity (*6*).

### 3.4.2. FVII, FIX, and FX Antigens

1. FVII and FIX antigens were measured using ELISA kits.
2. FX antigen was also measured by sandwich ELISA using $Ca^{2+}$-dependent anti-FX MAb and peroxidase-conjugated rabbit anti-FX polyclonal antibodies. Anti-human FX MAb was developed by immunizing BALB/c mice with plasma-derived FX using the method described by Sugo et al. (*7*).

### 3.4.3. SDS-PAGE Analysis

1. SDS-PAGE analysis (12.5% acrylamide concentration) was performed using Laemmli's buffer system (*8*).

**Table 3**
**Removal Efficacy of Spiked Viruses by Nanofiltration (Planova 15 N)**

| Virus (diameter, nm)[b] | Total virus titer[a] | | |
|---|---|---|---|
| | Prefiltration (A) | Postfiltration (B) | LRV log (A/B) |
| PPV (18–26) | $10^{7.1}$ | $<10^{2.5}$ | >4.6 |
| Polio virus (25–30) | $10^{8.5}$ | $<10^{2.5}$ | >6.0 |
| JEV (40–50) | $10^{5.8}$ | $<10^{2.5}$ | >3.3 |
| BVDV (46–57) | $10^{6.0}$ | $<10^{2.5}$ | >3.5 |
| CMV (150–200) | $10^{5.2}$ | $<10^{2.1}$ | >3.1 |

[a]Total CMV titer is expressed as pfu/mL × volume (mL), and other titers are expressed as $TCID_{50}$/mL × volume (mL).
[b]Particle sizes of the viruses are shown as the diameter (nm) of these viruses.

2. Proteins were visualized by staining with CBB.
3. The intensities of the visualized bands were measured using a densitometer.
4. The amount of FVII and FVIIa was calculated as the intensities of the 50-kDa band and sum of the intensity of the light and heavy chains of FVIIa (molecular masses of 25 and 35 kDa, respectively) in SDS-PAGE under reducing conditions.

### 3.4.4. Gel Permeation Chromatography Analysis

1. Gel permeation chromatography analysis was performed using an LC-10AS system and a TSK-GEL G3000SW$_{XL}$ column ($\varnothing$ 7.8 mm × 30 cm) in 50 m$M$ phosphate, pH 5.5, including 300 m$M$ NaCl at 0.5 mL/min, and the effluent was monitored by its absorbance at 280 nm ($A_{280}$).

## 4. Notes

1. A Q-FF column (200-L gel volume) was used to produce a vitamin K-dependent protein-rich fraction. If it is necessary to know the details of this process, please contact Kazuhiko Tomokiyo (Blood Products Research Department, Kaketsuken, 1-6-1, Okubo, Kumamoto 860-8568, Japan).
2. Mouse immunoglobulin G (IgG) that leaks into the FVII preparation from the immuno-affinity resin can be removed by DEAE-FF chromatography of the passed-through fraction. Mouse IgG was not detected by ELISA in the FVIIa concentrate.
3. Virus-spiking tests were performed on the following three processes: (1) immunoaffinity chromatography, (2) nanofiltration (15 nm), and (3) dry heating.
   Titration of spiked viruses was performed by measuring the median tissue culture-infective dose ($TCID_{50}$/mL) or plaque-forming units (pfu/mL). The removal efficacy of spiked viruses was estimated by comparing virus titers or total virus titers ($TCID_{50}$ or pfu/mL × vol) in the samples between pre- and postprocedures and calculating the logarithm reduction value (LRV; *see* **Tables 3** and **4**).
4. To test the efficacy of virus removal by immunoaffinity chromatography, a virus solution containing sindbis virus or sabin type I polio virus was added to the vitamin K-dependent protein-rich fraction at a volume ratio of 1:9, and the mixture was then applied to an anti-FVII MAb gel column ($\varnothing$ 2.5 cm × 2.0 cm). The validity of the scale-down of immunoaffinity chromatography for the spiking test was shown by determining the

**Table 4**
**Inactivation of Spiked Viruses by Dry-Heat Treatment at 65°C**

| Virus | (A) Prelyophilization | Postlyophilization | Total virus titer* Time-course | | | | (B) Postheat treatment | LRV log (A/B) |
|---|---|---|---|---|---|---|---|---|
| CMV | $10^{4.7}$ | $10^{1.6}$ | $<10^{0.7}$ (16 h) | $<10^{0.7}$ (32 h) | $<10^{0.7}$ (64 h) | $<10^{0.7}$ (96 h) | $<10^{0.7}$ | >4.0 |
| HIV | $10^{7.0}$ | $10^{6.0}$ | $<10^{0.5}$ (12 h) | $<10^{0.5}$ (48 h) | | | $<10^{0.5}$ | >6.5 |
| BVDV | $10^{4.8}$ | $10^{4.5}$ | $<10^{0.5}$ (12 h) | $<10^{0.5}$ (24 h) | $<10^{0.5}$ (48 h) | $<10^{0.5}$ (96 h) | $<10^{0.5}$ | >4.3 |
| Polio virus | $10^{6.2}$ | $10^{1.0}$ | $<10^{0.5}$ (16 h) | $<10^{0.5}$ (32 h) | $<10^{0.5}$ (64 h) | $<10^{0.5}$ (96 h) | $<10^{0.5}$ | >5.7 |
| PPV | $10^{5.5}$ | $10^{5.2}$ | $10^{4.8}$ (24 h) | $10^{3.8}$ (48 h) | $10^{3.7}$ (72 h) | $10^{3.3}$ (96 h) | $10^{3.3}$ | 2.2 |

*CMV titer is expressed as the value of pfu/mL, and the other titers are expressed as the value of $TCID_{50}$/mL.

recovery and purity of FVII in the eluate using SDS-PAGE analysis and gel permeation chromatography with TSK-GEL G3000 $SW_{XL}$. The recovery and purity of FVII were comparable to those of FVII in the manufacturing process. The efficacies of sindbis and polio viruses removal by immunoaffinity chromatography were 4.4 and 2.8 LRV, respectively, indicating a high degree of reduction.

5. In the FVII nanofiltration process, porcine parvovirus (PPV), polio virus, Japanese encephalitis virus (JEV), cytomegalovirus (CMV) or bovine viral diarrhea virus (BVDV) was added to the FVII fraction at a volume ratio of 1:9. It was eluted from the anti-FVII MAb gel, and the mixture was then applied to Planova® 15 N (30 $cm^2$). The efficacy of polio virus, JEV, BVDV, and CMV removal, which have various particle sizes, was tested by a nanofiltration process, resulting in excellent removal of these viruses with over 6.0, 3.3, 3.5, and 3.1 LRV, respectively. Using this virus elimination system, the LRV of PPV, which had the smallest particle size ($\varnothing$ 18–26 nm), was also over 4.6 LRV (*see* **Table 3**).

6. To optimize the conversion of zymogen FVII to FVIIa, the following experiment was performed. Seventy-five milliliters of purified FVII solution ($A_{280}$ = 0.260) was applied to a DEAE-FF column ($\varnothing$ 0.5 cm × 5.0 cm, 1/324 of the production scale), and after washing the column, the bound FVII was eluted with 50 m$M$ Tris-HCl, pH 7.8, containing 30 m$M$ NaCl and 1.75 m$M$ $CaCl_2$, at varying flow rates. The eluted FVII was immediately mixed with 500 m$M$ acetic acid, pH 3.0, at a volume ratio of 1:9 to stop further activation. To determine the activation rate of FVII, the eluate was then subjected to SDS-PAGE. As shown in **Fig. 3A**, 70% and 60% of FVII (calculated by the method described in **Methods**) was converted to FVIIa, which consists of two chains (heavy and light chains), when FVII on a DEAE-FF column was eluted at 0.4 and 0.6 mL/min, respectively. The activation rate of FVII decreased at a relatively high flow rate, showing that the elution flow rate determines the degree of FVII activation. A small amount of Gla-domainless (GD-less) FVII/FVIIa, which is cleaved from the light chain, also appeared during this activation procedure at low flow rates (*see* **Fig. 3A**). However, generation of these degradation products could be controlled by suppressing FVII activation to approx 50% (*see* **Fig. 3B**).

7. To optimize the incubation time for full activation of the FVII and FVIIa mixture (eluted fraction from DEAE-FF), a partially activated FVII solution (1.5 mg/mL total protein concentration of the FVIIa/FVII mixture), in which 25% of the FVII had been activated in advance, was converted to FVIIa in the presence of 1.75 m$M$ $CaCl_2$ for 20 h at 10°C without anion exchangers or other activators. FVII was activated to FVIIa in a time-dependent manner and was completely activated after 20 h (*see* **Fig. 4**).

8. To study the virus-inactivating effect of the dry heating of the FVIIa preparation, a virus-inactivation test was performed using the following viruses: human immunodeficiency virus (HIV), polio virus, CMV, and BVDV. After mixing the FVIIa preparation and virus solution (1:9 volume ratio), the mixture was lyophilized using methods similar to those for large-scale processes (e.g., glass vials, rubber stoppers, flip caps, filled volume, temperature for freezing, vacuum level, and shelf temperatures for primary and secondary drying), and the lyophilized samples were heated in a water bath at 65°C for 96 h. HIV was titrated using cytopathic effect in the SKT-1B cell line as previously described (*10*).

Except for PPV, the inactivation efficacy of these spiked viruses was over 4.0 LRV (*see* **Table 4**). PPV was not sufficiently inactivated by the dry-heating treatment. However, considering the efficacy of removing PPV by nanofiltration (LRV of PPV > 4.6) (*see* **Table 3**), our FVIIa concentrate should have a high potential for decreasing the transmission risk of human parvovirus B19 as well as other viruses.

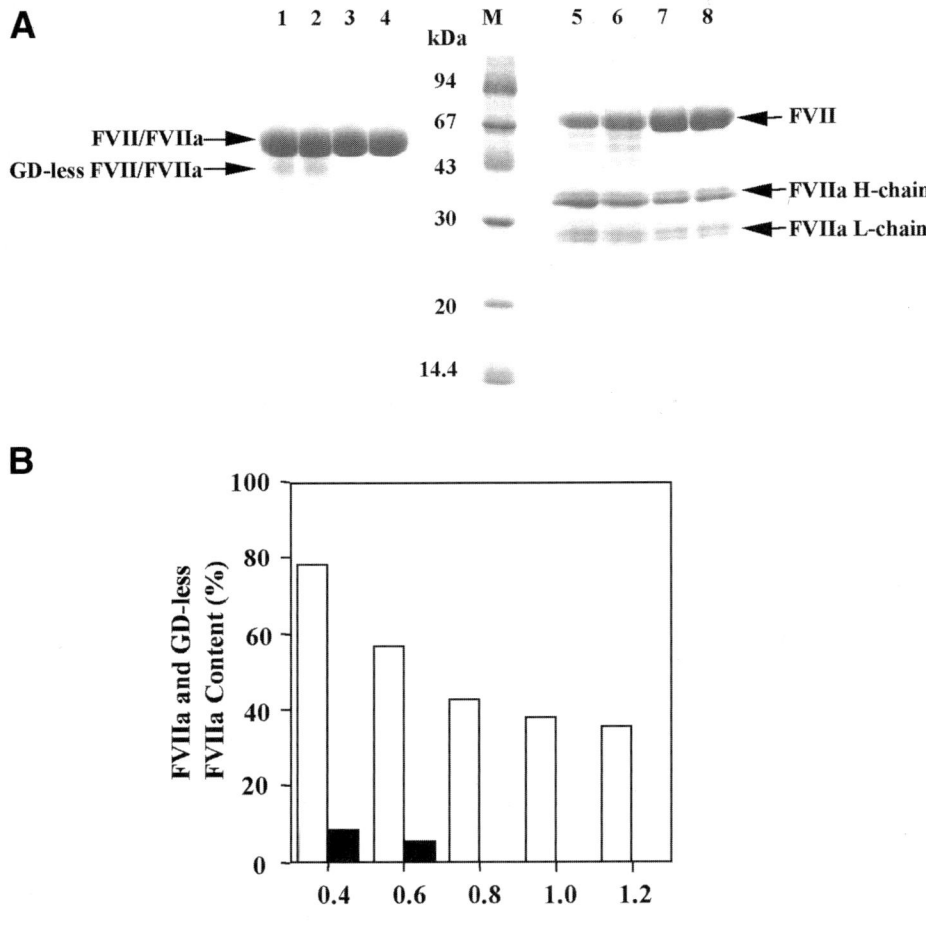

Fig. 3. Partial activation of FVII by small-scale anion exchange chromatography. (A) SDS-PAGE analyses of FVIIa under nonreducing (lanes 1–4) and reducing (lanes 5–8) conditions. Seventy-five milliliters of the filtrate with Planova® 15 N ($A_{280}$ = 0.260) were applied to a DEAE-FF (∅ 0.5 cm × 5.0 cm) column at 10°C. After washing with 10-column volumes of equilibration buffer, 50 m$M$ Tris-HCl, pH 7.8, containing 30 m$M$ NaCl, at 1.0 mL/min, proteins were eluted with the buffer containing 1.75 m$M$ CaCl$_2$ at various flow rates. (A) The results of SDS-PAGE analyses (12.5% acrylamide): lanes 1 and 5, 0.4 mL/min; lanes 2 and 6, 0.6 mL/min; lanes 3 and 7, 0.8 mL/min; lanes 4 and 8, 1.0 mL/min. (B) Amounts of FVIIa and GD-less FVIIa/FVII generated were quantified by densitometry of the light (25 kDa) and heavy (35 kDa) chains of SDS-PAGE under reducing conditions and GD-less FVIIa/FVII (47 kDa) under nonreducing conditions, respectively. The generation of FVIIa (□) is expressed as the percent ratio of the sum of light and heavy chains and GD-less FVIIa/FVII (■) is expressed as the percent ratio of the 47-kDa band.

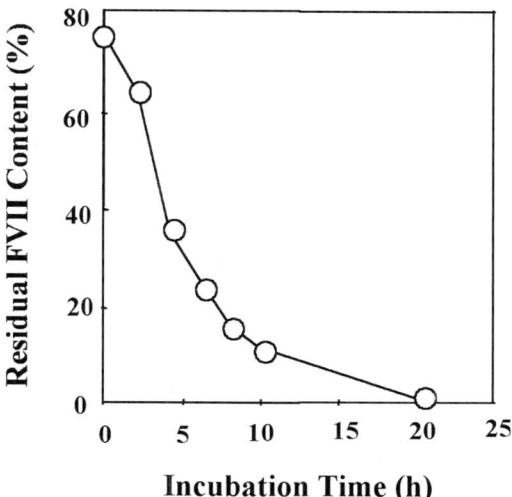

**Incubation Time (h)**

Fig. 4. Full activation of residual FVII in the FVII and FVIIa mixture. To optimize full activation of FVII in the FVII and FVIIa mixture, the FVIIa solution was mixed with the FVII solution at a percent ratio of 25%, and the mixture at a total protein concentration of 1.5 mg/mL was then incubated at 10°C for 20 h. The residual FVII content (% of the ratio of FVII and FVIIa) in the mixture was determined of the specified incubation times by densitometry on SDS-PAGE under reducing conditions as described in the legend of **Fig. 3**.

## Acknowledgments

This work was supported by the Blood Research Department (The Chemo-Sero-Therapeutic Research Institute, Kaketsuken, Japan). We gratefully acknowledge Dr. Sadaaki Iwanaga for his invaluable suggestions regarding our research and review of the text.

## References

1. Lusher, J. M., Roberts, H. R., Davignon, G., Joist, J. H., Smith, H., Shapiro, A., et al. (1998) A randomized, double-blind comparison of two dosage levels of recombinant factor VIIa in the treatment of joint, muscle and mucocutaneous haemorrhages in persons with haemophilia A and B, with and without inhibitors. rFVIIa Study Group. *Haemophilia* **4,** 790–798.
2. Bjoern, S. and Thim, L. (1986) Activation of coagulation factor VII to VIIa. *Research Disclosure* **269,** 564–565.
3. Pedersen, A. H., Lund-Hansen, T., Bisgaard-Frantzen, H., Olsen, F., and Petersen, L. C. (1989) Autoactivation of human recombinant coagulation factor VII. *Biochemistry* **28,** 9331–9336.
4. Tomokiyo, K., Yano, H., Imamura, Y., Nakano, Y., Nakagaki, T., Ogata, Y., et al. (2003) Large-scale production and properties of human plasma-derived activated Factor VII concentrate. *Vox Sang.* **84,** 54–64.
5. March, S. C., Parikh, I., and Cuatrecasas, P. (1974) A simplified method for cyanogen bromide activation of agarose for affinity chromatography. *Anal. Biochem.* **60,** 149–152.

6. Neuenschwander, P. F. and Morrissey, J. H. (1992) Deletion of the membrane anchoring region of tissue factor abolishes autoactivation of factor VII but not cofactor function. Analysis of a mutant with a selective deficiency in activity. *J. Biol. Chem.* **267,** 14477–14482.

7. Sugo, T., Mizuguchi, J., Kamikubo, Y., and Matsuda, M. (1990) Anti-human factor IX monoclonal antibodies specific for calcium ion-induced conformations. *Thromb. Res.* **58,** 603–614.

8. Laemmli, U. K. (1970) Cleavage of structural proteins during the assembly of the head of bacteriophage T4. *Nature* **227,** 680–685.

9. Bajaj, S. P., Rapaport, S. I., and Brown, S. F. (1981) Isolation and characterization of human factor VII. Activation of factor VII by factor Xa. *J. Biol. Chem.* **256,** 253–259.

10. Koito, A., Shirono, K., Suto, H., Matsushita, S., Hattori, T., and Takatsuki, K. (1987) Evaluation of the safety of blood products with respect to human immunodeficiency virus infection by using an HTLV-I-infected cell line (SKT-1B). *Jpn. J. Cancer. Res. (Gann)* **78,** 365–371.

# 16

## Purification of Clinical-Grade Monoclonal Antibodies by Chromatographic Methods

### Alberto L. Horenstein, Ilaria Durelli, and Fabio Malavasi

### 1. Introduction

Monoclonal antibodies (MAbs)—reagents mass-produced in the laboratory to recognize a specific molecular target *(1)*—are currently the most widely used protein-based therapeutic and diagnostic molecules in clinical trials *(2)*. This chapter presents the most effective ready-to-use protocol for MAb purification of clinical-grade standard. Samples of high purity are often produced via a highly selective single step. However, an appropriate combination of chromatographic materials in at least two orthogonal (i.e., based on two different mechanisms) steps is mandatory for obtaining clinical-grade MAbs for the in vitro functional analysis of ligand signals on cell-based assays and protein expression changes on arrayed MAb biochips *(3)*, or in vivo preclinical and therapeutic applications *(4)*.

Murine MAbs are obtained from tissue culture supernatants harvested from the hybridoma culture or from ascitic fluid produced by hybridoma cells grown as tumors in syngenic mice *(5)*. Cell culture systems have the advantage of producing material in unlimited volume and quantity, whereas production from ascitic fluid is limited in certain countries because of legal restrictions. Of the various factors that should be taken into consideration when planning purification, such as production yield, physicochemical characteristics of MAb, establishing the degree of purity required for the intended application, and scale of purification, is of utmost importance.

The clinical-grade purification of MAbs, ranging from milligram-to-gram scale per batch, involves several separation tasks, which can be clustered into two phases: upstream and downstream processing (**Fig. 1**). Quality-control tests of purified MAbs are carried out as a final stage. The protocols outlined yield an efficient amount of product in a reasonable turnaround time. Assuming that an high-performance liquid chromatography (HPLC; **Subheading 3.2.1.**) is available and operating properly, the timings required for the main steps of purification are:

Clarification (**Subheading 3.1.1.**): 2 h
Concentration (**Subheading 3.1.2.**): 3 h

From: *Methods in Molecular Biology, vol. 308: Therapeutic Proteins: Methods and Protocols*
Edited by: C. M. Smales and D. C. James © Humana Press Inc., Totowa, NJ

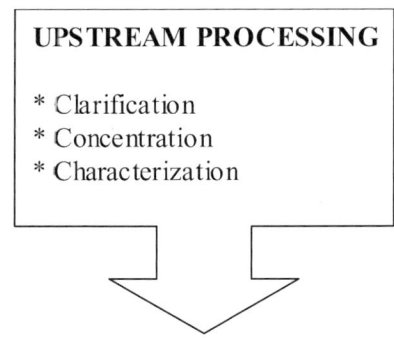

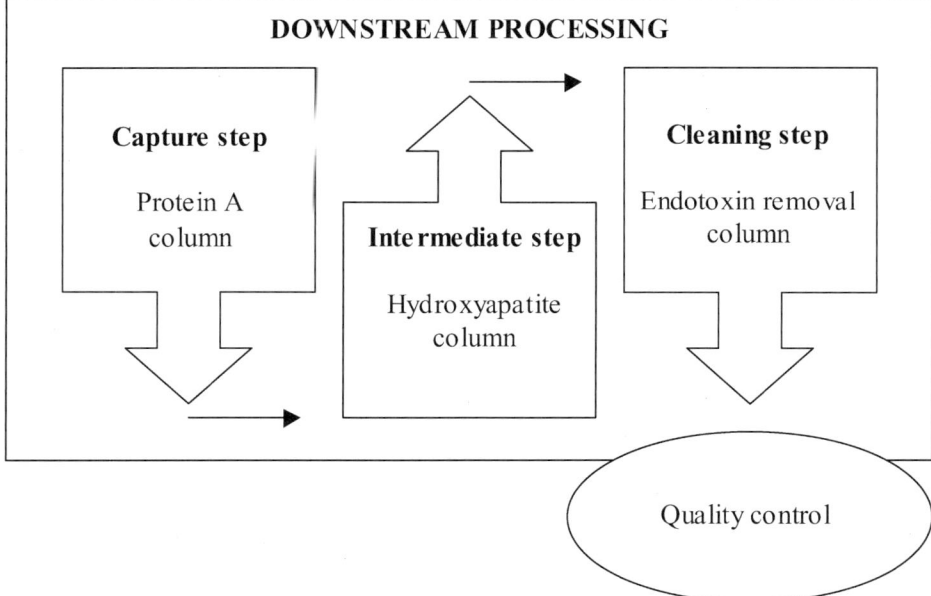

Fig. 1. Chromatographic purification of clinical-grade MAb. MAb harvested from supernatants of murine hybridomas and purified to clinical grade. The details for upstream and downstream procedures are outlined in the text.

Purification by protein A chromatography (**Subheading 3.2.2.**): 3 h
Diafiltration (**Subheading 3.2.2.4.**): 1 h
Purification by hydroxyapatite chromatography (**Subheading 3.2.3.**): 3 h
Purification by endotoxin removal gel chromatography (**Subheading 3.2.4.1.**): 3 h

## 2. Materials

### 2.1. General Reagents and Materials

1. Ultra-pure distilled water (e.g., Milli-Q generating system).
2. Sterile, pyrogen-free water, double-distilled (dd) ($H_2O$).

3. Methanol of HPLC grade.
4. E-Toxa-Clean (Sigma).
5. Salts and other chemicals of analytical grade.

## 2.2. General Equipment

1. Chromatographic columns (e.g., Glass Bio-Scale MT-10 column, Bio-Rad; 150-mm height × 10-mm internal diameter).
2. Chromatographic equipment (HPLC system; **Subheading 3.2.1.**).
3. Ultraviolet (UV) Spectrophotometer and quartz cuvets.
4. Class 2 laminar flow workstation.
5. Peristaltic pumps (1–2.5 L/min flow capability).
6. Centrifuge to spin up to 15,000$g$ and 40,000$g$ with fixed angle and swing-out rotors.

## 2.3. Upstream Processing

1. Sterile apyrogenic glass bottles (500 and 1000 mL).
2. Sterile apyrogenic 250-mL bottles (e.g., Nalgene bottles).
3. Sterile top-bottle filters: 0.22 and 0.45 μm.
4. Sterile syringe filters: 0.22, 0.45, and 1.20 μm.
5. 1 $M$ NaOH.
6. Sterile test tubes for ascitic fluid collection.
7. Sterile 18-gauge needles.
8. 50-mL tubes (e.g., Nalgene tubes).
9. 10.4-mL tubes (e.g., Beckman polycarbonate tubes). Soak tubes with 1% E-Toxa-Clean solution overnight. After depyrogenization, rinse the tubes with sterile, pyrogen-free ddH$_2$O and autoclave.
10. Pasteur pipets.
11. Clarified cell culture supernatant (**Subheading 3.1.1.1.**).
12. Hollow fiber membrane cartridge (30–50-kDa NMWC, e.g., Amersham Biosciences, NJ). Alternatively, select a UF cassette (30–50-kDa NMWC, e.g., Ultrasette, Filtron, Pall). Choose a membrane that is 3–5 times smaller in NMWC than the molecular weight of the MAb.
13. 150 m$M$ NaCl.
14. Glass microscopy slides.
15. Phosphate-buffered saline (PBS), pH 7.2, with 0.1% sodium azide.
16. 1% Agarose in PBS.
17. Cell culture supernatant (overgrown or 10X concentrated) or ascitic fluid (undiluted).
18. Specific isotyping anti-immunoglobulin (Ig) murine antibodies.
19. Polyethylene glycol (PEG) 6000.

## 2.4. Downstream Processing

All buffers are prepared with sterile and pyrogen-free ddH$_2$O, filtered through a 0.22-μm filter, and degassed prior to use. Store at 4°C for up to 1–2 wk after filtration through sterile units.

1. Bottles (500 and 1000 mL).
2. Sterile, pyrogen-free tubes (15 and 50 mL; e.g., Corning).
3. Top-bottle and syringe filters (0.22 and 0.45 μm; Corning).
4. Recombinant protein A-Sepharose Fast Flow (IPA 400 HC) resin, designed for low-ligand leakage (<5 ng protein A in 1.0-mg/mL antibody product), MabSelect, or other commercial sources may be used as bulk support media (**Subheading 3.2.2.**).

5.  Binding buffer: 1.5 $M$ glycine/NaOH and 3 $M$ NaCl, pH 8.9. Add 112.6 g glycine (free base; $M_r$ 75.07), 175.3 g NaCl ($M_r$ 58.44), and 4.0 g NaOH to 800 mL ddH$_2$O. If necessary, titrate with 5 $M$ NaOH to pH 8.9. Make up final volume to 1 L with ddH$_2$O.

6.  Elution buffer: 0.1 $M$ Na-citrate buffer, pH 3.5. Add 16.05 g citric acid (anhydrous; $M_r$ 192.1), 4.41 g trisodium citrate (dihydrate; $M_r$ 294.1), 2.0 g NaOH, and enough ddH$_2$O to make 1 L.

7.  Neutralization buffer: 1 $M$ Tris-HCl buffer, pH 8.0. Add 141 g Tris(hydroxymethyl)-aminomethane to 800 mL ddH$_2$O. Add 100 mL 5 $M$ HCl. Mix well and make up final volume to 1 L with ddH$_2$O.

8.  Sterile PBS, pH 7.2.

9.  Centricon Plus-80, 30-kDa molecular weight cut-off (MWCO; Millipore, Bedford, MA).

10. 10 m$M$ Sodium phosphate, pH 6.7 (P10).

11. Sterile pyrogen-free pipets: 1, 5, and 10 mL.

12. Sterile syringe filters: 0.22 and 0.45 μm.

13. BioGel HT Hydroxyapatite or Macro-Prep ceramic Hydroxyapatite-Type I, 20 μm (Bio-Rad Labs, Hercules, CA).

14. Stock solution of phosphate buffer: 1.0 $M$ sodium phosphate, pH 6.7. Dissolve 137.99 g NaH$_2$PO$_4$·H$_2$O in approx 1 L ddH$_2$O. When fully dissolved, bring the volume to 1.5 L, and slowly add 268.07 g Na$_2$HPO$_4$·7H$_2$O with stirring. When fully dissolved, make up to 2.0 L with ddH$_2$O, and filter through a 0.22-μm membrane filter. Upon dilution, the pH should be 6.7 ± 0.05.

15. 400 m$M$ Sodium phosphate, pH 6.7 (P400).

16. 200 m$M$ Sodium phosphate, pH 9.0 (P200).

17. 0.1 $M$ PBS, pH 7.2.

18. Chromatographic column (e.g., fast protein liquid chromatography [FPLC] glass column, 150-mm height × 15-mm internal diameter).

19. Chromatographic supports: Actigel ALD (Acticlean Etox, Sterogene Bioseparations Inc., CA), Affi-Prep Polymyxin Matrix (Bio-Rad), Detoxi-Gel (Pierce, Rockford, IL).

20. 150 m$M$ NaCl.

21. 0.1 and 1.0 $M$ NaOH.

22. 1% Deoxycholate.

23. Single-test glass vials of "Pyrotell" test kits (0.250-endotoxin unit [EU]/mL sensitivity; Bio-Whittaker Inc., Walkersville, MD).

24. Endotoxin-free water (*Limulus* Amebocyte Lysate [LAL] Reagent Water, Bio-Whittaker).

25. MAb sample (1 mg/mL concentration) in endotoxin-free water.

26. Hydrophilic polyvinylidene fluoride membrane (e.g., Posydine syringe filter, Pall, Ann Arbor, MI; or Sartobind Membrane Adsorber, Sartorius AG, Goettingen, Germany). Membrane filtration may be used manually with a Luer Lock apyrogenic syringe or an FPLC chromatographic system.

## 3. Methods

### *3.1. Upstream Processing*

The upstream process involves the separation of cells and suspended solids from MAb sourcing and consists of (1) clarification, (2) concentration, and (3) characterization of Igs in the source material.

### 3.1.1. Clarification

Hybridoma culture supernatant and ascitic fluid clarification (removal of visible particulates >1 μm) involves standard laboratory techniques, such as centrifugation and filtration, to minimize particulate clogging and to maintain the integrity of the liquid chromatography (LC) columns. Prefiltration for the removal of large particulates (>5 μm) that may clog smaller pore-size filters is also recommended.

#### 3.1.1.1. HYBRIDOMA CULTURE SUPERNATANT

1. Soak glassware or other materials not guaranteed to be endotoxin-free in 1.0 *M* NaOH overnight. After depyrogenization, rinse the bottles with sterile, pyrogen-free ddH$_2$O and autoclave (*see* **Notes 1–3**).
2. Harvest hybridoma culture supernatant once the cells (density range of 3–10 × 10$^6$ cell/mL) produce MAb at a concentration of 50–100 mg IgG/L.
3. Pool the harvested supernatant (2–4 L) to obtain a single batch for purification.
4. Centrifuge at 10,000*g* for 15 min to remove cell debris, cells, and particulate matter.
5. Filter cell culture supernatant through a 0.22-μm filter before concentration (**Subheading 3.1.2.**) or purification (**Subheading 3.2.**).
6. Store the solution at 4°C overnight, and filter again by 0.22 μm before use.

#### 3.1.1.2. ASCITIC FLUID

1. Collect ascitic fluids from mice by tapping with an 18-gauge needle. Add CaCl$_2$ to 1 m*M* final concentration to convert fibrinogen into fibrin.
2. After 2-h incubation at room temperature, carefully detach the fibrin clot from the side of the vessel and incubate 12 h further at 2–8°C.
3. Remove the clot and cell debris by centrifugation for 30 min at 10,000*g* at 2–8°C.
4. Collect the supernatant and centrifuge for 2 h at 40,000*g* at 2–8°C.
5. Remove lipids that form a layer after centrifugation by aspiration with a Pasteur pipet.
6. Filter ascitic fluids through a 0.45-μm filter, then a 0.22-μm sterile syringe filter prior to purification (**Subheading 3.2.**), or freeze the sample at –80°C.

### 3.1.2. Concentration

#### 3.1.2.1. HOLLOW FIBER SYSTEM AND TANGENTIAL FLOW ULTRAFILTRATION

For larger volumes with low MAb concentrations (e.g., cell culture supernatant), the sample can be concentrated by circulating it through a hollow fiber membrane cartridge (**Fig. 2**), allowing the slight pressure of circulation to force water out through the membrane, causing rapid concentration of the MAb up to 50-fold. As filtrate (permeate) is removed, the supernatant in the retentate vessel becomes more concentrated in the components retained by the membrane according to the MWCO selected. The advantage of this method is that very large volumes can be processed using sterile and endotoxin-free components (*see* **Notes 4–8**).

1. Set up equipment according to the manufacturer's directions, connecting the peristaltic pump and hollow fiber or ultrafiltration (UF) cassette.
2. Sanitize and depyrogenate the equipment at a flow rate of 1 L/min by sequentially washing the device with ddH$_2$O, 150 m*M* NaCl, 1 *M* NaOH, and pyrogen-free ddH$_2$O.
3. Measure the pH. If it is above 9, then wash with 150 m*M* NaCl until neutral.

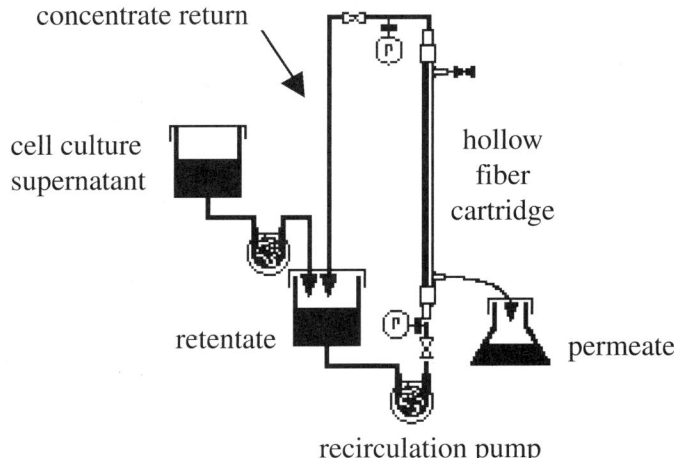

Fig. 2. Concentration system utilizing ultrafiltration. The cell culture supernatant is fed to a reservoir, which is then circulated through a hollow fiber cartridge. Supernatant flow continuously, both via a reject line in the retentate vessel and a shunt in the permeate vessel. Pump, valves, and pressure gauge are shown.

4. Concentrate the clarified cell culture supernatant according to the manufacturer's instructions.
5. Filter the cell culture supernatant in the retentate vessel (20–40X concentrated) through a 0.22-μm filter.
6. Sanitize the device (**step 2**), and store it at 4°C.

## 3.1.3. Immunoglobulin Characterization and Sourcing Storage

### 3.1.3.1. MAB ISOTYPING: OUCHTERLONY TECHNIQUE

MAbs are members of the Ig family of molecules that constitute the humoral branch of the immune system. The biochemical properties of Igs differ and determine how to purify the MAb. Thus, before proceeding with antibody purification, establish the Ig class and subclass or isotype, as defined by the sequence of amino acids found in the constant region. Igs are classified according to their heavy-[IgG ($\gamma$), IgA ($\alpha$), IgM ($\mu$), IgD ($\delta$), and IgE ($\epsilon$)] and light-chain components ($\kappa$ and $\lambda$). Murine IgG isotypes are designated as $IgG_1$, $IgG_{2a}$, $IgG_{2b}$, and $IgG_3$.

Several methods currently available can determine the isotypes. The Ouchterlony technique is relatively inexpensive and can be easily performed using standard laboratory equipment. The assay visualizes the antigen–antibody complexes formed after the diffusion of the MAb sample and the anti-IgG isotype antisera in a semisolid support.

1. Overgrow an aliquot of cells for 24 h in a 1-mL culture. Let the cells die, and spin them out.
2. Bring the agarose solution to 100°C, and allow to cool to 60°C. Add PEG 6000 at 1% final concentration.

**Table 1**
**MAb Purification Procedures**

| Methods | Ig | Yield | Purity | Ref. |
|---|---|---|---|---|
| Affinity chromatography | | | | |
| Protein A | $IgG_{2a/2b/3}$ | >95% | >95% | 7 |
| Protein G | $IgG_1$ | >95% | >95% | |
| Mixed-mode chromatography | | | | |
| Hydroxyapatite | IgG/IgM | 90% | | 12 |
| Abx | IgG/IgM | 80–90% | | 18 |
| Physicochemical-based chromatography | | | | |
| Hydrophobic charge induction | IgG | 76–90% | 69–90% | 19 |
| Cation exchange | IgG | 80% | 90% | 20 |
| Anion exchange | IgG | 90% | 90% | 21 |
| Size exclusion | IgM/IgG | 90% | 50–80% | 8 |

3. Working at 60°C, prepare a thin layer of agarose solution on a glass slide (~3 mL per glass slide).
4. Allow to harden in a humid chamber. Punch several wells in a circular pattern around a center well using a Pasteur pipet connected to a water pump.
5. Add the MAb-containing sample to the central well and isotyping antiserum to the peripheral wells.
6. As they diffuse, specific isotyping antibody will meet the Ig sample and form a precipitate. The precipitation line that occurs between the sample well and an antiserum well indicates the isotype.
7. Wash with PBS, changing the buffer every 2–3 h for 12 h, and stain for proteins using Coomassie blue.

### 3.1.3.2. STORAGE OF MAB SOURCING

Culture supernatants and ascitic fluid should be kept frozen in small aliquots at –20°C. Avoid repeated freeze/thawing, which may reduce biological activity. To minimize bacterial growth and protease activity, keep refrigerated at 4°C in a closed sterile vessel. After 24 h at 4°C, add a preserving agent if possible (e.g., 0.01% merthiolate). Sodium azide can interfere with some biological assays and can be a health hazard.

## 3.2. Downstream Processing

Large-scale MAb purification from clarified cell culture supernatants or ascitic fluids requires the use of LC. The various retention modes of LC, including affinity, mixed-mode, and physicochemical chromatographic mechanisms, have been in use for decades (**Table 1**). To obtain clinical-grade MAb, two or more of these techniques need to be combined effectively. A three-step strategy is outlined as a purification protocol with (1) a capture step—the MAb is separated from other sample components; (2) an intermediate step—the MAb is isolated from contaminants similar in size or other biochemical properties; and (3) a cleaning step—the complete removal of trace contaminants (**Fig. 1**). These steps use the resolving power, precision, and speed of HPLC.

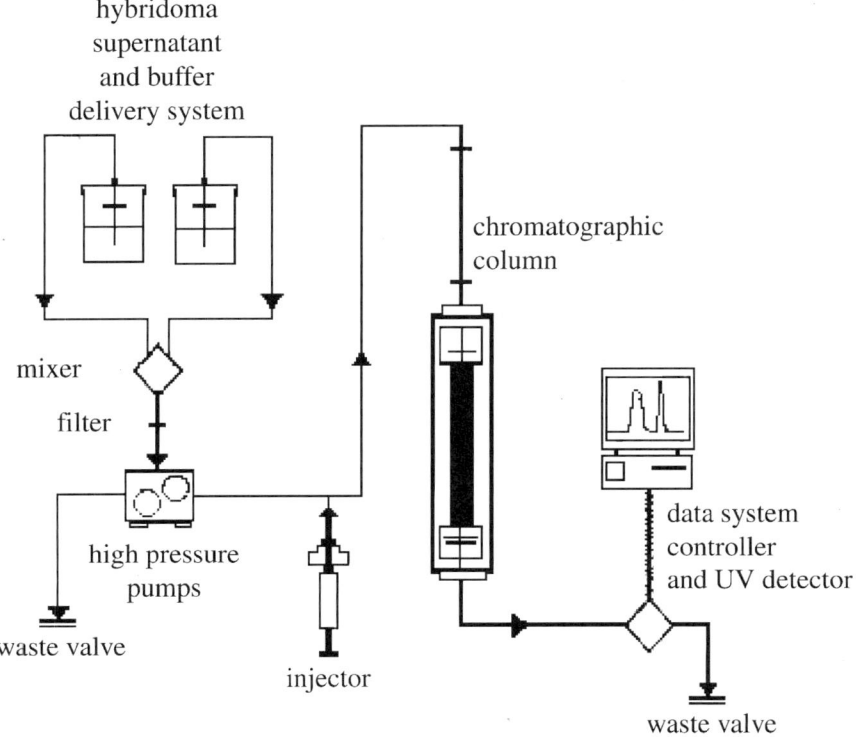

Fig. 3. Schematic diagram of the HPLC system.

### 3.2.1. General Principles for HPLC

Most purification procedures can be automated and improved by HPLC technology, which reduces purification time, increased resolution, and better data acquisition. More importantly, HPLC is a useful system to optimize a semipreparative MAb purification process (gram amounts of purified product), relying on the use of rigid, small-particle matrices at high pressure. Various HPLC equipments are commercially available: the ÄKTA system (Amersham Biosciences, NJ), Beckman system (Beckman Instruments, Fullerton, CA), and the Water system (Waters Corporation, Milford, MA). We employ a Beckman HPLC system (**Fig. 3**), consisting of a Programmable Solvent Module 166 (outfitted with two pumps able to work simultaneously and at high pressure), a Programmable Detector Module 166 NM set at 280 nm that allows the purification data to be obtained in real time, and a Solvent delivery system with an automatic analytical injector and a prep-load pump accessory for rapid, large-volume sample loading. Chromatographic data is collected and processed using System Gold software. The computer-based data acquisition system allows modifications to the conditions of the chromatographic run (flow speed and percentage of buffers injected through pumps) and scaling and replotting of data.

**Table 2**
**Characteristics of Protein A Affinity Media**

| Sorbent | Matrix | Manufacturer |
| --- | --- | --- |
| rProtein A Sepharose FF | Crosslinked agarose | Amersham Biosciences, Uppsala, Sweden |
| Protein A Sepharose 4 FF | Crosslinked agarose | Amersham Biosciences |
| MabSelect | Crosslinked agarose | Amersham Biosciences |
| IPA-400 | Crosslinked agarose | RepliGen Corp., Waltham, MA |
| Protein A Ceramic HyperD | Ceramic polyacrylamide | Biosepra, Cedex, France |
| Prosep-A/rA High Capacity | Porous glass | Bioprocessing (Millipore) Watford, UK |
| Poros 50 A High Capacity | Polystyrene divinylbenzene | PerSeptive Biosystems, MA |
| UltraLink Immobilized ProtA | Polymeric | Pierce, Rockford, IL |
| Affi-Gel Protein A Gel | Crosslinked agarose | Bio-Rad, Hercules, CA |
| Affi-Prep Protein A Support | Polymeric | Bio-Rad |
| Protein A Agarose 4XL | Crosslinked agarose | Affinity Chromat. Ltd., Freeport, UK |
| Protein A Cellthru | Not available | Sterogen Biosep., Carlsbad, CA |
| AF-Protein A Toyopearl 650M | Polymethacrylate | Tosoh Biosep, Stuttgart, Germany |

The general operating procedure for HPLC chromatography involves: column matrix and buffer selection; matrix equilibration for column packing; equipment setup; sample loading; column washing; elution of antibodies; cleaning of the column and system; and storage (*see* **Notes 9–11**).

### 3.2.2. Capture Step: Protein A Affinity Chromatography

Protein A chromatography is a type of affinity chromatography that relies on the specific interaction of protein A with the Fc region of Ig. The affinity of protein A (originally isolated from bacterial surface proteins) for IgG depends on source species, antibody subclass, and changes in pH.

Protein A is available from several commercial suppliers coupled to different matrices (**Table 2**). The choice of the protein A resin depends on a compromise between loading capacity, flow characteristics, and eluate purity, with crosslinked agarose or polyacrylamide beads as the best choice for HPLC purification. Some resins are unsuitable for semipreparative applications, for their residual bovine serum albumin and IgG (aggregated/fragmented), difficulty in regeneration, or low-binding capacity (**7**).

The protocol applied as a capture step for clinical-grade MAb purification includes three stages: (1) adsorption of the MAb to the affinity matrix at neutral or slightly basic pH values; (2) removing nonspecifically adsorbed proteins by washing; and (3) elution of the MAb by lowering the pH. To retain the native conformation of the MAb, the pH of the solution is adjusted to a neutral value immediately after elution (*see* **Notes 12–18**).

### 3.2.2.1. COLUMN PACKING AND SANITIZATION

1. Prepare the column according to the manufacturer's instructions.
2. Alternatively, we find it much easier to insert the bottom frit and distribution screen and to place the internal glass column into a 50-mL tube containing enough PBS to form a cushion of about 1–2 mL. Ensure no air is trapped under the column net.
3. Resuspend the degassed protein A and, after inverting and rotating the bottle, fill the column with the required volume of slurry.
4. Immediately fill the column with PBS, and allow the slurry to settle. Do not allow the resin to run completely dry. If necessary, add more slurry to the desired level.
5. Without inserting the upper frit and distribution screen, assemble the column. Mount the column bottom and top pieces and connect to a pump. Begin the flow rate at 1 mL/min for at least 3 column volumes (cv) after a constant bed height of 9–10 cm is obtained.
6. Stop the pump, then close the column outlet. Remove the top piece, and carefully pipet a 0.5-cm layer of buffer to form an upward meniscus at the top of the gel.
7. Insert the upper frit and distribution screen, ensuring that no air is trapped under the net. Assemble the column.
8. Pass 5 cv PBS through the column at a flow rate of 2.5 mL/min.
9. If necessary, stop the flow and lower the upper bed support via the lock nut to remove any head space that might have formed between the frit and the top of the bed.
10. Equilibrate the column by pumping binding buffer (**Subheading 2.4.**) at 1 mL/min. A column is equilibrated when HPLC pressure and detector $A_{280}$ nm are constant.
11. For column cleaning in place (CIP) wash sequentially with: 5 cv sterile, pyrogen-free ddH$_2$O, 5 cv 1 $M$ NaOH, and 5 cv sterile, pyrogen-free ddH$_2$O.
12. For long storage, wash the column with 20% methanol (5 cv for packed media), and store at 4°C.

### 3.2.2.2. MAB PURIFICATION

1. Bring all materials to room temperature.
2. Equilibrate the column with 5 cv binding buffer or until the baseline is flat.
3. Adjust the clarified cell culture supernatant (**Subheading 3.1.1.**) to the pH and ionic strength of the binding buffer by 1:3 dilution in binding buffer or the addition of 112.6 g/L glycine, 175 g/L NaCl, and 3 g/L NaOH to give a pH of 8.9. (Store the solution at 4°C overnight without stirring and filter through a 0.22-µm filter before loading in the column.) If using ascitic fluid, dilute 1:3 in binding buffer, and filter through a 0.45-µm filter, then a 0.22-µm sterile syringe filter prior to loading in column.
4. Load sample (10 mg Ig/mL resin), setting the flow rate during the initial stage at 2 mL/min, followed by a lower flow rate (1 mL/min). Continue with 5 cv binding buffer wash to remove impurities and unbound material (e.g., *see* **Fig. 4**). Collect unbound proteins (flow-through) until absorbance at 280 nm drops to baseline. Always retain the flow-through material for possible reprocessing.
5. Elute Ig with 1–3 cv elution buffer, separating eventual peaks (**Fig. 4**). Because elution conditions are quite harsh, collect fractions into a neutralization buffer (1–2 mL 1 $M$ Tris-HCl, pH 8.0 per 50-mL fraction), so that the final pH of the fractions will be approximately neutral.
6. Store the eluted Ig sample at –20°C or diafiltrate against 10 m$M$ sodium phosphate buffer (hydroxyapatite column starting buffer) using a Centricon Plus-80 (**Subheading 3.2.2.4.**).

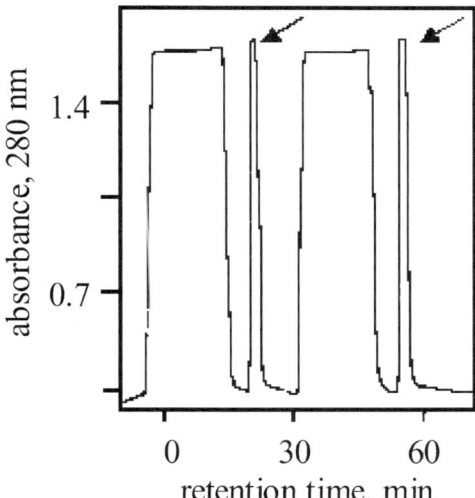

Fig. 4. HPLC elution profile of MAb purification on protein A affinity chromatography. A concentrated hybridoma culture supernatant is loaded and eluted at a flow rate of 2 mL/min using 1.5 $M$ glycine/3 $M$ NaCl, pH 8.9 as binding buffer. The capture-step chromatogram shows the flow-through peaks and sharp elution peaks of MAb (arrows) with sodium citrate buffer, pH 3.5.

7. Regenerate the column with 5–10 cv binding buffer. Check pH of effluent to ensure that the column is equilibrated.
8. Proceed with CIP column and storage if no more runs are to be performed, washing sequentially with 2 cv sterile, pyrogen-free ddH$_2$O, 2 cv 1 $M$ NaOH, and 2 cv sterile, pyrogen-free ddH$_2$O.

### 3.2.2.3. PROTEIN A LEAKAGE

Protein A affinity chromatography poses the risk of ligand leakage from the matrix, but at laboratory scale, this is not a significant problem. However, for purification of clinical-grade MAb, the final product must be free of even trace amounts (>10 ppm/dose) of ligand, as protein A may cause immunogenic or other physiological responses in humans *(9)*.

Most of the protein A that might leak is removed from the sample during the intermediate step of purification with hydroxyapatite *(10)*. Thus, the levels of protein A remaining to be detected are quite low. The enzyme-linked immunosorbent assay kit from RepliGen (RepliGen Corporation, Waltham, MA) is simple and accurate *(8)*.

### 3.2.2.4. SAMPLE DIAFILTRATION

Immunoglobulins eluted from the capture step on protein A are diafiltrated (simultaneous dialysis and concentration) against the hydroxyapatite column starting buffer using a Centricon concentrator.

1. Add the antibody solution to the sample filter cup of a Centricon C-80. Centrifuge at 3500*g* for 15–20 min.
2. Remove the filtrate, and add P10 solution to the sample filter cup. Retain the filtrate to detect the presence of MAb.
3. Centrifuge at 3500*g* for 10 min.
4. Recover the concentrate in the minimum volume of P10. Invert the unit and spin at no more than 1000*g* for 5 min.
5. Remove the sample with a sterile, pyrogen-free pipet.
6. Filter on a 0.22-μm membrane and store at 4°C.

## 3.2.3. Intermediate Step: Hydroxyapatite Chromatography

The specific MAb present in the purified IgG pool (**Subheading 3.2.2.2.**) can be isolated from other proteins by a subsequent passage through an hydroxyapatite column *(11)*. Hydroxyapatite (HA), a calcium phosphate-based matrix, provides excellent resolution of proteins using gentle separation conditions. The HA:MAb interactions are complex. Significant features are the attraction of amino groups of basic proteins to the phosphate moieties (P sites), while acidic proteins are retained by the calcium cations (C sites) of the HA. Associated to the geometric distribution of charges, this mechanism imparts a unique stereochemical property to the resin that endows HA chromatography with the ability to discriminate light-chain idiotypes from MAb mixtures with common heavy chains *(12)*. HA chromatography also supports both viral clearance and endotoxin removal. Furthermore, HA is capable of removing aggregation or degradation products *(13)*.

The use of HA—whether crystalline or ceramic—is a simple and straightforward process. The sample is loaded in low-ionic strength sodium phosphate buffer (1–10 m*M*) near neutral pH. Elution is normally achieved with a gradient of sodium phosphate buffer (10–400 m*M*) of the same pH (*see* **Notes 19–22**).

### 3.2.3.1. COLUMN PACKING AND SANITIZATION

1. Prepare the column according to the manufacturer's instructions.
2. Alternatively, insert only the bottom frit and distribution screen, and put the internal glass column into a 50-mL tube containing enough buffer to form a cushion of about 1–2 mL. Ensure no air is trapped under the column net.
3. Set up the column making a 20% (v/v) slurry of HA ceramic type I particles in P200 or BioGel HA HT in P10.
4. Degas the solution and, after gentle stirring, fill the column with the required volume of slurry.
5. Once the slurry has settled, carefully pipet the 0.5-cm layer of buffer to form an upward meniscus at the top of the gel. Do not allow the resin to run completely dry. If necessary, add more slurry to the desired level.
6. Insert the upper frit and distribution screen, ensuring that no air is trapped under the net, and assemble the column.
7. Set the HPLC-operating conditions at a flow rate of 2 mL/min for approx 2 h after a constant 9–10-cm bed height is obtained.
8. If necessary, stop the flow and lower the upper bed support via the lock nut to remove any head space that might have formed between the frit and the top of the bed.
9. Equilibrate the column by pumping P10-binding buffer at 2 mL/min. A column is equilibrated when pressure and $A_{280}$ nm are constant.

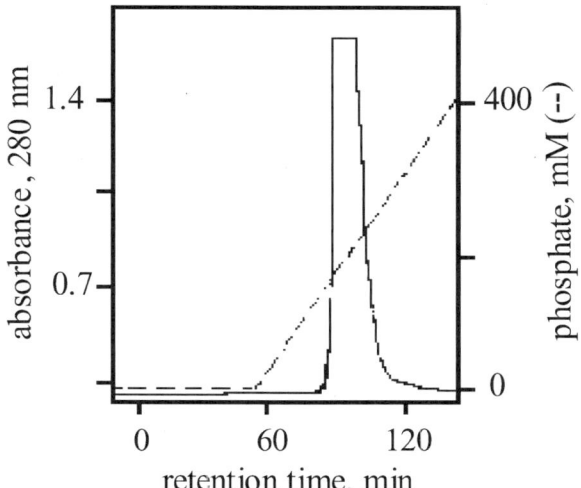

Fig. 5. HPLC elution profile of MAb purification on HA chromatography. The second purification step performed on a hydroxyapatite column at a flow rate of 4 mL/min. MAb is eluted with a 10 cv of phosphate gradient (10–400 m*M*), pH 6.7.

10. For CIP column and storage, wash sequentially with 5 cv sterile, pyrogen-free P10, 5 cv 0.1 *M* NaOH, and 5 cv sterile, pyrogen-free P10, and store at 4°C.
11. The results of column packing are evaluated by protein analysis of the nonretained peak, i.e., the flow-through of the column.

### 3.2.3.2. MAB PURIFICATION

1. Equilibrate the chromatographic column at room temperature with 5 cv starting buffer P10.
2. Load the column with 100 mL sample at an approx 2 mg/mL concentration at a flow rate of 4 mL/min. At this loading protein concentration and injection flow rate, the total MAb amount loaded must all be bound to the HA matrix.
3. Wash with 5 cv P10—the point at which the UV baseline should be stable.
4. Elute the MAb with 10 cv gradient from P10 to P400 at a flow rate of 4 mL/min (e.g., *see* **Fig. 5**).
5. Throughout the aforementioned procedures, IgG concentration is monitored and measured by $A_{280}$ nm. Calculations regarding retention times, peak heights, and relative peak areas are made using computer software that integrates the antibody peak obtained and displays the results immediately after the purification run has finished.
6. Regenerate the HA column upon completion of the run with 5 cv P400 buffer.
7. For sanitizing chromatographic media and systems, rinse the column with 5 cv 0.1 *M* NaOH with a contact time of 30 min at a flow rate of 1 mL/min.
8. The peak corresponding to pure MAb is diafiltrated against 0.1 *M* PBS, pH 7.2, using a Centricon Plus-80 (**Subheading 3.2.2.4.**).
9. Filter the concentrate sample through a 0.22-μm disposable membrane syringe filter.

### 3.2.4. Cleaning Step

The purification of clinical-grade MAb addresses the removal of endotoxins (lipopolysaccharide, peptidoglycans, muramyl peptides, and other still unidentified substances) and viral particles *(5,14)*.

### 3.2.4.1. Endotoxin Removal Gel Chromatography

Endotoxins are highly toxic to mammalian cells and potent modulators of the immune system. A variety of methods are available for the removal of endotoxins, such as ultrafiltration, anion-exchange chromatography, hydrophobic interaction chromatography, sucrose gradient centrifugation, and extraction with detergents. However, all these methods have distinct disadvantages *(15)*. The purification protocol described previously provides a certain degree of endotoxin clearance, and sodium hydroxide is used to destroy endotoxins in laboratory equipment. However, clinical-grade quality is achieved only by endotoxin-selective affinity sorbents chromatography (*see* **Notes 23** and **24**).

1. Pack a 5 mL-bed volume chromatographic column with the endotoxin-selective affinity sorbent under FPLC conditions using the peristaltic pump.
2. Render pump, tubing, connecting pieces, and pyrogen-free column by treating with 3 cv 0.1 *M* NaOH, and allow to stand at 4°C overnight. Wash with pyrogen-free ddH$_2$O until neutral.
3. Equilibrate the column with pyrogen-free PBS, pH 7.2. Check that the eluate is neutral and pyrogen-free using LAL gel-clot test (**Subheading 3.2.4.2.**).
4. Pass the 0.22-µm filtered MAb preparation through the column at a flow rate of 1 mL/min under sterile FPLC conditions.
5. Repeat the process at least twice to increase endotoxin-binding efficiency.
6. Regenerate the column at the end of the procedure by perfusion with 20 cv 1 *M* NaOH (for Acticlean Etox), 1% deoxycholate in pyrogen-free ddH$_2$O (for Detoxi-gel), or 0.1 *M* NaOH (for Affi-Prep Polymyxin Matrix). Allow to stand at 4°C overnight, and wash with sterile pyrogen-free ddH$_2$O until neutral.
7. Test for the presence of residual endotoxin using the LAL gel-clot test (**Subheading 3.2.4.2.**).

### 3.2.4.2. Endotoxin Evaluation by LAL Test

The term "endotoxin unit" (EU) is used to describe the activity of endotoxins. The presence or absence of endotoxin is determined in the final product and buffers used in the purification procedures with the LAL gel-clot test *(16)*. The endotoxin content of a MAb specimen can be evaluated with a test of a given sensitivity, so-called "Pass/Fail test," which indicates whether or not the MAb sample has more or less endotoxin than the limit set by Food and Drug Administration (FDA) guidelines *(5)*. The limit for endotoxin is set at 50 EU/mg for MAb destined for cell-based assays, whereas endotoxins should always be less than 0.650 EU/mg for clinical-grade MAb (*see* **Note 25**).

1. Add 0.2 mL test sample and controls to single-glass vials containing the LAL reagent.
2. Vortex tubes 1–2 s.
3. Place the reaction tube in a 37 ± 1°C dry bath for 60 ± 2 min.
4. Remove and read reaction tubes one at a time, inverting them 180 degrees in one smooth motion.
5. A positive test is indicated by the formation of a gel that does not collapse when the tube is inverted.

**Table 3**
**Quality-Control Tests for Clinical-Grade MAb Evaluation**

| Analytical technique | MAb evaluation |
|---|---|
| $A_{280}$ nm absorbance | Protein concentration |
| Ouchterlony | Isotype |
| IEF[a] and SDS-PAGE[a] | Charge heterogeneity/denatured size |
| HPLC-SEC[b] | Aggregation/native size |
| Bioassays[a,b] | Biological activity/potency |
| Dot-blot[b] | Residual DNA |
| LAL test | Endotoxins |
| EIA[b] | Protein A leakage |

[a]Technical procedures carried out using standard methods *(22)*.
[b]Technical procedures carried out as described elsewhere *(8)*.

### 3.2.4.3. DNA CLEARANCE

The hybridoma producing the MAb may harbor viral particles. Protein A chromatography, followed by an elution step attained at low (3.5) pH and HA chromatography, provides some virus inactivation and clearance. Thus, an adequate clearance technique is important to the purification process. As it is a relatively innocuous method, filtration is included to remove viral particles during the purification regime using hydrophilic PVDF or ion-exchange membrane filtration. After this procedure, no DNA should be detected in the purified product *(17)*.

The detection of residual murine DNA in purified samples is performed using a dot-blot apparatus as described elsewhere *(6)*. The sensitivity of the assay allows the detection of 5 pg total murine DNA/mg protein.

## 3.3. Quality Control for Clinical-Grade MAb

A clinical-grade MAb is formulated as a protein solution containing $2.5 \pm 0.5$ mg/mL in PBS, pH 7.2, with no stabilizers or preservatives, aliquoted under sterile conditions and stored at –80°C. The analytical characterization of this sample typically includes the following parameters.

1. Evaluation of MAb concentration by UV absorbance. The $A_{280}$ of a 1-mg/mL sample of a solution of IgG MAb is 1.37 in a UV quartz cuvet of 1-cm path length.
2. Identification by sodium dodecyl sulfate-polyacrylamide gel electrophoresis (SDS-PAGE) of the migration pattern and isoelectric focusing (IEF) profile.
3. Analysis of purity and structural integrity by size-exclusion chromatography (SEC-HPLC).
4. Biological immunoreactivity, specificity, and potency of MAb binding to the target molecule.

**Table 3** summarizes quality-control evaluation of the purified clinical-grade MAb.

## 4. Notes

1. Wherever possible, use sterile, specifically endotoxin-free disposable plastics.
2. All materials that are not compatible with NaOH should be washed with a 1% E-Toxa-Clean solution overnight, rinsed with ddH$_2$O, and autoclaved.

3. Similarly, assume that any source of water or buffers is contaminated unless checked by standard LAL assay. All buffers should be less than 1 EU/mL (**Subheading 3.2.4.2.**).

4. MAb are more stable in relatively concentrated solutions (1–10 mg/mL).

5. The laboratory target process is generally a 20-fold concentration of 2–4 L cell culture supernatant containing 100–200 mg MAb.

6. Retain the permeate to analyze the presence of the MAb.

7. The hollow fiber cartridge (10-kDa NMWC subunit) is suited to depyrogenation of purified water, buffers, and solutions of small molecules.

8. Concentration of the Ig fraction from ascitic fluid or cell culture supernatant by ammonium sulfate, followed by centrifugation and extensive dialysis, is often used prior to LC purification procedures (**6**). However, like most precipitation techniques, it is unsuitable for large-scale preparation. Ammonium sulfate precipitation frequently results in preparations containing higher pyrogen levels in the concentrated cell culture supernatants, precluding its use.

9. Minimal backpressure (BP) prolongs the useful life of HPLC components and columns. A flow rate of 2–4 mL/min provides a compromise between time spent, effort, and purification efficiency.

10. The most common problem in HPLC involves a BP increase, usually owing to blockage in the column upper frit (**Subheading 3.2.2.1.**). A gradual increase in BP should be followed by a standard CIP procedure (**Subheading 3.2.2.1.**). If the BP suddenly increases, the column should be disconnected from the system and reverse-flushed with adequate buffer or ddH$_2$O. Alternatively, the column may be disassembled and the frit replaced.

11. Normal column performance involves prevention:
    a. CIP the column regularly.
    b. Use the same buffer for the MAb sample as for column equilibration.
    c. Filter samples and buffers through a 0.22-µm filter.
    d. Wash column briefly with ddH$_2$O before starting column equilibration buffer.
    e. Never allow a column to dry out.
    f. Rinse system thoroughly with 20% methanol or ddH$_2$O before storage.

12. All materials should be sterile and pyrogen-free. For this purpose, NaOH is widely accepted for cleaning, sanitization, and storage of chromatography media and systems.

13. Interactions between protein A and the Fc of the MAb are primarily hydrophobic. Direct addition of high-salt concentrations to the cell culture supernatant increases the binding capacity and retention of mouse IgG$_1$ and also averts dilution of the loading material.

14. The volume of sample to be loaded can usually be calculated on the basis of the binding capacity of the protein A resin for IgG (10–20 mg antibody/mL wet beads) and amount of Ig in the starting material. However, semipreparative HPLC-protein A columns must be overloaded to purify as much MAb as possible in a single run. Column overloading can be performed either by concentration overloading through increased sample concentration at constant sample volume or by volume overloading through applying more sample volume of the same concentration onto the column.

15. A two-step loading strategy—a higher flow rate during the initial stage of loading (2 mL/min), followed by a lower flow rate (1 mL/min)—increases the productivity in protein A chromatography.

16. For practical reasons, elution is carried out in a single step with citrate buffer, pH 3.5. Mouse IgG subclasses can be eluted sequentially: IgG$_1$ with 0.1 $M$ Na-phosphate, pH 6.0; IgG$_{2a}$ with 0.1 $M$ Na-citrate, pH 4.5; IgG$_{2b}$ with 0.1 $M$ Na-citrate, pH 3.5; and IgG$_3$ with 0.1 $M$ glycine, pH 2.7.

17. To prevent cross-contamination, the protein A column should be reutilized only with identical samples unless the column has been washed sequentially with 0.5 $M$ NaOH and ddH$_2$O at the end of the purification procedure.
18. If the starting material is ascitic fluid, the final preparation will contain endogenous host IgG and specific antibodies.
19. Most protein samples will bind to the HA column in P10 buffer. However, some lightly binding samples may require loading in 1 m$M$ phosphate—the lowest recommended buffer concentration.
20. P400 buffer is generally sufficient to desorb most MAb and regenerate the HA column for subsequent use. If necessary, buffer concentrations up to 1 $M$ or the use of 1 $M$ NaCl can be used to remove tightly bound MAb.
21. The nonretained peak is tested for the presence of residual MAb by collecting and reinjecting the solution onto a protein A column (**Subheading 3.2.2.**).
22. MAb recoveries below 80% may be caused by: column overloading, improper column equilibration, or nonoptimal starting buffer conditions.
23. Lower final endotoxin levels can be achieved by increasing residence times on the column or increasing resin-to-sample ratios.
24. A centrifugal ion-exchange membrane spin column (Mini Q, Vivapure) offers an alternative to traditional chromatographic methods for the removal of endotoxin from research-grade MAb solutions. The centrifugal devices reduce the levels to approx 1.2–1.5 EU/mg (**Subheading 3.2.4.2.**).
25. The average endotoxin concentration in the clarified cell culture supernatant is less than 100 EU/mL (20–100-EU/mL range).

## Acknowledgments

This work was supported by grants from AIRC, Biotechnology (CNR/MURST), and FIRB special project (MIUR, Rome, Italy). The Compagnia di SanPaolo and International Foundation for Research in Experimental Medicine (FIRMS) provided valuable financial contributions.

## References

1. Köhler, G. and Milstein, C. (1975) Continuous cultures of fused cells secreting antibody of predefined specificity. *Nature* **256,** 495–497.
2. Gura, T. (2002) Magic bullets hit the target. *Nature* **417,** 584–586.
3. Stoll, D., Bachman, J., Templin. M. F., and Joos, T. (2004) Microarray technology: an increasing variety of screening tools for proteomic research. *DDT:TARGETS* **3,** 24–31.
4. Funaro, A., Horenstein, A. L., Santoro, P., Cinti, C., Gregorini, A., and Malavasi, F. (2000) Monoclonal antibodies and therapy of human cancers. *Biotechnol. Adv.* **18,** 385–401.
5. Center for Biologics Evaluation and Research. *Points to Consider in the Manufacture and Testing of Monoclonal Antibody Products for Human Use.* Food and Drug Administration, Rockville, MD, 1997.
6. Mariani, M., Bonelli, F., Tarditi, R., Calogero, R., Camagna, M., Seccamani, E., and Scassellati, G. A. (1989) Purification of monoclonal antibodies for in vivo use: a two-step HPLC shows equivalent performance to immunoaffinity chromatography. *BioChromatography* **4,** 149–155.
7. Hahn, R., Schlegel, R., and Jungbauer, A. (2003) Comparison of protein A sorbent. *J. Chromat. B* **790,** 35–51.

8.  Horenstein, A. L., Crivellin, F., Funaro, A., Said, M., and Malavasi, F. (2003) Design and scaleup of downstream processing of monoclonal antibodies for cancer therapy: from research to clinical proof of principle. *J. Immunol. Methods* **275,** 99–112.
9.  Gagnon, P. (1996) *Purification Tools for Monoclonal Antibodies.* Validated Biosystems, Tucson, AZ.
10. Horenstein, A. L., Crivellin, F., Durelli, I., and Malavasi, F. Application of Hydroxyapatite and Protein A in the Downstream Processing of MAb for Clinical Use. Presented at the Third International Hydroxyapatite Conference, Lisbon, November 3–5, 2003.
11. Stanker, L. H., Vanderlaan, M., Juarez-Salinas, H. (1985) One-step purification of mouse monoclonal antibodies from ascites fluid by hydroxylapatite. *J. Immunol. Methods* **76,** 157–169.
12. Juarez-Salinas, H., Ott, G. S., Chen, J. C., Brooks, T. L., and Stanker, L. H. (1986) Separation of IgG idiotypes by high-performance hydroxylapatite chromatography. *Methods Enzymol.* **121,** 615–622.
13. Franklin, S. G. (2003) Removal of aggregate from an IgG4 product using CHT Ceramic Hydroxyapatite. *BioProcess International* **1,** 50–51.
14. *European Pharmacopea*, 3rd ed. Biological Tests. 2.6.8 Pyrogens; 2.6.14 Bacterial Endotoxins, Council of Europe, Strasburg, 1997.
15. Petsch, D. and Anspach, F. B. (2000) Endotoxin removal from protein solutions. *J. Biotechnol.* **76,** 97–119.
16. Pearson, F. C. *Pyrogens.* Marcel Dekker, New York, 1985, p. 119.
17. Brandwein, H. and Aranha-Creado, H. (2000) Membrane filtration for virus removal. *Dev. Biol. Stand.* **102,** 157–163.
18. Nau, D. R. (1986) A unique chromatographic matrix for rapid antibody purification. *Biochromatography* **1,** 82–94.
19. Boschetti, E. (2002) Antibody separation by hydrophobic charge induction chromatography. *Trends Biotechnol.* **20,** 333–337.
20. Weaver, L. E. Jr. and Carta, G. (1996) Protein adsorption on Cation Exchangers: Comparison of Macroporous and Gel-Composite Media. *Biotech. Prog.* **12,** 342–355.
21. Follman, D. K. and Fahrner, R. L. (2004). Factorial screening of antibody purification processes using three chromatography steps without protein A. *J. Chromatogr. A.* **1024,** 79–85.
22. Harlow, E. and Lane, D. *Using Antibodies: A Laboratory Manual.* New York, Cold Spring Harbor Laboratory Press, 1999.

# 17

# Virus Elimination and Validation

## Nicola Boschetti and Anna Johnston

## 1. Introduction

Biological products carry a risk of transmitting viruses (or pathogens), which may come from the source material (e.g., cell banks of human or animal origin, human blood, and human or animal tissues) or as adventitious agents introduced by the production process (e.g., the use of animal sera and growth supplements in cell culture or protein stabilizers in formulation). A low-risk level of viral contamination persists in cell culture processes that harness viral replication which also promotes the replication of adventitious viruses despite a high level of cell bank characterization. Products derived from blood have a higher risk of contamination of blood-borne viruses regardless of stringent and high-sensitivity tests of donor blood. In both cases, it is mandatory to introduce steps within the manufacturing process that pose to remove and/or inactivate human pathogenic viruses and thus provide a "virally safe" therapeutic product. In such cases as blood and plasma, which can harbor known blood-borne viruses, it is common to incorporate two distinct effective viral reduction steps that compliment each other in their mode of action so that any virus passing the first step would be effectively eliminated (through inactivation and removal) by the second.

Designing viral elimination steps is not a straightforward task. (General guidance has to be taken from **ref. _1_**, for cell-derived products **ref. _2_** and for plasma-derived products from **ref. _3_**). Each elimination step has to contribute to the net safety of the product but should also be economically viable. Hence, be conscious of the properties of the biological protein or vaccine and the suspected contaminating viruses (relevant viruses). Such considerations lead to the design of the appropriate virus elimination steps, which eliminate viruses efficiently without irreversible damage to the therapeutic and with as little yield loss as possible. The efficiency of an elimination step is not only judged by the log reduction factor (LRF; $\geq 4$ LRF is required for a dedicated elimination step) but also on the robustness (e.g., inactivation kinetics and process variations).

Advice from virologists is necessary for the initial design of a production process to choose the appropriate source materials, perform proper cell bank characterization,

From: _Methods in Molecular Biology, vol. 308: Therapeutic Proteins: Methods and Protocols_
Edited by: C. M. Smales and D. C. James © Humana Press Inc., Totowa, NJ

and decide what are the most effective and appropriate virus elimination procedures. After the manufacturing process is outlined, some process steps may be assessed experimentally for their elimination capacity to avoid unfortunate surprises. The results may influence the choice of process.

After a process has been developed and process robustness and reproducibility has been established, either on a pilot or a manufacturing scale, viral validation can commence. This begins by designing a scaled-down version of the manufacturing step. In our experience, it is best if those working closely on the process design do this. After accurate down-scaling is proven by the preservation of process and product parameters, virus-spiking experiments may start. The choice of which viruses are to be spiked into the starting material of manufacturing quality has to be justified. If possible relevant viruses have to be used, but if these viruses cannot be cultivated to the necessary high titers or cannot be assayed in vitro, it is more appropriate to use specific model viruses (viruses closely related to relevant viruses). Model viruses that may be used are suggested by the guidelines *(1–3)*. To appreciate the general elimination potential of the particular elimination step, it is strongly suggested to use additionally nonspecific model viruses (not related to relevant viruses but with different physicochemical properties).

For virus inactivation steps, the kinetics of the inactivation have to be reported; for removal steps, a mass balance (input virus = sum of output virus) has to be established. The formula here is used for the calculation of the LRFs:

$$LRF = \log \left( \frac{\text{totalvirus}_{\text{input}}}{\text{totalvirus}_{\text{output}}} \right)$$

These results may be used to support an application to relevant regulatory agencies to commence clinical trials. To support an application to commence commercial production and sale of the product, further viral validation data demonstrating the robustness of the elimination steps to process variations is often required. This may require testing viral elimination at the lower limits of the pasteurization temperature or using the lowest concentration of the viral agent, for example, one must always be cognizant of the fact that "all these studies are an approximation of the true capacity of the process to eliminate viruses" *(1)*.

Changes in process parameters may require revalidation of the elimination steps, which are downstream of the change. New threats (e.g., emerging viruses) arise, and standards become more stringent. The draft guideline: Note for guidance on assessing the risk for virus transmission—new Chapter 6 of the *Note for Guidance on Plasma-Derived Medicinal Products (4)* requests "investigational" studies for each emerging pathogen if it is capable of being handled in the laboratory. The emerging issue of new variant Creutzfeldt-Jakob Disease (vCJD) is one such case, in which the requirement to demonstrate some level of vCJD removal has become a global challenge.

Furthermore, research is never finite. Scientists are always investigating new techniques that are more efficient and more applicable to a wide range of viral classes or that are gentler to therapeutic products. For example, irradiation techniques, such as γ irradiation and ultraviolet (UV) irradiation, are gaining considerable momentum, particularly in the field of biotech products.

We have chosen to describe three methods (two inactivation steps and one removal step) that are commonly used and widely accepted by regulatory authorities as effective viral elimination steps. The methods are readily performed in a laboratory and can be scaled appropriately to mimic the conditions presented at a manufacturing scale.

Solvent detergent treatment involves incubating a solution with a solvent, such as tri-n-butyl phosphate (TNBP) combined with a nonionic detergent, e.g., Triton X-100, which together can inactivate enveloped viruses. It has been developed since the mid-1980s and is a well-established, gentle, and robust technique *(5,6)*. It works on the principal of removing/stripping the membrane of an enveloped virus, thus rendering the virus incapable of infecting. It can be readily incorporated into a manufacturing process, as it does not require specialized equipment. It is necessary to have an additional purification step after this treatment to remove the solvent and detergent because these agents, at the concentrations used, are adverse to human administration.

Low pH involves inactivating viruses sensitive to acidic pH by reducing the pH of the solution to below pH 5 and incubating at a defined temperature (usually between room temperature and 37°C). It has been used in the plasma fractionation industry since the late 1980s *(7)*. The technique can be applied only to therapeutic proteins that are acid stable but has the advantage of being very simple to adopt, as it requires no special equipment or additional viral agents. Generally, only enveloped viruses are susceptible to low-pH conditions. Acidification is a trigger for the fusion with the host cell membrane during the route of infection and induces a conformational change of the viral membrane proteins. However, we know that the nonenveloped B19 virus is inactivated after 2 h at pH 4 *(8)*; therefore, it is reasonable that low-pH conditions also have impact on the conformation of viral capsid proteins and can thus inactivate some nonenveloped viruses.

Viral filtration involves the use of specially designed filters to remove viruses that are marginally bigger in size than the desired biological protein or therapeutic. It has been used since the mid-1990s *(9)*, and there are now three major suppliers of these filters—Asahi Chemical Industry (Planova), Millipore (Viresolve), and Pall BioPharmaceuticals (DV20)—supplying a range of volumetric capacities and pore sizes to suit different applications. The filter's efficiency is reliant on many parameters, both of an operational and a physicochemical nature. It is crucial to optimize parameters, such as operating pressure, cross-flow rates, and filtered volume per filter area, in advance to ensure a sufficient log reduction of virus. Parameters such as solution pH, protein concentration, and protein composition are also influential in establishing the filter capacity, fouling rate of the filters, and ultimately the efficiency of the step *(10)*. Owing to the complex nature of these filters, the three suppliers often work with the process technologists and virologists to design and optimize this removal step to ensure the best possible operational parameters.

The methods described next are down-scaled manufacturing process steps, which were used for virus validation studies. It should be kept in mind that accurate down-scaling is crucial for the validity of the obtained results, and each process step will have its own scaled-down version and will certainly differ from the examples described. This section also specifies some routine cell culture techniques, as well as interference and cytotoxicity assays that are usually done in advance using appropriate biological material. **Table 1** lists some of the viruses, cell lines, and appropriate media used in these assays.

## 2. Materials

### 2.1. Solvent Detergent Treatment

1. Relevant protein solution, e.g., a factor IX solution, obtained from plasma fractionation.
2. TNBP neat solution.
3. Tween-80 (polysorbate) or Triton X-100 neat solution.
4. Heating bath/heating unit, insulated.
5. Vortex/mixer.
6. Measuring cylinder.
7. A standard coagulometer to measure activated partial thromboplastin times (APTT; MLA Electra 800, Medical Laboratories Automatic).
8. Sindbis virus (MRM 39 strain, American Type Culture Collection [ATCC]) titer more than $10 \log_{10} TCID_{50}/mL$, kept at $-40°C$ or less.
9. Baby Syrian Hamster Kidney cells (BHK-21) cells (ATCC, cat. no. CCL 10).
10. Viral assay medium (VAM): 2% fetal calf serum (FCS) and minimal essential medium (MEM) buffer, stored at 4°C.
11. 96-Well flat-bottom microtiter plates.

### 2.2. Low-pH Treatment

1. Starting material, such as an immunoglobulin or antibody preparation.
2. A humidified incubator (37°C) supplied with a mixture of 5% $CO_2$ (v/v) in air.
3. 0.2 $M$ HCl.
4. 0.2 $M$ NaOH.
5. 100 m$M$ Phosphate buffer, pH 7.4.
6. Phosphate-buffered saline (PBS) solution.
7. Laminar flow cabinet.
8. pH meter.
9. Waterbath, thermostatically controlled.
10. Thermometer.
11. Cell proliferation assay (Promega CellTitre 96, cat. no. G5430).
12. 96-Well flat-bottom microtiter plates.
13. 0.5% Crystal violet: 5 g crystal violet, 1000 mL methanol, and 30 mL formaldehyde.
14. 0.1-, 0.2-, and 0.45-µm filters.
15. Appropriate virus and cell line (**Table 1**).

### 2.3. Cytotoxicity, Interference, and Tissue Culture Infectious Dose 50% (TCID₅₀) Assay

1. Appropriate virus and cell line (**Table 1**).
2. A humidified incubator (37°C) supplied with appropriate mixture of $CO_2$ in air.
3. Laminar flow cabinet.
4. Cell proliferation assay (Promega CellTitre96, cat. no. G5430).
5. 96-Well flat bottom microtiter plates.
6. 175 cm$^2$ cell culture flasks.
7. 0.5% Crystal violet: 5 g crystal violet, 1000 mL methanol, and 30 mL formaldehyde.

**Table 1**
**Viruses, Their Appropriate Cell Lines, and Media**

| Viruses | Model virus | Cell line | VAM | Cell culture medium | Cell culture conditions |
|---|---|---|---|---|---|
| Hepatitis C virus, West Nile virus | BVDV | BT cells (ATCC, cat. no. CRL-1390, Rockville, MD) | Earle's MEM; 2% horse serum | Earle's MEM; 10% horse serum | 37°C; 5% $CO_2$ |
| Large complex DNA virus | PRV | BT cells (ATCC, cat. no. CRL-1390, Rockville, MD) | Earle's MEM; 2% horse serum | Earle's MEM; 10% horse serum | 37°C; 5% $CO_2$ |
| Retroviruses | X-MuLV | Moloney sarcoma virus-transformed cat brain cells: PG-4 (ATCC, cat. no. CRL-2032) | McCoy's 5A medium; 2% heat-inactivated FCS | McCoy's 5A medium; 10% heat-inactivated FCS | 37°C; 5% $CO_2$ |

BVDV, Bovine Viral Diarrhea Virus; BT, bovine turbinate; PRV, Pseudorabies Virus; X-MuLV, Xenotropic Murine Leukemia.

## 2.4. Nanofiltration

1. Factor IX concentrate, obtained from plasma fractionation or recombinantly derived.
2. 35 N Asahi Filter, 0.01 $m^2$ (Asahi Chemical Industry).
3. 15 N Asahi Filter, 0.001 $m^2$.
4. Filtration buffer: 5 m$M$ phosphate-NaCl buffer, pH 6.0.
5. BVDV (NADL strain, ATCC) grown to achieve a titer of greater or equal to 8 $\log_{10}$ $TCID_{50}$/mL and stored at –40°C.
6. BT cells (ATCC).
7. Fetal bovine serum (FBS).
8. PBS.
9. VAM: 2% FBS in Eagle's MEM supplemented with nonessential acids, glutamine, penicillin/streptomycin, and HEPES buffer.
10. 96-Well flat-bottom microtiter plates.
11. A humidified incubator (37°C), 5% $CO_2$ (v/v).
12. An inverted microscope.

## 3. Methods

Viruses are commonly assayed using an infectivity assay in which a virus-spiked solution is added to the cell culture, and the virus infects the cells. Visible effects on the cell culture, such as cell lysis, plaque formation, or focus formation, manifest the infection; these effects can be quantifiable. Cells can also die if the solution is cytotoxic, and the virus assay sensitivity consequently will be compromised. Similarly, a solution, component, or characteristic may interfere in some manner with the virus assay. Thus, to determine an accurate reading of cell death, attributable to the presence of virus alone, cytotoxicity of the test material and interference of the test material with the virus titration has to be assessed. This is described in the next section.

### 3.1. Cytotoxicity and Interference

1. Before performing the virus validation experiments, the lowest dilution of the test material, which is not cytotoxic to the detector cells (e.g., BT cells), has to be determined. This is achieved by simply distributing the amount of test material at different dilutions in cell culture medium and analyzing the effect on the cells.
2. The cell viability is determined after 7 d by a cell proliferation assay (Promega CellTitre96, cat. no. G5430) or by visual inspection under the microscope.
3. The concentration with no presence of cytotoxicity is then used in the validation experiments and interference assay (primary dilution).
4. Virus interference is assessed by direct comparison of the virus titers obtained in interference assays to the virus titers obtained in the standard $TCID_{50}$ assays of stock virus performed in culture medium. Interference assays are conducted by serially diluting stock virus into the primary dilution of the test material.

### 3.2. Virus Assay (TCID$_{50}$)

#### 3.2.1. Day 1

1. Appropriate cells are grown to confluency in 175 $cm^2$ cell culture flasks ($10^6$–$10^7$ cells/flask).
2. After trypsinization, cells are resuspended in 150 mL final volume of cell culture medium.
3. Resuspended cells are distributed into 96-well microtiter plates (100 µL/well).

## 3.2.2. Day 2

1. Samples to be assayed are diluted (usually, they have already undergone some neutralization or predilution in VAM during the experiment) with cell culture medium to the concentration previously determined as noncytotoxic (primary dilution).
2. Serial dilution is then carried out; 1/3 or 1/10 dilutions are often selected, depending on the aimed precision of the assay. For example, if the sample is expected to have a viral titer of $10^5$, then a dilution of 10-fold carried out five times will be sufficient to ensure that no virus remains in the last dilution and that a negative score for cytopathic effect (CPE) can be observed.
3. The microtiter plate is typically plated out with 50 µL sample into eight wells, each dilution representing an entire column. Some columns of the plate should contain 50 µL assay medium as controls for cell growth.
4. The microtiter plates are then incubated for about 30 min in the incubator before 200 µL cell culture medium is added to each well.
5. The microtiter plates are then incubated for 6–9 d.

## 3.2.3. Days 8–11

1. If CPE is to be measured with the help of a microscope, the monolayers are visually inspected, and evidence of CPE is scored with "+," whereas absence of CPE is scored with a "–."
2. If measuring CPE on a microtiter plate reader, the medium is aspirated from each well. Optionally, the wells are washed with 100 µL PBS/well.
3. To stain the cells, 50 µL crystal violet solution is added to each well, then they are covered and incubated for 20 min at room temperature.
4. The crystal violet solution is removed, and the plate is rinsed with cold water and then allowed to dry.
5. Using a microtiter plate reader (560-nm wavelength), CPE is scored as before.
6. The virus titers and their errors are calculated using the Spearman-Kärber method *(11,12)* as shown in **Note 1**.

## 3.3. Solvent Detergent Treatment

1. The protein solution (e.g., factor IX [FIX]) is placed in the water bath, set at 25°C, and allowed to equilibrate for 10 min.
2. An aliquot of Sindbis virus is thawed the day of the experiment and kept on ice until used. The amount of virus used varies according to the stock titer and minimum scale-down volume achievable. Ideally, a minimum of 4 log reduction is required (*see* **Note 2**).
3. To 50 mL protein solution, add sufficient Tween neat solution and TNBP neat solution over a period of 7 min to achieve 1% Tween and 0.3% TNBP solution. This solution represents the protein control sample, and a 50-µL aliquot of this solution is taken at the beginning of the incubation time, and a further aliquot taken at the end of the incubation time (at least 8 h) to establish if there is any loss of protein activity because of the incubation.
4. To another 50 mL protein solution, add 500 µL Sindbis stock virus.
5. Mix gently as before, and add a 10-µL sample immediately to 1 mL VAM.
6. Store the aliquot on ice to stop any further viral inactivation. This is the virus spike sample to confirm the initial viral load at zero time.
7. To the spiked protein solution, add sufficient Tween and TNBP over a period of 7 min to achieve 1% Tween and 0.3% TNBP solution.
8. Continue to incubate, and stir gently both 50 mL protein solutions (spiked and unspiked) for at least 8 h, by which time complete inactivation would have occurred.

9. To establish the kinetics of the viral inactivation, sample from the contaminated solution at several time intervals, diluting 100-μL aliquots into 1 mL VAM, and store on ice to stop inactivation from occurring.

10. After incubation, test the two protein control solutions for factor IX activity using a standard coagulometer to measure APTT.

11. Assay the virus-containing samples by end-point dilution assays on BHK 21 cells (*see* **Subheading 3.2.**) to determine the extent of viral elimination over time. This experiment is usually carried out on different production lots of protein to ensure that the step is robust and reproducible.

12. If the extent of viral inactivation is less than 4 logs, it will be necessary to try other conditions to increase the effectiveness of this step. Some different parameters are included in **Note 3**.

    Following the viral inactivation treatment, it is then necessary to remove the solvent and detergent from the protein of interest. There are several ways to do this.

    a. Chromatography can be useful if further purification of the protein of interest is sought and if chromatography is an appropriate tool by which to do this.

    b. Extraction with soybean oil and reverse-phase chromatography on insolubilized C18 resin *(13)*. This does not necessarily change the protein composition, which is desirable, e.g., if one is inactivating plasma for use in plasma exchange.

## *3.4. Low-pH Treatment*

1. Starting material (30 mL) is equilibrated at the desired temperature (37°C), and the pH is adjusted with 0.2 *M* HCl.

2. The virus spike is filtered according to virus size (BVDV, 0.1 μm; PRV, 0.45 μm; X-MuLV, 0.2 μm) to disrupt viral aggregates.

3. Virus is spiked into the starting material (maximal ratio: 1/10 v/v; *see* **Note 2**) and thoroughly vortexed. If the starting material has limited buffer capacity, the spike may influence the previously adjusted pH. In any case, pH should be monitored throughout the experiment.

4. Immediately after spiking, the first sample is withdrawn. Subsequent time points for sample withdrawal are usually determined experimentally in pre-experiments, because the inactivation kinetics may vary depending on the virus and starting material.

5. After withdrawal, the acidity of the samples has to be neutralized with 0.2 *M* NaOH and buffered to stop inactivation.

6. Samples are sterile-filtered (0.45 μm) and diluted to the primary dilution with cell culture medium. Samples expected to be of high titer are then further diluted to be in the range of the $TCID_{50}$ assay.

If the viral reduction is insufficient, then **Note 4** suggests some methods to improve the kinetics or overall reduction.

## *3.5. Nanofiltration*

1. The 35 N Planova filter (0.01 $m^2$ filter area) is tested for leakage and preconditioned with a phosphate-NaCl buffer, pH 6.0, according to the manufacturer's directions. It is then configured in a dead-end mode.

2. An aliquot of BVDV is thawed on the day of the experiment, filtered (according to virus size; BVDV requires a 0.1-μm filter) to break up aggregates, and kept on ice until used.

3. 40 mL FIX solution is spiked with 0.4 mL BVDV stock virus (i.e., a 1:100 spike).

4. 50 μL BVDV stock virus is removed into 5 mL VAM and maintained at 2–8°C until assayed. This confirms the titer of BDVD stock.

5. A 0.5-mL aliquot of the spiked FIX solution is removed into 4.5 mL VAM and maintained at 2–8°C until assayed. This confirms the initial titer of BVDV in the protein solution.
6. The remaining product is passed through the 35 N filter at constant pressure (~5 psi).
7. To flush out the remaining solution in the lines and filter, 20 mL filtration buffer is used and added to the filtrate.
8. A 0.5-mL aliquot of this filtrate is removed into 4.5 mL VAM and held at 2–8°C until assayed. This measures the log reduction possible after this 35 N prefilter.
9. The 15 N Planova filter ($0.001$ m$^2$ filter area) is leakage-tested and preconditioned with the filtration buffer according to the manufacturer's instructions. It is then configured in dead-end mode.
10. The 35 N filtrate is passed through the 15 N filter at constant pressure (4 psi). A 30-mL flush with filtration buffer is performed, and the flushed volume is added to the filtrate.
11. A 0.5-mL sample of this final filtrate is taken and mixed with 4.5 mL VAM and held at 2–8°C until assayed.
12. All samples are titrated for cytopathic BVDV, including the unspiked material to test for interference according to the methodology in **Subheadings 3.1.** and **3.2.** using confluent BT cells and 100-µL sample volumes.
13. The spent filters are then integrity-tested according to the manufacturer's instructions to ensure that there was no breach in the capillary structures, i.e., that no virus passed through the filter purely as a result of an undesirable hole in the structure.

To improve the efficiency of virus removal, several factors can be investigated, which are considered in **Note 5**.

# 4. Notes

1. Virus assay calculations: Formula for the calculation of log TCID$_{50}$ values are shown. These can be used in Excel spreadsheets:

$$\text{LOG (TCID}_{50}/\text{mL}) = (Xo - d/2 + d * \text{sum } [pi]) + \log(1/v) - \log(pd)$$
$$SE = d * \sqrt{(\text{sum } [pi(1 - pi)/(ni - 1)])}$$

$pd$ = primary dilution factor
$Xo$ = -$\log_{10}$ of the concentration, where all wells were positive (virus present).
$d$ = $\log_{10}$ of the serial dilution factor.
$pi$ = ratio: positive wells/n wells, which are $> 0$.
$v$ = inoculation volume (mL/well).
$ni$ = number of wells/dilution (usually 8).
$SE$ = standard error.

Example: primary dilution = 2; inoculation volume = 0.05 mL

| Dilution | + | – | n | pi | | sum[pi] |
|---|---|---|---|---|---|---|
| Primary dilution | 8 | 0 | 8 | 1 | – | |
| $10^{-1}$ | 8 | 0 | 8 | 1 · | 1 | 2.25 |
| $10^{-2}$ | 6 | 2 | 8 | 0.75 | 0.75 | 2.25 |
| $10^{-3}$ | 3 | 5 | 8 | 0.375 | 0.375 | 2.25 |
| $10^{-4}$ | 1 | 7 | 8 | 0.125 | 0.125 | 2.25 |
| $10^{-5}$ | 0 | 8 | 8 | 0 | 0 | |

$$\text{LOG(TCID}_{50}/\text{mL}) = Xo - \frac{d}{2} + d * \sum pi + \log(\frac{1}{v}) - \log(pd) = 1 - \frac{1}{2} + 1 * 2.25 + \log(\frac{1}{0.05}) - \log(2) = 3.75$$

$$SE = \sqrt{\sum \frac{pi(1-pi)}{ni-1}} = \sqrt{\frac{1*0 + 0.75*0.25 + 0.375*0.625 + 0.125*0.875}{8-1}} = 0.275$$

If no virus is detected, the Poisson distribution may be used:

If sample volume is less than total volume, then $c = (LN\ p)/-LNv$

$c$ = virus concentration ($TCID_{50}/mL$).

$p$ = probability of the sample containing no virus ($p = 0.05$).

2. Virus titers and spiked volumes. A 1:100 spike is usually appropriate, provided that the initial titer of virus is high. Regulatory guidances allow a maximum spike of 1/10 to not alter the milieu. Typically, the sensitivity of a viral assay is $10^1$–$10^2$. Thus, for example, if the initial titer is $10^8$, and the assay sensitivity is $10^2$, a spike of 1:100 will give you an initial starting titer of $10^6$, and the maximum log reduction possible to show would be 4 log or greater. In some cases, it is possible to demonstrate greater than 6 log reduction.

3. Methods to improve inactivation include (1) reducing lipid content of the protein solution. Lipids in the solution will compete with the lipids on the viral membranes and reduce the effective concentration of the solvent and detergent combination. Lipids can be removed by filters (e.g., Cuno's delipidation filter). (2) Increase concentrations of TNBP to up to 1%, and (3) increase temperature to a maximum that the therapeutic protein can bear without having to include thermal stabilizers.

4. Efficiency of the low-pH step may be enhanced (at least against some viruses) by the additional use of trace amounts of pepsin or nonionic detergent. The appropriate protease concentration to be used has to be determined experimentally to not degrade the therapeutic protein. Nonionic detergents should be used above the critical micelle concentration. However, the use of additives necessitates a subsequent purification step to remove them.

5. The major parameters affecting the efficiency of virus removal with Asahi filters are volume and pressure. Excess volume passing through the filter may result in virus particles being inadvertently forced through the void and capillary structures, whereas the low pressure results in less clogging in the voids, which may allow small viruses to escape through. Protein concentration and pH can also affect volume throughput and thus capacity of the filters. A high-protein concentration can cause early clogging of the filter voids and ultimately block all flow; pH can affect the physical shape of the biological, which can affect passage through the membrane. These opposing attributes need to be considered before deciding how and when to use a virus filter.

## References

1. Committee for Proprietary Medicinal Products (CPMP) of the European Agency for the Evaluation of Medicinal Products (EMEA). CPMP/BWP/268/95. Note for Guidance on Quality of Biotechnological Products: Viral Safety Evaluation of Biotechnology Products Derived from Cell Lines of Human or Animal Origin. Section 5.2, Virus Validation Studies: The Design, Contribution and Interpretation of Studies Validating the Inactivation and Removal of Viruses.

2. Committee for Proprietary Medicinal Products (CPMP) of the European Agency for the Evaluation of Medicinal Products (EMEA). CPMP/ICH/295/95. Quality of Biotechnological Products: Viral Safety Evaluation of Biotechnology Products Derived from Cell Lines of Human or Animal Origin. Section 6, Evaluation and Characterization of Viral Clearance Procedures.

3. Committee for Proprietary Medicinal Products (CPMP) of the European Agency for the Evaluation of Medicinal Products (EMEA). CPMP/BWP/269/95 rev. 3. Note for Guidance on Plasma-Derived Medicinal Products. Section 5.2, Validation Studies.

4. Committee for Proprietary Medicinal Products (CPMP) of the European Agency for the Evaluation of Medicinal Products (EMEA). CPMP/BWP/5180/03. Note for guidance on

assessing the risk for virus transmission—new chapter 6 of the *Note for Guidance on Plasma-Derived Medicinal Products.*

5. Neurath, A. R. and Horowitz, B. Undenatured virus-free trialkyl phosphate treated biologically active protein derivatives. US patent 4,820,805.
6. Horowitz, B., Prince, A. M., Horowitz, M. S., and Watklevicz, C. (1993) Viral safety of solvent-detergent treated blood products. *Dev. Biol. Stand.* **81,** 147–161.
7. Mitra, G. and Mozen, M. M. Preparation of retrovirus-free immunoglobulins. US patent 4,762,714.
8. Boschetti, N., Niederhauser, I., Kempf, C., Stühler, A., Löwer, J., and Blümel, J. (2004) Different susceptibility of B19 virus and mice minute virus to low pH treatment. *Transfusion* **44,** 1079–1086.
9. Burnouf, T. and Radosevich, M. (2003) Nanofiltration of plasma-derived biopharmaceutical products. *Haemophilia* **9,** 24–37.
10. Ensuring regulatory compliance: Validation of regulatory compliance. Millipore publication
11. Spearman, C. (1908) The method of right and wrong cases ("constant stimuli") without Gauss's formulae. *Br. J. Psychol.* **2,** 227–242.
12. Kärber, G. (1931) Beitrag zur kollektiven Behandlung pharmakologischer Reihenversuche. *Arch. Exp. Pathol. Pharmakol.* **162,** 480–483.
13. Strancar, A., Raspor, P., Schwinn, H., Schutz, R., and Josic D. (1994) Extraction of Triton X-100 and its determination in virus-inactivated human plasma by the solvent—detergent method. *J. Chromatogr. A.* **658,** 475–481.

# 18

# Virus Removal by Nanofiltration

## Marina Korneyeva and Scott Rosenthal

## 1. Introduction

Virus removal filters (nanofilters) are specifically designed to remove viruses and other biomolecules through a size-exclusion mechanism. Nanofilters can be broadly classified as either direct (or normal) flow filtration (DFF) or tangential flow filtration (TFF). In the TFF mode, the material to be filtered is flowed across the membrane in a direction roughly parallel to the filtration matrix. This operational mode is analogous to that used with conventional ultrafiltration/diafiltration membranes. Product passage is facilitated by restricting the flow of the retentate relative to the feed, thereby applying backpressure to the system, which results in product migration through the membrane. In the DFF mode, the product is driven through the filter perpendicular to the filtration matrix. The main advantage to the TFF format is that a constant sweeping current (across the membrane surface) minimizes the opportunity for the membrane to be clogged by impurities in the product being nanofiltered. Nanofilter operation in the DFF mode offers several advantages over TFF, including flow rates that are generally higher, easier integration into a production environment, a simpler unit operation, and lower capital expenditure.

Nanofiltration is a promising technology for significantly improving the safety of any protein product small enough to pass through its pores. It has been shown that nanofiltration in the DFF configuration provides robust retention of large viruses in the size range of 80–110 nm (Reovirus 3, retroviruses) (*1*). Removal of small viruses in the size range from 18 to 30 nm (hepatitis A virus, polio virus, and parvoviruses) by filtration has proven to be more challenging. The difficulty encountered is because of the conflicting demands of virus retention and protein passage (*2*).

A panel of recommended model viruses for recombinant products is described in ICH Q5a, 1998. Rodent endogenous retroviruses are a nonadventitious contaminant that could potentially be present in harvests from murine hybridoma, Chinese hamster ovary (CHO), or baby hamster kidney (BHK) cell cultures. Thus, evaluation of rodent retrovirus removal using Xenotropic Murine Leukemia virus or a similar murine

From: *Methods in Molecular Biology, vol. 308: Therapeutic Proteins: Methods and Protocols*
Edited by: C. M. Smales and D. C. James © Humana Press Inc., Totowa, NJ

retrovirus are expected by regulatory agencies prior to clinical use of mammalian cell culture-derived biopharmaceuticals *(3)*.

Viruses used in studies to evaluate pathogen clearance across various process steps employed in the production of plasma-derived products are chosen to model known pathogens with the potential to infect the blood supply *(4)*. The smallest known pathogenic human virus is parvovirus B19, with an average diameter of 18–24 nm. Several manufacturers (Asahi/Planova, PALL, Millipore, and Amersham/AG Technology) produce nanofiltration products, which are reported to be capable of removing viruses of this class size. Owing to its mechanism of action (size-based discrimination), nanofiltration should have little impact on product activity, stability, or chemical composition. Because B19 cannot be cultured efficiently in the laboratory, porcine parvovirus (PPV, mean diameter of 18–22 nm) can be used as a substitute challenge virus for the validation of viral clearance.

There are no specific regulatory requirements for the acceptance criteria for the validation of viral clearance by nanofiltration. This criteria should be selected based on the product, manufacturing process, and filter performance. The important element of the virus filter selection is the consideration of the virus clearance and protein transmission requirements, availability of a sensitive in-process integrity test, and ability to validate virus clearance at a small scale. Because of the robustness of the technique, the targeted removal of viruses by nanofiltration is often four logs. However, for those processes in which additional virus clearance is sought, and a four-log clearance step is not absolutely required, the target may be lower. In addition, adequate protein passage is essential for implementation of nanofiltration to the purification process with a typical protein recovery more than 90%. Factors affecting viral clearance and protein passage are protein concentration, impurity content, pH, viscosity, and ionic strength *(5)*. If the nanofiltration step is placed immediately after a chromatography column, filter flux may be an important parameter to consider given that the nanofilter must be capable of keeping up with column flow rates *(6)*.

## 2. Materials

1. 0.001 m$^2$ Asahi Planova 15 N nanofilter (Asahi Kasei Corporation, Planova Division, cat. no. 15NZ-001).
2. 0.1 µ filter (e.g., Millipak Filter Unit, Millipore, cat. no. MPVL10CA3 or similar).
3. Table-top style centrifuge (e.g., Sorvall RT 6000D; Beckman GS-6R; Jouan MR1812, or similar) Beckman ultracentrifuge (TLX or similar) with rotor (TLA-100.3 or similar).
4. Variable-speed peristaltic pump, 60–600 rpm (e.g., Easy-Load Tubing Pump, Millipore, cat. no. XX80EL000 or similar).
5. Pressure gauge, 0–30 psi for nanofiltration.
6. PPV, strain NADL-2 (American Type Culture Collection [ATCC], Bethesda, MD).
7. Mini Pig Kidney cells (ATCC).
8. Cell culture medium: Dulbecco's modified Eagle's medium (DMEM) (Gibco/Life Technologies), 0.1 m$M$ nonessential amino acids (Gibco/Life Technologies), 10 m$M$ HEPES (Sigma), 0.05 mg/mL gentamicin (Gibco/Life Technologies), 2.5 µg/mL fungizone (Amphotericin B; Gibco/Life Technologies).
9. Phosphate-buffered saline (PBS), pH 7.2 (1X), liquid (Gibco/Life Technologies).
10. Hank's balanced salt solution (HBSS; Gibco/Life Technologies).

## 3. Methods

The methods described outline the preparation of a monodispersed virus stock *(1)*, assay for virus infectivity *(2)*, and execution of a filtration run *(3)*.

### 3.1. Monodispersed Virus Stock Preparation

Crude viral stocks (clarified cell lysates) are not suitable for use in nanofilter challenge studies. The majority of virus present in crude tissue culture lysate preparations is aggregated. This leads to a significant overestimation of the clearance ability of the nanofilter—a result that is unacceptable from a product-safety standpoint. Therefore, it is important to employ procedures that minimize the presence of these aggregates. The procedure described here can be used to produce purified monodispersed viral suspensions using crude cell lysates from infected cells as the starting material.

1. Infect subconfluent monolayers of minipig kidney (MPK) cells at a low multiplicity of infection (PPV MOI ranged from 0.1 to 0.01).
2. Propagate PPV in MPK cells grown in cell culture medium containing 10% fetal bovine serum (FBS) at 36–38°C in a humidified $CO_2$ incubator until advanced cytopathic effect (CPE) is observed.
3. Disrupt the infected cells by freezing and thawing 3 times, and store the cell lysates at ≤ –68°C until used. Freeze–thaw cycles can be carried out directly in the cell culture flasks, if desired.
4. Thaw cell lysates at room temperature or 37°C.
5. Clarify cell lysates by centrifuging at 4100$g$ for 15 min at 4°C using the Sorvall SLA 600TC rotor.
6. Decant the supernatant into a sterile container. Filter clarified cell culture supernatant stock solution through a 0.22-μm rated membrane. This is the crude virus stock. Pool supernatants to ensure homogeneity prior to aliquoting.
7. Aliquot the crude virus stock solution into ultracentrifugation tubes. Using a permanent marker, place a single small mark on the opening of the tubes to help orient the location of the virus pellet after centrifugation (*see* **step 10**).
8. Place the tubes of crude virus into an appropriate rotor (e.g., BeckmanTLA-100.3) with the marked side of the tube opening facing outward.
9. Place the rotor into an appropriate ultracentrifuge (e.g., Beckman TLX) and centrifuge at 300,000$g$ for 1.5 h at 4°C to pellet virions.
10. Carefully decant the supernatant to avoid disruption of the pellet. Using the mark on the tube to orient the location of the pellet, resuspend the resulting virus pellet in 1/20 vol (of the original virus solution) of PBS and allow to equilibrate overnight at 2–8°C.
11. Following overnight incubation at 2–8°C, add another 1/20 vol PBS to the resuspended virus, and mix well.
12. Pass through a 0.1-μm membrane filter to remove large aggregates. Aliquot, label, and store at –70°C until ready to use. Determine the titer of the newly prepared monodispersed virus stock using the virus infectivity assay, and record the stock titer. Electron microscopy may be used to confirm the virus stock is monodispersed.

This technique produces suspensions of virus that are largely monodispersed and works well for small nonenveloped viruses (e.g., porcine parvovirus). Nanofilter performance may decrease dramatically if virus spike concentration exceeds 5% of the total feed volume. As such, it is desirable to maintain the virus spike concentration between 1% and 2% of the total filter feed volume *(7)*.

## 3.2. Virus Infectivity Detection and Quantification

Virus infectivity can be detected using cell and tissue culture–based assays in which CPEs are observed. Using these types of assay, virus titer can be determined in feed, filtrate, and control samples by serially diluting the samples, plating the dilutions on an appropriate indicator cell line, and assessing for CPE after a predetermined incubation period. Titers can be calculated from the CPE data using the Spearman–Kärber method *(8)*. This method calculates a titer for the sample expressed as tissue culture infectious dose at 50% infectivity ($TCID_{50}$). The method described here can be used to assay for and quantify PPV infectivity.

1. Harvest MPK cells, and resuspend in cell culture medium supplemented with 7.5% FBS to a concentration of $2 \times 10^4$ cells/mL. Transfer 0.1 mL cell suspension into each well of a 96-well plate.
2. Use cells for virus infectivity assay when they are approx 50% confluent. Remove the medium from the cells, and replace with fresh cell culture medium with 7.5% FBS prior to virus titration.
3. Titrate virus by diluting the samples of interest in HBSS using half-log serial increments (i.e., 1:3.2, volume to volume) and transferring 0.1 mL of each dilution to a single row of cells on the 96-well plate.
4. A positive control can be prepared by spiking HBSS with the same lot of virus stock that is used to spike the nanofiltration feed. The positive control should be spiked at the same volumetric ratio used to spike the feed material. Include unspiked HBSS as a negative control, which this can be accomplished by adding 0.1 mL HBSS to the last row of each plate.
5. Incubate the sample inoculum, 0.1 mL per well on the cells for approx 60–75 min at 37°C, and then remove all liquid from each well.
6. Add 0.2 mL fresh cell culture medium supplemented with 2% FBS to each well, and incubate plates at 37°C in a humidified $CO_2$ incubator for 7 d. Observe cell monolayers for the presence of viral infection as indicated by viral CPE.
7. Score each well as a positive or negative, and convert the results into a titer ($log_{10}$ tissue culture infective dose per milliliter; $TCID_{50}$/mL) by the method of Spearman and Karber as described.

The Spearman and Kärber formula for calculating virus titers is

$$m = X_k + (d/2) - d \, \Sigma \, p_1$$

where: $m$ = logarithm of the titer relative to the test volume.
$X_k$ = logarithm of the smallest dosage that induces viral CPE in all cultures.
$d$ = logarithm of the dilution factor.
$p_i$ = proportion of positive results at dilution $i$.

The values were converted to $TCID_{50}$/mL using a sample inoculum of 0.1 mL.

The standard deviation, $s_m$, was calculated using the following formula:

$$s_m^2 = d_f^2 \, \Sigma \, [p_i \, (1 - p_i)/(n_i - 1)]$$

where: $d_f$ = logarithm of the dilution factor.
$p_I$ = proportion of positive results at dilution $i$.
$s_m$ = standard deviation.
$n_i$ = number of replicates at dilution $i$.

The 95% confidence limit is $m \pm 1.96 \, s_m$

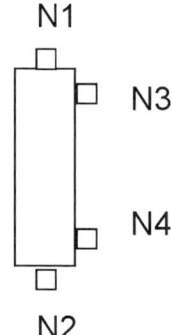

N1

N3

N4

N2

Fig. 1. Schematic drawing of Asahi 15 N filtration unit (vertical position).

### 3.3. Filtration Run

The evaluation of viral clearance by nanofiltration consists of numerous individual studies designed to demonstrate the effectiveness, consistency, and robustness of the manufacturing process step to reduce the level of viable viruses in the feed stream. A relatively large number of studies are performed at laboratory scale (virus challenges) to determine the appropriate filter sizing for a manufacturing scale process and to evaluate the effect of feedstream variability on nanofilter performance (*see* **Note 1**).

In this chapter, the performance of a small-scale Asahi Planova 15 N nanofiltration unit (0.001 m², Asahi Kasei Corporation, Planova Division, Tokyo, Japan) is described (*see* **Fig. 1**). This filter can be used to evaluate the capacity of virus filtration to remove infectious virus and prions (*9*). Owing to the small surface area of the filtration unit, research can be carried out with a relatively small amount of feed material. PPV is often selected as a worst-case scenario virus for filter performance validation (*see* **Note 2**).

### 3.3.1. Filter Integrity Testing

The preuse leakage test consists of applying process air to an operational pressure range of 10–14 psi (1.0 kg/cm² = 1 bar) to the nanofilter with all outlets blocked off and visual observation of the filter stored in sterile water for any bubbles migrating from the filters that indicate a gross defect in the filter. This is not a validated test by the manufacturer; however, it allows the user to screen the filter for gross defects prior to use (*see* **Subheading 3.3.4.**). The postuse, destructive, gold particle test is the validated test from the manufacturer. This test uses the monodisperse colloidal gold particle suspension to confirm the distribution of pore size in the filter and filter grade differentiation (*see* **Note 3**).

### 3.3.2. Virus Infectivity Control Samples

The virus-spiked product sample passes through the nanofiltration unit. Nanofiltrate is collected into several fractions and titrated individually along with the rinse fraction to evaluate virus removal/partitioning. As a titration control, an aliquot of HBSS should be spiked with virus at the same ratio as the filter feed material. A titration control sample should be divided into two portions. The first portion should be titrated immediately while the remainder of the sample will be held at 5°C and titrated at the end of the nanofiltration run. The results of the virus infectivity assay for titration control samples should not differ for more than 1 log $TCID_{50}$. If the results vary by more than 1 log, assessment of virus stability may need to be evaluated.

### 3.3.3. Nanofilter Preparation

1. Remove sterile bag containing the nanofilter from outer packaging, and inspect bag for tears or holes. Do not use nanofilter if the sterile bag has been compromised.
2. Remove nanofilter from its packaging, and visually inspect it for any cracks on the casing or ports that may compromise its integrity. Do not use the nanofilter if the casing has been compromised.
3. Record the nanofilter material number and batch number for future references. The Leakage Test must be performed with water in the outer jacket of the filter. Hollow fibers within casing must be fully submerged in water storage solution. Perform steps listed in the next section.

### 3.3.4. Nanofilter Leakage Test

1. Assemble a nitrogen or compressed air gas line with the gas valve and pressure regulator *closed*.
2. Attach the filter vertically to the stand and secure it with a clamp (*see* **Fig. 1**). First, remove the cap from nozzle N2, connect one end of the gas line tubing to N2, and close the line with forceps or clamp.
3. Remove the cap from nozzle N1, connect the other end of the gas line tubing to N1, and close the line with forceps.
4. Remove the cap from nozzle N3, and connect N3 to the silicone tubing, which leads into the discharge liquid flask.
5. Perform pressurization.
   a. Open the gas valve (with the pressure regulator still closed), and open the lines to nozzles N1 and N2.
   b. Slowly adjust the air regulator (over a 10-s period) until the pressure gauge connected to the filter retentate port reads 10–14 psi. Do not raise pressure above 14 psi. Higher pressures may affect the filter integrity.
   c. Turn the filter from the vertical position to the horizontal position with nozzles N3 and N4 pointing upward (**Fig. 2**).
   d. Permit water to discharge from nozzle N3 into a waste container under the pressure of the expanding hollow fibers. *Note:* The amount of water discharged from the 15 N filter is not sufficient to drip from the tubing—it will only enter into the tubing by a few inches.
   e. Beginning at about 10 s after the attainment of 12 psi, observe over 2 min the inside of the filter from all sides for any internal bubbling that continues for more than several seconds. (A short noncontinuous bubbling may occur in a normal filter following pressurization due to the release of residual gas within the filter and is not a sign of loss of integrity.) Do not pressurize the filter for more than 5 min. Do not use the nanofilter if continuous bubbling occurs (*see* **Note 4**).
6. If the filter is observed to be free from bubbling for a period of 20 s or more, it is assessed to have *passed* the leakage test.

Fig. 2. Schematic drawing of Asahi 15 N filtration unit (horizontal position).

*Note:* Although no maximum time limit is necessary, it is generally preferable to avoid prolonged pressurization of the Planova much beyond the total period of about 60 s as described.

7. Clamp off nozzles N1 and N2, and disassemble the N2 line. Place the filter back in the vertical position. Drain liquid from the nanofilter housing by uncapping upper the permeate port. When housing is empty, recap the permeate port.

### 3.3.5. Product Preparation

1. Prepare feed solution just prior to filtration. Spike 1 mL virus into 99 mL product for a 1:100 dilution.
2. Remove sample to titrate as 0.1-μm prefilter feed.
3. Prefilter the feed solution with a 0.1-μm prefilter. Remove sample after prefiltration. Titrate as 0.1-μm postfilter feed.
4. Prepare an infectivity assay positive control. Record dilution. For example, 1 mL virus + 9 mL DMEM/serum ($10^{-1}$ dilution).
5. Store the infectivity assay positive control, 0.1-μm prefilter feed and 0.11-μm postfilter feed samples at 4°C until ready to titrate.

### 3.3.6. Filtration Procedure

1. Drain the water from the outside of the hollow fiber in the filter (open ports N3 and N4).
2. Completely fill all prefilter tubing (dashed lines in **Fig. 3**) with buffer before connecting to the filter (to reduce air introduced into filter).
3. Wash filters with buffer solution. The buffer solution should be the same as used in the product manufacturing process.
4. Add the buffer solution into the feed flask. Pump through buffer to replace water and any air trapped from the inside of the hollow fibers. No pressure should be measurable on the monitor. Collect discharge from port N2 (port N4 should be capped, and port N3 should be clamped).
5. The volume of the solution to be discharged should be determined based on the surface area of Planova filters (*see* **Table 1**).
6. Immediately, before switching the tubing to the feed solution flask, briefly stop the pump, and switch the tubing into the feed solution flask. Make the transfer quickly to minimize pressure change across the membrane during the switch. Be careful not to pump any air into the tubing when switching from the buffer flask to the feed solution flask.
7. Change the pump speed during the filtration process to maintain the pressure at 12 psi (no more than 14 psi). This will require frequent rate reductions throughout the collection of the first filtrate.
8. Collect the filtrate in several fractions, each representing 20–25% of the total volume of the feed solution for the run. Record the volume, elapsed time, pressure, and pump setting for each filtrate collected.

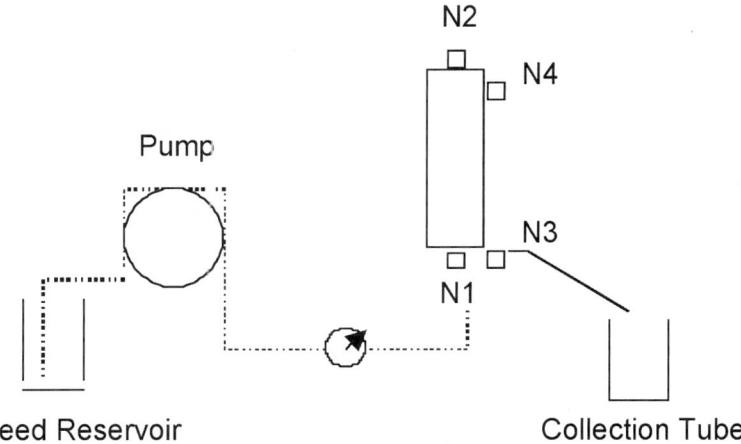

Fig. 3. Filtration unit connections to the feed reservoir and filtrate collection tubes.

**Table 1**
**Suggested Prewashing and Discharged Volume**

| Unit | Surface area (m²) | Volume discharged (mL) | Prewashing volume (mL) |
|------|------|------|------|
| Asahi/Planova 15 N | 0.001 | 10 | 10 |
|  | 0.01 | 10 | 30 |
| Asahi/Planova 35 N | 0.001 | 10 | 10 |
|  | 0.01 | 10 | 30 |
| Asahi/Planova 20 N | 0.001 | 10 | 10 |
|  | 0.01 | 10 | 30 |

9. When the feed solution is exhausted, add 1 mL buffer solution to rinse the feed flask. When this material is exhausted, briefly stop the pump, and switch the feeding tubing into a flask-containing buffer. Once again, be careful not to pump any air into the tubing when switching the tubing, moving quickly to minimize pressure change.

10. Collect the wash in two fractions, each representing 1/10 of the total volume of the feed solution for the run, or 2 mL, whichever is larger. (A minimum of 2 mL is required for protein concentration assay and viral titration.) Record the volume, elapsed time, pressure, and pump setting for each wash collected.

11. Measure and record protein concentration values of all samples: initial product, initial product plus virus (0.1-μm prefeed), 0.1 μm postfeed, all filtrates, and all washes (**Table 2**). Store all samples at 4°C until titrations are completed.

12. Dismantle and enclose filter in a Zip-lock bag. Store at 4°C until results and analysis are completed. If viral titration indicates loss of filter integrity, perform leakage test.

**Table 2**
**Suggested Samples of Nanofiltration Using Asahi Planova 15 N Filters**

| Sample | Wt of container + sample (g) | Wt of container (g) | Wt of sample (g) | Time min/s | Pressure (psi) | Flow rate mL/min | Protein conc. | Suggested virus titration dilutions |
|---|---|---|---|---|---|---|---|---|
| Positive control | NA[b] | NA | NA | NA | NA | NA | NA | $3.2^{-4}$–$3.2^{-14}$ |
| 0.1-μm prefeed | | | | NA | NA | | | $3.2^{-4}$–$3.2^{-14}$ |
| 0.1-μm postfeed | | | | NA | | | | $3.2^{-4}$–$3.2^{-14}$ |
| Buffer purge | | | | | | | | $3.2^{0}$–$3.2^{-10}$ |
| Filtrate 1 | | | | | | | | $3.2^{0}$–$3.2^{-10}$ |
| Filtrate 2 | | | | | | | | $3.2^{0}$–$3.2^{-10}$ |
| Filtrate 3 | | | | | | | | $3.2^{0}$–$3.2^{-10}$ |
| Filtrate 4 | | | | | | | | $3.2^{0}$–$3.2^{-10}$ |
| Pooled filtrate[a] | | | | | | | | $3.2^{0}$–$3.2^{-10}$ |
| Wash 1 | | | | | | | | $3.2^{0}$–$3.2^{-10}$ |
| Wash 2 | | | | | | | | $3.2^{0}$–$3.2^{-10}$ |

[a]Pooled filtrate = 10% from each of the filtrates.
[b]NA, not applicable.

## 4. Notes

1. Virus filter retention capability may be altered during the course of filtration. Protein aggregates may deposit on the surface of the filter during the filtration process, reducing the effective pore size of the filter and adding an additional retentive layer. Viruses can be also held on the filter surface by adsorptive sites through a variety of forces, such as charge, van der Waals forces, hydrogen bonding, steric entrapment, and so on. These forces may act alone or in an additive manner. Adsorptive retention also depends on virus diffusion and virus size relative to the effective membrane pore size *(10)*.

2. A typical virus filter provides consistent high virus clearance for viruses larger than the filter rating. For viruses with sizes below the filter rating, clearance may be dependent on fluid and process conditions, especially for small viruses significantly below the filter's rating. A virus size rating has been assigned by the manufacturer based on the retention of a particular model (e.g., bacteriophage) at a given LTR, or an average pore-size rating is established from a mathematical model for the permeability.

3. The virus retention capability of nanofilters is assessed by challenging samples of the specific membrane under the vendor-specific challenge organism and test conditions, as well as analyzing the retention efficiency of the filter. Generally, all virus retention integrity tests, e.g., forward flow tests, pressure hold tests, and particle challenge tests, have an intrinsic limit point where defects cannot be detected even if the defects are comparably large. This occurs when the number of the defects per unit membrane surface area is small. To overcome this disadvantage, it is recommended that these integrity tests be performed in combination with a leakage test.

4. For nano-grade membrane filters, the integrity test value established by vendors is linked to the titer reduction (i.e., $\log_{10}$ reduction values) afforded by the filter. Filter manufacturers use different microorganism challenges to release filter lots. *See* examples listed here.

| Test type | Integrity test principle | Test characteristics | Virus/phage retention correlation |
|---|---|---|---|
| Gold Particle Test | Use the monodisperse colloidal gold particle suspension to confirm the distribution of pore size in the filter. | Destructive test. May not detect large defects (pinholes). JEV, size 40–50 nm. | Correlation is established and validated. Virus markers: PPV, size 18–24 nm; poliovirus, size 30 nm; |
| Leakage Test | Modification of the Bubble Point method. | The outside of the hollow fibers in a test filter is filled with water, and the pressure to the inside of the fibers is raised to a specified value. Continuous stream of bubbles is fail criterion. | To assess for gross defects, it is recommended to use in conjunction with a more sensitive integrity test (Gold Particle test, Liquid Forward Flow test, and Pressure hold test) |
| Forward/ Diffusive Flow | Migration of test gas molecules through the wetted filter. The rate of diffusion depends on the solubility and of the test gas. | The test is sensitive to temperature fluctuations. Suitable for confirming proper installation and absence of gross damage. | Correlation is established and validated. Virus markers: Phage Phi 6; size 78 nm; Phi X174, size 28 nm |

## Acknowledgments

The authors thank Dr. Peter Charles and Dr. Hiroo Nakano for their technical expertise in performing the evaluation tests.

## References

1. Brough, H., Antoniou, C., Carter, J., Jakubik, J., Xu, Y., and Lutz, H. (2002) Performance of a novel Viresolve NFR virus filter. *Biotechnol. Prog.* **18,** 782–795.
2. Aranha, H. Viral Clearance Strategies for biopharmaceutical safety. (2001) Part 2: Filtration for viral clearance. *BioPharm* **14,** 32–43.
3. ICH Q5A. Viral Safety Evaluaticn of Biotechnology Products Derived from Cell Lines of Human or Animal Origin (1998) Available at www.ifpma.org/ich5q.html.
4. CPMP Note for Guidance on Virus Validation Studies: The design, contribution and interpretation of studies validating the inactivation and removal of viruses. CPMP/BWP/268/95.
5. Levy, R. V., Phillips, M. W., and Luntz, H. Filtration and the removal of viruses from biopharmaceuticals, in *Filtration in the Biopharmaceutical Industry* (Meltzer, T. and Jornitz, M., eds.), Marcel Dekker, New York, Basel, Hong Kong, 1998, 619–646.
6. Van Holten, R. W., Ciavarella, D., Oulundsen, G., Harmon, F., and Riester, S. (2002) Incorporation of an additional viral-clearance step into a human immunoglobulin manufacturing process. *Vox Sang* **83,** 227–233.
7. Korneyeva, M., Rosenthal, S., Charles, P., Trejo, S., Pifat, D., and Petteway, S. (2004) Important parameters for optimal target virus clearance: virus spike selection issues. *Am. Pharm. Rev.* **7,** 38–42.
8. Cavalli-Sforza, L., ed. Biometri Grundzuge Biologisch-Medizinischer Statistik, Gustav Fischer Verlag Stuttgart, 1974.
9. Tateshi, J., Kitamoto, T., Mohri, S., Satoh, S., Sato, T., Shepherd, A., and Macnaughton, M. R. (2001) Scrapie removal usirg Planova virus removal filters. *Biologicals* **29,** 17–25.
10. Grant, D. C., Liu, B. Y. H., Fischer, W. G., and Bowling, R. A. (1989) Particle capture mechanisms in gases and liquids: an analysis of operative mechanisms. *J. Environ. Sci.* **42,** 43–51.

# 19

## Determining Residual Host Cell Antigen Levels in Purified Recombinant Proteins by Slot Blot and Scanning Laser Densitometry

**Aaron P. Miles, Daming Zhu, and Allan Saul**

## 1. Introduction

The production of recombinant proteins for therapeutic and vaccine uses necessitates the analysis of purified products for residual host cell antigens. Various assays to quantify these antigens in both in-process streams and purified bulk solutions have been described *(1–3)*. Process-specific assays utilize those host cell antigens most likely to copurify with the protein of interest, whereas generic assays use all recoverable host cell antigens, thereby testing for the presence of any antigen during protein production and that might therefore be recovered with the purified protein. With limited utility of commercially available assay kits, we developed a fast, simple, quantitative, and highly sensitive method based on immunoblotting to accurately determine residual host cell antigen levels. Samples are applied directly to a membrane, and signals are measured by densitometry. This assay results in very low-background signals, providing accuracy and precision at the low nanogram/milliliter level, comparable to reported enzyme-linked immunosorbent assay (ELISA)-based methods.

## 2. Materials

1. Microfluidized, soluble *Escherichia coli* antigens: sham fermentation run using null (no gene insert in the vector) BL21 (DE3) *E. coli* (*see* **Note 1**).
2. Purified test protein: BL21 (DE3) *E. coli*-produced $MSP1_{42}$-3D7 *(4)*.
3. Pools of rat antisera to antigens.
4. Slot-blot apparatus with nitrocellulose membrane.
5. 96-Well plate.
6. High-performance liquid chromatography (HPLC)-grade water (*see* **Note 2**).
7. 1X Tris-buffered saline (TBS): 25 m$M$ Tris-HCl, pH 7.4, 137 m$M$ NaCl, and 3 m$M$ KCl.
8. Blocking buffer: 3% skim milk in 1X TBS.
9. TBS/Tween: 1X TBS and 0.05% Tween-20.

From: *Methods in Molecular Biology, vol. 308: Therapeutic Proteins: Methods and Protocols*
Edited by: C. M. Smales and D. C. James © Humana Press Inc., Totowa, NJ

10. Goat anti-rat immunoglobulin G (IgG) secondary antibody conjugated to alkaline phosphatase.
11. Alkaline phosphatase substrate: 5-bromo-4-chloro-3-indolyl-phosphate/nitroblue tetrazolium (BCIP/NBT) membrane phosphatase substrate system (Kirkegaard & Perry Labs, Gaithersburg, MD).
12. Personal Densitometer SI with ImageQuant software (Molecular Dynamics Inc., Sunnyvale, CA; *5*).

## 3. Methods

The methods described here outline the loading of standard, test, and "spiked" samples onto a membrane using a slot-blot apparatus, followed by blocking, incubation in primary and secondary antibody solutions, and development. Also, this chapter describes scanning and analyzing the membrane using laser densitometry and ImageQuant software, along with graphing software, to determine the level of host cell antigen impurities.

### 3.1. Slot Blot

1. Block all wells of a 96-well plate by adding 400 µL blocking buffer and incubating at room temperature for 30 min.
2. Wash the plate with five changes of TBS/Tween and dry.
3. Dilute the soluble *E. coli* antigens (standard) to 1.56 µg/mL and the purified test protein (MSP1$_{42}$-3D7) to 0.1 mg/mL. In the plate, prepare 12 twofold serial dilutions of each.
4. Assemble the slot-blot apparatus. Add 300 µL 1X TBS to all wells, and draw through the membrane with a gentle vacuum.
5. Into wells 1–12 of one column of the assembled slot-blot apparatus, load 60 µL of each dilution of one of the antigen solutions (with the highest concentration at the top and lowest at the bottom). Add 60 µL 1X TBS to the remaining wells (13–24) of that column. In the same manner, repeat with the other antigen solution to fill another column. Then add 60 µL of each dilution of both antigen solutions to the remaining column of the apparatus. This column then contains standard *E. coli* antigens spiked with the purified MSP1$_{42}$-3D7 in amounts equivalent to those in the first two columns, which tests for interference.
6. Draw the solutions in all of the wells through the membrane with gentle vacuum.
7. Wash all wells with 2 × 300 µL 1X TBS, then disassemble the apparatus.
8. Cut the membrane horizontally between rows 12 and 13, and label each piece appropriately.
9. Place the membrane pieces in a suitable container, and add enough blocking buffer to cover them. Incubate for 1 h at room temperature with gentle agitation.
10. Pour off the blocking buffer, and add enough pooled rat anti-*E. coli* soluble antigens sera at an appropriate dilution in blocking buffer to cover the membranes (*see* **Notes 3–6**). Incubate for 1 h at room temperature with gentle agitation.
11. Pour off the primary antibody solution (can be frozen and reused several times), and wash the membranes by agitating for approx 10 min three times at room temperature in plenty of TBS/Tween.
12. Pour off the TBS/Tween, and add enough goat anti-rat IgG alkaline phosphatase-conjugated secondary antibody solution at an appropriate dilution (e.g., 1:3000 in blocking buffer) to cover the membranes. Incubate for 1 h at room temperature with gentle agitation.
13. Repeat the wash as in **step 11**.

14. Prepare the BCIP/NBT substrate solution according to the manufacturer's instructions. Pour off the TBS/Tween, and add enough substrate solution to cover the membranes. Incubate with gentle agitation for 40 min at room temperature (*see* **Note 7**). Add water and gently agitate at room temperature for 20 min with two changes of water to stop color development.

## 3.2. Scanning and Data Analysis

1. Scan the membranes with the densitometer while still wet (*see* **Note 8**) according to the manufacturer's instructions, ensuring that the lanes of the membranes run straight from the top to the bottom of the scanning bed. **Figure 1** shows a representative membrane with soluble *E. coli* antigens in column A, $MSP1_{42}$-3D7 in column B, and a mixture of the two (soluble *E. coli* antigens spiked with $MSP1_{42}$-3D7 to test for interference; *see* **Note 9**) in column C. The membranes loaded with only 1X TBS should be scanned and analyzed to control for background signal.

2. Draw an encompassing line for analysis down each lane of interest, and widen the lines appropriately. Include in the analyzed area of each lane enough length from top to bottom and enough width to envelope all areas on the membrane that had samples loaded onto them through the wells. **Figure 2** illustrates this using the same membrane as shown in **Fig. 1**.

3. Create a line graph for each line drawn. Examples are shown in **Fig. 3**, which gives representations (in optical density units vs pixel) of the serial dilutions in each column on the membrane, where lines 1, 2, and 3 of the graphs correspond to columns A, B, and C, respectively.

4. Adjust all relevant settings to give the most accurate representation of the baseline signal. Options for this within the ImageQuant software may include "automatic" (the baseline is detected by the software based on manually adjusted settings to appropriately identify peaks), "lowest point" (the most transparent part of the lane is taken as the baseline), and point-by-point manipulation. Other software programs are available that allow more customized identification and tuning of the baseline and peaks (e.g., PeakFit). It is then usually necessary to redefine the peaks by hand, rather than letting the software do it, to fine-tune where each peak begins and ends. Once the peaks are defined, they are integrated, i.e., the area under each peak is calculated by the software. In this way, one quantifies the signal arising from each diluted sample loaded onto the membrane.

6. Using the areas generated from the integration of each soluble *E. coli* standard peak, create a graph of area counts vs known antigen load using the appropriate software (e.g., Excel and SigmaPlot). Determine the linear range, fit a line for the data points of that range, and obtain an equation for the line. One or more peaks in the test protein lane should then be identified as both falling within that range and giving rise to a linear signal response. The amount of host cell antigen in each of those peaks is then obtained from the equation for the line. It is best to obtain amounts for several peaks (several different dilutions of test sample) so that a mean can be determined. A typical standard curve, generated in Excel, is shown in **Fig. 4**. The curve demonstrates that this assay is linear over seven twofold dilutions or from 12 to 0.2 ng total host cell antigen loaded on the membrane. Peaks 2, 3, and 4 from **Fig. 3** fall within the linear range and give rise to a linear response of signal vs sample load. By plugging test sample peak signals into the equation of the line for the standard curve, one obtains the level of host cell antigen present per unit of test protein solution. In the case of $MSP1_{42}$-3D7, the impurity level was found to be 1.06 ng host cell antigen per microgram of protein or 0.106%, based on

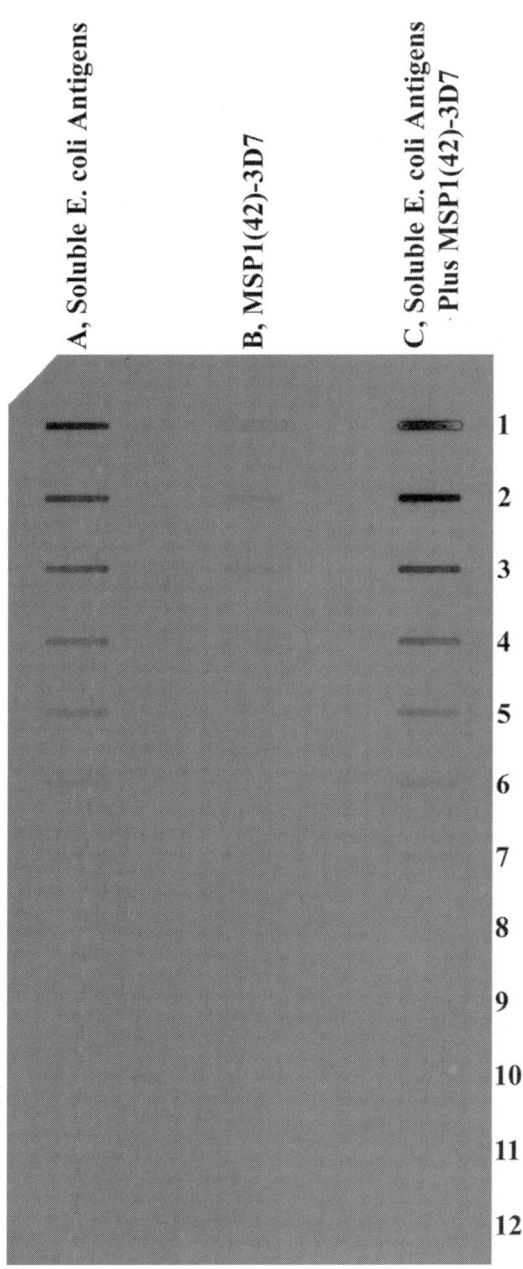

Fig. 1. Typical membrane following development. Each column contains serial twofold dilutions of the indicated samples, beginning with 94 ng (column A), 6 μg (column B), and 94 ng (A) + 6 μg (B) (column C).

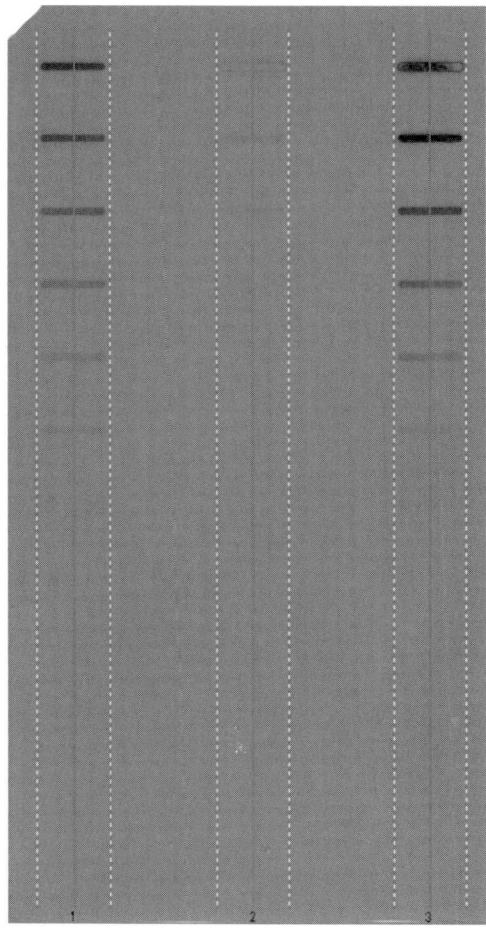

Fig. 2. Membrane with encompassing lines for analysis. The same membrane in **Fig. 1** is shown with lines drawn and widened in ImageQuant to envelope all sample loads.

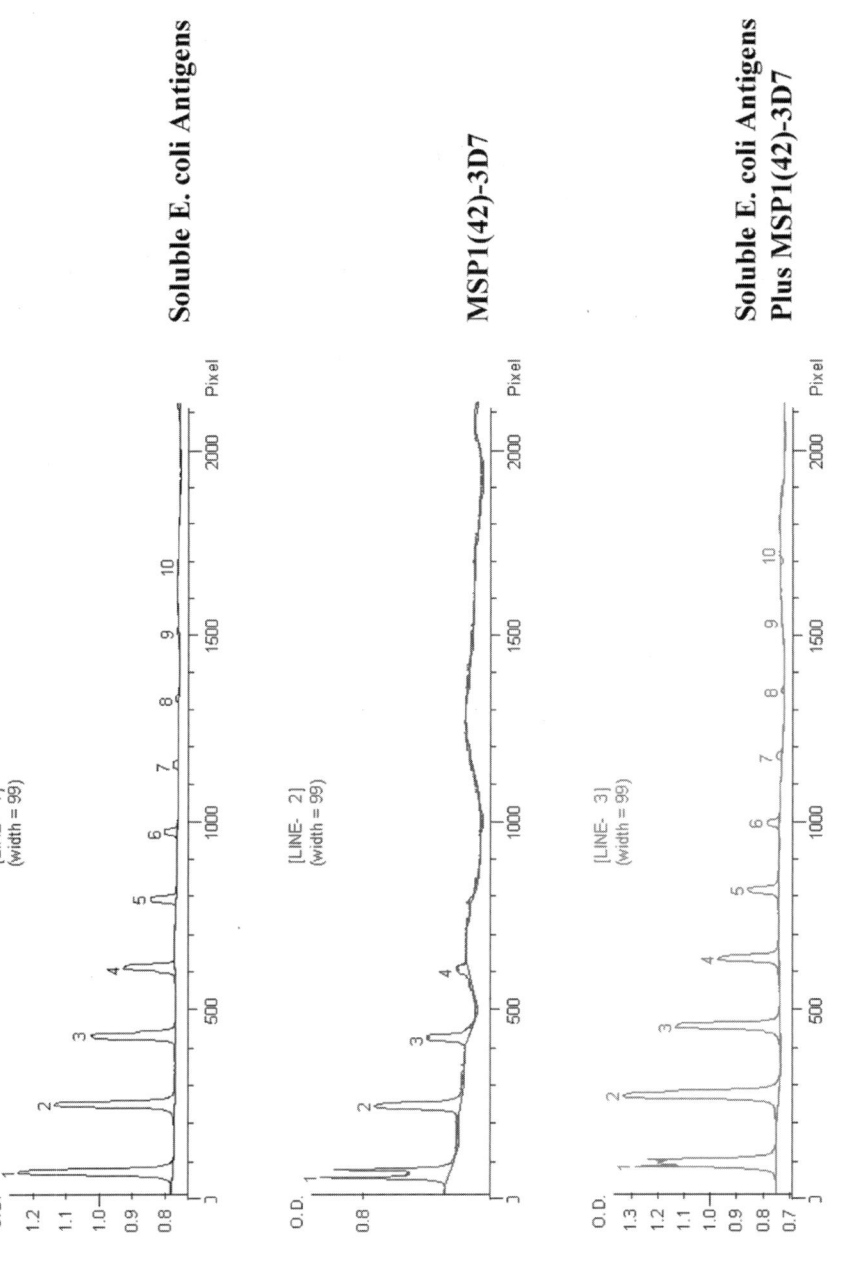

Fig. 3. Line graphs generated in ImageQuant. Graphical representations of the lines in **Fig. 2** are shown, left to right lines 1, 2, and 3 corresponding to the top to the bottom of columns A, B, and C, respectively. Optical density vs pixel (numbered from top to bottom of each line) is plotted.

*238*

## Standard Curve of *E. coli* Soluble Antigens

$y = 0.2663x + 0.0201$ (linear fit)

$R^2 = 0.9992$

| X Sample Load (ng) | Y Peak Area | X Back Calc. (ng) |
|---|---|---|
| 93.8 | 9.59 | 35.92 |
| 46.9 | 7.53 | 28.20 |
| 23.4 | 5.17 | 19.34 |
| **11.7** | **3.11** | **11.60** |
| 5.9 | 1.64 | 6.08 |
| 2.9 | 0.80 | 2.93 |
| 1.5 | 0.39 | 1.39 |
| 0.7 | 0.18 | 0.60 |
| 0.4 | 0.12 | 0.38 |
| 0.2 | 0.08 | 0.22 |

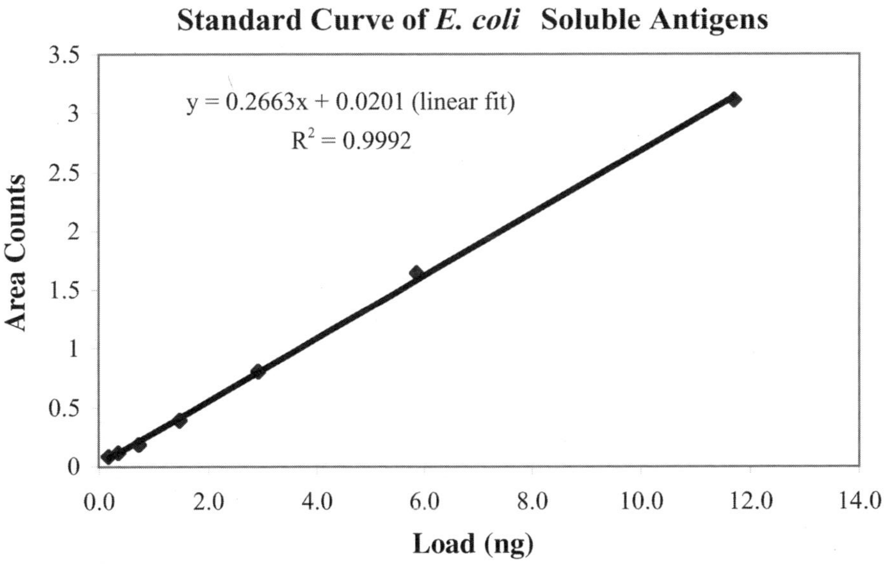

Fig. 4. Standard curve of *E. coli*-soluble antigens. A representative standard curve (in area counts *vs* ng loaded) is shown of dilutions of *E. coli*-soluble antigens generated in Excel. The equation for the line (based on a linear fit) and correlation coefficient ($R^2$) are given. Known and back-calculated *E. coli* antigen sample loads and the observed peak areas are tabulated.

**Table 1**
**Interassay Variation of *E. coli*-Soluble Antigen Standard Curve**

| Sample load (ng) | Assay no. | | | | | | | | | | | Mean area | SD |
|---|---|---|---|---|---|---|---|---|---|---|---|---|---|
| | 1 | 2 | 3 | 4 | 5 | 6 | 7 | 8 | 9 | 10 | 11 | | |
| | Area* | | | | | | | | | | | | |
| 93.8 | 6.54 | 8.99 | 6.59 | 7.39 | 7.35 | 7.91 | 9.59 | 8.20 | 8.44 | 8.44 | 9.36 | 8.07 | 1.03 |
| 46.9 | 6.00 | 6.96 | 5.13 | 5.35 | 5.66 | 6.33 | 7.53 | 6.17 | 6.43 | 7.24 | 7.19 | 6.36 | 0.80 |
| 23.4 | 4.07 | 5.57 | 4.19 | 3.65 | 4.36 | 4.97 | 5.17 | 4.33 | 4.57 | 5.20 | 5.16 | 4.66 | 0.59 |
| 11.7 | 2.11 | 2.84 | 2.11 | 2.08 | 2.30 | 3.11 | 3.11 | 2.58 | 2.73 | 3.12 | 3.17 | 2.66 | 0.45 |
| 5.9 | 1.16 | 1.46 | 1.13 | 0.97 | 1.16 | 1.77 | 1.64 | 1.37 | 1.66 | 1.60 | 1.69 | 1.42 | 0.27 |
| 2.9 | 0.51 | 0.86 | 0.58 | 0.51 | 0.54 | 0.65 | 0.80 | 0.69 | 0.76 | 0.84 | 0.87 | 0.69 | 0.14 |
| 1.5 | 0.25 | 0.42 | 0.27 | 0.22 | 0.28 | 0.30 | 0.39 | 0.33 | 0.34 | 0.45 | 0.38 | 0.33 | 0.07 |
| 0.7 | 0.12 | 0.23 | 0.14 | 0.13 | 0.15 | 0.18 | 0.18 | 0.20 | 0.21 | 0.22 | 0.18 | 0.18 | 0.04 |
| 0.4 | 0.07 | 0.12 | 0.08 | 0.08 | 0.09 | 0.07 | 0.12 | 0.11 | 0.12 | 0.10 | 0.09 | 0.09 | 0.02 |

*Units are "area counts." SD, standard deviation.

Results for 11 replicates (performed by a single analyst on multiple days) are shown. The sample load is the known amount of *E. coli* soluble antigen applied to a given well and bound to the membrane. All assays were performed with pools of rat anti-*E. coli* soluble antigen sera. It is generally found that 0.2 (not shown) or 0.4 ng is the lower limit of quantitation, and 11.7 ng is the upper limit (7–190 ng/mL, based on 60-μL loads).

means of three peaks per assay and three assays. **Table 1** shows the interassay variation among 11 replicate assays in producing standard curves.

## 4. Notes

1. $MSP1_{42}$-3D7 is overexpressed in *E. coli* and packed into inclusion bodies. The sham fermentation run (which uses the host cells containing the expression plasmid without the gene insert) does not result in the formation of inclusion bodies but generates those *E. coli* host cell antigens that are most likely to be present at any given stage of product purification. These antigens are then used for immunizations, as well as standards on the membrane (*see* below). Aside from the induction step that results in expression of the protein of interest, the sham fermentation run is identical to that used for protein production.

2. TBS, TBS/Tween, and blocking buffer must be prepared using HPLC-grade water. This assay was initially performed using sterile, distilled, cell culture–grade water. But with the long development times needed, strong signals in the blank (TBS) lanes developed. These were eliminated when we switched to HPLC-grade water.

3. Initially for this assay, antisera were raised against host cell antigens in rabbits. We found that preimmune bleeds of these sera showed reactivity to some of our recombinant proteins; thus, we generated antisera in mice and rats. These sera were characterized by ELISA and immunoblotting, and we detected broader and stronger reactivity with the rat sera, which were therefore chosen for development of this assay.

4. It is necessary to assay the antisera by ELISA and immunobloting. Although the mouse and rat antisera showed only low-level antibodies by ELISA, the rat sera exhibited much broader and higher reactivity to the host cell antigens on immunoblots than the mouse sera. All five immunized rats generated sera that showed similar reactivity. Therefore, we pooled them for the assay.

5. A study was performed with varying concentrations of primary and secondary antibodies in an attempt to increase the signal-to-noise ratio, but all combinations showed a similar ratio. We therefore settled on a primary antibody dilution of 1:1000 and a secondary of 1:3000. We later found that the noise in the assay arose from the use of non-HPLC–grade water.

6. As a check for information purposes only, this assay should also be performed using sera from rats immunized with *E. coli*-insoluble cellular material (microfluidized and centrifuged, then resuspended in 1X phosphate-buffered saline or suitable buffer). This will ensure that the recovery/purification process does not result in unacceptably high levels of any *E. coli* antigen. However, the method is only quantitative when using sera generated against soluble antigens, since the standard curve is derived from these.

7. Highly purified proteins can require long incubation times to develop a signal that is within the linear range of the assay.

8. A wet membrane improves the signal-to-noise ratio by making the membrane more transparent, allowing the scanner to detect small peaks that it otherwise could not. This improves the sensitivity of the assay.

9. We have observed only negligible interference in spiked samples when performing this assay with several different purified proteins. The signals of the standard lane and spiked lane are similar within the linear range of the assay.

## Acknowledgments

The authors wish to thank Anthony Stowers, Michael Whitmore, Sanjay Singh, Richard Shimp, Vu Nguyen, Jacqueline Glen, and Jin Wang for fermentations, refolding, and purification of $MSP1_{42}$-3D7.

# References

1. Dagouassat, N., Haeuw, J.-F., Robillard, V., Damien, F., Libon, C., Corvaïa, N., et al. (2001) Development of a quantitative assay for residual host cell proteins in a recombinant subunit vaccine against human respiratory syncytial virus. *J. Immunol. Methods* **251,** 151–159.

2. Wan, M., Wang, Y., Rabideau, S., Moreadith, R., Schrimsher, J., and Conn, G. (2002) An enzyme-linked immunosorbent assay for host cell protein contaminants in recombinant PEGylated staphylokinase mutant SY161. *J. Pharm. Biomed. Anal.* **28,** 953–963.

3. Rathore, A. S., Sobacke, S. E., Kocot, T. J., Morgan, D. R., Dufield, R. L., and Mozier, N. M. (2003) Analysis for residual host cell proteins and DNA in process streams of a recombinant protein product expressed in *Escherichia coli* cells. *J. Pharm. Biomed. Anal.* **32,** 1199–1211.

4. Singh, S., Kennedy, M. C., Long, C. A., Saul, A. J., Miller, L. H., and Stowers, A. W. (2003) Biochemical and immunological characterization of bacterially expressed and refolded *Plasmodium falciparum* 42-kilodalton C-terminal merozoite surface protein 1. *Infect. Immun.* **71,** 6766–6774.

5. ImageQuant User's Guide Version 5.0. Molecular Dynamics, Inc., Sunnyvale, CA, 1998.

# 20

## Principles of Biopharmaceutical Protein Formulation

### An Overview

### Scott P. Sellers and Yuh-Fun Maa

### 1. Introduction

Ideally, formulation development of biopharmaceutical protein therapeutics will provide a final dosage form that offers sufficient ex vivo stability during processing, handling, and long-term storage and also provide adequate in vivo stability in terms of bioavailability that meets the pharmacokinetics/pharmacodynamics (PK/PD) therapeutic requirements. This chapter focuses on ex vivo stability of protein therapeutics and is targeted for the novice researcher who struggles with the inherent instabilities of biological molecules. More important, this chapter provides the inexperienced formulators with fundamental understanding and the basic tools for formulation development of biopharmaceutical proteins.

Because of the marginal bioavailability of protein therapeutics via noninjectable delivery routes (e.g., oral or inhalation), parenteral administration is often preferred. Yet, the toxicological and safety requirements for parenteral formulations are much more stringent. As most protein therapeutics cannot survive terminal sterilization procedures currently available, e.g., γ-irradiation or ethylene oxide treatment, considerations for formulation component/excipient selection should include microbiological requirements in addition to biological, chemical, and physical functions. The presence of measurable quantities of endotoxin, pyrogens, or objectionable organisms (transmissible spongiform encephalopathies [TSEs]) in raw materials can be difficult and costly to remove during product manufacturing and should be avoided. **Subheading 2.** provides a list of commonly used excipients typically evaluated during formulation development. Throughout this chapter, the function of these excipients in specific dosage forms is discussed.

The final pharmaceutical dosage form should remain stable both physically and chemically upon room temperature storage for at least 2-yr shelf-life; analytical methodologies used to detect chemical and physical instabilities are discussed in

From: *Methods in Molecular Biology, vol. 308: Therapeutic Proteins: Methods and Protocols*
Edited by: C. M. Smales and D. C. James © Humana Press Inc., Totowa, NJ

**Subheading 3.** Equally important, the final pharmaceutical formulation must also provide sufficient stability of the active pharmaceutical ingredient (API) during shipping and handling to ensure that the dosage and safety label claims are met at the time of administration. Owing to the inherent chemical and physical instability of therapeutic proteins, the aforementioned stability requirements for the final biopharmaceutical product are often difficult to achieve with a simple liquid formulation. Therefore, development or codevelopment of a frozen or dry solid-state formulation, such as a lyophilized or spray-dried preparation, is often necessary. **Subheading 4.** specifies the principal steps used to develop each dosage form, along with the common physical and chemical degradation mechanisms encountered within each formulation type. Lastly, **Subheading 5.** reviews the primary considerations and concerns of containers and closures used for liquid and solid-state lyophilized formulations.

## 2. Materials

Available for the formulation scientist is a long list of excipients that have been previously approved for use in commercial pharmaceutical products and are generally regarded as safe (GRAS). However, the time and cost for developing an injectable product can be significantly reduced if all selected formulation excipients have been previously approved for parenteral administration. Preferably, the presence and quantity of all excipients used in the final product should be accompanied by supportive data, and only the minimum quantities required to elicit the desired protective effect should be used. Additionally, the chosen excipients should be inert toward chemical degradation–reducing sugars, and surfactants that contain or can produce oxidative byproducts should be avoided from the possibility of chemically altering the protein upon storage. The following list of materials is not comprehensive but does include some typically used excipients that are evaluated during initial formulation development based on the specific product requirements and the major degradation pathways for the therapeutic protein.

### 2.1. Buffers

The following buffers are often used to help regulate the pH of the formulation solution to minimize chemical degradation mechanisms, such as hydrolysis or deamidation upon storage, which are accelerated at low and high pH, respectively (*see* **Subheading 4.**). Each buffer is usually evaluated at a concentration between 10 and 100 m$M$ and within 1 pH unit of its respective pKa: acetate, pKa 4.76; citrate, pKa 6.4; phosphate, pKa 7.2; and Tris hydroxylaminomethane (Tris), pKa 8.1.

### 2.2. Tonicifiers and Bulking Agents

The excipients that are typically added to liquid formulations to achieve isotonicity are sodium chloride and dextrose. Inorganic salts, e.g., sodium chloride, can also stabilize proteins at certain concentrations by nonspecifically binding to the native-folded state, increasing thermal stability and inhibiting denaturation and inactivation. However, sodium chloride is not ideal for solid-state formulations because of the low-eutectic freezing temperatures and concentration effect that occurs during freezing—

topics that are discussed further in **Subheading 4.** Mannitol or glycine are commonly used as bulking agents to facilitate freeze drying of small protein quantities and improve cake appearance. Dextrans and hydroxyethyl starch (HES) have also been used as bulking agents to improve lyophilized cake appearance and in tonicity adjustment.

## 2.3. Disaccharide Stabilizers

Sucrose is perhaps the most common excipient that can stabilize proteins in liquid or frozen solutions via preferential exclusion (*see* **Subheadings 4.1.** and **4.2.**). For solid-state formulations, nonreducing amorphous disaccharides, such as sucrose or trehalose, are generally required to achieve adequate storage stability, acting as a water surrogate for the hydration sphere of the protein and also significantly reducing protein mobility in the dried state by forming a glassy matrix (*see* **Subheading 4.3.**). However, some sugars should be avoided in protein formulations, including reducing sugars like lactose and maltose as the most notable examples. By definition, a reducing sugar contains reactive aldehyde or ketone groups, which are prone to react with amine groups of amino acids and proteins by the Maillard reaction (*1*).

## 2.4. Surfactants

Nonionic surfactants, such as polysorbates (Tween 80) or polyethylene oxide polypropylene oxide block copolymers (Pluronic F68), can be added to liquid formulations to prevent surface denaturation at air–water interfaces, surface adsorption to water–glass interfaces, and intermolecular interactions between exposed hydrophobic residues of the protein (*see* **Subheading 4.1.**). The role of surfactants in stabilizing proteins through freezing has also been recognized (*see* **Subheading 4.2.**). Ionic surfactants, e.g., sodium dodecyl sulfate (SDS), tend to denature proteins at relatively low concentrations (<0.1% w/v) and should be avoided.

## 2.5. Chelating Agents

Metal ions are possible contaminants in several raw materials, and their presence in protein formulations should be avoided if possible, as they can catalytically degrade/ oxidize the protein. If removing the source of metal ion impurities is not possible, or the source cannot be readily identified, the addition of a chelator (e.g., EDTA) can generally minimize the effect of metal-catalyzed oxidation. Citric acid can also be added as a chelating agent for trace metal and can be used as a buffering system in a dual role.

## 2.6. Antioxidants

Methionine and ascorbic acid are two frequently used reducing agents that can act as oxygen scavengers that prevent oxidation of the protein. These additives are evaluated when physical measures (removing oxygen from the headspace of the container) are not effective. Antioxidants (e.g., butylated hydroxyl toluene [BHT], propyl gallate, and vitamin E) have also been used to reduce oxidative damage in protein formulations.

## *2.7. Preservatives*

For multidose liquid protein formulations, antimicrobial preservatives are often needed to prevent microbial growth after the seal is punctured for first-dose withdrawal. Sometimes such preservatives are incorporated into single-dose liquid formulations to ensure sterility throughout the product shelf-life. Aromatic organic compounds (phenol, benzyl alcohol, and meta-cresol) are common preservatives used in marketed protein products. However, formulation scientists should use caution when dealing with these preservatives because of their incompatibility with certain proteins, including recombinant human growth hormone, human insulin-like growth factor, and recombinant human interferon-γ. These proteins have been shown to precipitate in the presence of these preservatives.

## 3. Analytical Methodologies

To some degree, formulation development must begin with analytical method development or method qualification to monitor the performance of each candidate formulation through a series of initial forced degradation experiments, excipient screening, and eventual long-term stability studies. The least stable component of the formulation—the protein—must be as fully characterized as possible from the initial stages of development to detect any perturbations that may lead to significant stability issues. Monitoring the performance of candidate formulations through forced degradation experiments by several complementary analytical methods is often necessary to elucidate the major mechanism(s) of degradation. These methods can eventually be developed into the stability-indicating methods used for analysis of long-term storage studies. Several common analytical techniques and their uses and limitations are now described.

## *3.1. Protein Characterization*

### *3.1.1. Protein Concentration*

Protein concentration can be measured by simple ultraviolet (UV) spectroscopy owing to the absorbance at 190–210 nm from the carbonyl functionalities of the backbone peptide bonds and absorbance at 240–320 nm from the aromatic residues (tyrosine, tryptophan, and phenylalanine). To a lesser extent, disulfide bonds (if present) also contribute to the UV signal at 250–300 nm. Protein concentration is typically quantitated at a wavelength around 280 nm using an extinction coefficient derived from an absolute independent method, such as Kjeldahl (nitrogen content) or amino acid analysis. Both can be time- and sample-consuming methods and are therefore not routinely performed in the research laboratory. If the primary sequence of the therapeutic protein is known, the molar extinction coefficient can be predicted from the known extinction coefficients of the individual amino acids. However, most therapeutic proteins are globular molecules that feature hydrophobic aromatic residues buried inside the folded structure, exposing only the hydrophilic residues to the polar aqueous solvent. Because the internal aromatic residues may not be fully accessible to UV light, thus affecting the overall absorbance signal, the calculated value from the individual extinction coefficients may not be suitable for the native folded protein (*see* **Note 1**).

Colorimetric methods, such as the bicinchoninic acid (BCA) and Lowry or Bradford assays, can also be used to routinely determine protein concentration with the use of a suitable reference standard (e.g., bovine serum albumin). However, as some proteins interact differently with the chromogenic reagent of the colorimetric methods, it is necessary to select a reference standard that is comparable to the analyte over the desired range of analysis *(2)*. Additionally, several commonly used excipients and surfactants (EDTA and reducing sugars and polysorbates) are known to interfere with colorimetric assays. For example, when measuring a blank protein-free solution containing only the formulation buffer, excipients, and Tween-80, the background signal was observed to increase markedly over time, as the stock blank solution was stored at 2–8°C, compromising the results from storage stability studies. Fresh preparations of the blank solution at each time point were much more consistent.

These methods for the determination of protein concentration are nonspecific and cannot be used to evaluate protein purity or heterogeneity, which must be determined by other methods.

### 3.1.2. Protein Conformation

Absorbance spectroscopy can be used to determine the thermal denaturation temperature (unfolding temperature or $T_m$), because the aromatic residues contributing to the absorbance signal are sensitive to their immediate environment. A change in protein conformation is observed as a slight shift in the maximum absorbance, as the environment surrounding the hydrophobic aromatic residues (typically buried deep inside the hydrophobic core) will be much different than that in the unfolded state in which all residues are exposed to the aqueous polar environment. As the shift in UV absorbance is generally quite small even for complete unfolding (1–2 nm), differential UV spectroscopy, utilizing a dual-beam UV spectrometer and a control sample, or analyzing the second-derivative UV absorbance spectrum, can help to identify conformational changes in proteins *(3,4;* **Note 2**).

Fluorescence spectroscopy is an effective tool for characterizing protein conformation, capitalizing on the unique properties of the aromatic residues of the peptide that fluoresce in the 300–400-nm range when excited at wavelengths between 250 and 300 nm. Compared to absorbance spectroscopy, fluorescence spectroscopy is more sensitive, partly owing to detection at 90 degrees from the incident beam; therefore, much less sample is required for analysis. Furthermore, the fluorescence emission signal is much more susceptible to the environment surrounding the chromophores than absorbance spectroscopy. For example, ribonuclease T1, when experiencing a conformational change from the native state to the unfolded random coil, has been monitored by a decrease in the fluorescence emission intensity at 320 nm because of the environmental change surrounding the tryptophan and tyrosine residues *(5)*.

Circular dichroism (CD) spectroscopy is another useful tool for analyzing the secondary and tertiary structure of proteins. The chiral amino acids that make up the polypeptide interact differently with left and right circularly polarized light—the difference between these two interactions giving rise to the CD signal. The percentages of the secondary structural elements that are present in a given protein conformation (e.g., α-helix, β-sheet and random coils) can be determined by analyzing the CD spectrum in the far-UV region (190–250 nm). CD is also capable of detecting changes in tertiary structure as a result of differences in the environment surrounding the aromatic residues in the near-UV region (240–320 nm). The CD spectrum can be monitored as a function of pH, temperature, buffer species/strength, and denaturant to determine the stability of the native state. With the transglutaminase enzyme recombinant human factor XIII, CD was used to characterize the secondary structure of the native protein and to monitor the changes in conformation through thermal denaturation *(6)*. As the changes in the optical activity of the protein are directly related to the concentrations of the native state and the unfolded state(s), thermodynamic properties, such as the equilibrium constant of folding and the free energy of folding (usually useful in formulation screening), can be obtained through CD characterization *(7)*.

Fourier transform spectroscopy (FTIR) is another popular tool that can be used to analyze the elements of secondary structure by deconvoluting the overlapping carbonyl stretch signals of the peptide bonds in the amide I band (1700–1600 cm$^{-1}$). To gain more detailed information about the secondary structural elements present in the sample, the secondary-derivative absorbance spectrum is compared to previously analyzed proteins *(8,9)*. FTIR has the additional facility of being able to detect structural elements in the solid-state, such as lyophilized formulations *(10,11)*. As discussed next, maintaining the native folded state of the protein through the lyophilization process and during storage is essential to preserving the activity of labile proteins. In the case of a model recombinant humanized monoclonal antibody (rhMAb HER2), the capability of formulation excipients to maintain the native aqueous secondary structure in the lyophilized solid was assessed by FTIR and monitored upon storage at elevated temperature. A specific sugar-to-antibody molar ratio (360:1) was found sufficient to preserve the native structure through the lyophilization process and maintained activity upon long-term storage *(12)*.

### 3.1.3. Gel Electrophoresis

Under reducing and nonreducing conditions, SDS-polyacrylamide gel electrophoresis (SDS-PAGE), can be used to separate protein mixtures based on molecular size. This technique is particularly useful in monitoring protein stability by observing the number and intensity of bands that develop in a sample formulation upon forced degradation. Larger covalently bound aggregates or smaller proteolytic fragments will migrate differently than the parent molecule, and the qualitative purity can be monitored as a function of the applied stress. The relative molecular weights of the degradents can be compared to molecular weight standards loaded on the same gel. Comparison between reduced and nonreduced conditions will provide information about the involvement of disulfide linkages or shuffling in protein degradation. The most common methods used for detection of protein on polyacrylamide gels are Comassie blue dye binding and silver staining. The dye-binding method has the disadvantage of variation in binding capacity between different proteins. Therefore, quantitation by densitometry for Comassie-stained gels is not straightforward. Silver staining can greatly improve sensitivity of detection and is hence effective in identifying trace levels of impurities. Unfortunately, the development of the silver stain on the gel is difficult to control and interacts nonstoichiometrically with the amount of protein in the sample. For these reasons, quantitative conclusions are difficult to draw for SDS gel-based assays and can add significant time and effort to method validation in this respect.

Proteins will migrate in the presence of an electric field based on their isoelectric point (pI). Using a pH gradient made up of stable ampholytes in agarose or acrylamide gels, proteins will separate when an electric field is applied and diffuse to the portion of the gel where the pH of the gel and the pI of the protein coincide to result in net zero charge. This method of separation is termed isoelectric focusing (IEF). Thus, a protein can be separated from other related protein impurities that differ in pI by as little as 0.02 pH units. IEF is therefore an excellent tool to monitor chemical degradation in protein formulations. In evaluating the thermal stability of interleukin 1β (IL-1β) formulations, IEF was used in conjunction with other methods to screen and optimize pH, buffer species, and the presence/quantity of stabilizing excipients *(13)*. At higher temperatures, an aggregation mechanism predominated the degradation, but at temperatures below 30°C, analysis by IEF was able to discern two distinct bands attributed to degradation by a possible deamidation mechanism. Analysis of these same samples by SDS-PAGE indicated that the apparent molecular weights of these degradation products were not distinguishable and would therefore go unnoticed. Yet, this method has the same drawbacks in the available detection methods as mentioned for SDS-PAGE.

### 3.1.4. High-Performance Liquid Chromatography (HPLC)

Gel filtration or size-exclusion chromatography (SEC-HPLC) can be used to determine the apparent molecular weight of proteins with the use of molecular weight standards. As the SEC column separates molecules based on their hydrodynamic or Stokes radius, rather than true molecular weight, accurately assigning molecular weights for nonspherical proteins can be problematic. However, SEC is a useful analytical tool for quantitating the relative amounts of soluble protein aggregates, dimers, trimers, and other multimers and the presence of any detectable protein fragments. This method is routinely used to monitor the status of the aggregation profile during forced degradation experiments, such as thermal treatment or agitation.

Reversed-phase chromatography (RP-HPLC) separates molecules based on hydrophobicity. The sample is typically loaded onto the hydrophobic stationary phase in a polar aqueous mobile phase and is then eluted off the column by applying an organic solvent gradient (e.g., acetonitrile). This method is particularly useful to determine chemical degradation of proteins and peptides that have an effect to change the surface hydrophobicity. The degree to which the hydrophobic character changes—and can therefore be distinguished or sufficiently resolved on the reversed-phase column— is inversely proportional to the size of the molecule (e.g., a single chemical change in one amino acid of 150-kD protein may go largely undetected). For this reason, chemical degradation (deamidation, isomerization, proteolysis, and oxidation) are much more easily detected for small peptides than for large globular proteins, which, in worst cases, can irreversibly bind to the hydrophobic stationary phase of the column. RP-HPLC has also found great utility in peptide mapping when used in conjunction with Edman degradation or trypsin digests.

Ion-exchange chromatography (cation- or anion-IEX) separates proteins based on differences in the protein's surface charge, which is dependent on the difference between the protein's pI and pH of the mobile phase. The sample is loaded onto the column at a low-ionic strength, then a salt gradient is applied to elute the varying species in the sample at the critical salt concentration for each species. This type of chromatography is efficient in monitoring changes that affect the surface charge of the protein, such as deamidation, isomerization, dimerization and any conformational changes that alter the electrostatic interaction between the protein and stationary phase. For this reason, ion exchange is an excellent stability-indicating method. However, developing the proper conditions to resolve the different species in a degraded sample can be time-intensive. In addition, ion exchange will only detect degradation products that are charged at the pH of the mobile phase. Therefore, multiple conditions and using both anion- and cation-exchange columns must be explored when developing this method.

### 3.2. Solid-State Characterization

#### 3.2.1. Karl Fisher Moisture Analysis

As many degradation mechanisms are facilitated by the presence of water, and the amount of water in solid-state formulations can have a direct impact on the storage stability of the product (*see* **Subheading 4.3.**), knowledge and control of the moisture content in solid-state formulations is essential. Karl Fisher coulometry is a titrimetric method used to quantitate the amount of water present in solids or nonaqueous solutions. The quantitation method is based on the stoichiometric reaction of water with iodine and can routinely measure the moisture content down to 0.01%. Owing to the sensitivity of this method, diligent sample preparation is required, and residual moisture that may be present on the surface of sample vials, stoppers, spatulas, and so on, cannot be discounted. Special care must be taken when preparing lyophilized or spray-dried powders, as these samples have large specific surface area and can be highly hygroscopic.

#### 3.2.2. Differential Scanning Calorimetry (DSC)

DSC can determine the thermal properties of the protein and/or excipients in a given dosage form. DSC measures heat flow as a function of temperature and is sensitive to thermal transitions, such as melting and crystallization of the formulation component(s). Importantly, DSC can be used to analyze liquid formulations to determine the $T_m$ of the protein—a complimentary method to the spectroscopic methods described previously. In addition, DSC can be used to determine subambient thermal transitions, e.g., the freezing point of ice, eutectic freezing temperature ($T_e$) of crystalline excipients, and the glass transition temperature of the amorphous phase in frozen formulations ($T_g'$). This feature is particularly useful in screening frozen or freeze-dried formulations as the Tg' defines the conditions at which freeze-drying must be conducted (*see* **Subheading 4.3.**). Moreover, DSC can be used to determine the glass transition temperature ($T_g$) of the amorphous phase containing the protein in solid-state formulations. If the product is stored near or above $T_g$, molecular mobility is greatly enhanced, and all degradation reactions are kinetically increased. For this reason, DSC is an excellent screening tool for solid-state formulations. Unfortunately, the presence of several components in a formulation, such as buffers, excipients, and surfactants, will complicate the DSC thermogram, and distinguishing the $T_g$ or $T_g'$ of the formulation can be difficult.

### 3.2.3. Moisture Vapor Adsorption

Measurement of the hygroscopicity of solid-state lyophilized powders is typically performed on a moisture balance that is composed of a cantilever microbalance contained inside a humidity-controlled vacuum chamber. The sample is placed on the balance pan, and the system is evacuated until an equilibrium sample weight is achieved at 0% relative humidity (RH). Water vapor is then introduced into the evacuated chamber to expose the sample to incremental RH. The conditions are held constant at each % RH step until the sample mass reaches equilibrium, and the amount of moisture uptake is recorded. The change in sample mass is measured as a function of increasing RH (absorption) and decreasing RH (desorption). Typically, for lyophilized amorphous powders, moisture uptake is minimal at low humidity (<20% RH), but then increases dramatically at higher humidity and can adsorb more than 30% (w/w) water from the surrounding atmosphere.

### 3.2.4. Scanning Electron Microscopy (SEM)

The morphology of solid-state formulations can be characterized by SEM. The sample is placed on conductive carbon tape, mounted onto an aluminum sample holder, and evacuated in the SEM chamber. Resolution can be improved for some nonconductive powders by sputter coating with a thin film of gold prior to analysis. As the sample is evacuated and bombarded with an electron beam during microscopy, significant sample degradation may occur during analysis. For this reason, samples with high-moisture content or fragile powders with high-protein content must be handled appropriately.

## 4. Formulation Development of Proteins

Although a major goal of successful formulation development is to provide an adequately stable product, an understanding of the mechanisms of protein degradation is essential. Degradation mechanisms can be divided into two main categories: chemical and physical. Chemical is defined as those mechanisms in which existing chemical bonds are broken and/or new chemical bonds are formed (e.g., hydrolysis, racemization, and oxidation), and physical is defined as no covalent bonds being broken or formed (e.g., conformational alterations, aggregation, and surface adsorption). The major pathways of protein degradation are aggregation, deamidation, and oxidation *(14)*. Aggregation is the association between two or more partially unfolded protein molecules to form dimers, trimers, and higher molecular weight multimers. Deamidation is the formation of a carboxylic acid following hydrolysis of the amide side-chain linkages of asparagine or glutamine. Oxidation can occur at methionine, cystine, histidine, tryptophan, and tyrosine; methionine is the most susceptible site *(15)*. The extent of damage caused by these mechanisms can be attenuated by optimizing the pH of the solution phase, reducing storage temperature, resourcing raw materials that may contain contaminants that catalyze oxidation, removing oxygen in the head-space by purging with nitrogen, or, in the case of photo-oxidation, protection from light with the use of amber containers or secondary packaging. If these measures still do not provide an adequately stable product, formulation with additional excipients that can provide further stability must be explored. Again, it must be stressed that the presence of each excipient in a formulation must be justified with data supporting its beneficial role.

## 4.1. Development of Aqueous Liquid Formulations for Biopharmaceutical Proteins

Liquid formulations for biopharmaceutical proteins are routinely developed because of the ease and low cost of production. Nevertheless, the marginal stability of most biopharmaceutical proteins in liquid formulations often prevents long-term storage at room temperature or refrigerated conditions; thus, frozen storage is often required. This drawback will add significant burden in validating cold-chain storage and shipping requirements and poses significant product risks in the event of temperature excursions during long-term storage, handling, or transportation.

Formulation development generally begins with preformulation studies to determine the most stable solution environment for the protein, as well as the major modes of protein degradation. The parameters typically explored during preformulation are pH, protein concentration (solubility), ionic strength, and the quantity and nature of the buffering species. A series of accelerated stability studies are conducted at elevated temperatures, and the rate of degradation under each condition is plotted as a function of the parameter under evaluation to determine the range that provides maximum stability. For example, to generate a pH profile, all other parameters are held constant, and only the buffer type and pH are varied. As most proteins are stable at physiologic pH of approx 7.4, the range of 4–9 is often evaluated. The progress of degradation as a percentage of initial protein concentration is plotted as a function of time, and the degradation rate constant is calculated for each condition. The rate constants are then plotted as a function of pH, revealing the optimal range of stabilization. In the case of recombinant DNA-derived human deoxyribonuclease (rhDNase), the rate of deamidation and aggregation was monitored by IEX and SEC-HPLC, respectively, in solutions buffered to pH 5, 6, 7, and 8. Lower pH effectively reduced the deamidation rate, whereas the aggregation rate was found to increase with precipitation occurring at pH 5 *(16)*. As is often the case, optimizing a formulation parameter requires balancing the rate of one degradation mechanism with another. For this reason, and the fact that the major degradation pathways have not yet been identified during preformulation, the importance of analyzing stability samples by multiple methods cannot be overemphasized.

Alternatively, a multi-variable factorial design of experiment (DOE) can be used to determine the most important parameters and optimum formulation excipient compositions. Because this may involve a huge set of experiments, it is prudent to perform preformulation studies to develop a rational formulation strategy by determining the major degradation pathways and elucidate the more critical parameters to focus on for further studies.

Stabilizing the protein's native folded structure during storage is essential to maintaining the biologic activity, but the native state of most naturally occurring proteins is only about 5–15 kcal more stable than the unfolded conformation *(17)*. Labile proteins that are not sufficiently stabilized during storage in the optimized liquid formulation can be further stabilized (to an extent) with the addition of excipients that are "preferentially excluded" from the protein's surface, e.g., sucrose, which has been shown to stabilize with this mechanism *(18)*. The reason for stabilization has been explained as an increased concentration of any excipient that is excluded from the protein surface will cause the protein molecules to adopt a more compact folded state to minimize the protein-excluded solvent interaction. The compact protein state is, in turn, less prone to chemical and/or physical degradation. The flexible native protein state is thereby stabilized against more diffuse conformations that would otherwise increase the surface contact of the protein with the excluded solute. The protein's response to adopt a more compact conformation also serves to conceal backbone amino acids, chemically reactive sites, and other hydrophobic residues that are further buried in the stabilized conformation *(19)*.

Even with the addition of an excluded solute like sucrose, the presence of hydrophobic residues on the surface of a protein may still lead to physical instability. Protein denaturation can occur as a result of interfacial stress caused by adsorption of exposed hydrophobic residues to hydrophobic surfaces, such as the air–water interface or glass–water interface at the container walls *(20)*. Glass-surface adsorption can be a major issue for proteins formulated at low concentrations, particularly at the low microgram/milliliter range, and specific attention must be directed toward maintaining the label concentration throughout the storage life of the product. Additionally, shipping and handling of nonfrozen aqueous formulations may render additional interfacial stress that arises from a substantial increase in the air–water interface caused by agitation *(21)*. This stress event can lead to an increase in surface-associated denaturation of proteins, as hydrophobic residues typically buried in the interior of the native protein conformation have an affinity for hydrophobic surfaces, e.g., glass– and air–water interfaces. As the native state is in constant flux, these hydrophobic residues have access to potential adsorption sites when the protein adopts a partially unfolded conformation. Depending on the nature and complexity of the protein's native conformation, the probability of refolding to the native state from the surface-induced denatured state may be very low. Furthermore, if surface-denatured proteins suffer unfolding, the attractive forces between hydrophobic residues on neighboring molecules can interact to form intermolecular noncovalent bonds, leading to protein aggregation and the formation of soluble aggregates or, in the worst cases, macroscopic protein precipitation.

Instabilities associated with interfacial stress can be neutralized with the addition of a surface-active agent (Tween-20 or Tween-80; *22,23*). One action of the amphiphilic surfactant is to noncovalently bind with the exposed hydrophobic sites on the protein, preventing protein–protein interaction and thereby reducing the formation of soluble and insoluble aggregates *(24,25)*. Excess surfactant can also interact with hydrophobic surfaces, such as the air–water interface and container surface–water interfaces, reducing the surface tension of the fluid, favoring the native protein-folded state, and reducing the likelihood interfacial adsorption and potential denaturation *(26,27)*.

A formulation's susceptibility to agitation stress can be evaluated by forced degradation studies in a vortex mixer or an orbital shaker at high rpm. The extent of the damage can be monitored by the increasing light-scattering signal (>300 nm) of the absorbance spectrum. However, special attention may be required for formulation components that may also absorb in this region. Alternatively, the aggregation profile can be monitored by SEC-HPLC. However, this method often underestimates the extent of damage posed by agitation stress, because larger insoluble aggregates are generally filtered from the sample with in-line column filters before detection or settle in the analytical sample vial prior to sample injection.

## 4.2. Formulation Development of Frozen Aqueous Formulations

As the stability of many protein biopharmaceuticals is typically marginal at room temperature and refrigerated conditions, a frozen formulation is often pursued as a continuation of the failed liquid formulation. In addition to the concerns mentioned in the previous section for liquid formulations, frozen solutions have the following additional considerations. Although substantially decreasing the storage temperature of a liquid formulation to well below the freezing point ($< -10°C$) alleviates agitation stress during shipping and handling and kinetically reduces the degree of chemical degradation, there are several stress events associated with the freezing process that must be specifically addressed with formulation. Many proteins are vulnerable to freeze-thaw cycling with the belief that acute protein denaturation at the ice–water interface is the major cause of freezing instability *(28,29)*. Interfacial stress is generated upon freezing because of the interaction between the protein and the forming ice crystals. An increase in protein concentration or a decrease in the freezing rate—leading to larger ice crystals and less specific interfacial surface area—can reduce the magnitude of interfacial denaturation upon freezing. Additionally, the presence of a cryoprotectant, such as polyethylene glycol (PEG), can stabilize the protein via the same preferential exclusion mechanism described previously *(30)*. Surfactants (e.g., Tween-80) have also been shown to alleviate the acute interfacial-freezing stress *(31)*. In the case of recombinant human interferon $\gamma$ (rhIFN-$\gamma$), addition of Tween-20 reduced the protein concentration at the hydrophobic ice–water interface, reducing protein aggregation *(32)*.

Along with the formation of ice crystals upon freezing, the eutectic freezing of other components in the formulation may also occur. This can be especially problematic when one component of the formulation buffer system crystallizes out of solution upon freezing, resulting in a large pH change for the remaining unfrozen solution containing the protein. Dibasic sodium phosphate ($Na_2HPO_4$) and monobasic potassium phosphate ($KH_2PO_4$) readily crystallize upon freezing, leading to a large decrease or increase in solution pH for sodium and phosphate buffers, respectively *(33)*. For these reasons, the use of these buffers should be limited to nonfrozen liquid formulations. Buffers systems without this tendency include citrate, succinate, histidine, and Tris.

The addition of inorganic salts (e.g., sodium chloride) has long been known to affect protein solubility. The salt effect is dictated by the nature of the salt (Hofmeister series) but is independent of the protein *(34)*. The salt concentration can have a dramatic effect on protein solubility and even "salt-out" proteins at high concentrations, giving rise to aggregation and precipitation. Thus, the nature and effect of a salt concentration on the protein should be well understood, especially for frozen formulations. As freezing will concentrate the salt species in the freeze concentrate until the salt's $T_e$ is reached, a significant increase in relative salt concentration will be experienced by the protein.

Owing to the constraints of the stress events associated with the freezing rate, the development of a freezing protocol capable of providing a consistent and homogeneous freezing rate to all vials/containers in a large-scale manufacturing setting is pivotal for the success of this dosage form. Also important is the establishment of a shipping protocol that allows reliable low-temperature ($< -10°C$) maintenance to prevent local melting of the frozen formulation. Certainly, this constraint adds significant manufacturing costs to the product.

### 4.3. Formulation Development of Lyophilized Formulations

Unless a solid protein formulation is specifically required by the route of administration or delivery technology (e.g., inhalable dry powders or microspheres), the development of a solid-state formulation often occurs only when it is known that the liquid or frozen formulation will not meet the product-stability requirements. Still, early development of a solid-state formulation is often a judicial choice, because the shelf-life stability typically far exceeds that of a liquid or frozen formulation. The improved stability aspect of solid-state formulations must certainly be weighed against the development time, manufacturing costs, and the added inconvenience of reconstitution prior to dose administration. The aspects of developing a lyophilized formulation will be discussed, and a discussion of solid-state dry-powder formation techniques, their process considerations, and formulation development will follow in the next chapter.

The goal of a lyophilized formulation is not only to provide a formulation that is stable upon handling, transportation, and long-term storage, but it is also to provide a cosmetically pristine cake of low-moisture content with no signs of shrinkage or collapse upon storage. Furthermore, the lyophilized cake must be easily and quickly reconstituted by the end-user at the time of administration. In terms of process considerations, the prelyophilized liquid formulation in the short term is susceptible to all chemical and physical degradation mechanisms discussed previously for both liquid and frozen formulations. After freezing, additional stress is experienced by the protein upon dehydration during both primary and secondary drying. Proteins can be stabilized through this dehydration process with the addition of an amorphous stabilizer (discussed further below). For the dried-solid product, many chemical and physical degradation mechanisms that govern protein destabilization in liquid and frozen formulation are kinetically slowed from the absence of water and the greatly reduced mobility in the dried state *(35–37)*. Preserving the protein's stability within the lyophilized cake upon long-term storage is dependent on preserving the native folded state upon storage and maintaining storage temperature below the $T_g$ or the temperature at which the lyophilized cake begins to soften *(38,39)*. When stored above $T_g$, mobility within the dried solid increases along with the reaction rates of all degradation mechanisms. In extreme cases, recrystallization of stabilizing excipients can occur, markedly reducing the product quality in a relatively short storage period *(40)*.

Lyophilization of proteins in the absence of amorphous stabilizers often leads to conformational denaturation as the minimal solvation energy of the native folded state is lost upon dehydration. Rehydration of the dried denatured protein often causes aggregation and low recovery of activity, except for rare cases of robust proteins (e.g., lysozyme, which can readily refold to the native structure and recover full activity upon reconstitution from a denatured dried state; *41)*. A lyophilized formulation that preserves the protein's native structure throughout the freezing and drying process has been shown to be the most effective approach to improving activity recovery upon reconstitution *(42)*. An amorphous disaccharide (sucrose or trehalose) can serve this purpose. This stabilizing effect has been explained in terms of the sugar acting as a replacement for the hydration sphere of the protein, fulfilling the protein's hydrogen-bonding requirements to maintain the native state *(43,44)*. Additionally, these sugars are known to form glassy matrices, significantly reducing protein mobility and greatly reducing the impact of chemical and physical degradation when stored below $T_g$ *(45)*. However, although capable of enhancing protein stability, glass formation alone is not sufficient to stabilize the protein *(46)*.

The lyophilization process consists of three defined steps: a freezing step, followed by primary drying, and then secondary drying. During the freezing step, ice crystals begin to form after a brief period of supercooling and phase-separate from the aqueous solution as the shelf temperature is reduced. As the shelf temperature is further reduced, other crystalline components in the formulation (if present) may crystallize at the $T_e$, thereby also separating from the remaining nonfrozen aqueous solution. Because ice crystals grow at the expense of nonfrozen water, all other noncrystalline components are concentrated during freezing. This concentration effect continues until the maximal freeze-concentration point is reached, and no further ice can crystallize. The remaining maximal freeze-concentrate containing the protein and other noncrystalline excipients exists as an amorphous phase in the interstitial regions between the ice crystals. This amorphous phase exhibits a transition between a glassy rigid state and a rubbery state at a critical temperature termed the glass transition temperature $(T_g')$. Primary drying must be conducted below $T_g'$ to avoid cake collapse during the process *(47)*. Primary drying must also be conducted below the $T_e$ of any crystalline components in the formulation to avoid cake meltback *(48)*. For this reason, salts with relatively low $T_e$, such as sodium and calcium chloride ($T_e$ of NaCl = $-21\,°C$; $T_e$ of CaCl$_2$ = $-51\,°C$), should be avoided (**Note 3**). $T_g'$ is formulation-dependent and can be measured by several available thermal analysis techniques, e.g., DSC. $T_g'$ and $T_e$ are the most commonly used excipients that have been published *(49–51)*, but the transition temperature is unique to the exact composition of the final formulation, as each component and its relative concentration will influence $T_g'$. Therefore, extensive thermal analysis of each lyophilized formulation is critical to maintaining good-cake quality throughout the primary drying process.

Following the successful completion of primary drying, which removes the frozen water by sublimation, the moisture content of the resultant cake can still be above 10% (w/w) and may collapse if brought to ambient pressure and temperature at this point. Secondary drying is then conducted at a higher shelf temperature to further remove the nonfrozen water from the product and reduce the final moisture content to approx 1–2% (w/w). The main reason to reduce the moisture content to such low levels is that water can serve to plasticize the amorphous phase containing the protein, thereby reducing the $T_g$. If the $T_g$ of the freeze-dried cake falls below the storage temperature or significantly high temperatures are experienced during shipping, there is a risk of cake collapse and increased protein mobility. For these reasons, ensuring a low-moisture level during the freeze-drying process and maintaining it upon storage is paramount to preserving product quality (*52,53*; **Note 4**).

To improve the structural and mechanical strength of the lyophilized cake, bulking agents (mannitol or glycine) are often used. Although mannitol is effective at enhancing the appearance of the cake partly because of its propensity to crystallize, the crystalline mannitol does not occupy the same phase as the protein and is therefore not available for protein stabilization by the previously mentioned mechanisms. Hence, bulking agents can be used in combination with amorphous stabilizers, especially in cases where the total solid content of the prelyophilized solution is low (<2%). Lyophilization of low-solid content solutions often leads to wispy, fragile cakes that are difficult to control, and some dried material may even escape from the vials during the drying process prior to stoppering. In the case of recombinant human granulocyte–macrophage colony-stimulating factor (Leukine), the protein is formulated with both sucrose as an amorphous stabilizer and mannitol as a bulking agent at a 4:1 ratio to sucrose *(54)*.

Amino acids can also exhibit the stabilizing functions observed for amorphous disaccharides. In the case of a fully human anti-IL-8 monoclonal antibody (ABX-IL-8), the use of histidine as a buffering component also afforded an additional protective effect in both liquid and lyophilized formulations *(55)*. In this report, increasing the histidine level in lyophilized formulations with mannitol, glycine, glutamic acid and Tween-20 was shown to reduce the formation of aggregates detectable by SEC-HPLC and to eliminate the formation of high-molecular-weight bands detectable by nonreducing SDS-PAGE during accelerated stability. Histidine was also shown to perform as well as sucrose in reducing the aggregation amount during freeze-thaw evaluations at equivalent molar concentrations and outperformed succinate and citrate buffers at equivalent concentrations in liquid formulations during accelerated storage.

## 5. Considerations for Containers and Closures

One aspect of formulation development often overlooked until the late stages is the selection of the final product: containers and closures. Owing to the different product requirements for liquid- and solid-dosage forms, there are varying considerations. For liquid-protein formulations, the primary concerns are surface adsorption of protein to the container, extractables from the containers/closures, and maintaining seal integrity. For solid-protein formulations, such as lyophilized products, important considerations are vial breakage during manufacturing, seal integrity, the presence of moisture in stoppers or moisture/oxygen ingress during product storage, and surface adsorption/extractables when the product is in liquid form and in contact with the final containers/closures (before lyophilization and after reconstitution). Adsorption to hydrophobic glass surfaces has been a known therapeutic protein formulation issue for many years. Early reports detail the prevention of insulin adsorption to glass surfaces with the addition of gelatin *(56)*. As mentioned already, adding surface-active agents (e.g., polysorbates) to the protein formulation has been shown to effectively reduce surface adsorption and surface-induced denaturation. Along with rational formulation strategies to specifically address protein adsorption, specialty glass coatings are available that can significantly reduce product interactions *(57)*.

Central regulatory issues are maintaining seal integrity to ensure the sterility of the parenteral product and the leaching of extractables from the containers or stoppers into liquid formulations during storage. Metal ions leached from glass containers at high or low pH can catalyze many degradation reactions, such as oxidation, but this is less of an issue at neutral pH, where most therapeutic proteins are formulated. Extractables from rubber stoppers, such as silicone oils or waxes added for lubricity, low-molecular-weight polymer fragments and/or particulates could have a dramatic impact on storage stability of liquid-protein formulations. To accelerate the observation of such incompatibilities, inverted vials can be placed on stability to magnify the contribution to product degradation caused by extractables from the closures.

In the case of lyophilized dosage forms, vials are often sealed under reduced pressure following freeze-drying. This measure helps to ensure that the stoppers will remain closed until secondary aluminum crimp-type seals are in place. Unfortunately, this practice can also increase the likelihood of moisture/oxygen ingress or the volatilization of gaseous extractables (e.g., moisture vapor) from the stoppers to dramatically impact product quality upon storage. Volatilized moisture from the stoppers will have a strong affinity for the hygroscopic lyophilized cake. Moisture absorption into the amorphous matrix could occur over a period of weeks or even a few hours if the vial is exposed to low temperatures, resulting in moisture condensation onto the cake surface. The increased moisture content would, in turn, impart a reduction in the glass transition temperature that could potentially trigger the onset of cake collapse even though the product passed quality specifications for moisture content following manufacture. Several manufacturers now offer coated stoppers to address the moisture stopper issue. The durability of these coatings, particularly under harsh sterilization procedures, must be considered. A detailed knowledge of the levels and nature of extractable materials from containers and closures will help to identify potential product incompatibilities.

## 6. Notes

1. Aqueous protein solutions should not have significant absorption above 320 nm. If a decaying background signal is observed in the 320–800 nm region, the sample may be scattering light from particulates, protein aggregates, or micelles in the solution. As the light-scattering signal increases significantly as wavelength decreases (increasing energy), the light-scattering signal will add to the absorbance signal at the wavelength used to quantitate the protein ($\lambda_{max}$). The magnitude of the light-scattering signal at the $\lambda_{max}$ can be extrapolated from the portion of the spectrum above 320 nm, where the protein does not have any notable absorption. Procedures for extrapolating the light-scatter signal at the wavelength are detailed in the literature *(58,59)*. The simplest method: Begin to multiply the light-scattering signal at 333 nm by 2 to obtain the light-scattering contribution to the signal at 280 nm. As the light-scattering signal increases as a function of the number of light-scattering particles in solution, forced degradation studies that cause aggregation (shear and agitation) are particularly susceptible to this interference. Sample preparation relating to filtration or centrifugation prior to measuring the absorbance spectrum can be used to help reduce the interference caused by aggregation. However, development of the sample preparation conditions (filtration membrane type, pore size, collection vial type, and so forth) should be performed on the initial protein solution to ensure that the native protein can be completely recovered. The magnitude of the light-scattering signal at any wavelength above the region where the protein normally absorbs can be used as a simple method to track increases in solution turbidity. This can be shown to be directly related to the amount of aggregation in solution. However, this technique has its limitations, because as degradation by the aggregation mechanism progresses, larger protein aggregates begin to settle out of solution.
2. The $T_m$ of a protein can be used to identify the optimum solution conditions for protein stability. $T_m$ can be measured by UV difference spectroscopy, but other spectroscopic methods (e.g., CD, fluorescence spectroscopy, and DSC are typically used more often. The $T_m$ is directly related to the solution environment that the protein is exposed to, as the simple act of unfolding a protein will expose a larger protein surface that increases solvent interactions. If the increasing protein surface has a negative interaction with the surrounding environment, the unfolded state will be destabilized, and more energy will be required to denature the protein, thereby increasing $T_m$ as a result. As the solution pH, buffer concentration, buffer species, ionic strength, and presence and concentration of excipients are varied, the temperature at which the protein unfolds can be measured and plotted as a function of each parameter. The conditions at which the $T_m$ has been maximized is directly associated with the physical stabilization of the native folded state. Therefore, conditions at which $T_m$ is increased have been found to increase physical storage stability, and conditions at which $T_m$ is decreased have been typically shown to decrease physical storage stability *(60,61)*. Thus, monitoring $T_m$ as a function of solution conditions during preformulation and formulation is extremely useful during screening and optimization studies.
3. An additional reason that ionic salts are not used for tonicity adjustment in frozen or lyophilized formulations is the freeze-concentrate phenomenon that occurs prior to

eutectic melting. This phenomenon, as previously mentioned, will concentrate all nonfrozen, noncrystalline components, typically in a single amorphous phase. Hence, on a physical level, the protein is experiencing a huge concentration increase for all components present in the formulation during freezing, which includes buffers and salts. Protein exposure to high levels of ionic strength or to a large pH change when one component of the buffer system eutectically freezes (as previously mentioned) can have drastic effects on protein stability.

4. What confounds the low-moisture content requirement is the fact that some excipients form metastable amorphous or hydrated crystalline phases that contain residual or crystalline-hydrated water during the freeze-drying process. By nature, amorphous phases are metastable and can crystallize under certain critical conditions. Researchers have observed the crystallization of sucrose, lactose, and trehalose above a critical storage humidity of approx 55%, 50%, and 50% RH, respectively *(62)*. Others have found the presence of a partial metastable amorphous phase of mannitol that can spontaneously crystallize when stored above its glass transition temperature of approx 45°C. Recently, a crystalline-hydrated form of mannitol has been observed to occur under certain lyophilization conditions *(63)*. Upon storage, these states can convert or recrystallize to more stable crystalline forms and release previously associated water as residual liquid moisture into the freeze-dried cake. This moisture can then be adsorbed into the hygroscopic amorphous protein matrix, reducing the $T_g$ and greatly deteriorating product stability.

## References

1. Hageman, M. J. Water sorption and solid-state stability of proteins, in *Stability of Protein Pharmaceuticals, Part A: Chemical and Physical Pathways of Protein Degradation* (Ahern, T. J. and Manning, M. C., eds.), Plenum Press, New York, NY, 1992, pp. 273–309.
2. Sapan, C. V., Lundblad. R. L., and Pace, N. C. (1999) Colorimetric protein assay techniques. *Biotechnol. Appl. Biochem.* **29,** 99–108.
3. Donovan, J. W. (1973) Ultraviolet difference spectroscopy—new techniques and applications. *Methods Enzymol.* **27,** 497–525.
4. Balestrieri, C., Colonna, G., Giovane, A., Irace, G., and Servillo, L. (1978) Second-derivative spectroscopy of proteins. A method for the quantitative determination of aromatic amino acids in proteins. *Eur. J. Biochem.* **90,** 433–440.
5. Thomson, J. A., Shirley, B. A., Grimsley, G. R., and Pace, C. N. (1989) Conformational stability and mechanism of folding of ribonuclease T1. *J. Biol. Chem.* **264,** 11614–11620.
6. Dong, A., Kendrick, B.. Kreilgard, L., Matsuura, J., Manning, M. C., and Carpenter, J. F. (1997) Spectroscopic study of secondary structure and thermal denaturation of recombinant human factor XIII in aqueous solution. *Arch. Biochem. Biophys.* **347,** 213–220.
7. Kelly, S. M. and Price, N. C. (2000) The use of circular dichroism in the investigation of protein structure and function. *Curr. Protein Pept. Sci.* **1,** 349–384.
8. Dong, A. and Caughey. W. S. (1994) Infrared methods for study of hemoglobin reactions and structures. *Methods Enzymol.* **232,** 139–175.
9. Dong, A., Huang, P., and Caughey, W. S. (1990) Protein secondary structures in water from second-derivative amide I infrared spectra. *Biochemistry* **29,** 3303–3308.
10. Dong, A., Prestrelski, S. J., Allison, S. D., and Carpenter, J. F. (1995) Infrared spectroscopic studies of lyophilization- and temperature-induced protein aggregation. *J. Pharm. Sci.* **84,** 415–424.
11. Carpenter, J. F., Prestrelski, S. J., and Dong, A. (1998) Application of infrared spectroscopy to development of stable lyophilized protein formulations. *Eur. J. Pharm. Biopharm.* **45,** 231–238.

12. Cleland, J. L., Lam, X., Kendrick, B., Yang, J., Yang, T. H., Overcashier, D., et al. (2001) A specific molar ratio of stabilizer to protein is required for storage stability of a lyophilized monoclonal antibody. *J. Pharm. Sci.* **90,** 301–321.

13. Gu, L. C., Erdos, E. A., Chiang, H. S., Calderwood, T., Tsai, K., Visor, G. C., et al. (1991) Stability of interleukin 1 beta (IL-1 beta) in aqueous solution: analytical methods, kinetics, products, and solution formulation implications. *Pharm. Res.* **8,** 485–490.

14. Cleland, J. L., Powell, M. F., and Shire S. J. (1993) The development of stable protein formulations: a close look at protein aggregation, deamidation, and oxidation. *Crit. Rev. Ther. Drug Carrier Syst.* **10,** 307–377.

15. Manning, M. C., Patel, K., and Borchardt, R. T. (1989) Stability of protein pharmaceuticals. *Pharm. Res.* **6,** 903–918.

16. Shire, S. J. Stability characterization and formulation development of recombinant human deoxyribonuclease I [Pulmozyme®, (Dornase Alpha)], in *Formulation, Characterization, and Stability of Protein Drugs* (Pearlman, R. and Wang, R. J., eds.), Plenum Press, New York, NY, 1996, pp. 393–426.

17. Pace, C. N. (1974) The stability of globular proteins. *CRC Crit. Rev. Biochem.* **3,** 1–43.

18. Lee, J. C. and Timasheff, S. N. (1981) The stabilization of proteins by sucrose. *J. Biol. Chem.* **256,** 7193–7201.

19. Kendrick, B. S., Chang, B. S., Arakawa, T., Peterson, B., Randolph, T. W., Manning, M. C., and Carpenter, J. F. (1997) Preferential exclusion of sucrose from recombinant interleukin-1 receptor antagonist: role in restricted conformational mobility and compaction of native state. *Proc. Natl. Acad. Sci.* **94,** 11917–11922.

20. Norde, W. and Lyklema, J. (1991) Why proteins prefer interfaces. *J. Biomater. Sci. Polym. Ed.* **2,** 183–202.

21. Sluzky, V., Tamada, J. A., Klibanov, A. M., and Langer, R. (1991) Kinetics of insulin aggregation in aqueous solutions upon agitation in the presence of hydrophobic surfaces. *Proc. Natl. Acad. Sci.* **88,** 9377–9381.

22. Bam, N. B., Cleland, J. L., Yang, J., Manning, M. C., Carpenter, J. F., Kelley, R. F., and Randolph, T. W. (1998) Tween protects recombinant human growth hormone against agitation-induced damage via hydrophobic interactions. *J. Pharm. Sci.* **87,** 1554–1559.

23. Charman, S. A., Mason, K. L., and Charman, W. N. (1993) Techniques for assessing the effects of pharmaceutical excipients on the aggregation of porcine growth hormone. *Pharm. Res.* **10,** 954–962.

24. Tandon, S. and Horowitz, P. M. (1989) Reversible folding of rhodanese. Presence of intermediate(s) at equilibrium. *J. Biol. Chem.* **264,** 9859–9866.

25. Cleland, J. L. and Randolph, T. W. (1992) Mechanism of polyethylene glycol interaction with the molten globule folding intermediate of bovine carbonic anhydrase B. *J. Biol. Chem.* **267,** 3147–3153.

26. Suelter, C. H. and DeLuca, M. (1983) How to prevent losses of protein by adsorption to glass and plastic. *Anal. Biochem.* **135,** 112–119.

27. Sato, S., Ebert, C. D., and Kim, S. W. (1984) Prevention of insulin self-association and surface adsorption. *J. Pharm. Sci.* **72,** 228–232.

28. Strambini, G. B. and Gabellieri, E. (1996) Proteins in frozen solutions: evidence of ice-induced partial unfolding. *Biophys. J.* **70,** 971–976.

29. Chang, B. S., Kendrick, B. S., and Carpenter, J. F. (1996) Surface-induced denaturation of proteins during freezing and its inhibition by surfactants. *J. Pharm. Sci.* **85,** 1325–1330.

30. Carpenter, J. F. and Crowe, J. H. (1988) The mechanism of cryoprotection of proteins by solutes. *Criobiology* **25,** 244–255.

31. Kerwin, B. A., Heller, M. C., Levin, S. H., and Randolph, T. W. (1998) Effects of Tween 80 and sucrose on acute short-term stability and long-term storage at −20°C of a recombinant hemoglobin. *J. Pharm. Sci.* **87,** 1062–1068.

32. Webb, S. D., Golledge, S. L., Cleland, J. L., Carpenter, J. F., and Randolph, T. W. (2002) Surface adsorption of recombinant human interferon-gamma in lyophilized and spray-lyophilized formulations. *J. Pharm. Sci.* **91,** 1474–1487.

33. van den Berg, L. and Rose, D. (1959) Effect of freezing on the pH and composition of sodium and potassium phosphate solutions; the reciprocal system $KH_2PO_4$-$Na_2$-$HPO_4$-$H_2O$. *Arch. Biochem. Biophys.* **81,** 319–329.

34. Arakawa, T. and Timasheff, S. N. (1985) Theory of protein solubility. *Methods Enzymol.* **114,** 49–79.

35. Carpenter, J. F. and Chang, B. S. Lyophilization of protein pharmaceuticals, in *Biotechnology and Biopharmaceutical Manufacturing, Processing, and Preservation* (Avis, K. E. and Wu, V. L., eds.), Interpharm Press, Buffalo Grove, IL, 1996, pp. 199–264.

36. Prestrelski, S. J., Pikal, K. A., and Arakawa, T. (1995) Optimization of lyophilization conditions for recombinant human interleukin-2 by dried-state conformational analysis using Fourier-transform infrared spectroscopy. *Pharm. Res.* **12,** 1250–1259.

37. Chang, B. S., Beauvais, R. M., Dong, A., and Carpenter, J. F. (1996) Physical factors affecting the storage stability of freeze-dried interleukin-1 receptor antagonist: glass transition and protein conformation. *Arch. Biochem. Biophys.* **331,** 249–258.

38. Pikal, M. J. (1990) Freeze-drying of proteins, part 1: process design. *BioPharm* **3,** 18–27.

39. Pikal, M. J. Freeze-drying of proteins, in *Formulation and Delivery of Proteins and Peptides* (Cleland, J. L. and Langer, R., eds.), ACS Symposium Series, vol. 567, 1994, pp. 120–133.

40. Carpenter, J. F., Pikal, M. J., Chang, B. S., and Randolph, T. W. (1997) Rational design of stable lyophilized protein formulations: some practical advice. *Pharm. Res.* **14,** 969–975.

41. Sellers, S. P., Clark, G. S., Sievers, R. E., and Carpenter, J. F. (2001) Dry Powders of stable protein formulations from aqueous solutions prepared using supercritical $CO_2$-assisted aerosolization. *J. Pharm. Sci.* **90,** 785–797.

42. Prestrelski, S. J., Arakawa, T., and Carpenter, J. F. (1993) Separation of freezing- and drying-induced denaturation of lyophilized proteins using stress-specific stabilization. II. Structural studies using infrared spectroscopy. *Arch. Biochem. Biophys.* **303,** 465–473.

43. Carpenter, J. F. and Crowe, J. H. (1989) An infrared spectroscopic study of the interactions of carbohydrates with dried proteins. *Biochemistry* **28,** 3916–3922.

44. Crowe, J. H., Carpenter, J. F., and Crowe, L. M. (1998) The role of vitrification in anhydrobiosis. *Annu. Rev. Physiol.* **60,** 73–103.

45. Franks, F., Hatley, R. H. M., and Mathias, S. F. (1991) Material science and the production of shelf stable biologicals. *BioPharm* **4,** 38–55.

46. Pikal, M. J., Dellerman, K. M., Roy, M. L., and Riggin, R. M. (1991) The effects of formulation variables on the stability of freeze-dried human growth hormone. *Pharm. Res.* **8,** 427–436.

47. Pikal, M. J. (1990) Freeze-drying of proteins, part II: formulation selection. *BioPharm* **3,** 26–30.

48. MacKenzie, A. P. Collapse during freeze-drying—qualitative and quantitative aspects, in *Freeze-drying and Advanced Food Technology* (Goldblith, S. A., Rey, L., and Rothmayr, W. W., eds.), Academic Press, NY, 1975, pp. 277–307.

49. Levine, H. and Slade, L. (1988) Principles of "cryostabilization" technology from structure/property relationships of carbohydrate/water systems—a review. *Cryoletters* **9,** 21–63.

50. Levine, H. and Slade, L. (1988) Thermomechanical properties of small-carbohydrate-water glasses and "rubbers": Kinetically metastable systems at subzero temperatures. *Pure Appl. Chem.* **60**, 1841–1864.

51. Chang, B. S. and Randall, C. S. (1992) Use of subambient thermal analysis to optimize protein lyophilization. *Cryobiology* **29**, 632–656.

52. Hancock, B. C. and Zografi, G. (1994) The relationship between the glass transition temperature and the water content of amorphous pharmaceutical solids. *Pharm. Res.* **11**, 471–477.

53. Franks, F. (1990) Freeze-drying: from empiricism to predictability. *Cryoletters* **11**, 93–110.

54. Geigert, J. and Ghrist, B. F. D. Development and shelf-life determination of recombinant human granulocyte-macrophage colony stimulating factor (Leukine®, GM-CSF), in *Formulation, Characterization, and Stability of Protein Drugs*, (Pearlman, R. and Wang, Y. J., eds.), Plenum Press, New York, NY, 1996, pp. 329–342.

55. Chen, B., Bautista, R., Yu, K., Zapata, G. A., Mulkerrin, M. G., and Chamow, S. M. (2003) Influence of histidine on the stability and physical properties of a fully human antibody in aqueous and solid forms. *Pharm. Res.* **20**, 1952–1960.

56. Hill, J. B. (1959) Adsorption of insulin to glass. *Proc. Soc. Exp. Biol. Med.* **102**, 75–77.

57. Schwarzenbach, M. S., Reimann, P., Thommen, V., Hegner, M., Mumenthaler, M., Schwob, J., and Guntherodt, H. J. (2002) Interferon alpha-2a interactions on glass vial surfaces measured by atomic force microscopy. *PDA J. Pharm. Sci. Technol.* **56**, 78–89.

58. Leach, S. J. and Scheraga, H. A. (1960) Effect of light scattering on ultraviolet difference spectra. *J. Am. Chem. Soc.* **82**, 4790–4792.

59. Pace, N. C., Vajdos, F., Fee, L., Grimsley, G., and Gray, T. (1995) How to measure and predict the molar absorption coefficient of a protein. *Protein Sci.* **4**, 2411–2423.

60. Gekko, K. and Timasheff, S. N. (1981) Thermodynamic and kinetic examination of protein stabilization by glycerol. *Biochemistry* **20**, 4677–4686.

61. Lee, L. L. and Lee, J. C. (1987) Thermal stability of proteins in the presence of poly(ethylene glycols). *Biochemistry* **26**, 7813–7819.

62. Fakes, M. G., Dali, M. V., Haby, T. A., Morris, K. R., Varia, S. A., and Serajuddin, A. T. (2000) Moisture sorption behavior of selected bulking agents used in lyophilized products. *PDA J. Pharm. Sci. Technol.* **54**, 144–149.

63. Yu, L., Milton, N., Groleau, E. G., Mishra, D. S., and Vansickle, R. E. (1998) Existence of a mannitol hydrate during freeze-drying and practical implications. *J. Pharm. Sci.* **88**, 196–198.

# 21

## Solid-State Protein Formulation

*Methodologies, Stability, and Excipient Effects*

### Yuh-Fun Maa and Scott P. Sellers

### 1. Introduction

An important message was delivered in Chapter 20 by Sellers and Maa—the protein is generally more stable in the solid state than in the liquid state. Obviously, this belief is related to protein mobility. Protein movement is restricted in the dry state, substantially prohibiting the surrounding influence on the protein. Then, the notion evolves that escalating the glass transition temperature ($T_g$) of the dry formulation will suppress protein mobility, thereby improving protein stability *(1)*. Yet, enhancing the $T_g$ alone is not sufficient for protein stabilization without a more necessary strategy— the use of appropriate protein stabilizers, such as amorphous sugars. Again, the prevailing mechanisms of these stabilizers have already been depicted in the previous chapter.

The suggestions described above are general solid-state formulation strategies/ guidelines for proteins. Still, there are other factors affecting protein stability, which have not yet been addressed. The most significant perhaps is the stress imposed on the protein during formulation processing. Each process dictates its own stress events and, in turn, presents unique influences to short-term and possibly long-term protein stability. Freeze-drying (FD) imparts two major stresses to the protein: freezing and dehydration. Nevertheless, more distinct, subset stress events might be associated with either of these two major stresses. In response to freezing, ice crystals of various sizes grow to intermingle with numerous microscopic liquid pockets—freeze concentrate (e.g., a highly viscous fluid phase containing noncrystalline components and the remaining nonfrozen water). First, this physical process creates ice/liquid interfaces that may present hydrophobic surface areas for protein adsorption and result in conformational changes to the protein. Second, some excipients crystallize upon freezing and phase separate from the freeze concentrate, causing composition changes and thereby rendering the protein instable. The most prominent example is the downshift

From: *Methods in Molecular Biology, vol. 308: Therapeutic Proteins: Methods and Protocols*
Edited by: C. M. Smales and D. C. James © Humana Press Inc., Totowa, NJ

of formulation pH because of the crystallization of sodium dibasic phosphate (2). Third, residual stress in the frozen glass at the ice/liquid interface might cause perturbations in the protein's secondary structure, which can be irreversible and cause aggregation upon drying (3,4). Therefore, as formulation processes will intimately dictate the stress posed to the protein, protein formulation development requires a full understanding of the processes employed, from which appropriate formulation strategies can be derived.

With the concept that proteins/peptides are more stable in the solid state, many protein products are prepared as solid-dosage forms, mainly by FD (also discussed in Chapter 20). However, these products are primarily delivered parenterally by subcutaneous or intramuscular route after being reconstituted. However, amid the host of protein delivery technologies that have emerged over the last decade, several are powder-based methods, such as microspheres for long-acting delivery (5), fine powders for pulmonary delivery (6), and biopharmaceutical/vaccine powders for intradermal delivery (7). Indeed, the efforts to identify appropriate powder formation methods are mounting.

Despite being a mature and well-established process for preparing marketed biopharmaceuticals, FD, or lyophilization, is not a direct particle formation method and requires a secondary procedure to pulverize the product cake generated by the FD process. Mechanical milling can be performed to break apart the cake into particles, but such a secondary process would significantly reduce the process yield and control over critical particle properties (e.g., particle size and size distribution, morphology, and porosity) is relatively poor.

With this limitation in mind, two dehydration methodologies are discussed—spray drying (SD) and spray freeze-drying (SFD)—both of which offer a direct particle formation capability. SD, the most common powder-generating methodology, allows a liquid formulation to be rapidly transformed into a dry-particulate form by atomizing into a hot-drying medium. SFD combines atomization to generate suitable size droplets and freezing followed by removal of ice by sublimation under vacuum (the latter two steps are conceptually similar to classic FD in a vial).

This chapter describes the respective dehydration mechanisms of each of these processes along with the influences to the powder's physical attributes and protein stability in response to the stress events specific to these mechanisms.

## 2. Materials

1. Trehalose:mannitol:dextran (10,000 Dalton; 3:3:4 weight ratio solution [30 g/100 mL, 30% solid content]).
2. Hepatitis B surface antigen vaccine.
3. Phosphate-buffered saline (PBS).
4. Recombinant human anti-immunoglobulin E monoclonal antibody (anti-IgE MAb).
5. Recombinant human deoxyribonuclease (rhDNase).
6. Recombinant human growth hormone (rhGH).
7. Amray 1810T scanning electron microscope (Amray, Bedford, MA).
8. Hummer JR Technics unit (Pergamon Corporation, King of Prussia, PA).
9. Time-of-flight particle size analyzer (Aerosizer, API, Minneapolis, MN).
10. Nova 2200 Gas Sorption Analyzer (Quantachrome Corporation).

11. Karl Fisher Coulometer (Model 737, Brinkmann) equipped with a drying oven (Model 707).
12. Bio-Select size-exclusion chromatography (SEC) 250-5 column (Bio-Rad).
13. 1090L High performance liquid chromatography (HPLC; Hewlett Packard) system.
14. Molecular weight standards (Bio-Rad) consisting of 670 kDa thyroglobulin, 158 kDa γ-globulin, 44 kDa ovalbumin, and 1.35 kDa cyanocobalamin.
15. Tosoh TSK2000SWXL column (7.8-mm inner diameter [ID] × 30-cm L; 5-μm particle size).
16. Equilibration buffer: 5 m$M$ HEPES, 150 m$M$ NaCl, and 1 m$M$ CaCl$_2$ at pH 7.0.

## 3. Methods

### 3.1. Powder and Protein Characterization

#### 3.1.1. Scanning Electron Microscopy (SEM)

The external morphology of coated particles was examined using an Amray 1810T scanning electron microscope (15). Powder samples were first sputter-coated with gold using a Hummer JR Technics unit prior to microscopy.

#### 3.1.2. Particle Size Analysis

The mean geometric/aerodynamic diameter of the particles in the volume distribution was determined using a time-cf-flight particle size analyzer (15). The mean volumetric size was calculated by the software using a density of 1.0 g/mL, and the particle population between 10% ($D_{10}$) and 90% ($D_{90}$) was reported for particle size distribution. Each analysis required approx 3–5 mg of the powder sample.

#### 3.1.3. Specific Surface Area

The specific surface area (m$^2$/g) of the powder samples was determined by the multipoint Brunauer–Emmett–Teller (BET) method from the adsorption of nitrogen gas at 77 K using a Nova 2200 Gas Sorption Analyzer. All samples were weighed and loaded into 9-mm large pellet cells and outgassed at 25°C for 16 h based on the gas flow method using nitrogen. For surface-area analysis, a five-point BET adsorption was performed with P/P$_0$ ranging from 0.1 to 0.3. For pore size distribution, 12-point adsorption (P/P$_0$ from 0.05 to 0.99) and 25-point desorption (P/P$_0$ from 0.99 to 0.01) were performed. The desorption branch was used to calculate pore diameter and pore volume based on the Barret–Joyner–Halenda (BJH) equation (*see* the operating manual for the theory). This model allowed the pore size of mesopores (>0.002 μm) to be accurately measured.

#### 3.1.4. Moisture Content Analysis

Approximately 4–5 mg of the powder sample was weighed into an aluminum-weighing vessel, and the weight was recorded. The sample was then loaded into a Karl Fisher Coulometer equipped with a drying oven. The sample was heated to 150°C for 150 s with a gas flow rate of 100 mL/min within the drying oven. The extraction length was set to 120 s with the drift level of 10 μg/min.

#### 3.1.5. Native SEC-HPLC

Soluble aggregation of recombinant human monoclonal antibody (rhMAb) was determined by SEC-HPLC on a Bio-Select SEC 250-5 column. The column was equilibrated and run in PBS at a flow rate of 0.5 mL/min using a HPLC system. Molecular weight standards were used to calibrate the column. The sample load was 25 μg, and protein was detected by monitoring the ultraviolet (UV) absorption at 214 nm.

Soluble aggregation of rhDNase was determined by SEC-HPLC on a silica-based TSK2000SWXL column. The column was equilibrated in an equilibration buffer at a flow rate of 1 mL/min using a 1090L HPLC system. The column load was 100 μg, and protein was detected by monitoring the UV absorption at 280 nm.

## 3.2. Spray Drying

SD consists of three steps of operation: atomization, dehydration, and powder collection. As shown in **Fig. 1A**, the protein solution feed is sprayed by an atomizer into a drying chamber. Aided by the large specific surface area of the droplets and the hot air, dehydration takes place in a matter of seconds in the laboratory-scale drying chamber. Finally, the dry particles are carried into the cyclone and settle in the product collector. The most important process parameters include drying air inlet temperature ($T_{inlet}$), drying air outlet temperature ($T_{outlet}$), drying air flow rate ($Q_{DA}$), and atomizing air flow rate ($Q_{AR}$). Of these parameters, $T_{outlet}$ is the dominating factor in controlling the drying rate and important particle characteristics, such as particle shape and moisture content (*see* **Subheadings 3.2.4.** and **3.2.5.**, respectively). Detailed analysis on stress events associated with SD are described in **Subheading 3.2.3.** Owing to the importance to process understanding, process and scale-up considerations are delineated in **Subheadings 3.2.1.** and **3.2.2.** Moreover, to explore an emerging technology as a viable alternative to SD, supercritical $CO_2$-assisted bubble drying will be depicted in **Subheading 3.2.6.**

### 3.2.1. Process Consideration

Many researchers studying SD of proteins have used bench-top spray dryers (e.g., Büchi dryers) for their ability to process a relatively small quantity of the protein—an important attribute if the availability of high-valued proteins is limited. Unfortunately, powder collection efficiency for this kind of dryer is relatively low at 20–50% *(8,9)*. For formulation scientists, it is hard to realize that with appropriate modifications (**Fig. 1B**), this bench-top dryer can be an efficient and useful manufacturing tool, at least for early phase clinical trials (*see* **Note 1**).

Yet, there is another limitation with the bench-top spray dryer—generating particle size of primarily less than 10 μm—the reason why dehydration requires only a few seconds. Certainly, larger dryers offer longer drying times for larger droplets to be dehydrated, but this comes at the expense of the small batch-size capability. To overcome this limitation, a custom-made spray dryer can be designed with a long, narrow drying chamber and a direct in-line filter collection system. The utility of a filter collection system also substantially benefits powder collection efficiency. Such a custom-made spray dryer can produce particles as large as 50 μm in median diameter.

### 3.2.2. Scale-up Consideration

Although SD has evolved into a mature technique for industry-scale production up to a few tons per day *(10)*, within the relatively young biotechnology industry,

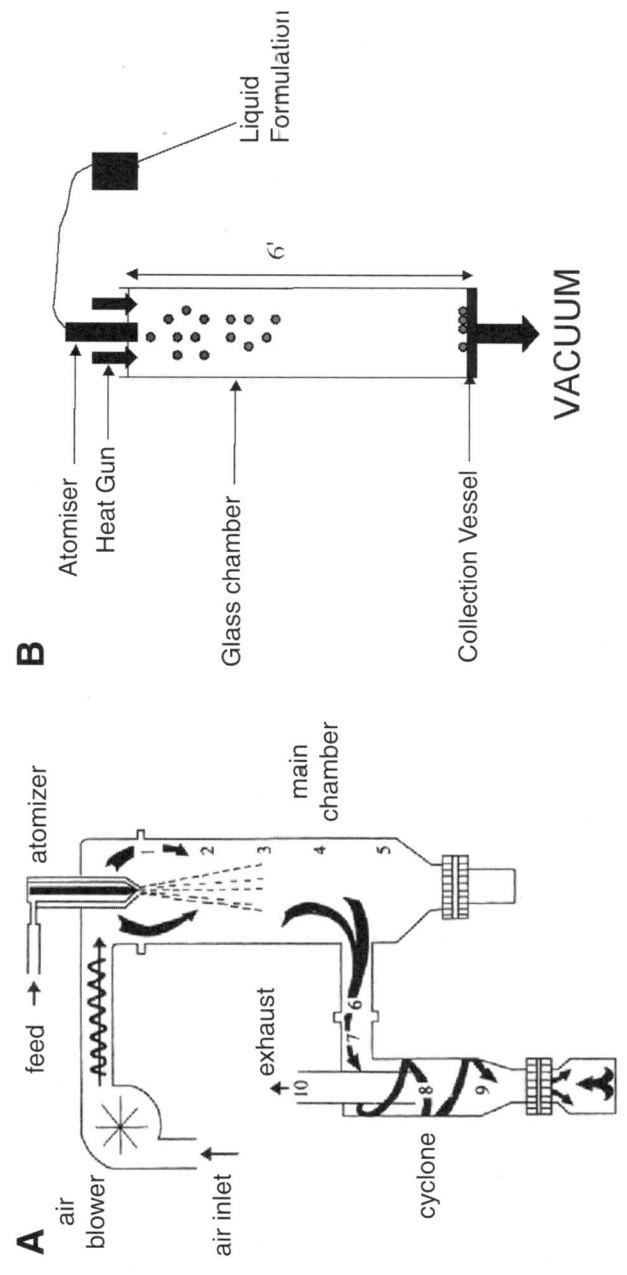

Fig. 1. The steps involved in SD. (**A**) The three steps of SD consist of atomization, dehydration, and powder collection. (**B**) Modifications to a bench-top dryer can be an efficient manufacturing tool.

the production of high-valued protein powders has been limited to the laboratory bench-top. There is very little information available for considering the scale-up issues of this powder formation process. However, an obvious challenge for the direct application of an industry-scale spray dryer is that such dryers might be prohibitively expensive because of the high-material costs. Again, scale-up strategies are available for the laboratory-scale spray dryer (*see* **Note 2**).

### 3.2.3. Process Stress on Protein Stability

A typical concern with SD of proteins/peptides is how thermally labile proteins can resist heat denaturation by hot air. This is a simple physical chemical phenomenon. During the early stage of drying, where the droplet surface remains moisture saturated (i.e., 100% relative humidity), the droplet surface temperature maintains at the wet-bulb temperature that is significantly lower than the hot air temperature. As drying continues, the droplet temperature begins to rise as water diffusion to the droplet surface cannot maintain 100% moisture. At this stage, the protein is primarily in the solid state, and the surrounding air temperature also decreases significantly due to moisture uptake. Thus, thermal denaturation is not typically observed in SD. However, it is a good practice to use a lower inlet air temperature to reduce the potential thermal stress to the protein. Nevertheless, as in FD, protein denaturation often occurs during dehydration; thus, it is necessary to incorporate a stabilizer (e.g., sugars or amino acids) into the protein formulation *(11,12)*. For some proteins, SD can alter the secondary structure (α-helix, β-sheet, and random coil; *13*) as observed in lyophilization *(14,15)*, which may irreversibly inactivate the protein. These alterations are attributed to the removal of hydration water molecules that are required to form hydrogen bonds to stabilize the protein's secondary structure *(16)*. Therefore, in developing a biochemically stable spray-dried protein product, it is judicious to dry the protein with a substance (e.g., sucrose or trehalose) that serves as a good water-replacing agent *(17,18)*.

Atomization and the air–water interface are the two other possible sources of stress during SD. It has been demonstrated previously that proteins can sustain shear rates as high as $10^5$ s$^{-1}$ *(19)*. Mathematical modeling estimates that the shear rate arising from atomization is in the range of $10^4$–$10^5$ s$^{-1}$ at most; therefore, it should not be a significant stress to the protein. However, when shear stress of this magnitude is combined with air–water interface, it may cause significant aggregation for air–water interface sensitive proteins, such as rhGH, bovine serum albumin (BSA), and lactate dehydrogenase (LDH) *(20–22)*.

The structure of most proteins is more or less amphiphilic, i.e., surfactant-like structure. The protein molecules tend to be adsorbed to the air–water droplet interface, where unusual surface energies may cause the protein to unfold, exposing hydrophobic regions. The unfolded protein may then undergo aggregation by the interaction of the exposed hydrophobic regions with other unfolded molecules until precipitation occurs *(23,24)*. This kind of surface denaturation has a great influence on spray-dried proteins because atomization generates fine droplets with an extremely high-specific surface area (*A*), that follows the relationship of $A = 6/D_{droplet}$ (e.g., 6000 cm$^2$/cm$^3$ for 10-μm droplets). A linear relationship between rhGH aggregation and $1/D_{droplet}$ suggests that aggregation is dominated by the total air–water interfacial area *(25)*.

Three options were found to be effective in minimizing the rhGH aggregation: the addition of a surfactant to prevent the formation of insoluble aggregates, addition of divalent zinc ions to prevent the formation of soluble aggregates, and increasing the rhGH concentration in the liquid feed.

### 3.2.4. Shape/Morphology of Spray-Dried Particles

Two factors predominate the shape and morphology of the particles upon SD: the rate of droplet evaporation and formulation composition. The speed of solvent evaporation dictates particle quality. Fast drying may cause deformed or defective particles, but slow drying may result in particles that are too wet and sticky to be collected effectively. Meanwhile, spray-dried composition (e.g., materials) prescribes the shape of the spray-dried particles. For example, some materials tend to form solid spherical particles, whereas others form hollow, deformed (shrivelled and cenospherical) or disintegrated particles (26). A hypothetical film, formed at the external surface of the droplets during drying, offers a possible explanation for both types of observed particles. Theoretically, if formed, such a film will encumber the outward diffusion of water and cause the water vapor pressure inside the droplet to increase. At a critical pressure, the film bursts, deforming the particle shape from its original sphericity. Certainly, the extent of such hindrance in diffusion will be dictated by the film properties, such as flexibility, mechanical strength, porosity, and so on.

Without doubt, formulation composition (i.e., the protein and the excipients) determines the film properties which, in turn, governs the observed particle morphology. However, the nature of the protein appears to have a stronger influence on particle morphology because of the protein's sensitive nature to the air–water interface. rhDNase tends to form spherical particles with a smooth-surface morphology (**Fig. 2A**). Recombinant human anti-IgE antibody typically forms donut-shaped or dimpled particles (**Fig. 2B**). However, proteins like rhGH and BSA, often form particles of raisin-like morphology (**Fig. 2C**).

The rate of droplet evaporation may also affect the film properties at the droplet's surface. A fast-drying rate will promote the formation of a more viscous film at the earlier phase of drying. Because water pressure increases rapidly inside the droplet, the film is prone to burst; therefore, the fast-drying conditions result in a large fraction of the particles containing dimples or holes. For example, based on the notion that a higher outlet temperature prompts a faster drying rate, rhDNase was spray-dried using two outlet drying air temperatures: 46°C and 88°C. Indeed, the particles collected with the outlet temperature adjusted to 88°C had extensive holes, whereas particles collected with an outlet temperature of 46°C were more spherical (**Fig. 3**).

### 3.2.5. Factors Affecting the Powder's Residual Moisture Content

The impact of moisture and temperature on solid-state protein stability is well-documented (27,28). Generally, a protein's chemical stability decreases with increasing moisture in the solid because of changes in either the dynamic activity or conformational stability of the protein or from water that serves as a reactant and/or medium for mobilization of reactants (28). In another perspective, residual water can serve as a plasticizer to lower the $T_g$ of the solid formulation—a physical attribute

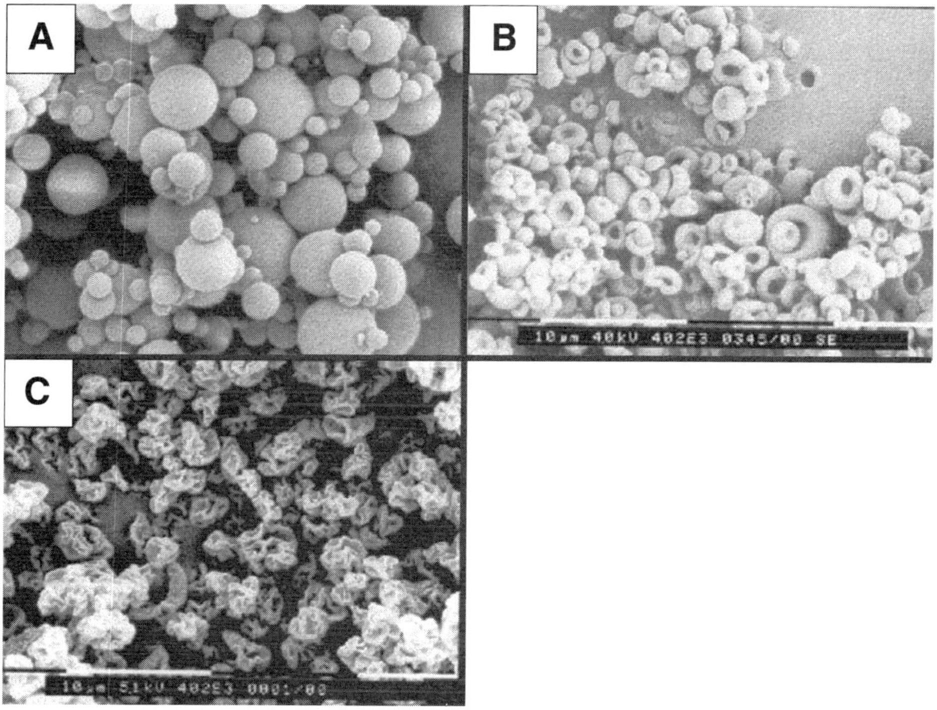

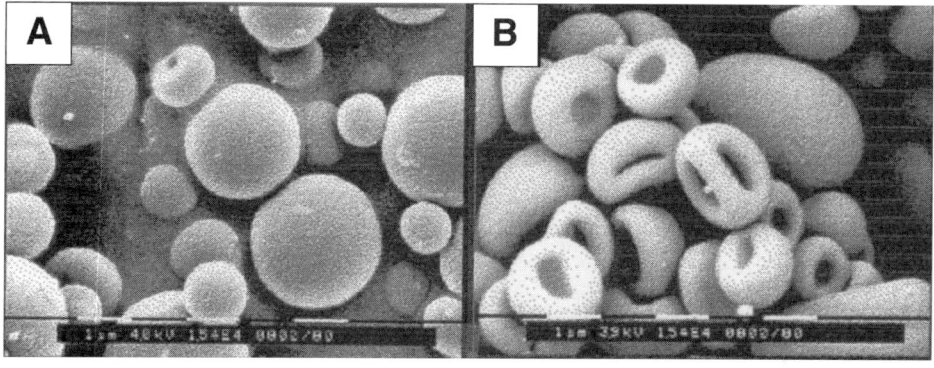

Fig. 2. SEM of pure protein spray-dried at an outlet temperature of 53°C: (A) rhDNase, (B) rhuMAb, and (C) BSA.

Fig. 3. SEM photograph of rhDNase:lactose (3:2 weight ratio) spray-dried at an outlet temperature of (A) 46°C and (B) 88°C.

unfavorable to protein stability. Furthermore, the level of moisture in the powder may affect particle size and promote excipient crystallization during long-term storage, thereby altering the properties and characteristics of the original powder. Thus, in the developmental stages of a spray-dried powder product, it is essential to understand how the SD conditions affect the powder's residual moisture content. By vapor pressure and mass transfer analysis, the effect of SD conditions on the powder's final moisture content can be elucidated (*see* **Note 3**). Overall, to reduce moisture content of the spray-dried powder, escalating the outlet temperature of the drying air is the most effective approach, but it is limited as a result of the possibility of thermally denaturing the protein.

Another important concept about maintaining the dry powder's moisture content is the humidity control on the powder-processing environment. If uncontrolled, residual moisture stored with the powder following SD may result in protein destabilization.

### 3.2.6. Supercritical CO₂-Aided Spray Drying

As an emerging particle formation technology, this innovative method enhances the ability of atomization (or nebulization/aerosolization) and dehydration posed by conventional SD *(29,30)*. Instead of establishing fine droplets by a two-fluid or pressurized atomizing mechanism, the protein/excipients-containing liquid formulation is mixed with supercritical $CO_2$ in a low dead-volume tee, and subsequent expansion out of a capillary flow restrictor into a drying chamber forms a fine emulsion, consisting of microbubbles (supercritical $CO_2$) and microdroplets. Thus, this technology is often referred to supercritical $CO_2$-assisted nebulization by a bubble dryer (CAN-BD).

There are two driving forces attributed to such emulsion formation. These are the sudden physical dispersion of the liquid solution by the rapid expansion of compressed $CO_2$ in the drying chamber and further breakup of microdroplets because of the sudden release of $CO_2$ dissolved in the liquid solution during their intimate contact in the mixing tee. Indeed, these physical mechanisms render added power to aerosolization, thereby readily producing a fine powder of less than 5 μm in diameter.

Owing to the large-specific surface area, the microdroplets can undergo rapid dehydration even using a low-temperature drying gas, e.g., dry nitrogen of 25–65°C. Certainly, such a drying temperature is much lower than that encountered in conventional SD, normally at 80–150°C. Therefore, the thermally induced stress by SD can be mitigated, which is one prominent advantage offered by CAN-BD.

There is a stress event specifically associated with CAN-BD, i.e., the interactions between protein and $CO_2$, occurring at the interface of the microdroplet and supercritical $CO_2$ and/or within the microdroplet where the protein encounters the dissolved $CO_2$. Although the number of proteins tested by CAN-BD or supercritical $CO_2$ methodologies in general remains low, several case studies have been reported that demonstrated the overall effect on protein stability is mild if the proteins are formulated with a proper buffer and pH, an amorphous disaccharide (e.g., trehalose or sucrose), and surfactant (e.g., Tween-20 or Tween-80). These proteins include LDH *(29)*, trypsinogen *(30,31)*, anti-CD4 *(31)*, and a-1-antitrypsin *(31)*.

Overall, CAN-BD is a viable technology capable of generating fine powders that greatly benefit formulation scientists in processing proteins of limited availability in early development work. Equally important is the demonstration of CAN-BD for powder manufacturing at a commercial scale *(32)*. Yet, a broader use of this method for producing protein powders for various biopharmaceutical and drug delivery applications requires a more extensive compilation of protein case studies and the ability of particle size manipulation to go beyond the 5-μm barrier.

## 3.3. Spray Freeze Drying

SFD is a relatively new method for biopharmaceutical powder preparation. This method combines the atomization and FD processes to present a potential advantage for fine-powder preparation. The principle is to atomize the protein solution into a cryogenic medium, such as liquid nitrogen, to quench freeze the droplets. The frozen droplets are then dried by lyophilization (**Fig. 4**). This process involves no heat for drying so that heat denaturation associated with the SD process can be avoided.

### 3.3.1. Advantages of the SFD Powder-Formation Method

In our view, the assessment of a protein powder process can be performed based on six criteria (*see* **Note 4**). Based on these criteria, SFD is considered a highly attractive process because it inherits the overall advantages of lyophilization, and it involves a viable particle formation mechanism. SFD is highly efficient (e.g., >90% yield) because all atomized droplets can be collected and dried in a confined space with a small-surface area. Also, control of particle characteristics can be achieved through atomization, formulation, and drying conditions. Furthermore, SFD is compatible with a wide range of excipients and biopharmaceuticals, whereas some methods (including SD) may be impossible or impractical, e.g., only possible with extremely low yield. In addition, the relatively mild stresses posed by SFD on biopharmaceuticals can be ameliorated by judicious process and formulation approaches. Moreover, SFD can be readily scaled up or down, the latter an important point when conducting screening studies with small quantities of high-valued biopharmaceuticals. Finally, SFD has been sufficiently developed to the point where it is employed in manufacturing a marketed biopharmaceutical (Nutropin Depot™).

### 3.3.2. Overview of SFD for Biopharmaceuticals Applications

The utility of SFD has been demonstrated in numerous proteins and other biopharmaceuticals (*see* **Note 5**). In each case, the process and formulation variables can be optimized based on the physicochemical properties of the protein and its susceptibility to the process stresses as well as the intended application.

The use of SFD began with microencapsulation *(5)*. For this application, spray freeze-dried protein particles are suspended in an organic solvent containing biodegradable poly(lactide-co-glycolide) (PLG). The suspension is spray frozen, and residual solvents are removed, e.g., by solvent exchange into ethanol followed by drying under vacuum. This process, referred to as ProLease®, is suitable for processing fragile drugs like protein, as the process is cryogenic, and there are no potentially destabilizing water–oil interfaces involved *(33)*. This technology is the basis for an FDA-approved, sustained-release form of rhGH: Nutropin Depot.

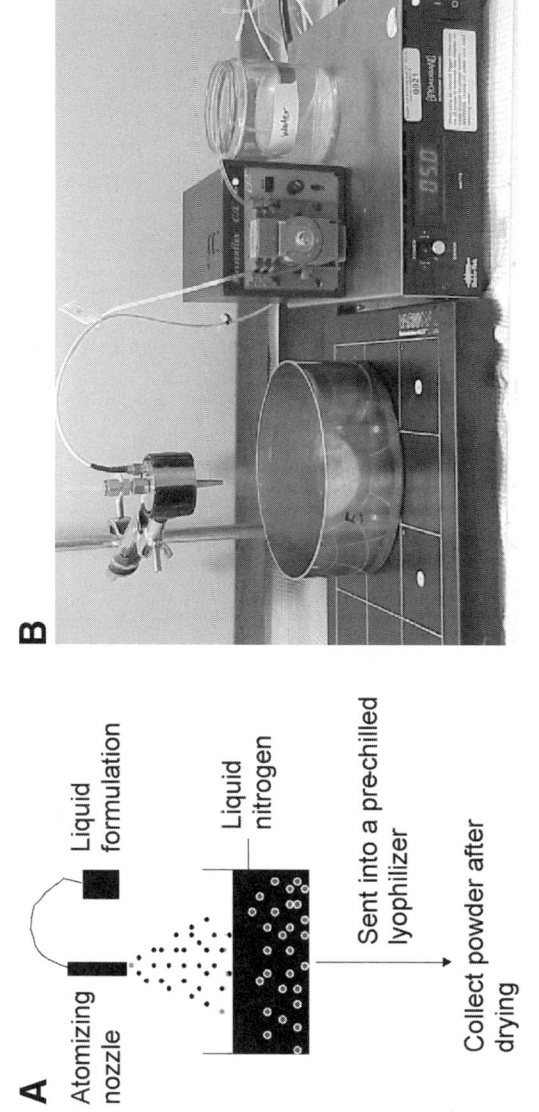

Fig. 4. SFD experimental set-up presented graphically (**A**) and in a real system (**B**).

275

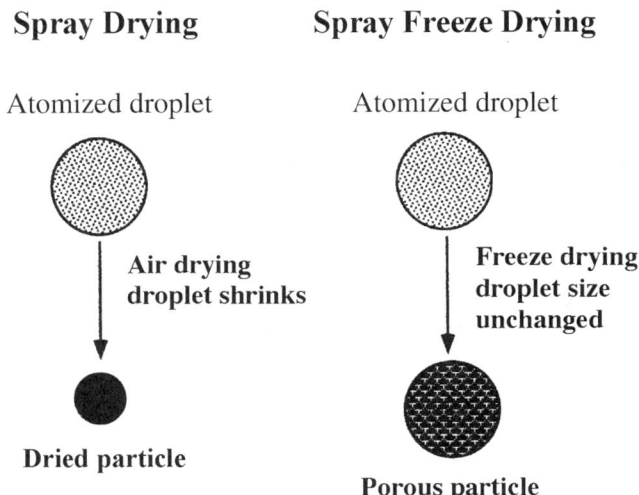

**Spray Drying**          **Spray Freeze Drying**

Atomized droplet          Atomized droplet

**Air drying**            **Freeze drying**
**droplet shrinks**       **droplet size**
                          **unchanged**

**Dried particle**

                          **Porous particle**

Fig. 5. Different drying mechanism by SD and SFD, resulting in distinct particle characteristics.

Another potential application for SFD proteins is the preparation of powders suitable for aerosol delivery. SFD is capable of producing highly porous particles suitable for pulmonary delivery *(6)*. Conceptually, the hydrodynamic diameter of a large porous particle would be comparable to that of a physically smaller, high-density particle. Physically larger porous particle would endure less surface energy, thereby suffering less cohesive forces upon inhalation.

Needle-free skin delivery utilizing high-density SFD particles represents a unique drug delivery application. More specifically, a powder-based intraepidermal delivery focusing on epidermal powder immunization for vaccine delivery has been successfully tested preclinically and in human clinical trials.

The high degree of porosity and high-specific surface area that is obtainable by SFD allows for other opportunities as well. Along these lines, yet another application that has been recently touted for SFD is to improve the dissolution of poorly water-soluble drugs *(34,35)*.

### 3.3.3. Particle Characteristics: SFD vs SD

SFD and SD produce powders of distinct physical properties. In general, the SFD powders have larger median particle size, larger specific surface area, and are more porous (i.e., less dense) than the SD powders. Whereas particle size is primarily dictated by the droplet size in both cases, the size of SD droplets shrinks during dehydration by hot air, and the particle shape is prone to change (*see* **Subheading 3.2.4.**). However, in the absence of hot air, atomized droplets during SFD maintained their spherical shape and size upon immediate freezing, and the subsequent drying process did not affect the shape or size (**Fig. 5**). Instead, the SFD process rendered the particles quite porous. The significant increase (approx 40-fold) in specific surface area for the SFD powder suggests a highly porous structure. Indeed, SEM analysis confirms that

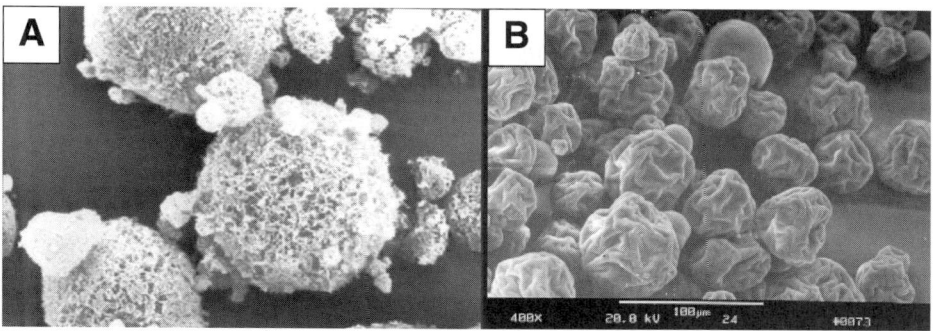

Fig. 6. Particle manipulation by SFD: **(A)** porous and **(B)** dense particles.

the SFD particles are spherical and porous (**Fig. 6A**). Such characteristics are particularly suitable for aerosol applications.

Despite this inherent nature, SFD can dictate particle density using one of the two following approaches: increasing the solid content of the spraying solution or employing a unique formulation composition. For example, a SFD powder formulation prepared from a solution containing hepatitis B surface antigen vaccine, along with a tertiary composition of trehalose:mannitol:dextran (3:3:4) at a solid content of 35%, results in particles with a "shrunken" morphology (**Fig. 6B**). Indeed, this powder has a much higher particle density, which has been attributed to the flow of freeze concentrates during drying to fill the voids left behind by ice crystals (*see* **Note 6**).

### 3.3.4. Stabilization of Proteins Upon SFD

A series of stress events are associated with SFD, i.e., atomization, freezing, and dehydration. Each stress event can create different stress sources, and each stress source potentially activates various mechanisms that might damage peptides and proteins. The first step in SFD, atomization, provides shear force that can potentially cause protein conformational changes *(36)* and an air–water interface that possibly promotes adsorption, conformational changes, and aggregation *(37,38)*. The stresses of the next two steps—freezing and drying—are conceptually the same as that posed by conventional lyophilization. For example, freezing can potentially cause damage to proteins via cold denaturation *(39)*, concentration, and pH changes *(2,40,41)*, as well as possible interaction at the ice interface *(42,43)*. In addition, the loss of hydrogen bonding upon removal of water can cause the structural rearrangement of proteins.

Despite such challenges, a mechanistic rational approach can be successfully employed to stabilize proteins upon SFD. To date, some studies on biopharmaceutical formulations, primarily on rhDNase, anti-IgE MAb, and interferon-γ, processed by SFD have appeared in the literature. In these studies, all potential stress sources were assessed, including using various excipients (e.g., amorphous sugars, surfactants, and so on), as well as processes like freeze/thaw, different freezing protocols, annealing, and so forth. Overall, in most cases, stresses posed by SFD were benign to the proteins. However, one particular stress appeared to intensify when the freezing rate exceeded a critical point. It was conjectured that a freezing-related stress event (probably due to shear stress associated with separation between ice- and freeze-concentrated phases) was responsible for protein denaturation upon SFD, ultimately leading to aggregation during drying and/or reconstitution. This freezing stress generated at the freezing front (ice–water interface) had an adverse effect on two therapeutic proteins, rhDNase and rhMAb, at a freezing rate beyond a critical point, which was estimated to be in the range of $8.4 \times 10^{-4}$ and $3.4 \times 10^{-3}$°C/s. When the ice front advances faster than this critical rate, it might alter protein conformation, leading to aggregation.

Yet, such an effect can be mitigated through stress relaxation by thawing (*see* **Note 7**). However, annealing may compromise the powder properties. Thus, a more effective approach might be to decelerate the freezing rate by making larger droplets. For instance, when three particle sizes were prepared (7 and 19 µm by two-fluid atomization and 32 µm by ultrasonic atomization), the aggregate content of anti-IgE MAb was 14.6%, 9.3%, and 0%, respectively. This is an important realization because SFD has not been considered benign to protein denaturation since the use of ultrasonic atomization to prepare powders with many of the proteins (**Note 4**).

### 3.4. Conclusions

SD and SFD present two important dehydration methodologies for biopharmaceutical formulations with unique particle formation attributes. In the authors' opinion, the use of these two methods side-by-side will enable researchers to engineer particles with a wide spectrum of properties. Furthermore, by understanding the stress posed by each process, appropriate formulation strategies can be taken to minimize detrimental effects on biomolecules. More importantly, these two dehydration methods can not only serve as an effective research tool, but they also offer large-scale manufacturing capabilities.

## 4. Notes

1. The most widely used powder collection equipment is a cyclone separator, in which the particle-laden gas enters a cylindrical or conical chamber tangentially at one or more points and leaves through a central opening at the top. Solid particles, by virtue of their inertia, move toward the wall of the separator from which they are fed into a receiver. Working essentially as a settling chamber, the cyclone uses centrifugal acceleration to replace gravitational acceleration as the separating force. The centrifugal-separating force can be as high as 2500g in very small, high-resistance cyclone units (*44*). Powder collection by the cyclone is governed by complex fluid dynamic behaviors in the cyclone and receiver (*45*). The fluid behavior is affected by cyclone design (*46*). Therefore, the study to improve powder recovery using different cyclone and receiver designs and their configurations yielded useful information in this area (*47*).

   Mass balance indicated that material loss in the bench-top SD system is due mostly to the attachment of sprayed droplets and dry powder to the wall of the apparatus and the cyclone's poor efficiency in collecting fine particles. Particle adhesion to the wall mainly occurs in the drying chamber, as well as in the cyclone, and is affected by the nature of the spray-dried materials and SD conditions. When the cyclone tends to retain a significant amount of the powder, recovering the powder from the cyclone as part of the product may be necessary and is only possible if the protein's stability and particle properties remain unchanged. Many researchers may find cooling down the cyclone helps maintain the protein's stability (unpublished observations). This may be true for very thermally labile proteins. However, despite the high temperature in the cyclone, most proteins investigated thus far are stable *(20,48,49)*. One possible drawback with reducing the cyclone temperature is that the relative humidity of the air in the cyclone will increase and result in higher moisture content of the powder.

The design of the bench-top dryer (Büchi) has limitations in drying air flow, thereby limiting the batch size for SD. This is because of a bag-filter unit located downstream of the system (**Fig. 2A**) through which a respirator pulls the drying air. This filter unit presents a major resistance to airflow. Unfortunately, the resistance increases during the drying process as fine particles slowly build on the bag to foul the filter. A report *(47)* described modifications to the dryer to improve airflow. The modifications (**Fig. 1C**) include removal of the bag-filter unit, relocation of the aspirator, and addition of a vacuum-filter unit. This modified system increased the capacity of drying air, which allowed droplets to be dried at a lower inlet air temperature, whereas the outlet air temperature remained unchanged. This is important to the SD of heat-sensitive proteins. Both the removal of the bag filter and addition of a vacuum-filter resulted in nearly a 100% increase in the airflow rate. This allows nonstop SD of 2-L batch volume (up to 60 g of solid). Design changes, such as using dual cyclones and dual receivers in different configurations and cyclones of different designs, were tested and their effects on powder collection are minor. Also, the effect of using an antistatic-treated cyclone on powder collection was found to be insignificant. This information suggests that the bench-top spray dryer can be a useful production tool for preparing high-valued, low-volume protein products if powder collection efficiency is acceptable. Powder collection is affected more by protein formulation than by system design.

2. By definition, scale-up of a SD process involves increasing the powder output while maintaining product quality comparable to the small-scale process. Several factors influence the rate of product output; these include liquid throughput, solid (protein/excipient) content in the solution/suspension formulation, and product recovery efficiency. Low recovery of aerosol particles upon SD by cyclone collection is a concern owing to the poor deposition characteristics and cohesive nature of the powder as a result of very small particles. Although the general concept favors improved recovery associated with large dryers, the published results are lacking.

When liquid throughput is increased, the process to produce desired powders is limited by two factors: atomization and heating capacity. Of many atomizing mechanisms, air (two-fluid) atomization, based on kinetic energy, should be one of the most effective methods for generating fine sprays. However, its efficiency will decrease as the liquid feed rate increases, because the atomizing air will be acting on the liquid in a less homogeneous fashion. Drying capacity is measured by the rate of water removal from the dryer in the form of vapor. Drying air with more thermal energy can be achieved with higher temperatures and higher flow rates depending on the power source of the system. Larger dryers are normally equipped with larger power supplies to boost heating capacity. However, because the outlet air temperature is the most critical parameter in the SD process *(50)*, heating capacity is limited by how high the inlet air temperature can go at a fixed outlet air temperature without compromising protein stability during the process.

A more direct way to increase powder production is to maximize the protein concentration or total solid content in the starting liquid formulation. The solubility of the active or inactive components limits the solid content of the feed solution. Nevertheless, the stability of the protein at high concentrations should be monitored and studied to ensure good stability prior to SD. The solid concentration can also affect particle characteristics, e.g., particle size, density, and morphology.

The dimensions of the drying chamber determines the scale of the dryer. The taller the chamber, the longer the residence times for the droplets and the larger the particles that can be produced. Under normal conditions, the bench-top spray dryers, such as Büchi-190, can only produce powders of less than 10 μm. The next scale-up is the laboratory-scale spray dryer that is commercially available in different sizes. The Yamato DL-41, Mobile Minor by Niro, and Tower-unit by Bowen Engineering represent the commonly used models at an increasing scale, which can produce powders with a particle size of typically 15, 20, and 30–50 μm, respectively. Dryers of larger scale are rarely used for therapeutic protein products because of limited production quantity.

3. In the drying chamber, the drying force for dehydration is the difference of partial water vapor pressure (or relative humidity) between the solid surface and the environment, $P_{droplet} - P_{DA}$. When there is enough water initially to keep the surface saturated, the surface relative humidity is 100%. This surface vapor pressure decreases, as the subsurface can no longer supply sufficient water for surface saturation due either because of a decease in liquid diffusion or to the reduced moisture level within the protein solid. Therefore, the final moisture level of the protein solid is determined by two factors: the nature of the material and the humidity of the drying conditions. The former is caused by the interactions of water molecules with protein and/or formulation excipient molecules *(51)*. The number and distribution of strong- and weak-binding sites on protein and the excipient molecules are among the intrinsic properties determining these interactions. Therefore, the moisture level of the protein solid can be primarily controlled by the humidity of the environment where the powder is manufactured, processed, and stored, resulting from the equilibration between the powder's residual moisture and the environmental humidity.

Under normal-drying conditions, the powders prepared by FD are drier than those prepared by SD, 1–4% vs 4–10% *(52)*. During FD, the final moisture content is determined by the secondary drying step, virtually a vacuum-drying process performed near room temperature. Therefore, a long-drying time is normally used to ensure that the solid is dried to its equilibrium moisture content, i.e., in equilibration with the environment. However, inside the drying chamber of the spray dryer, the droplets encounter a continuously changing environment where the drying air temperature decreases and the relative humidity (%RH) increases along the chamber from moisture uptake. Thus, the driving force for heat and mass transfer decreases, as does the rate of water removal. If, during the SD process, it is assumed that the powder reaches equilibrium with the surrounding drying air, the final moisture content of the powder will be determined by the %RH of the air inside the collection vessel where the powder resides for the most time. Therefore, to produce drier spray-dried powders, higher inlet (outlet) air temperatures and lower liquid feed rates are required. Indeed, many spray-dried powders with moisture contents of less than 3% have been reported using inlet air temperatures of 140°C or higher (or outlet air temperatures of 90°C or higher) or liquid feed rates of 2 mL/min or lower *(53)*. This represents an undesirable manufacturing condition that decreases the overall production rate and may impose potential adverse effects on protein denaturation *(8,52)*. Therefore, subjecting the powder to a secondary vacuum-drying process might be a better alternative to reduce the moisture content of the spray-dried powder. Nevertheless, if this powder is exposed to a humid environment, it will pick up moisture until its moisture content equilibrates with the %RH of the surrounding environment. Regardless of the drying method, the final moisture of the powder is determined by the environment where the powder was further processed or stored.

Another approach to improve moisture removal is to decrease $P_{DA}$ by dehumidifying the air prior to entering the chamber, as mass transfer is affected by the driving force of $P_{droplet} - P_{DA}$. As the %RH of ambient air was reduced to 5% or lower by a dehumidifier, this additional dehumidification step does not further reduce the moisture content of the powder *(52)*. However, the dehumidified drying air can improve the drying capacity, i.e., removing more water per unit of time.

4. The six criteria in the assessment of the protein powder process are (1) process efficiency/yield; (2) ability to control particle properties, such as size/size distribution, powder flowability, and density; (3) compatibility with a wide range of excipients and biomolecules; (4) mildness of process stress and impact on stability of biopharmaceuticals; (5) flexibility for both scale-up and scale-down; and (6) suitability of manufacturing, i.e., ability to conduct steps in a straightforward manner aseptically.

5. Over the years, a host of macromolecules have been tested by SFD: rhDNase *(6)*, rhMAb *(6)*, recombinant human interferon-$\gamma$ *(3,4)*, rhGH *(5,54)*, recombinant human vascular endothelial growth factor *(55)*, recombinant human nerve growth factor *(56)*, recombinant human insulin-like growth factor-I *(57)*, recombinant humanized monoclonal antibody *(58)*, BSA *(43)*, trypsinogen *(58)*, lysozyme and lactate dehydrogenase *(60)*, catalase *(61)*, insulin *(62)*, and vaccines, including hepatitis B surface antigen, diphtheria toxoid, and tetanus toxoid *(63,64)*.

6. It remains unclear which factor affects particle shrinkage more—the solid content of the spraying solution, drying conditions, or formulation composition. Because increasing the solid content is limited by each excipient's solubility maximum and by the solution viscosity maximum where atomization ceases, we attempted to increase particle density by inducing particle shrinkage via formulation composition. As previously reported *(59,60)*, formulations containing trehalose alone or the binary mixture of trehalose/mannitol still produce lower density particles even at high levels of solid content (>25%), whereas the mixture of three excipients, trehalose/mannitol/dextran, resulted in significant particle shrinkage. The working hypothesis is that either the freeze-concentrate structure collapses during ice sublimation (or possibly during secondary drying) when the freeze-concentrate mixture softens and can flow to fill the void left behind by ice crystals. Certainly, it must be related to the maximally frozen freeze-concentrate's glass transition temperature ($T_g'$), which drops below the drying temperature. Yet, the collapse occurs only at the intraparticle level, because the powder would lose particle characteristics if the collapse took place at the interparticle level. More specifically, the composition of trehalose: mannitol:dextran at the 3:3:4 weight ratio was particularly effective in plasticizing the mixture and promoting particle shrinkage, thereby resulting in a higher tap density.

7. To test this hypothesized mechanism, a series of experiments were performed to anneal the spray-frozen powder of anti-IgE MAb:trehalose (60:40) at temperatures (−5, −10, and −15°C) higher than the primary drying temperature (−25°C) for 1 h before the lyophilization cycle began. As the annealing temperature increased, the surface area and aggregate content was reduced: 74.6 m²/g (nonannealing), 15.4 m²/g (−15°C), 11.3 m²/g (−10°C), and 8.3 m²/g (−5°C). Annealing temperatures near the melting point caused large ice crystals to grow at the expense of more energetically unfavorable small crystals (a migratory recrystallization process), thereby reducing the surface area of the particle. Indeed, such annealing appeared to relax the stress as the protein's aggregate content decreased as well: 15.4% (nonannealing), 5.6% (15°C), 4.2% (−10°C), and 2.1% (−5°C).

# References

1. Dudu, S. P., Zhang, G., and Dal Monte, P. R. (1997) The relationship between protein aggregation and molecular mobility below the glass transition temperature of lyophilized formulations containing a monoclonal antibody. *Pharm. Res.* **14,** 596–600.

2. Pikal-Cleland, K. A., Cleland, J. L., Anchordoquy, T. J., and Carpenter, J. F. (2003) Effect of glycine on pH changes and protein stability during freeze-thawing in phosphate buffer systems. *J. Pharm. Sci.* **91,** 1969–1979.

3. Webb, S. D., Golledge, S. L., Cleland, J. L., Carpenter, J. F., and Randolph, T. W. (2002) Surface adsorption of recombinant human interferon-gamma in lyophilized and spray-lyophilized formulations. *J. Pharm. Sci.* **91,** 1474–1487.

4. Webb, S. D., Cleland, J. L., Carpenter, J. F., and Randolph, T. W. (2003) Effects of annealing lyophilized and spray-lyophilized formulations of recombinant human interferon-gamma. *J. Pharm. Sci.* **92,** 715–729.

5. Tracy, M. A. (1998) Development and scale-up of a microsphere protein delivery system. *Biotechnol. Prog.* **14,** 108–115.

6. Maa, Y.-F., Nguyen, P.-A., Sweeney, T. D., Shire, S. J., and Hsu, C. C. (1999) Protein inhalation powder: Spray drying vs. spray freeze drying. *Pharm. Res.* **16,** 249–254.

7. Chen, D., Maa, Y.-F., and Haynes, J. (2002) Needle-free epidermal powder immunizations. *Expert Vaccine Rev.* **1,** 265–276.

8. Broadhead, J., Edmond Rouan, S. K., Hau, I., and Rhodes, C. T. (1994) The effect of process and formulation variables on the properties of spray-dried—galactosidase. *Pharm. Pharmacol.* **64,** 458–467.

9. Labrude, P., Rasolomanana, M., Vigneron, C., Thirion, C., and Chaillot, B. (1989) Protective effect of sucrose on spray drying of oxyhemoglobin. *J. Pharm. Sci.* **78,** 223–229.

10. Masters, K. Spray-drying Handbook, 4th ed. Longman (UK) and J. Wiley & Sons (US), 1985.

11. Broadhead, J., Edmond Rouan, S. K., and Rhodes, C. T. (1992) The spray-drying of pharmaceuticals. *Drug Dev. Ind. Pharm.* **18,** 1169–1206.

12. Labrude, P., Rasolomanana, M., Vigneron, C., Thirion, C., and Chaillot, B. (1989) Protective effect of sucrose on spray drying of oxyhemoglobin. *J. Pharm. Sci.* **78,** 223–229.

13. Tzannis, S. T., Meyer, J. D., and Prestrelski, J. D. Secondary structure consideration during protein spray drying. Presented at 213th ACS National Meeting, BIOT-297, April 13–17, San Francisco, CA, 1997.

14. Costantino, R. H., Nguyen, T. H., and Hsu, C. C. (1996) Fourier transform infrared spectroscopy demonstrates that lyophilization alters the secondary structure of recombinant human growth hormone. *Pharm. Sci.* **2,** 229–232.

15. Prestrelski, S. J., Pikal, K. A., and Arakawa, T. (1995) Optimization of lyophilization conditions for recombinant human interlukin-2 by dried-state conformational analysis using fourier transform infrared spectroscopy. *Pharm. Res.* **12,** 1250–1259.

16. Carpenter, J. F. and Crowe, J. H. (1989) An infrared spectroscopic study of interactions of carbohydrates with dried proteins. *Biochemistry* **28,** 3916–3922.

17. Arakawa, T. and Timsheff, S. N. (1982) Stabilization of protein structure by sugars. *Biochemistry* **21,** 6536–6544.

18. Sarciaux, J.-M. E. and Hageman, M. (1997) Effect of bovine somatotropin (rbSt) concentration at different moisture levels on the physical stability of sucrose in freeze-dried rbSt/sucrose mixture. *J. Pharm. Sci.* **86,** 365–371.

19. Maa, Y.-F. and Hsu, C. C. (1996) Effect of high shear on proteins *Biotech. Bioeng.* **51,** 458–465.

20. Maa, Y.-F. and Hsu, C. C. (1997) Protein denaturation by combined effect of shear and air-liquid interface. *Biotech. Bioeng.* **54,** 503–512.
21. Adler, M. and Lee, G. (1999) Stability and surface activity of lactate dehydrogenase in spray-dried trehalose. *J. Pharm. Sci.* **88,** 199–208.
22. Faldt, P. and Berganstahl, B. (1994) The surface composition of spray-dried protein-lactose powders. *Colloids and Surfaces* **90,** 183–190.
23. MacRichie, F. (1978) Proteins at interfaces. *Adv. Protein Chem.* **32,** 283–311.
24. Thurow, H. and Geisen, K. (1984) Stabilization of dissolved proteins against denaturation at hydrophobic interfaces. *Diabetologia* **27,** 212–218.
25. Maa, Y.-F., Nguyen, P.-A., and Hsu, C. C. (1998) Spray drying of air-sensitive recombinant human growth hormone. *J. Pharm. Sci.* **87,** 152–159.
26. Masters, K. (1991) Spray drying handbook, 5th ed. John Wiley and Sons, New York, 1991.
27. Hageman, M. J. (1988) The role of moisture on protein stability. *Drug. Dev. Ind. Pharm.* **14,** 2047–2070.
28. Bell, L. N., Hageman, M. J., and Muraoka, L. M. (1995). Thermally induced denaturation of lyophilized bovine somatotropin and lysozyme as impacted by moisture and excipients. *J. Pharm. Sci.* **84,** 707–712.
29. Sellers, S. P., Clark, G. S., Sievers, R. E., and Carpenter, J. F. (2001) Dry Powders of Stable Protein Formulations from Aqueous Solutions Prepared Using Supercritical CO2-Assisted Aerosolization. *J. Pharm. Sci.* **90,** 785–797.
30. Villa, J. A, Sievers, R. E., and Huang, E. T. S. Bubble drying to form fine particles from solutes in aqueous solutions. Proceedings of the 7th Meeting on Supercritical fluids: Particle design—Materials and natural products processing (Perrut, M. and Reverchon, E., eds.), Antibes/Juan-Les-Pins, France, December 6–8, 2000, pp. 83–88.
31. Cape, S. P., Villa, J. A., Huang, E. T. S., Yang, T. H., Carpenter, J. F., and Sievers, R. E. Preparation of active protein fine powders using supercritical or near-critical carbon dioxide—Protein Formulation, micronization and delivery. Manuscript in preparation.
32. Sievers, R. E., Huang, E. T. S., Villa, J. A., and Walsh, T. Process for rapidly forming and drying fine particles. Proceedings of the 8th Meeting on Supercritical fluids: Chemical reactivity and material processing in supercritical fluids (Besnard, and cansell, F., eds.), Bordeaus, France, April 14–17, 2002, pp. 73–78.
33. Gombotz, W. R., Healy, M. S., and Brown, L. R. (1991) Very low temperature casting of controlled release microspheres. U.S. Patent no. 5,019,400.
34. Hu, J., Rodgers, T. L., Brown, J. N., Young, T., Johnson, K. P., and Williams III, R. O. (2002) Improvement of dissolution rates of poorly water soluble APIs using novel spray freeze into liquid technology. *Pharm. Res.* **19,** 1278–1284.
35. Rodgers, T. L., Nelson, A., Hu, J, Brown, J. N., Sarkari, M., Young, T., et al. (2002) A novel particle engineering to enhance dissolution rates of poorly water soluble drugs: spray freeze into liquid. *Eur. J. Pharm. Biopharm.* **54,** 271–280.
36. Singh-Zocchi, M., Hanne, J., and Zocchi. G. (1999) Plastic deformation of protein monolayers. *Biophys. J.* **83,** 2211–2218.
37. Maa, Y.-F., Nguyen, P.-A., and Hsu, C. C. (1998) Spray drying of air-sensitive recombinant human growth hormone. *J. Pharm. Sci.* **87,** 152–159.
38. Sluzky, V., Tamada, J. A., Klibanov, A. M., and Langer, R. (1991) Kinetics of insulin aggregation upon agitation in the presence of hydrophobic surfaces. *Proc. Natl. Acad. Sci. USA* **88,** 9377–9381.

39. Tsai, C. J., Maizel, J. V., and Nussinov, Jr., R. (2002) The hydrophobic effect: a new insight from cold denaturation and a two-state water structure. *Crit. Rev. Biochem. Mol. Biol.* **37,** 55–69.

40. Lam, X. M., Costantino, H. R., Overcashier, D. E., Nguyen, T. H., and Hsu, C. C. (1996) Replacing succinate with glycolate buffer improves the stability of lyophilized interferon-γ. *Int. J. Pharm.* **142,** 85–95.

41. Pikal-Cleland, K. A. and Carpenter, J. F. (2001) Lyophilization-induced protein denaturation in phosphate buffer systems: monomeric and tetrameric beta-galactosidase. *J. Pharm. Sci.* **90,** 1255–1268.

42. Hsu, C. C., Hguyen, H. M., Yeung, D. A., Brooks, D. A., Koe, G. S., Bewley, T. A., and Pearlman, R. (1995) Surface denaturation at solid-void interface—a possible pathway by which opalescent particulates form during the storage of lyophilized tissue-type plasminogen activator at high temperatures. *Pharm. Res.* **12,** 69–77.

43. Costantino, H. R., Firouzabadian, L., Hogeland, K., Wu, C., Beganski, C., Carrasquilla, K. G., et al. (2000) Protein spray freeze drying. Effect of atomization conditions on particle size and stability. *Pharm. Res.* **17,** 1374–1383.

44. Perry, R. H. and Green, D. (1984) Chemical Engineering Handbook, 6th ed. McGraw-Hill, New York, 1984, pp. 77–89.

45. Bradley, D. (1965) The Hydrocyclone. Pergamon Press Ltd., Oxford, UK, 1965, pp. 88–98.

46. Bloor, M. I. G. and Ingham, D. B. (1973) On the efficiency of the industrial cyclone. *Trans. Instn. Chem. Engrs.* **51,** 173–176.

47. Maa, Y.-F., Nguyen, P.-A., Sit, K., and Hsu, C. C. (1998) Spray-drying performance of a bench-top spray dryer for protein aerosol powder preparation. *Biotech. Bioeng.* **60,** 301–309.

48. Mumenthaler, M., Hsu, C. C., and Pearlman, R. (1994) Feasibility study on spray-drying protein pharmaceuticals: Recombinant human growth hormone and tissue-type plasminogen activator. *Pharm. Res.* **11,** 12–20.

49. Costantino, H. R., Andya, J. D., Nguyen, P.-A., Dasovich, N., Sweeney, T. D., Shire, S. J., et al. (1998) The effect of mannitol crystallization on the stability and aerosol performance of a spray-dried pharmaceutical protein, recombinant humanized anti-IgE monoclonal antibody. *J. Pharm. Sci.* **87,** 1406–1411.

50. Maa, Y.-F., Costantino, H. R., Nguyen, P.-A., and Hsu, C. C. (1997) The effect of operating and formulation variables on the morphology of spray-dried protein particles. *Pharm. Dev. Technol.* **2,** 213–223.

51. Hageman, M. J. Stability of protein pharmaceuticals. Part A, in Chemical and Physical Pathways of Protein Degradation (Ahern, T. J. and Manning, M. C., eds.), Plenum Press, New York, 1994, pp. 273–309.

52. Maa, Y.-F., Nguyen, P.-A., Andya, J. D., Dasovich, N., Sweeney, T. D., Shire, S. J., and Hsu, C. C. (1998) Effect of spray drying and subsequent processing conditions on residual moisture content and physical/biochemical stability of protein inhalation powders. *Pharm. Res.* **15,** 768–775.

53. Tzannis, S. T., Meyer, J. D., and Prestrelski, J. D. Secondary structure consideration during protein spray drying. Presented at 213th ACS National Meeting, BIOT-297, San Francisco, CA, April 13–17, 1997.

54. Johnson, O. L., Jaworowicz, W., Cleland, J. L., Bailey, L., Charnis, M., Duenas, E., et al. (1997) The stabilization of human growth hormone into biodegradable microspheres. *Pharm. Res.* **14,** 730–735.

55. Cleland, J. L., Duenas, E. T., Park, A., Daugherty, A., Kahn, J., Kowalski, J., and Cuthbertson, A. (2001) Development of poly(D,L-lactide-co-glycolide) microsphere

formulations containing recombinant human vascular endothelial growth factor to promote local angiogenesis. *J. Control. Release* **72,** 13–24.

56. Lam, X. M., Duenas, E. T., and Cleland, J. L. (2001) Encapsulation and stabilization of nerve growth factor into poly(lactic-co-glycolic) acid microspheres. *J. Pharm. Sci.* **90,** 1356–1365.

57. Lam, X. A., Duenas, E. T., Daugherty, A. L., Levin, N., and Cleland, J. L. (2000) Sustained release of recombinant human insulin-like growth factor-I for treatment of diabetes. *J. Control. Release* **67,** 281–292.

58. Mordenti, J., Thomsen, K., Licko, V., Berleau, L., Kahn, J. W., Cuthbertson, R. A., et al. (1999) Intraocular pharmacokinetics and safety of a humanized monoclonal antibody in rats after intravitreal administration of a solution or a PLGA microsphere formulation. *Toxicol. Sci.* **52,** 101–106.

59. Sonner, C., Maa, Y.-F., and Lee, G. (2002) Spray-freeze-drying for protein powder preparation: particle characterization and a case study with trypsinogen stability. *J. Pharm. Sci.* **91,** 2122–2139.

60. Sonner, C. Protein loaded powders produced by spray freeze drying. Ph. D. Thesis. Department of Pharmaceutics, University of Erlangen, Erlangen, Germany, 2002.

61. Rochelle, C. and Lee, G. Spray-freeze-drying of proteins: Inactivation of catalase on atomization, freezing in liquid nitrogen, and freeze-drying steps. Presented at the CRS German Chapter Meeting, Munich, Germany, March 2003.

62. Yu, Z., Rogers, T. L., Hu, J., Johnson, K. P., and Williams III, R. O. (2002) Preparation and characterization of microparticles containing peptide produced by a novel process: Spray freezing into liquid. *Eur. J. Pharm. Biopharm.* **54,** 221–228.

63. Maa, Y.-F., Lu, Z., Payne, L. G., and Chen, D. (2003) Stabilization of alum-adjuvanted vaccine dry powder formulations: Mechanism and application. *J. Pharm. Sci.* **92,** 319–332.

64. Maa, Y.-F., Shu, C., Ameri, M., Zuleger, C., Che, J., Osorio, J. E., et al. (2003) Optimization of alum-adjuvanted vaccine powder formulation for epidermal powder immunization. *Pharm. Res.* **20,** 969–977.

# 22

## Stabilization of Therapeutic Proteins by Chemical and Physical Methods

### Ken-ichi Izutsu

### 1. Introduction

Proteins are complex molecules composed of numerous reactive chemical groups and delicate three-dimensional structures. The chemical (oxidation, deamidation, hydrolysis) and physical (unfolding, aggregation) changes of proteins during the formulation process and storage not only reduce biological activity, but they can cause adverse reactions, such as immune responses, even when the total protein population is at a low level (**Table 1**) *(1–5)*. Developing new formulations that maintain the integrity of the purified protein during the manufacturing process, delivery, and 1–2 yr of storage is a minimum requisite for bringing protein pharmaceuticals to the market. Aqueous solutions that are ready for injection are often the most convenient and preferable dosage forms for therapeutic proteins, whereas many purified proteins are not stable enough in the solutions to meet the shelf-life requirements, even when they are stored at low temperatures.

Altering the protein structure (amino acid alternation and chemical modification) and altering the environment (e.g., pH change and dehydration) are two major methods to stabilize various proteins. Optimizing the protein environments by changing the physical state and/or use of excipients usually attains the stability of therapeutic proteins. Protein engineering methods are employed more often to alter the clinical properties of the protein, including slower disposition by covalent attachment of polyethylene glycol. This chapter focuses on the strategies to optimize stability of protein therapeutics. Understanding the various stresses and relevant stabilizing mechanism of excipients in each phase would facilitate rational formulation design. The protein-stabilizing techniques can also be applied in protein maintenance for general use.

### 2. Materials

The required reagents and equipments vary depending on the protein and formulation types. Typical excipients and their concentrations used for protein formulations are listed in **Table 2**.

From: *Methods in Molecular Biology, vol. 308: Therapeutic Proteins: Methods and Protocols*
Edited by: C. M. Smales and D. C. James © Humana Press Inc., Totowa, NJ

**Table 1**
**Typical Protein Degradation Pathways**

|  | Examples | Analytical methods |
|---|---|---|
| Physical instability | Denaturation, aggregation, precipitation, and adsorption | Size-exclusion HPLC, light scattering, SDS-PAGE, FT-IR, Raman, UV, and fluorescence |
| Chemical instability | Deamidation, β-elimination, oxidation, disulfide exchange, and hydrolysis | Reversed-phase HPLC, ion-exchange HPLC, SDS-PAGE, mass spectroscopy, and peptide mapping |

**Table 2**
**Excipients for Protein Formulations**

| Class | Purpose | Examples | Typical concentration |
|---|---|---|---|
| Buffer | pH buffer | Sodium phosphate, potassium phosphate, histidine, and Tris-HCl | 10–50 m$M$ |
| Salt | Tonicity modifier and solubilizer | NaCl and CaCl$_2$ | 10–100 m$M$ |
| Sugar | Protein-stabilizer, bulking agent, and glass former | Sucrose and trehalose | 10–100 mg/mL |
| Polyol | Tonicity modifier and bulking agent | Mannitol and sorbitol | 10–100 mg/mL |
| Amino acid | Tonicity modifier, bulking agent, and stabilizer | Glycine and arginine | 10–100 mg/mL |
| Polymer | Bulking agent and glass former | Hydroxyethyl starch | 10–50 mg/mL |
| Surfactant | Solubilizer, stabilizer, and aggregation inhibitor | Tween-80, Tween-40, and SDS | <1 mg/mL |
| Preservative | Antimicrobial preservation | Benzyl alcohol and phenol | 1–10 mg/mL |
| Antioxidant | Antioxidant | Ascorbic acid | 1–10 mg/mL |
| Other | Protein-specific stabilization | Ligands | <1 mg/mL |

1. High-performance liquid chromatography (HPLC) (ion-exchange, size-exclusion, reversed-phase, and affinity), circular dichroism, mass spectrometry, fluorescence spectroscopy, FT-IR, sodium dodecyl sulfate polyacrylamide gel electrophoresis (SDS-PAGE), and other apparatus for protein analysis.
2. System to measure protein biological activity.
3. Freeze drier, shaker, and chambers (temperature, light, and humidity) to apply storage- or process-related stresses.
4. Differential-scanning calorimeter and Karl Fischer coulometer to analyze the physical property of dried formulations.

## 3. Methods

### 3.1. Preformulation Studies

Preformulation studies are designed to elucidate the physical and chemical properties of the object protein molecules to determine the stabilization strategies.

1. Obtain the protein solubility and stability profile in aqueous solutions of different pH and ionic strength.
2. Clarify the possible ligands that affect protein stability.
3. Establish analytical methods to identify the chemical, physical, and biological changes of the particular protein.
4. Clarify the chemically liable residues from experimental studies and/or *in silico* calculations.
5. Clarify the relative importance of the changes on the pharmaceutical qualities: e.g., pharmacology, pharmacokinetics, and toxicity.

### 3.2. Formulation Design

Formulation types and excipients are selected to maximize the pharmaceutical quality of the product depending on the protein stability, clinical needs, and pharmaceutical acceptability. Aqueous formulations are the preferred choice because of the convenience of manufacturing and using them. Optimization of the protein formulations is attained by a combination of the following relevant experiments.

1. Obtain the protein stability profile through the analysis of protein solutions exposed to possible process- or storage-related stresses (e.g., high temperature and agitation).
2. Apply protein solutions containing excipient or excipient combinations (pH buffer, tonicity modifier, and protein stabilizer) to "the stresses." Select the candidate formulations from the stability data of the protein, focusing on the physical and major chemical changes.
3. Optimize the protein stability by attenuating the candidate formulations.
4. Perform a long-term real-time stability test at a larger scale. Determining the formulation sometimes requires a parallel study of the protein stability.

Lyophilized formulation is a default option to foster the long-term stability of proteins that are not stable enough to fill the required shelf-life in aqueous solutions. Conformation changes by dehydration and interface stresses are major reasons for the reduction of protein activity.

1. Freeze-dry the protein with and without the conformation stabilizers, then analyze the protein in rehydrated solutions to elucidate the susceptibility of the protein against the dehydration stresses.
2. Select formulations that retain the protein secondary structure (FT-IR study) and favorable freeze-dried cake structure (high-glass transition temperature ($T_g$) in thermal analysis).
3. Apply the candidate freeze-dried formulations to high-temperature and high-humidity conditions (50°C, 75% relative humidity), then analyze the protein.
4. Optimize the protein stability, and run the longer-time stability test.

Suspension of protein crystal or amorphous aggregates is another formulation choice to overcome stability problems of certain proteins, such as insulin zinc suspension. Specific application and delivery may demand other types of protein formulation, including spray-dried powder for pulmonary delivery.

### 3.3. Stabilizing Excipients

#### 3.3.1. Salts and Buffers

Protein stability, solubility, and pharmaceutical acceptance (low-irritative stimulation at the administration site) are the major factors in selecting buffer and salt components. The salts and buffers have significant effects on protein stability, especially in frozen solutions and freeze-dried solids because of the increased concentrations and possible pH changes. For example, the freezing of neutral sodium phosphate buffer and potassium phosphate buffer often results in acidification and alkalization in the freeze-concentrate, respectively. It is better to keep the concentrations of salts and buffers to a minimum to avoid protein denaturation and other physical changes during freeze-drying (e.g., cake collapse and meltback).

#### 3.3.2. Sugars

Various saccharides (sugars) protect the conformation of proteins in aqueous solutions and during freeze-drying. Unfavorable interaction of the sugar molecules with the protein surface (preferential exclusion) in aqueous solutions makes the native protein conformation with the least surface area thermodynamically favored against the unfolded state (*6*). The preferentially excluded cosolutes also reduce conformational fluctuation of the protein at lower temperatures, which results in a decreased risk for aggregation and chemical changes during storage. The sugars stabilize the protein conformation against dehydration stresses by substituting surrounding water molecules through hydrogen bonding. They also prevent a chemical reaction kinetically by keeping the protein in highly viscose glass-state freeze-dried matrices with limited molecular mobility. Nonreducing disaccharides (e.g., sucrose and trehalose) are the most potent and useful excipients to protect protein conformation in aqueous solutions and freeze-dried solids. Sucrose has been widely used in various pharmaceutical formulations, whereas the application of trehalose for lyophilization is increasing because of the advantages in its physical properties, such as a relatively high $T_g$ and a resistance to humidity. Reducing sugars (maltose and lactose) can degrade proteins through Maillard reactions during storage. The sugar concentrations required to protect proteins in aqueous solution and during freeze-drying varies because of the different stabilizing mechanisms. Effective stabilization of protein conformation in aqueous solutions requires relatively high concentrations (approx > 0.3 $M$) of disaccharide. An approx 1:1 weight–concentration ratio of disaccharides is needed to freeze-dry proteins without structural changes. The sugars also stabilize protein conformation in other dehydrating formulations (spray-drying).

### 3.3.3. Sugar Alcohols and Amino Acids

Some sugar alcohols and amino acids (mannitol and glycine) tend to crystallize in frozen solutions and, subsequently, freeze-dried solids. The crystallizing excipients are good bulking agents with a favorable cake structure, and they can be used in combination with the protein-stabilizing amorphous excipients. They also serve as tonicity modifiers in place of salts. Some amorphous polyols and amino acids, such as inositol and arginine hydrochloride, protect the protein conformation against the dehydration stresses.

### 3.3.4. Surfactants

Many nonionic surfactants protect proteins from structural changes at various interfaces (e.g., air–liquid, container–liquid, and ice–freeze-concentrated phase) by competing adsorption. These surfactants also modify the protein conformational stability by binding directly to the protein surface. The direct binding increases protein structural stability under certain combinations and concentration ratios, whereas it damages the protein conformation in many other cases. Some nonionic surfactants also increase the recovery of native protein from a partially unfolded state by assisting refolding as a chemical chaperon and/or by preventing aggregation.

### 3.3.5. Other Excipients

Some water-soluble polymers (e.g., hydroxyethyl starch) maintain the integrity of multimeric proteins from subunit dissociation and resulting inactivation in frozen solutions. The high $T_g'$ (glass transition temperature of maximally freeze-concentrated solutes) and $T_g$ values of various polymers also provides greater cake stability to sugar-based freeze-dried solids. Certain excipients, such as enzyme substrates, inhibitors, and metal ions, that bind specifically to proteins possess another category of protein stabilizer. These ligands affect the free energy of the native protein conformation state (polyanions in acidic fibroblast growth factor formulations). Chemical preservatives in some multidose liquid formulations often adversely affect the protein stability.

## 4. Notes

1. It is desirable to determine appropriate formulations in the early stages of product developments, because changing a formulation in the latter stages would require many more supporting studies.
2. Freezing an aqueous solution with appropriate excipients is a potent method for protein stabilization, whereas the frozen solution formulations are limited to a few products (e.g., denileukin difitox) because of the difficulty to ensure the cold chain during delivery.
3. Protein concentrations are an important factor in the stability of the formulations. Lower concentration protein solutions (~ <1 mg/mL) are more susceptible to protein adsorption and unfolding at various interfaces.

4. It is preferred to select excipients that have been used in approved parenteral products in major markets. The use of other excipients with insufficient safety records would require significant risk assessments in the course of product development.

5. The quality of the excipients should be carefully evaluated to avoid unnecessary complications in the stability profiles. The use of excipients derived from animals or humans should be avoided owing to the risks associated with transmissible diseases.

6. Physiologically isotonic total solute concentrations are desirable to avoid irritation at the administration site.

7. Excess removal of water damages the protein conformation during freeze-drying. Alternatively, residual water molecules in the freeze-dried formulations reduce the protein storage stability as a medium for chemical reactions and/or as a plasticizer of amorphous solids. It is necessary to control the moisture level during the freeze-drying process and to prevent moisture uptake during storage.

8. The cooling speed in the freeze-drying process influences the protein structural stability. Rapid freezing often produces smaller ice crystals and the accompanying larger total ice surface area for possible protein denaturation.

9. The composition of freeze-dried solids determines the component molecular mobility that affects the protein stability. Increasing the sugar–protein mass ratio will raise the $T_g$ level of the freeze-dried solids and improve protein stability.

10. Using surface-active excipients requires multiple studies, as the agents show distinctly different effects on each protein.

## References

1. Nayar, R. and Manning, M. C. (2002) High throughput formulation: strategies for rapid development of stable protein products. *Pharm. Biotechnol.* **13,** 177–198.

2. Chang, B. S. and Hershenson, S. (2002) Practical approaches to protein formulation development. *Pharm. Biotechnol.* **14,** 1–25.

3. Pikal, M. J. Mechanism of protein stabilization during freeze-drying and storage: The relative importance of thermodynamic stabilization and glassy state relaxation dynamics, in *Freeze-drying/Lyophilization of Pharmaceutical and Biological Products* (Rey, L. and May, J. C., eds.), Marcel Dekker, New York, 1999, pp. 161–198.

4. Arakawa, T., Prestrelski, S. J., Kenney, W. C., and Carpenter, J. F. (2001) Factors affecting short-term and long-term stabilities of proteins. *Adv. Drug Deliv. Rev.* **46,** 307–326.

5. Wang, W. (2000) Lyophilization and development of solid pharmaceuticals. *Int. J. Pharm.* **203,** 1–60.

6. Lee, J. C. and Timasheff, S. N. (1981) The stabilization of proteins by sucrose. *J. Biol. Chem.* **256,** 7193–7201.

# 23

## Extraction and Characterization of Vaccine Antigens from Water-in-Oil Adjuvant Formulations

### Aaron P. Miles and Allan Saul

### 1. Introduction

Water-in-oil adjuvants are currently undergoing experimental testing in human vaccine trials *(1–10)*. Two such adjuvants are the squalene-based Montanide® ISA 720 and the mineral oil-based Montanide ISA 51 *(11)*. Vaccines containing these adjuvants are intended to provide a lasting depot effect, with vaccine persisting at the injection site for many months *(12)*. Because it is often convenient to formulate these vaccines well ahead of use, and because of the extended residence at the injection site, their stability must be established. Therefore, studies should be designed that will address stability at 4°C (intended storage temperature) and 37°C (body temperature) to ensure that the antigens remain intact and the vaccines consequently remain potent. With the addition of benzyl alcohol, the emulsion is broken, and the antigens are released and recoverable in the resulting aqueous layer. Storage at 4°C and 37°C of fresh formulations of ISA 720 and ISA 51 containing either of two recombinant malarial proteins, followed by extraction at various time points and analysis by sodium dodecyl sulfate-polyacrylamide gel electrophoresis (SDS-PAGE), N-terminal sequencing, and Western blotting will be used to illustrate the methods involved in monitoring the stability of protein antigens in these water-in-oil formulations.

### 2. Materials

1. AMA1 in ISA 720 and Pfs25 in ISA 720 and ISA 51 vaccine formulations (ISA 720 formulations are 30:70 aqueous:oil, and ISA 51 formulations are 50:50 aqueous:oil).
2. Benzyl alcohol.
3. SDS-PAGE and Western blotting equipment.
4. Coomassie blue-staining solution: 40% methanol, 10% glacial acetic acid, and two PhastGel Blue R tablets (Amersham Biosciences, Piscataway, NJ) per liter.
5. Coomassie blue-destaining solution: 5% methanol and 10% glacial acetic acid.

From: *Methods in Molecular Biology, vol. 308: Therapeutic Proteins: Methods and Protocols*
Edited by: C. M. Smales and D. C. James © Humana Press Inc., Totowa, NJ

6. Silver stain fixing solution: 40% methanol and 10% glacial acetic acid.
7. 30% Ethanol.
8. Distilled water.
9. 0.25 g/L Dithionite.
10. 0.2% Silver nitrate and 0.75 mL/L 37% formaldehyde.
11. Silver stain developer: 6% sodium carbonate, 0.5 mL/L 37% formaldehyde, and 4 mg/L sodium thiosulfate.
12. Silver stain stop solution: 3.5% glacial acetic acid.
13. Penta-histidine monoclonal antibody (MAb; Qiagen, Valencia, CA).

## 3. Methods

The methods described here outline the storage of water-in-oil vaccines followed by antigen extraction and SDS-PAGE, N-terminal sequencing, and immunoblotting as part of a program to monitor vaccine stability.

### 3.1. Vaccine Storage, Antigen Extraction, and SDS-PAGE

1. Place a sufficient number of vials of each vaccine formulation at 4°C and 37°C for extractions and subsequent analysis at 1 d, 1 wk, and 3 wk, or any other desired time points. Samples should also be stored long term (i.e., at least 1–2 yr) at 4°C so that stability at the intended storage temperature can be followed.
2. Reserve one vial of each formulation for d-0 samples.
3. To each d-0 sample, add benzyl alcohol to a final concentration of 10%.
4. Vortex for 20 min at maximum speed using a bench-top vortex.
5. Centrifuge at 16,100g for 10 min to form three layers. Protein is recovered in the aqueous (middle) layer, and this can be frozen (preferably stored at –80°C to maximize stability) for later analysis so that all samples can be run together (*see* **Note 1**).
6. At each time point in the study, perform an extraction as in **steps 3–5**, and store aqueous samples frozen until analyzed.
7. Analyze samples by both nonreduced and reduced SDS-PAGE, loading equal quantities of protein in the lanes for best results. The maximum load of the lowest concentration formulation generally defines the loads for all other concentrations. One or more purified standards should also be run on each gel for comparison.
8. Silver stain the gels as follows: Gently shake the gels in fixing solution for at least 1 h. Shake for 45 min in 30% ethanol (change solution twice). Shake in distilled water for 25 min (change solution once). Shake in dithionite solution for 1 min. Rinse/shake with distilled water for 2 min. Shake in silver nitrate/formaldehyde solution for 25 min. Rinse with distilled water for 1 min. Develop the gels by shaking in sodium carbonate/formaldehyde/sodium thiosulphate solution until new minor bands have ceased to appear (approx 5–7 min). Stop the development by adding 3.5% acetic acid and shaking for 20 min (change solution twice). Examples of nonreduced and reduced gels containing extracts of Pfs25 *(13)* from ISA 720 (formulated at 2, 9, 35, and 150 µg/mL) after storage at 37°C for 3 wk are shown in **Fig. 1** (*see* **Notes 2–4**).
9. If samples are available that have been stored for extended periods at 4°C, perform extractions and analyze by SDS-PAGE. **Figure 2** shows a Coomassie blue-stained SDS-PAGE gel containing nonreduced samples of AMA1 *(14)* extracted from ISA 720 following storage at 4°C for 9 mo. This gel demonstrates that the degradation seen at 37°C also occurs at 4°C but at a slower pace (*see* **Note 5**).

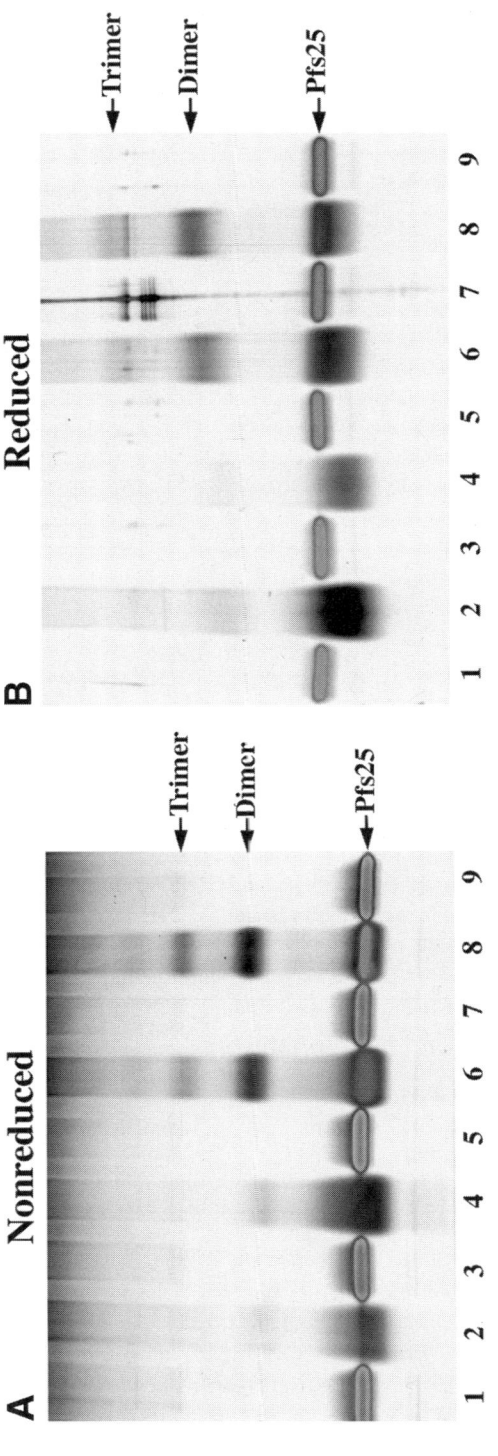

Fig. 1. SDS-PAGE of extracts of Pfs25 in ISA 720 stored at 37°C. Nonreduced (**A**) and reduced (**B**) silver-stained SDS-PAGE gels containing Pfs25 extracted from 2, 9, 35, and 150 µg/mL ISA 720 formulations following storage at 37°C for 3 wk. Lanes 1–9 (both gels) are loaded as follows: 1, Pfs25 standard; 2, 2 µg/mL extract; 3, Pfs25 standard; 4, 9 µg/mL extract; 5, Pfs25 standard; 6, 35 µg/mL extract; 7, Pfs25 standard; 8, 150 µg/mL extract; and 9, Pfs25 standard. Approximately 90 ng protein is loaded in all lanes. Multimers are indicated. Note that the multimers are not disulfide bond-dependent, because reduction by 50 m$M$ dithiothreitol (a concentration that fully reduces Pfs25, indicated by a shift in mobility) does not obliterate them.

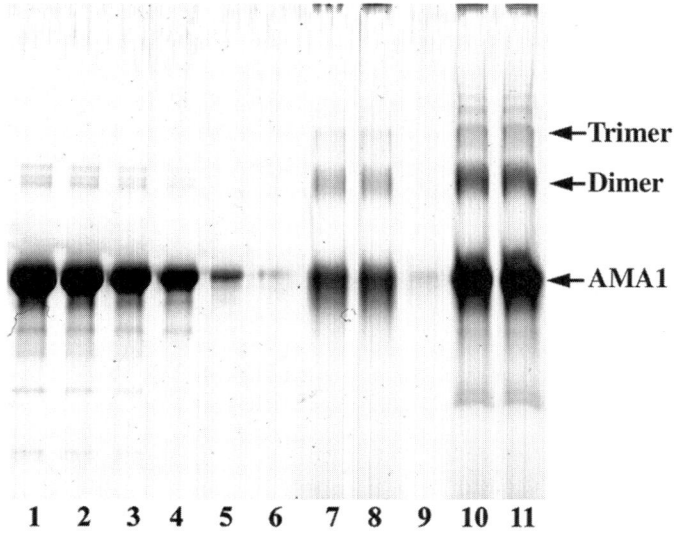

Fig. 2. SDS-PAGE of extracts of AMA1 in ISA 720 stored at 4°C. Nonreduced Coomassie blue-stained SDS-PAGE analysis of extracts of 40 and 160 µg/mL formulations of AMA1 in ISA 720 following storage at 4°C for 9 mo. Lanes 1–5 are 5, 4, 3, 2, or 1 µg, respectively, reference AMA1; lane 6, blank; lanes 7 and 8, duplicate 40 µg/mL extracts with approx 0.9 µg loaded per lane; lane 9, blank; lanes 10 and 11, duplicate 160 µg/mL extracts with approx 3.4 µg loaded per lane. Note that the degradation pattern in the extracts resembles that of samples stored at 37°C for 3 wk (*see* **Fig. 1**).

10. The previous techniques can be applied without modification to vaccine formulations of proteins in ISA 51 (mineral oil-based). **Figure 3** shows nonreduced and reduced silver-stained SDS-PAGE gels of extracts of two formulations (3 and 15 µg/mL) of Pfs25 in ISA 51 stored at either 4°C or 37°C for 10 d. Some protein modification occurs upon storage in ISA 51 but at a reduced rate when compared to formulations containing ISA 720.

## 3.2. N-Terminal Sequencing and Immunoblotting

1. It is useful to analyze extracted samples by N-terminal sequencing to verify that no proteolytic degradation detectable by sequencing is occurring. Run SDS-PAGE gels and transfer to a polyvinylidene difluoride membrane, and stain according to manufacturer's instructions. Cut out samples where there is any indication of modification (multimers and smearing bands), and sequence them (*see* **Note 6**).
2. As the N-terminus may be blocked, or degradation could be occurring from the C-terminal end, if possible, it is important to show that the C-terminus is intact. Pfs25 has a C-terminal 6X histidine tag used in a nickel-nitrilo triacetic acid agarose initial capture step. There is a commercially available antipenta-histidine MAb (Qiagen) that is used to identify this tag. **Figure 4** is a Western blot showing reactivity between this MAb and extracted Pfs25 following storage at 37°C for 2 wk. The reactivity of the main Pfs25 band to the MAb is similar between the reference and extract. The multimers formed during storage also react strongly with the MAb, demonstrating that, in the case of Pfs25, the modification(s) occurring during storage in ISA 720 do not involve C-terminal cleavage.

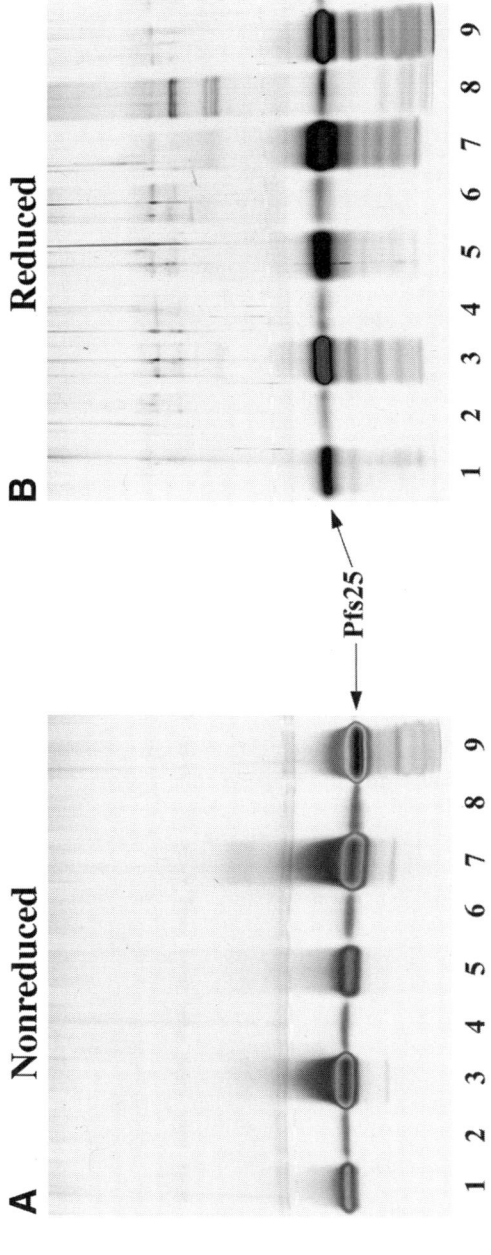

Fig. 3. SDS-PAGE of extracts of Pfs25 in ISA 51 stored at 4°C or 37°C. Nonreduced (**A**) and reduced (**B**) silver-stained SDS-PAGE gels containing Pfs25 extracted from 3 and 15 μg/mL ISA 51 formulations following storage at either 4°C or 37°C for 10 d. Gels are loaded identically as follows: lane 1, 0.1 μg of 3 μg/mL at 4°C extract; lane 2, blank (the protein in the blank lanes is carryover from neighboring lanes), lane 3, 0.4 μg of 15 μg/mL at 4°C extract; lane 4, blank; lane 5, 0.1 μg of 3 μg/mL at 37°C extract; lane 6, blank; lane 7, 0.4 μg of 15 μg/mL at 37°C extract; lane 8, blank; lane 9, 0.8 μg of reference (unextracted) Pfs25. Most modifications seen with proteins stored in ISA 720 for a similar length of time are absent from these samples, with only slight dimerization and main-band broadening apparent in the samples stored at 37°C (A and B, lane 7).

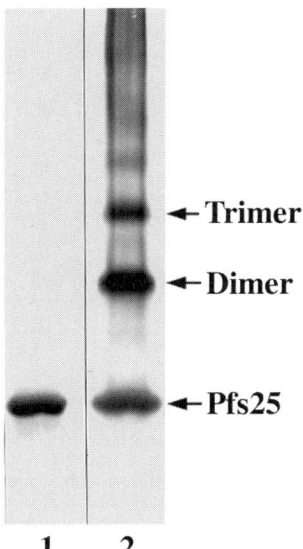

Fig. 4. Western blot of an extract of Pfs25 in ISA 720 stored at 37°C. Western blot showing reactivity of an anti-hisitidine MAb with Pfs25 (nonreduced) extracted from ISA 720 after storage for 2 wk at 37°C. Lane 1, 3 µg reference (unextracted) Pfs25; lane 2, 7.5 µg extracted Pfs25. The antibody is directed against the C-terminal polyhistidine tag on the protein. Much of the protein has undergone multimerization, reducing the amount of Pfs25 at the expected size.

## 4. Notes

1. When extracting, recoveries vary from protein to protein in an as yet poorly understood way. The recovery rate is consistent for any given protein and possibly correlates to molecular weight, but not hydrophobic index. To date, we have been unable to recover proteins from the residual oil layer, even when recoveries in the aqueous layer are only approx 50%. This is probably attributable to aggregation (sometimes visible in one or more phases after extraction) or the general inconvenience of working with oils.

2. The degradation occurs in a protein concentration-dependent manner, with lower concentrations tending to undergo main-band broadening and higher concentrations more susceptible to multimer formation. It is important to load equal amounts of protein so that relatively small differences in degradation patterns are not obscured by over- or underloading any samples. Although, for simplicity, early samples are not shown, the analysis of d-0 samples (along with other time points) is important in establishing that the modifications are not due to the extraction procedure itself. This procedure does slightly modify the way proteins migrate by SDS-PAGE, causing a small amount of smearing, but comparison to later samples makes clear that substantial modification is occurring upon storage.

3. We have seen a similar pattern of modification when analyzing other vaccine formulations (e.g., oil-in-water, as opposed to water-in-oil, formulations consisting of adjuvants other than ISA 720 that also contain squalene). Squalene consists of numerous double bonds that have the potential to react with proteins, and these reactions seem the most likely cause of the modifications.

4. Gels should usually be silver-stained because Coomassie blue is not sensitive enough to detect much of the degradation that occurs, especially with lower concentrations (e.g., 5 µg/mL) and at lower protein loads. Coomassie blue is useful if enough protein can be loaded.

5. Only AMA1 has been analyzed following 9 mo at 4°C, but we have worked with many recombinant proteins at 37°C, and all show the same pattern of degradation, suggesting that this phenomenon is universal with proteins in ISA 720.

6. We have not detected anything other than the expected sequence in samples that are heavily modified when extracted from ISA 720 formulations.

## Acknowledgments

The authors wish to thank Holly McClellan, Daming Zhu, and Kelly Rausch for excellent technical assistance.

## References

1. Genton, B., Al-Yaman, F., Anders, R., Saul, A., Brown, G., Pye, D., et al. (2000) Safety and immunogenicity of a three-component blood-stage malaria vaccine in adults living in an endemic area of Papua New Guinea. *Vaccine* **18,** 2504–2511.

2. Lawrence, G., Cheng, Q., Reed, C., Taylor, D., Stowers, A., Cloonan, N., et al. (2000) Effect of vaccination with 3 recombinant asexual-stage malaria antigens on initial growth rates of *Plasmodium falciparum* in non-immune volunteers. *Vaccine* **18,** 1925–1931.

3. Lawrence, G., Saul, A., Giddy, A., Kemp, R., and Pye, D. (1997) Phase I trial in humans of an oil-based adjuvant SEPPIC MONTANIDE ISA 720. *Vaccine* **15,** 176–178.

4. Saul, A., Lawrence, G., Smillie, A., Rzepczyk, C. M., Reed, C., Taylor, D., et al. (1999) Human phase I vaccine trials of 3 recombinant asexual stage malaria antigens with Montanide ISA720 adjuvant. *Vaccine* **17,** 3145–3159.

5. Toledo, H., Baly, A., Castro, O., Resik, S., Laferté, J., Rolo, F., et al. (2001) A Phase I clinical trial of a multi-epitope polypeptide TAB9 combined with Montanide ISA 720 adjuvant in non-HIV-1 infected human volunteers. *Vaccine* **19,** 4328–4336.

6. Yamshchikov, G. V., Barnd, D. L., Eastham, S., Galavotti, H., Patterson, J. W., Deacon, D. H., et al. (2001) Evaluation of peptide vaccine immunogenicity in draining lymph nodes and peripheral blood of melanoma patients. *Int. J. Cancer* **92,** 703–711.

7. Pinto, L. A., Berzofsky, J. A., Fowke, K. R., Little, R. F., Merced-Galindez, F., Humphrey, R., et al. (1999) HIV-specific immunity following immunization with HIV synthetic envelope peptides in asymptomatic HIV-infected patients. *AIDS* **13,** 2003–2012.

8. Carr, A., Rodríguez, E., del Carmen Arango, M., Camacho, R., Osorio, M., Gabri, M., et al. (2003) Immunotherapy of advanced breast cancer with a heterophilic ganglioside (NeuGcGM3) cancer vaccine. *J. Clin. Oncol.* **21,** 1015–1021.

9. Slingluff, C. L., Jr., Yamshchikov, G., Neese, P., Galavotti, H., Eastham, S., Engelhard, V. H., et al. (2001) Phase I trial of a melanoma vaccine with gp100$_{280-288}$ peptide and tetanus helper peptide in adjuvant: immunologic and clinical outcomes. *Clin. Cancer Res.* **7,** 3012–3024.

10. van Driel, W. J., Ressing M. E., Kenter, G. G., Brandt, R. M. P., Krul, E. J. T., van Rossum, A. B., et al. (1999) Vaccination with HPV16 peptides of patients with advanced cervical carcinoma: clinical evaluation of a phase I-II trial. *Eur. J. Cancer* **35,** 946–952.

11. Aucouturier, J., Dupuis, L., Deville, S., Ascarateil, S., and Ganne, V. (2002) Montanide ISA 720 and 51: a new generation of water in oil emulsions as adjuvants for human vaccines. *Expert Rev. Vaccines* **1,** 111–118.

12. Herbert, W. J. (1968) The mode of action of mineral-oil emulsion adjuvants on antibody production in mice. *Immunology* **14,** 301–318.

13. Zou, L., Miles, A., Wang, J., and Stowers, A. W. (2003) Expression of malaria transmission-blocking vaccine antigen Pfs25 in *Pichia pastoris* for use in human clinical trials. *Vaccine* **21,** 1650–1657.

14. Kennedy, M. C., Wang, J., Zhang, Y., Miles, A. P., Chitsaz, F., Saul, A., et al. (2002) In vitro studies with recombinant *Plasmodium falciparum* apical membrane antigen 1 (AMA1): production and activity of an AMA1 vaccine and generation of a multiallelic response. *Infect. Immunity* **70,** 6948–6960.

# 24

## Probing Reversible Self-Association of Therapeutic Proteins by Sedimentation Velocity in the Analytical Ultracentrifuge

### Bernardo Perez-Ramirez and John J. Steckert

## 1. Introduction

Proteins associate into two types of complex: (1) non-native (irreversible) aggregates, generated by stressing the native form of the protein to unfold, followed by assembly of partially unfolded forms *(1,2)* or assembly of misfolded forms *(3)*; and (2) complexes consisting of reversibly associated native proteins, e.g., tobacco mosaic virus protein *(4,5)* and tubulin *(6)*. Both types may be encountered in the development of formulations for therapeutic proteins. From a quality and safety perspective, drug manufacturers and regulatory agencies view the presence of non-native protein aggregates or reversible protein complexes in the final drug product as undesirable. Protein drug manufacturers are concerned about aggregation or self-association because it may lead to the following:

1. A decrease in protein solubility and, hence, to lower active drug concentrations and doses.
2. An undesirable appearance and opalescence in the drug product *(7)*.
3. A potential immunological response *(8–12)*.
4. Modification of bioactivity *(13)* or drug absorption rate *(14)*.
5. Obstruction of filtration equipment and drug delivery devices *(15)*.

Thus, evaluating the state of association of therapeutic proteins and distinguishing non-native aggregates from reversible complexes are critical to developing stable, potent, and safe protein drug products. Currently, the biopharmaceutical industry employs various methods (denaturating gel electrophoresis, size exclusion chromatography, turbidity, light scattering, and analytical ultracentrifugation) to evaluate the associative state of therapeutic proteins in solution. The application of these methods depends on the type of protein complex to be evaluated. Sodium dodecyl sulfate-polyacrylamide-gel electrophoresis (SDS-PAGE) and size exclusion chromatography (SEC) are methods used to evaluate non-native aggregates (crosslinked proteins or

From: *Methods in Molecular Biology, vol. 308: Therapeutic Proteins: Methods and Protocols*
Edited by: C. M. Smales and D. C. James © Humana Press Inc., Totowa, NJ

noncovalent protein complexes consisting of very tightly coupled protein molecules) are generally not suitable for evaluating reversible protein complexes. Reversible complexes dissociate in the presence of protein denaturants, such as SDS, and SEC may yield uncertain complex sizes (molecular masses) due to interactions between the protein and chromatographic medium or to a change from the native buffer to the SEC mobile-phase solvent. Complex size, based on molecular weight markers run on size exclusion columns or in SDS polyacrylamide gels, may be uncertain because of hydrodynamic differences between the protein complex and markers. By contrast, analytical ultracentrifugation comprises a set of biophysical methods that are suitable for evaluating protein complexes in solution under native conditions.

## 1.1. Analytical Ultracentrifugation (AUC) of Proteins

There are two experiments that can be run in an analytical ultracentrifuge: sedimentation equilibrium and sedimentation velocity. Sedimentation equilibrium is a method for determining the molecular weights of proteins and protein complexes in solution and establishing the reaction stoichiometries and equilibrium constants of reversibly interacting proteins *(16)*. As its name implies, sedimentation equilibrium enables the investigator to analyze systems of interacting proteins while they are at true thermodynamic equilibrium in the analytical ultracentrifuge. On the other hand, sedimentation velocity is a mass transport method with which the investigator can evaluate protein distribution in solution, along with their size, shape, and association properties *(17)*. Although sedimentation velocity experiments are not carried out at thermodynamic equilibrium, they can nevertheless be employed to analyze systems of reversibly interacting proteins *(18,19)*. Unlike chromatographic or gel electrophoretic methods, sedimentation equilibrium and sedimentation velocity allows the investigator to analyze proteins under native solution conditions without the need for separation media or molecular weight markers. The basic theory and practice of both, including descriptions of modern instrumentation and data analysis methods and sample applications to complex macromolecular systems and resources available to the AUC community (*see* **Note 1**), are discussed in **refs.** *17,20–25*. Incidentally, sedimentation equilibrium and sedimentation velocity can also be used to analyze other types of macromolecules (nucleic acids, polysaccharides, synthetic polymers, surfactant micelles, viruses, and chromatin) and low-molecular-weight solutes, in addition to proteins.

Sedimentation equilibrium is used primarily to analyze the chemical equilibria in systems of reversibly interacting proteins. If the data for these systems contain contributions from irreversible aggregates or other types of protein heterogeneity, then the investigator should expect to obtain large uncertainties in the thermodynamic parameters extracted from these data *(18,26)*. Therefore, proteins to be analyzed by sedimentation equilibrium must be both very stable and homogeneous. Alternatively, sedimentation velocity is much better suited for analyzing complex protein mixtures, such as those that contain irreversible aggregates or reversible complexes. Using sedimentation velocity, the investigator can readily determine the number of protein species in solution and their concentration and size. Moreover, sedimentation velocity carried out as a function of protein concentration or different solution variables (e.g., temperature, pH, and ionic strength) is an efficient method to distinguish irreversible aggregates from reversible complexes. As discussed in **ref. *18***, sedimentation velocity is preferred over sedimentation equilibrium for:

1. Analyzing unstable proteins.
2. Analyzing proteins that self-associate in a highly cooperative manner.
3. Analyzing proteins that self-associate into large, slowly diffusing complexes.
4. Identifying protein intermediates along assembly pathways.
5. Analyzing proteins that self-associate very slowly.
6. Analyzing proteins or protein complexes that undergo conformational transitions in the absence of self-association.

In practice, sedimentation velocity should be used to evaluate the composition, size, and stability of a protein before launching a series of time-consuming sedimentation equilibrium experiments. One practical advantage of sedimentation velocity experiments is that they can be performed in less time than sedimentation equilibrium experiments. Thus, sedimentation velocity is better not only for analyzing unstable proteins *(18)* but also for performing survey or screening type experiments, such as those in which protein solubility is studied as a function of different solution variables. Because of its ability to rapidly analyze complex mixtures of proteins, it is highly relevant to preformulation development.

### 1.2. Recombinant Human Bone Morphogenetic Protein-2 (rhBMP-2)

rhBMP-2 is an osteoinductive protein that is a member of the transforming growth factor-$\beta$ (TGF-$\beta$) superfamily *(27–29)*. rhBMP-2 stimulates bone growth by signaling osteoprogenitor cells to differentiate into osteoblasts *(30–33)*. rhBMP-2, manufactured for therapeutic purposes in humans *(34,35)*, is expressed in Chinese hamster ovary (CHO) cells *(36)*. Translation of the gene for rhBMP-2 produces a prepropeptide containing 396 amino acids. Subsequent intracellular proteolytic processing of the prepropeptide yields a mature rhBMP-2 monomer, 114 amino acids in length, with an NH$_2$-terminus of glutamine (Q$_{283}$), and an extended form of rhBMP-2, 131 amino acids in length, with an NH$_2$-terminus of threonine (T$_{266}$). (The numbers for these amino acids indicate their position in the prepropeptide. Of course, Q$_{283}$ and T$_{266}$ are at position 1 in the mature and extended monomers, respectively.) In solution, Q$_{283}$ at the NH$_2$-terminus of mature rhBMP-2 cyclizes to pyroglutamate (<Q$_{283}$) until equilibrium is reached. The result is three possible rhBMP-2 monomer forms: mature, extended, and mature cyclized. rhBMP-2 purified from CHO cell culture is predominantly found as a dimer in which the two monomers are linked by a single disulfide bond. The remaining six cysteine residues in each monomer comprise an intrachain cystine knot—a structure characteristic of TGF-$\beta$ proteins. Each monomer in CHO-expressed rhBMP-2 contains a single N-linked glycosylation site that is occupied by a high mannose oligosaccharide *(36)*. The crystal structure of rhBMP-2 expressed in *Escherichia coli* has been solved to a resolution of 2.7 Å *(37)*. As a result of the NH$_2$-terminal structural heterogeneity in the monomers, rhBMP-2 comprises several isoforms denoted as: <Q$_{283}$/<Q$_{283}$, <Q$_{283}$/Q$_{283}$, Q$_{283}$/Q$_{283}$, <Q$_{283}$/T$_{266}$, Q$_{283}$/T$_{266}$, and T$_{266}$/T$_{266}$ *(38)*. The theoretical molecular masses of <Q$_{283}$/<Q$_{283}$, <Q$_{283}$/T$_{266}$, and T$_{266}$/T$_{266}$ are 29.5 kDa, 31.5 kDa, and 33.5 kDa, respectively.

Abbatiello and Porter *(39)* have shown that rhBMP-2 in a buffer of low-ionic strength becomes increasingly insoluble as the pH is raised above 4.5. Their results are consistent with an affinity precipitation mechanism that requires anion-dependent formation of protein complexes. Understanding the nature of these complexes was critical to the development of a stabilizing formulation buffer for rhBMP-2. In particular, we were interested in determining whether these rhBMP-2 complexes comprised irreversible aggregates or reversibly dissociable assemblies. To achieve this goal, we employed sedimentation velocity in the analytical ultracentrifuge to evaluate the association state of rhBMP-2 in the low-ionic strength drug product formulation buffer as a function of increasing pH.

## 2. Materials

1. Protein: rhBMP-2 was expressed in CHO cell culture *(36)* and purified at Wyeth BioPharma (Genetics Institute Campus, Andover, MA).
2. Drug product formulation buffer: 25 m$M$ L-glutamic acid, 2 m$M$ sodium chloride, 2.5% (w/v) glycine, 0.5% (w/v) sucrose, and 0.01% (w/v) polysorbate 80, pH 4.5–5.8. Sodium hydroxide was used to adjust the pH of the buffer.
3. Equipment: Sedimentation velocity experiments were performed using a Beckman Coulter Optima XL-A analytical ultracentrifuge equipped with absorption optics and an An-60 Ti rotor. The XL-A was connected to a personal computer, and software supplied by Beckman Coulter was used for instrument control and data acquisition. Radius and wavelength calibrations were performed frequently as described in the instrument and service manuals and in the service menu of the user interface. Optical surfaces, including the windows on the xenon lamp, monochromator, and photomultiplier tube and the slit and lenses on the detector assembly were periodically cleaned to reduce stray light and improve the precision of the absorbance data *(40)*.
4. PD-10 desalting columns: prepacked with Sephadex G-25 medium (Amersham Biosciences).

## 3. Methods

In a sedimentation velocity experiment, a centrifuge cell containing a protein solution is spun at high speed in a rotor. The ensuing centrifugal force field causes the protein to sediment radially away from the center of the rotor. As the protein sediments, a concentration gradient, known as a boundary, forms in the centrifuge cell. In the XL-A analytical ultracentrifuge, a scanning absorption optical system is used to record the concentration gradient of the protein. Using this system, the absorption profiles of the protein can be measured at many time points throughout the course of the sedimentation velocity run. At the end of the run, the absorption profiles, often referred to as scans, are analyzed to determine the sedimentation coefficients and distribution of species present in the solution. The sedimentation coefficient of a protein is defined as the rate at which the protein sediments per unit of centrifugal force field strength. Sedimentation coefficients are denoted as $s$ and are usually expressed in Svedberg units: 1 Svedberg (S) = $10^{-13}$ s. The sedimentation coefficient of a protein depends on its molecular size and shape in solution. These molecular properties are related to the sedimentation coefficient through the Svedberg equation:

$$s = [M(1 - v\rho)] \div [Nf]$$

where $s$ is the sedimentation coefficient of the protein; $M$ is the molar mass of the protein; $f$ is the frictional coefficient of the protein; $N$ is Avogadro's number; $v$ is the partial specific volume of the protein; and $\rho$ is the density of the solvent. The molecular size of the protein is reflected by the value of $M$, whereas the shape of the protein affects the value of the frictional coefficient, $f$.

The methods described below outline the (1) assembly and filling of the centrifuge cells, (2) installation of the rotor and monochromator in the XL-A, (3) XL-A control parameters used for the sedimentation velocity experiments, (4) computer methods for analyzing the sedimentation velocity data, (5) distribution of complexes formed by rhBMP-2 at pH 4.5–5.8, (6) dissociation of rhBMP-2 complexes, and (7) distribution of complexes formed by three rhBMP-2 isoforms at pH 4.5–5.5.

### 3.1. Centrifuge Cell Assembly and Filling

The centrifuge cells were assembled with 12-mm charcoal-filled Epon double-sector centerpieces and window assemblies consisting of a window holder and a quartz window. A window gasket and liner were placed in each window holder before inserting the quartz window. Each centrifuge cell consisted of an aluminum cell housing, a centerpiece, two window assemblies, a screw ring washer, and a screw ring. The two window assemblies and centerpiece were inserted into a cell housing in the following order: bottom window assembly (quartz window facing up), centerpiece, and upper window assembly (quartz window facing down). A screw-ring washer was then placed on the upper window holder, and a screw ring was screwed into the threads of the cell housing. A thin film of Spinkote lubricant was applied to the washer and threads of the screw ring before placing them in the cell housing. Finally, the screw ring was tightened to 120 inch-pounds using a torque wrench and torque stand assembly (Beckman Coulter). This procedure presses the quartz windows against the flat surfaces of the centerpiece to form a leak-tight seal. The cells were then filled by loading 0.41 mL buffer into the reference sector and 0.40 mL rhBMP-2 into the sample sector. By convention, the reference sector is on the left, and the sample sector on the right as the cell is viewed with the filling holes facing up and the screw ring facing the analyst. A Pipetman P200 (Gilson) equipped with 250-µL plastic gel-loading tips was used for filling the cells. The cell-filling holes were sealed with plastic plug gaskets that were held in place by brass plugs that screw into the cell housing. Following this step, the filled cells and a counterbalance were weighed on a balance to ensure that their weights matched within 0.5 g or less of each other. The filled cells and counterbalance were then placed in the rotor and aligned with the scribe lines at the bottom of the rotor holes. The cells were placed in rotor positions 1–3, and the counterbalance was placed in position 4, which is the designated rotor position for the counterbalance. When placing the cells in the rotor, the analyst must ensure that the screw ring faces up and that the filling holes face toward the center of the rotor. The analyst must also check that the setscrew on the counterbalance faces up and that the arrow faces away from the center of the rotor. All these procedures are discussed in detail in **ref. 41**.

### 3.2. Installation of the Rotor and Monochromator in the XL-A

The rotor holding the filled cells and counterbalance was placed on the drive spindle at the bottom of the rotor chamber, and the monochromator was inserted into its mounting receptacle as described in the XL-A instruction manual. After installing the rotor and monochromator, the vacuum pumps were started to evacuate the air in the rotor chamber. The rotor was then equilibrated at 20°C for 2 h prior to the start of each run (*see* **Note 2**).

### 3.3. XL-A Control Parameters for Sedimentation Velocity Experiments

Instrument control parameters were entered in the user interface as described in the XL-A instruction manual. Different rotor speeds were used depending on the state of association of rhBMP-2 in the samples (*see* **Note 3**), and the run time varied depending on the rotor speed (*see* **Note 4**). The rotor temperature was controlled at 20°C in all runs. Absorbances were measured at 280 or 250 nm based on the rhBMP-2 concentration in the samples (*see* **Note 5**). Absorbance data were measured from 5.8 to 7.2 cm using a radial step size of 0.003 cm and one replicate in the continuous scan mode. In almost all cases, the An-60 Ti rotor was run with three cells in place, each cell containing a different rhBMP-2 sample. In these runs, the time between scans was set to 5 min. The delayed start was set at zero in all runs. The number of scans were dependent on the run time and interval of time between scans (*see* **Note 6**).

### 3.4. Analysis of the Sedimentation Velocity Data

Sedimentation coefficient distributions, denoted as $g(s^*)$, were computed from the time derivative of the absorbance profiles as described by Stafford *(42,43)*. The computer program DCDT version 3.10s (*see* **Note 7**) was employed for computing $g(s^*)$. Direct boundary-fitting of the absorbance profiles *(44,45)* was performed using the computer program SVEDBERG version 6.39 (*see* **Note 8**). Sedimentation coefficients were corrected to standard conditions of 20°C and water as described by Laue et al. *(46; see* **Note 9**). The density and viscosity of the formulation buffer were measured at 20°C using a pycnometer and glass capillary viscometer, respectively. These measurements yielded 1.0109 g·mL$^{-1}$ for the buffer density and 1.055 poise for the buffer viscosity. The partial specific volume of rhBMP-2 at 20°C was calculated *(46)* to be 0.717 mL·g$^{-1}$ based on the known amino acid and carbohydrate composition for rhBMP-2 (*see* **Note 10**).

### 3.5. Distribution of Complexes Formed by rhBMP-2 at pH 4.5–5.8

rhBMP-2 samples for the sedimentation velocity experiments were prepared as follows. Aliquots of a concentrated solution of rhBMP-2 were buffer-exchanged into the drug product formulation buffer at pH 4.5–5.8 by desalting on PD-10 columns. Aliquots of the desalted rhBMP-2 samples were then diluted into known volumes of the drug product formulation buffer at the same pH as the desalted sample to achieve a final protein concentration of 1.5 mg/mL. The concentrations of rhBMP-2 in these samples were determined by absorption at 280 nm using an absorptivity value of 1.43 [mL/(mg·cm)].

**Figure 1** shows that rhBMP-2 in the formulation buffer at pH 4.5 yields an asymmetric $g(s^*)$ pattern that displays a shoulder on its leading edge. This shoulder is replaced by a distinct $g(s^*)$ peak that grows in area and whose maximum shifts to higher sedimentation coefficients as the pH of the formulation buffer increases. As a result of these changes, the $g(s^*)$ patterns at pH values above 4.5 are bimodal in

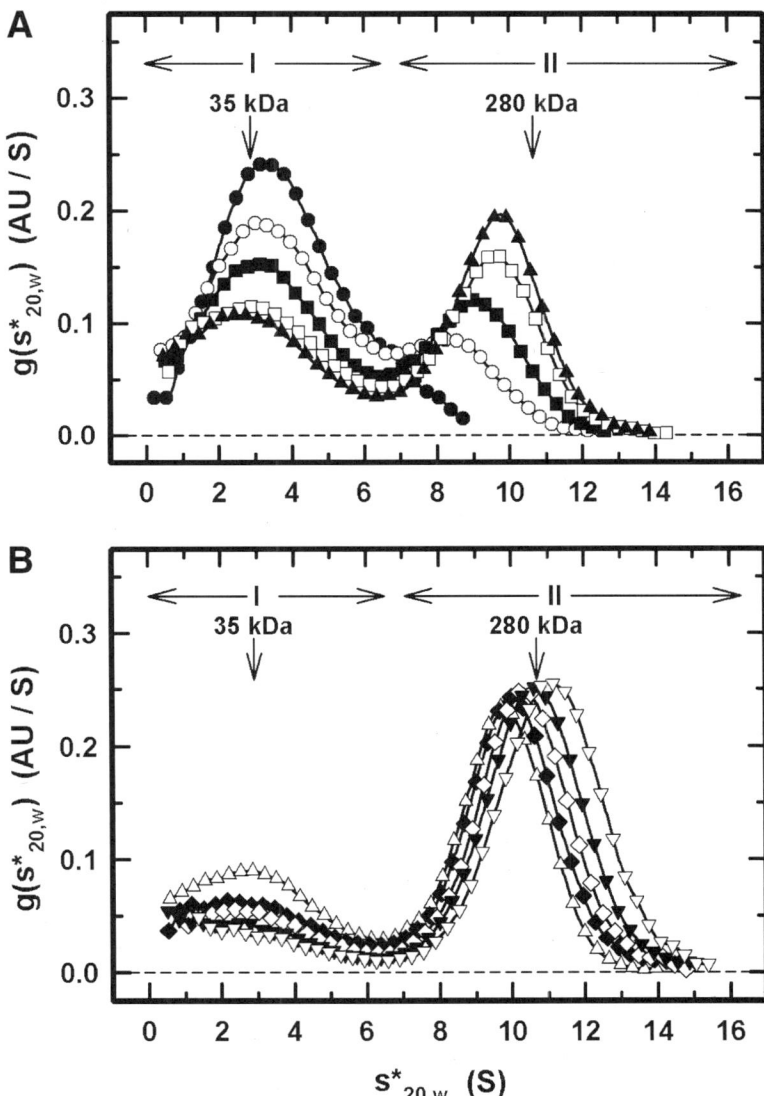

Fig. 1. Distribution of complexes formed by rhBMP-2 at pH 4.5–5.8. Samples of rhBMP-2 in the drug product formulation buffer at pH 4.5–5.8 were analyzed by sedimentation velocity. The protein concentration of each sample was 1.5 mg/mL. The absorbance profiles of all the samples were measured at 250 nm. The samples were spun at the following rotor speeds: (a) 40,000 rpm, $1.2 \times 10^5 g$ for pH 4.5–4.7; (b) 35,000 rpm, $8.9 \times 10^4 g$ for pH 4.8–5.0; (c) 32,000 rpm, $7.4 \times 10^4 g$ for pH 5.2–5.8. The sedimentation coefficient distributions shown in the figure were determined based on the time derivative of the absorbance profiles *(42,43)*. (A) ●, pH 4.5; ○, pH 4.6; ■, pH 4.7; □, pH 4.8; ▲, pH 4.9. (B) △, pH 5.0; ◆, pH 5.2; ◇, pH 5.4; ▼, pH 5.6; ▽, pH 5.8.

**Table 1**
**Composition of rhBMP-2 Samples at pH 4.5–5.8**

| pH[a] | Mass fraction unassociated protein[b] | Mass fraction complexes[c] |
|---|---|---|
| 4.5 | 0.87 | 0.13 |
| 4.6 | 0.68 | 0.32 |
| 4.7 | 0.55 | 0.45 |
| 4.8 | 0.46 | 0.54 |
| 4.9 | 0.40 | 0.60 |
| 5.0 | 0.35 | 0.65 |
| 5.2 | 0.30 | 0.70 |
| 5.4 | 0.26 | 0.74 |
| 5.6 | 0.22 | 0.78 |
| 5.8 | 0.18 | 0.82 |

[a]pH value of the drug product formulation buffer.
[b]Mass fraction of unassociated protein that sediments in the range labeled I in **Fig. 1**.
[c]Mass fraction of complexes that sediment in the range labeled II in **Fig. 1**.

shape. The leading $g(s^*)$ peak reflects a population of fast-sedimenting rhBMP-2 complexes, whereas the slow peak reflects unassociated protein. Complexes sedimenting at 8–11 S were formed in the drug product formulation buffer at pH 4.6–5.8. Integration of the areas under the two peaks yielded mass fraction values for the rhBMP-2 complexes and unassociated protein (*see* **Table 1** and **Note 11**). As indicated in **Table 1**, the fraction of complexes increased as the pH of the drug product formulation buffer was adjusted from 4.5 to 5.8. The complexes yielded a distinct $g(s^*)$ peak, suggesting that each complex has a definite molecular mass. To determine the molecular masses of the complexes, we used the direct boundary-fitting method developed by Philo *(44,45)* to analyze the absorbance profiles of the rhBMP-2 samples at pH 4.6–5.8. However, the models for multiple (2–4) noninteracting species were poorly fitted to these data (results not shown). Thus, the molecular masses for the rhBMP-2 complexes could not be determined by this method. The molecular mass values indicated in **Fig. 1** were determined by direct boundary-fitting of sedimentation velocity data for two purified dimeric isoforms of rhBMP-2 (*see* **Subheading 3.7.**). The 35-kDa value is consistent with the known molecular masses for the unassociated dimeric isoforms of rhBMP-2, and the 280-kDa value is consistent with a complex consisting of eight or nine dimeric rhBMP-2 molecules. Therefore, the $g(s^*)$ peaks for the 10–11-S species represent 280-kDa complexes of rhBMP-2, and the $g(s^*)$ peaks for the 2–3-S species represent unassociated dimeric forms of rhBMP-2.

### 3.6. Dissociation of the rhBMP-2 Complexes at pH 4.5

To determine whether the 8–11-S rhBMP-2 complexes formed at pH 4.6–5.8 are dissociable, we dialyzed samples from the pH series (*see* **Fig. 1**) back into the formulation buffer at pH 4.5. **Figure 2** shows that all the dialyzed samples yielded $g(s^*)$

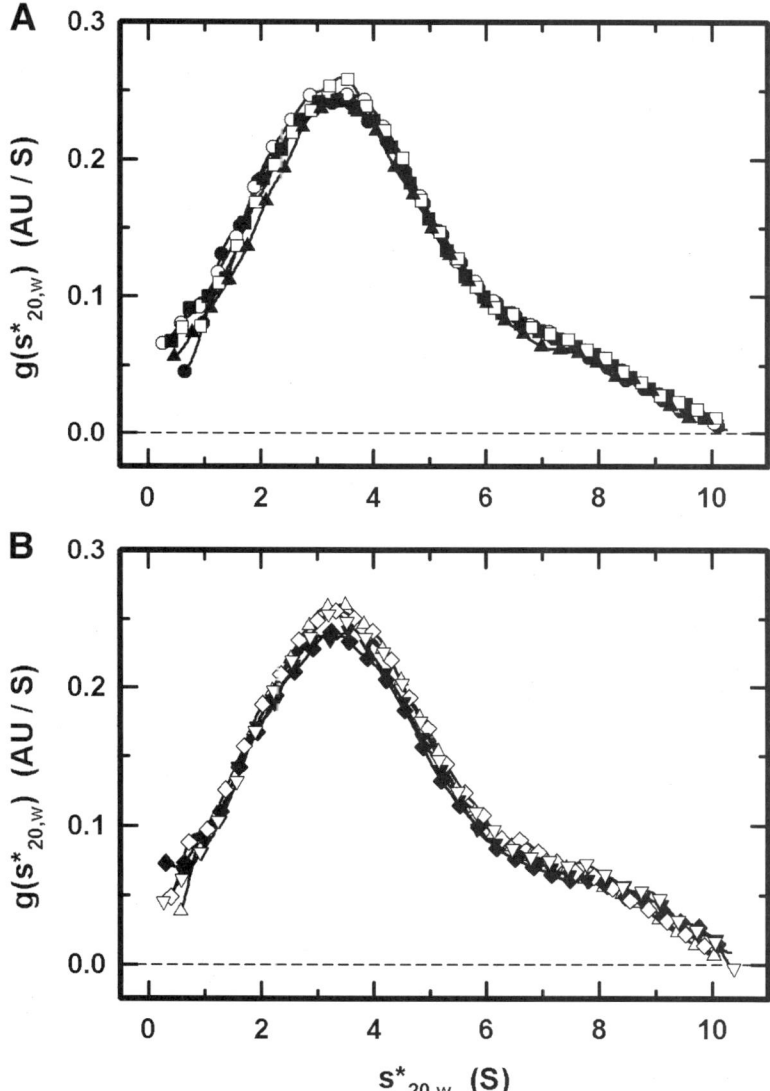

Fig. 2. Dissociation of rhBMP-2 complexes at pH 4.5. Samples in the pH series (*see* **Fig. 1**) were dialyzed back into the drug product formulation buffer at pH 4.5 and then analyzed by sedimentation velocity. The absorbance profiles of all the dialyzed samples were measured at 250 nm. All the dialyzed samples were spun at a rotor speed of 40,000 rpm, $1.2 \times 10^5 g$. The sedimentation coefficient distributions shown in the figure were determined based on the time derivative of the absorbance profiles *(42,43)*. The pH values indicated in the figure refer to those of the sample before dialysis. (A) ●, pH 4.5; ○, pH 4.6; ■, pH 4.7; □, pH 4.8; ▲, pH 4.9. (B) △, pH 5.0; ◆, pH 5.2; ◇, pH 5.4; ▼, pH 5.6; ▽, pH 5.8.

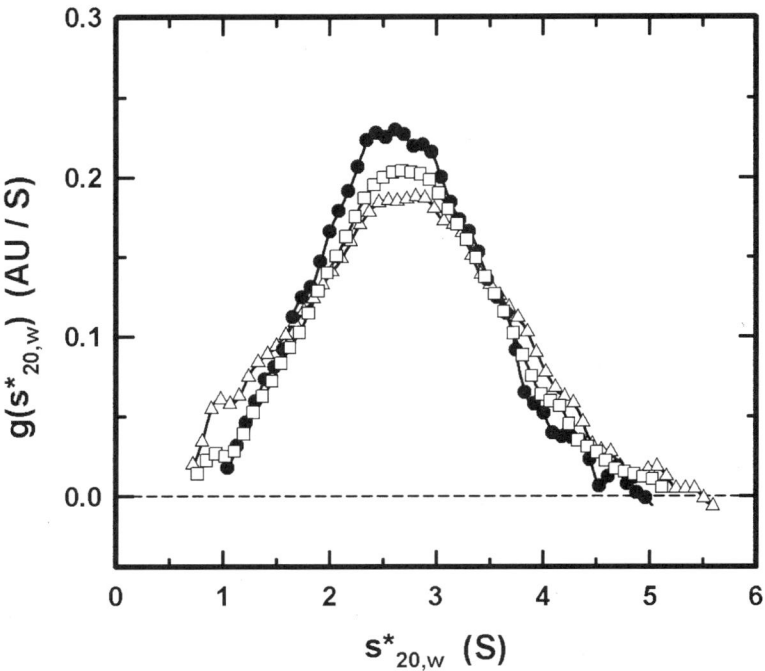

Fig. 3. Distribution of $T_{266}/T_{266}$ at pH 4.5, 5.1, and 5.5. Samples of $T_{266}/T_{266}$ in the drug product formulation buffer at pH 4.5, 5.1, and 5.5 were analyzed by sedimentation velocity. The protein concentration of each sample was 0.24 mg/mL. The absorbance profiles of all the samples were measured at 280 nm. The samples were spun at the following rotor speeds: (a) 40,000 rpm, $1.2 \times 10^5 g$ for pH 4.5 and 5.5; (b) 35,000 rpm, $8.9 \times 10^4 g$ for pH 5.1. The sedimentation coefficient distributions shown in the figure were determined based on the time derivative of the absorbance profiles *(42,43)*. △, pH 4.5; ●, pH 5.1; □, pH 5.5.

patterns consistent with that for the dialyzed control at pH 4.5. These results confirm that the 8–11-S complexes are dissociable assemblies of rhBMP-2 and not irreversible aggregates. When viewed together, the results shown in **Figs. 1** and **2** indicate that rhBMP-2 undergoes reversible pH-dependent self-association in the drug product formulation buffer at pH 4.5–5.8.

### 3.7. Distribution of Complexes Formed by $T_{266}/T_{266}$, $<Q_{283}/T_{266}$, and $<Q_{283}/<Q_{283}$

As indicated in the **Introduction**, rhBMP-2 comprises six different $NH_2$-terminal isoforms. However, the interactions that take place between these dimeric isoforms, as well as their distribution in the 8–11-S complexes (*see* **Fig. 1**), are unknown. As a first step in probing these interactions, we purified the $<Q_{283}/<Q_{283}$, $<Q_{283}/T_{266}$, and $T_{266}/T_{266}$ dimeric isoforms by cation exchange chromatography *(47,48)*, then employed sedimentation velocity in the XL-A to examine the tendency to self-associate. Each of the purified dimeric isoforms was studied at a concentration of 0.24 mg/mL in the drug product formulation buffer at pH 4.5, 5.1, and 5.5.

Table 2
Sedimentation Coefficients and Molecular Masses of $T_{266}/T_{266}$, $<Q_{283}/T_{266}$, and $<Q_{283}/<Q_{283}$[a]

| Isoform | pH[b] | $s_{20,w}$ (S)[c] | M (kDa)[d] |
|---|---|---|---|
| $T_{266}/T_{266}$ | 4.5 | 2.704 | 35.02 |
| | | [2.701, 2.708] | [34.82, 35.21] |
| $T_{266}/T_{266}$ | 5.1 | 2.831 | 35.19 |
| | | [2.827, 2.836] | [34.96, 35.39] |
| $T_{266}/T_{266}$ | 5.5 | 2.759 | 30.49 |
| | | [2.756, 2.764] | [30.25, 30.66] |
| $<Q_{283}/T_{266}$ | 4.5 | 2.920 | 32.03 |
| | | [2.917, 2.923] | [31.85, 32.20] |
| $<Q_{283}/<Q_{283}$ | 5.1 | 10.779 | 288.05 |
| | | [10.762, 10.820] | [278.94, 299.74] |
| $<Q_{283}/<Q_{283}$ | 5.5 | 10.661 | 274.82 |
| | | [10.623, 10.697] | [266.46, 286.11] |

[a]Sedimentation coefficients and molecular mass values were determined by direct boundary fitting (*44,45*) of the sedimentation velocity data for the isoforms.

[b]pH value of the drug product formulation buffer.

[c]Sedimentation coefficient at standard conditions of 20°C and water (*see* **Note 9**). The bracketed values are 95% confidence limits for the fitted value of $s_{20,w}$. The fitted $s_{20,w}$ values are expressed in Svedberg units (S).

[d]Molecular mass expressed in kilodaltons. The bracketed values are 95% confidence limits for the fitted value of M.

### 3.7.1. $T_{266}/T_{266}$

**Figure 3** shows the $g(s^*)$ patterns for $T_{266}/T_{266}$ in the drug product formulation buffer at pH 4.5, 5.1, and 5.5. These $g(s^*)$ patterns are very similar in shape and display a single peak, indicating that $T_{266}/T_{266}$ sediments as a single 2.7-S species under these conditions. Direct boundary-fitting of the sedimentation velocity data for $T_{266}/T_{266}$ revealed that it behaves as an unassociated dimer at pH 4.5, 5.1, and 5.5 (*see* **Table 2**). Thus, $T_{266}/T_{266}$ displayed no tendency to self-associate in the formulation buffer at any of these three pH values. However, these results do not rule out the possibility that $T_{266}/T_{266}$ may self-associate at concentrations higher than 0.24 mg/mL under these conditions.

### 3.7.2. $<Q_{283}/T_{266}$

**Figure 4** shows that the shape of the $g(s^*)$ pattern for $<Q_{283}/T_{266}$ changes progressively from a single peak at pH 4.5 to a bimodal pattern at pH 5.5. The $g(s^*)$ pattern at pH 4.5 is consistent with the sedimentation of a single 2.5-S species. Indeed, direct boundary-fitting of the sedimentation velocity data for $<Q_{283}/T_{266}$ at pH 4.5 confirmed that it behaves as an unassociated dimer at this pH value (*see* **Table 2**). The bimodal pattern shows that $<Q_{283}/T_{266}$ forms a mixture that contains the unassociated 2.5-S dimer and a 9-S complex at pH 5.5. When examined at pH 5.1, $<Q_{283}/T_{266}$ yielded a

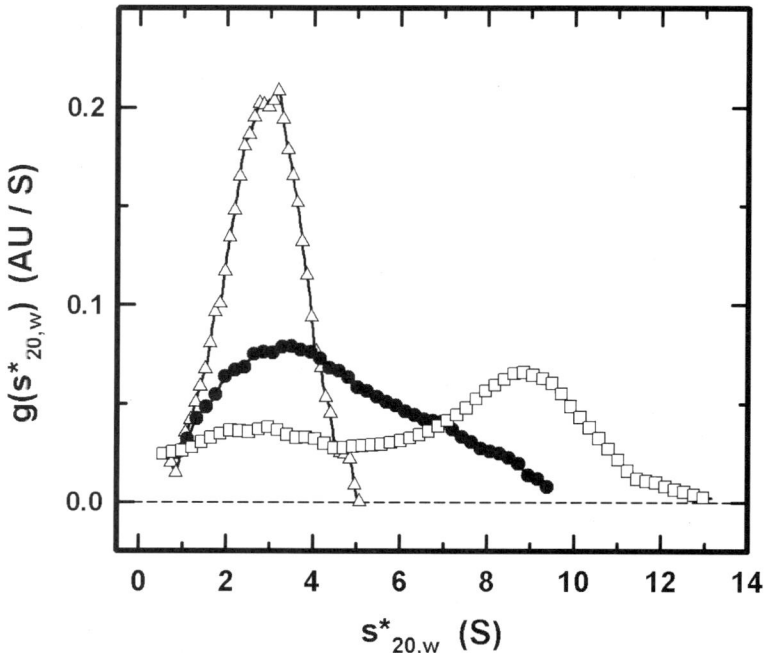

Fig. 4. Distribution of <$Q_{283}/T_{266}$ at pH 4.5, 5.1, and 5.5. Samples of <$Q_{283}/T_{266}$ in the drug product formulation buffer at pH 4.5, 5.1, and 5.5 were analyzed by sedimentation velocity. The protein concentration of each sample was 0.24 mg/mL. The absorbance profiles of all the samples were measured at 280 nm. The samples were spun at the following rotor speeds: (a) 40,000 rpm, $1.2 \times 10^5 g$ for pH 4.5; (b) 35,000 rpm, $8.9 \times 10^4 g$ for pH 5.1 and 5.5. The sedimentation coefficient distributions shown in the figure were determined based on the time derivative of the absorbance profiles *(42,43)*. △, pH 4.5; ●, pH 5.1; □, pH 5.5.

highly skewed $g(s^*)$ pattern that displays a broad leading edge. This reflects the sedimentation of oligomeric forms of <$Q_{283}/T_{266}$ that are larger in size than the unassociated 2.5-S dimers but smaller than the 9-S complex. Analyzing the sedimentation velocity data for <$Q_{283}/T_{266}$ at pH 5.1 and 5.5 in terms of multiple noninteracting species yielded poor fits (results not shown). These findings suggest that the unassociated 2.5-S dimer, intermediate-size oligomers, and 9-S complex cannot be treated as independent noninteracting species at pH 5.1 and 5.5. At both pH values, the mass transport of these species during the sedimentation process may not be fully uncoupled from the chemical equilibria that describe the formation of the oligomers and 9-S complex from the 2.5-S dimer (*see* **Note 12**). This point provides a plausible explanation for why direct boundary-fitting of the sedimentation velocity data for unfractionated rhBMP-2 at pH 4.6–5.8 yielded poor fits. Overall, the results shown in **Fig. 4** indicate that <$Q_{283}/T_{266}$ exhibits a stronger tendency to self-associate at pH 5.5 than at 5.1 and no tendency to self-associate at pH 4.5. Additional sedimentation velocity studies are required to determine whether <$Q_{283}/T_{266}$ self-associates at concentrations higher than 0.24 mg/mL in the formulation buffer at pH 4.5.

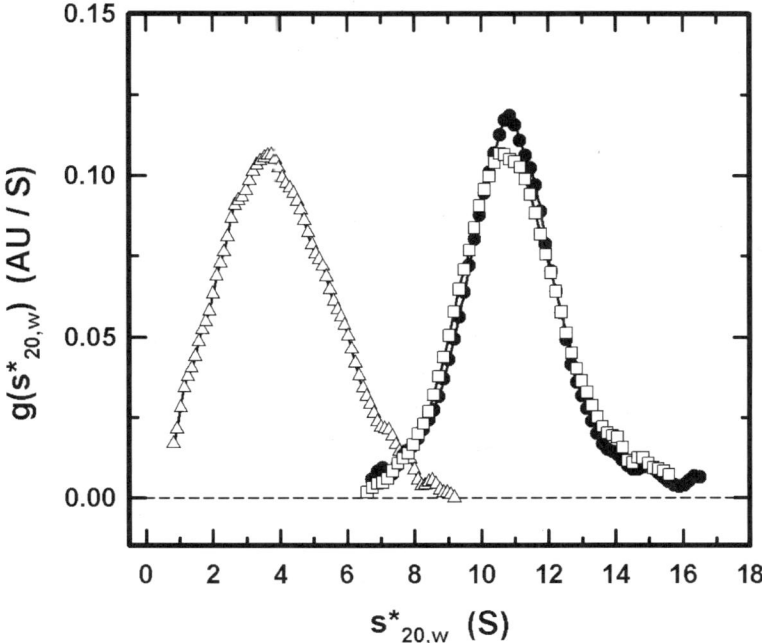

Fig. 5. Distribution of $<Q_{283}/<Q_{283}$ at pH 4.5, 5.1, and 5.5. Samples of $<Q_{283}/<Q_{283}$ in the drug product formulation buffer at pH 4.5, 5.1, and 5.5 were analyzed by sedimentation velocity. The protein concentration of each sample was 0.24 mg/mL. The absorbance profiles of all the samples were measured at 280 nm. The samples were spun at the following rotor speeds: (a) 40,000 rpm, $1.2 \times 10^5 g$ for pH 4.5; (b) 35,000 rpm, $8.9 \times 10^4 g$ for pH 5.1 and 5.5. The sedimentation coefficient distributions shown in the figure were determined based on the time derivative of the absorbance profiles *(42,43)*. △, pH 4.5; ●, pH 5.1; □, pH 5.5.

### 3.7.3. $<Q_{283}/<Q_{283}$

The $g(s^*)$ pattern for $<Q_{283}/<Q_{283}$ at pH 4.5 displays a broad leading edge, illustrating that weakly associated oligomers of this isoform sediment ahead of the unassociated dimer (*see* **Fig. 5**). These $<Q_{283}/<Q_{283}$ oligomers are likely to account for the leading shoulder that is evident on the $g(s^*)$ pattern for unfractionated rhBMP-2 at pH 4.5 (*see* **Figs. 1** and **2**). Direct boundary-fitting of the sedimentation velocity data for $<Q_{283}/<Q_{283}$ at pH 4.5 yielded poor fits, suggesting that the unassociated dimer and oligomers cannot be treated as independent noninteracting species (*see* **Note 13**). **Figure 5** shows that $<Q_{283}/<Q_{283}$ quantitatively self-associates at pH 5.1 and 5.5 into a 10.7-S complex with an average molecular mass of 280 kDa (*see* **Table 2**). These conclusions support the view that the assembly of the 280-kDa complex involves very strong cooperative interactions between the $<Q_{283}/<Q_{283}$ molecules.

### 3.7.4. Summary

When viewed together, the results shown in **Figs. 3–5** indicate that the dimeric isoforms ($<Q_{283}/<Q_{283}$, $<Q_{283}/T_{266}$, and $T_{266}/T_{266}$) exhibit widely different pH-depen-

dent tendencies to self-associate in the formulation buffer (*see* **Note 14**). As the pH of the drug product formulation buffer increases, $<Q_{283}/<Q_{283}$ exhibits a much stronger tendency to self-associate than $<Q_{283}/T_{266}$, whereas $T_{266}/T_{266}$ exhibits no tendency to self-associate. These findings imply that the extended $NH_2$-terminal sequence on the $T_{266}$ monomer has a moderating effect on the $<Q_{283}$ monomer to self-associate. Thus, the solubility of these isoforms is expected to decrease in the following order: $(T_{266}/T_{266}) >$ $(<Q_{283}/T_{266}) > (<Q_{283}/<Q_{283})$. Additional sedimentation velocity studies are required to determine whether the solubilities of the other dimeric isoforms $(<Q_{283}/Q_{283}, Q_{283}/Q_{283}$, and $Q_{283}/T_{266})$ follow this same trend. The methodology described here should provide sufficient experimental background to enable the biopharmaceutical scientist to address similar questions in different systems of reversibly interacting proteins.

## 4. Notes

1. Resources available to the AUC community, including training, data analysis software, consultations, collaborations, and literature, can be found at the Reversible Associations in Structural and Molecular Biology (RASMB) website (www.bbri.org/RASMB/rasmb.html) and at several other sites accessible through the RASMB homepage.
2. Sedimentation coefficients are affected by variations in viscosity owing to changes in sample temperature (*45*). Therefore, the rotor and centrifuge cells must reach temperature equilibrium before starting a run. A hold time of 2 h is sufficient for equilibrating the rotor and cells.
3. Using a high rotor speed is necessary if the protein sediments slowly and a low speed if the protein sediments rapidly. However, note that the recommended maximum speed for the 12-mm charcoal-filled Epon double-sector centerpieces is 42,000 rpm. The analyst may have to perform a series of pilot sedimentation velocity runs to establish an appropriate rotor speed for the protein of interest. The absorbance scans used for time-derivative analysis must be closely spaced in time (*see* **Note 7**). Therefore, the rotor speed must be chosen with this requirement considered.
4. The run time can be estimated by calculating a clearing time for the protein. The clearing time is the total time required for all the protein to sediment from the sample meniscus to the bottom of the centrifuge cell. The clearing time is calculated according to the following formula:

$$t = [\ln(r_b/r_m)/(s\omega^2)]/(3600 \text{ sec/h})$$

where $t$ is the clearing time in hours; ln is the natural logarithm; $r_b$ is the radius at the bottom of the centrifuge cell; $r_m$ is the radius of the protein sample meniscus; $s$ is the sedimentation coefficient of the protein expressed in seconds; and $\omega$ is the angular velocity of the rotor (expressed in $s^{-2}$). Note that $\omega = (/30 \text{ sec}) \times \text{rpm}$, where rpm is the rotor speed expressed in revolutions per minute. One can use values of 6.0 and 7.2 cm for $r_m$ and $r_b$, respectively. These values are typical for the centrifuge cells and rotor used in these experiments. Therefore, the clearing time for a 3-S protein centrifuged at 40,000 rpm is 9.6 h, and the clearing time for an 11-S protein centrifuged at 32,000 rpm is 4.1 h. Suppose the sedimentation coefficient for the protein of interest is not known, but the molecular weight is known or can be estimated. In that case, a sedimentation coefficient for the protein can be calculated, assuming that it can be modeled as a hydrated sphere (*49*). The calculated sedimentation coefficient may not be accurate, but it provides a basis for estimating a clearing time.

5. Sample concentrations should be adjusted to achieve absorbances between 0.1 and 1.0. The analyst should verify that absorbances in this range vary linearly with concentration. Note that the centerpieces used in these experiments have an optical pathlength of 1.2 cm. The absorption spectrum of rhBMP-2 exhibits a maximum at 280 nm and a minimum at 250 nm. The absorptivity values for rhBMP-2 at these wavelengths are 1.43 [mL/(mg·cm)] and 0.52 [mL/(mg·cm)], respectively. Therefore, the highest rhBMP-2 concentrations that can be measured at 280 and 250 nm without exceeding an absorbance of 1.0 are 0.6 and 1.6 mg/mL, respectively.

6. The number of scans to be collected can be calculated by dividing the run or clearing time (*see* **Note 4**) by the time interval between scans. For example, 115 scans can be collected in a run of 9.6 h if the interval between scans is 5 min.

7. The DCDT program can be downloaded from the RASMB analytical ultracentrifugation software archive (http://www.bbri.org/RASMB/rasmb.html). Unnormalized $g(s^*)$ patterns were computed in these experiments and expressed in absorbance units (AU) per Svedberg (S). The $g(s^*)$ patterns were computed based on an even number (12–16) of absorbance scans. The scans chosen for time-derivative analysis were closely spaced in time at the rotor speeds used in the sedimentation velocity runs. Apparent sedimentation coefficients are denoted as $s^*$. The sedimentation coefficients are apparent values because the distributions were not corrected for the effect of diffusion. In general, the value of $s^*$ corresponding to the maximum in the $g(s^*)$ pattern represents the sedimentation coefficient of the protein while the area under the $g(s^*)$ peak is directly proportional to concentration.

8. The SVEDBERG program is a commercial product that can be ordered at John Philo's software homepage (www.jphilo.mailway.com).

9. Sedimentation coefficients corrected to standard conditions of 20°C and water are denoted as $s_{20,w}$.

10. The amino acid sequence of the $T_{266}$ *(38)* monomer is the same as that for the $<Q_{283}$ monomer *(37)*, except for the additional 17 amino acid residues at the $NH_2$-terminus. For calculating the value of v, the high mannose oligosaccharide was assumed to contain nine residues of mannose and two residues of *N*-acetylglucosamine.

11. Mass fractions were calculated by dividing the areas under the $g(s^*)$ peaks for the unassociated protein and complexes by the total area. Calculating the fractions in this manner assumes that the unassociated protein and complexes have the same absorptivity value at 250 nm.

12. The self-association of $<Q_{283}/T_{266}$ at pH 5.1 and 5.5 could be studied in greater detail by analyzing sedimentation equilibrium data *(16)* or by analyzing weight average sedimentation coefficient data *(18,19)*. In these experiments, the concentration of $<Q_{283}/T_{266}$ should be varied over a wide range.

13. The self-association of $<Q_{283}/<Q_{283}$ at pH 4.5 could be studied in greater detail by analyzing sedimentation equilibrium data *(16)* or analyzing weight average sedimentation coefficient data *(18,19)*. In these experiments, the concentration of $<Q_{283}/<Q_{283}$ should be varied over a wide range.

14. Fitting sedimentation equilibrium data for unfractionated rhBMP-2 at pH 4.5–5.8 to different models of reversible homogeneous associations *(16)* would have been fruitless, as the dimeric isoforms ($<Q_{283}/<Q_{283}$, $<Q_{283}/T_{266}$, and $T_{266}/T_{266}$) exhibit widely different tendencies to self-associate. Analyzing these purified dimeric isoforms individually by sedimentation velocity was valuable for revealing these different tendencies. Using sedimentation equilibrium to analyze the potential two-component or three-

component heteroassociations *(16)* between the dimeric isoforms (e.g., $T_{266}/T_{266}$ + $<Q_{283}/T_{266}$; $T_{266}/T_{266}$ + $<Q_{283}/<Q_{283}$; $<Q_{283}/T_{266}$ + $<Q_{283}/<Q_{283}$; $T_{266}/T_{266}$ + $<Q_{283}/T_{266}$ + $<Q_{283}/<Q_{283}$) would be very difficult because both $<Q_{283}/<Q_{283}$ and $<Q_{283}/T_{266}$ reversibly self-associate.

# References

1. Schein, C. H. (1990) Solubility as a function of protein structure and solvent components. *Bio/Technology* **8,** 308–317.
2. Wetzel, R. Protein aggregation *in vivo*. Bacterial inclusion bodies and mammalian amyloid, in *Stability of Protein Pharmaceuticals, Part B*. In vivo *pathways of degradation and strategies for protein stabilization* (Ahern, T. J. and Manning, M. C., eds.), Plenum Press, New York, 1992, pp. 43–88.
3. Dobson, C. M. (2003) Protein folding and misfolding. *Nature* **426,** 884–890.
4. Schuster, T. M., Scheele, R. B., and Khairallah, L. H. (1979) Mechanism of self-assembly of tobacco mosaic virus protein. I. Nucleation-controlled kinetics of polymerization. *J. Mol. Biol.* **127,** 461–485.
5. Shire. S. J., Steckert, J. J., and Schuster, T. M. (1979) Mechanism of self-assembly of tobacco mosaic virus protein. II. Characterization of the metastable polymerization nucleus and the initial stages of helix formation. *J. Mol. Biol.* **127,** 487–506.
6. Perez-Ramirez, B., Andreu, J. M., Gorbunoff, M. J., and Timasheff, S. N. (1996) Stoichiometric and substoichiometric inhibition of tubulin self-assembly by colchicine analogues. *Biochemistry* **35,** 3277–3285.
7. Eckhardt, B. M., Oeswein, J. Q., Yeung, D. A., Milby, T. D., and Bewley, T. A. (1994) A turbidimetric method to determine visual appearance of protein solutions. *J. Pharm. Sci. Technol.* **48,** 64–70.
8. Hanson, L. A., Roos, P., and Rymo, L. (1966) Heterogeneity of human growth hormone preparations by immuno-gel filtration and gel filtration electrophoresis. *Nature* **212,** 948–949.
9. Lewis, U. J., Cheever, E. V., and Seavey, B. K. (1969a) Aggregate-free human growth hormone. I. Isolation by ultrafiltration. *Endocrinology* **84,** 325–331.
10. Lewis, U. J., Parker, D. C., Okerlund, M. D., Boyar, R. M., Litteria, M., and Vanderlaan, W. P. (1969b) Aggregate-free human growth hormone. II. Physico-chemical and biological properties. *Endocrinology* **84,** 332–339.
11. Schwartz, P. L. and Batt, M. (1980) The aggregation of [$^{125}$I]human growth hormone in response to freezing and thawing. *Endocrinology* **92,** 1795–1798.
12. Moore, M. V. and Leppert, P. (1980) Role of aggregated human growth hormone (hGH) in development of antibodies to hGH. *J. Clin. Endocrinol.* **51,** 691–697.
13. Becker, G. W., Bowsher, R. R., Mackellar, W. C., Poor, M. L., Tackitt, P. M., and Riggin, R. M. (1987) Chemical, physical, and biological characterization of biosynthetic human growth hormone. *Biotech. Appl. Biochem.* **9,** 478–487.
14. Brange, J., Ribel, U., Hansen, J. F., Dodson, G., Hansen, M. T., Havelund, S., et al. (1988) Monomeric insulins obtained by protein engineering and their medical implications. *Nature* **333,** 679–682.
15. Sluzky, V., Tamada, J. A., Klibanov, A. M., and Langer, R. (1991) Kinetics of insulin aggregation in aqueous solutions upon agitation in the presence of hydrophobic surfaces. *Proc. Natl. Acad. Sci. USA* **88,** 9377–9381.
16. Laue, T. M. (1995) Sedimentation equilibrium as a thermodynamic tool. *Methods Enzymol.* **259,** 427–452.

17. Stafford, W. F., III (1997) Sedimentation velocity spins a new weave for an old fabric. *Curr. Opin. Biotech.* **8,** 14–24.
18. Correia, J. J. (2000) Analysis of weight average sedimentation velocity data. *Methods Enzymol.* **321,** 81–100.
19. Stafford, W. F. (2000) Analysis of reversibly interacting macromolecular systems by time derivative sedimentation velocity. *Methods Enzymol.* **323,** 302–325.
20. Ralston, G. Introduction to analytical ultracentrifugation, in *Analytical Ultracentrifugation, vol. 1,* Beckman Instruments, Palo Alto, CA, 1993.
21. McRorie, D. K. and Voelker, P. J. Self-associating systems in the analytical ultracentrifuge, in *Analytical Ultracentrifugation, vol. 2,* Beckman Instruments, Palo Alto, CA, 1993.
22. Hansen, J. C., Lebowitz, J., and Demeler, B. (1994) Analytical ultracentrifugation of complex macromolecular systems. *Biochemistry* **33,** 13155–13163.
23. Cole, J. L. and Hansen, J. C. (1999) Analytical ultracentrifugation as a contemporary biomolecular research tool. *J. Biomol. Tech.* **10,** 163–176.
24. Laue, T. M. and Stafford, W. F III (1999) Modern applications of analytical ultracentrifugation. *Annu. Rev. Biophys. Biomol. Struct.* **28,** 75–100.
25. Lebowitz, J., Lewis, M. S., and Schuck, P. (2002) Modern analytical ultracentrifugation in protein science: A tutorial review. *Protein Sci.* **11,** 2067–2079.
26. Yphantis, D. A., Correia, J. J., Johnson, M. L., and Wu, G.-M. Detection of heterogeneity in self-associating systems, in *Physical Aspects of Protein Interactions* (Catsimpoolas, N., ed.), Elsevier/North Holland, New York, 1978, pp. 275–303.
27. Wozney, J. M., Rosen, V., Celeste, A. J., Mitsock, L. M., Whitters, M. J., Kriz, R. W., Hewick, R. M., and Wang, E. A. (1988) *Science* **242,** 1528–1534.
28. Wozney, J. M. (1989) Bone morphogenetic proteins. *Progress in Growth Factor Research* **1,** 267–280.
29. Wang, E. A., Rosen, V., D'Alessandro, J. S., Bauduy, M., Cordes, P., Harada, T., Israel, D. I., et al. (1990) Recombinant human bone morphogenetic protein induces bone formation. *Proc. Natl. Acad. Sci. USA* **87,** 2220–2224.
30. Takuwa, Y., Ohse, C., Wang, E. A., Wozney, J. M., and Yamashita, K. (1991) Bone morphogenetic protein-2 stimulates alkaline phosphatase activity and collagen synthesis in cultured osteoblastic cells, MC3T3-E1. *Biochem. Biophys. Res. Commun.* **174,** 96–101.
31. Katagiri, T., Yamaguchi, A., Komaki, M., Abe, E., Takahashi, N., Ikeda, T., et al. (1994) Bone morphogenetic protein-2 converts the differentiation pathway of C2C12 myoblasts into the osteoblast lineage. *J. Cell Biol.* **127,** 1755–1766.
32. Chaudhari, A., Ron, E., and Rethman, M. P. (1997) Recombinant human bone morphogenetic protein-2 stimulates differentiation in primary cultures of fetal rat calvarial osteoblasts. *Mol. Cell. Biochem.* **167,** 31–39.
33. Lecanda, F., Avioli, L. V., and Cheng, S. L. (1997) Regulation of bone matrix protein expression and induction of differentiation of human osteoblasts and human bone marrow stromal cells by bone morphogenetic protein-2. *J. Cell. Biochem.* **67,** 386–396.
34. Li, R. H. and Wozney, J. M. (2001) Delivering on the promise of bone morphogenetic proteins. *Trends Biotechnol.* **19,** 255–265.
35. Geiger, M., Li, R. H., and Friess, W. (2003) Collagen sponges for bone regeneration with rhBMP-2. *Adv. Drug Deliv. Rev.* **55,** 1613–1629.
36. Israel, D. I., Nove, J., Kerns, K. M., Moutsatsos, I. K., and Kaufman, R. J. (1996) Expression and characterization of bone morphogenetic protein-2 in Chinese hamster ovary cells. *Growth Factor* **13,** 139–150.
37. Scheufler, C., Sebald, W., and Hülsmeyer, M. (1999) Crystal structure of human bone morphogenetic protein-2 at 2. 7 Å resolution. *J. Mol. Biol.* **287,** 103–115.

38. Friess, W. *Drug Delivery Systems Based on Collagen.* Shaker Verlag, Aachen, 2000.

39. Abbatiello, S. E. and Porter, T. J. (1997) Anion-mediated precipitation of recombinant human bone morphogenetic protein (rhBMP-2) is dependent upon the heparin binding N-terminal region. *Protein Sci.* **6(suppl. 2),** 99.

40. Laue, T. M. *Solution Interaction Analysis: Choosing Which Optical System of the Optima XL-I Analytical Ultracentrifuge to Use.* A-1821A, Beckman Instruments, Fullerton, CA, 1996.

*41. An-60 Ti Analytical Rotor, Cells, and Counterbalance for Use with the Optima XL-A Analytical Ultracentrifuge,* LXL/A-TB-003, Beckman Instruments, Palo Alto, CA, July 1991.

42. Stafford, W. F. (1992) Boundary analysis in sedimentation transport experiments: A procedure for obtaining sedimentation coefficient distributions using the time derivative of the concentration profile. *Anal. Biochem.* **203,** 295–301.

43. Stafford, W. F. III (1994) Boundary analysis in sedimentation velocity experiments. *Methods Enzymol.* **240,** 478–501.

44. Philo, J. S. Measuring sedimentation, diffusion, and molecular weights of small molecules by direct fitting of sedimentation velocity concentration profiles. In: *Modern Analytical Ultracentrifugation: Acquisition and Interpretation of Data for Biological and Synthetic Systems* (Schuster, T. M. and Laue, T. M., eds.), Birkhäuser, Boston, 1994, pp. 156–170.

45. Philo, J. S. (1997) An improved function for fitting sedimentation velocity data for low-molecular-weight solutes. *Biophys. J.* **72,** 435–444.

46. Laue, T. M., Shah, B. D., Ridgeway, T. M., and Pelletier, S. L. Computer-aided interpretation of analytical sedimentation data for proteins, in *Analytical Ultracentrifugation in Biochemistry and Polymer Science* (Harding, S. E., Rowe, A. J., and Horton, J. C., eds.), The Royal Society of Chemistry, Cambridge, UK, 1992, pp. 90–125.

47. Rathore, S., Hammerstone, K. M., Dansereau, S., and Porter, T. J. (1995) N-terminal isoforms of recombinant human bone morphogenetic protein (rhBMP-2) are active in vitro and in vivo. *Protein Sci.* **4,** 140.

48. Morin, R. D. Interactions between recombinant human bone morphogenetic protein-2 (rhBMP-2) and a collagen matrix. M. S. dissertation, Tufts University, Medford, MA, 2001.

49. Waxman, E., Laws, W. R., Laue, T. M., Nemerson, Y., and Ross, J. B. A. (1993) Human factor VIIa and its complex with soluble tissue factor: Evaluation of asymmetry and conformational dynamics by ultracentrifugation and fluorescence anisotropy decay methods. *Biochemistry* **32,** 3005–3012.

# 25

## Biological Characterization of Pegylated Interferons

### A Case Study

**Lei Xie, Constance Cullen, Sheri Bradshaw,
Marc DeLorenzo, and Michael J. Grace**

## 1. Introduction

Type I interferon-α (IFN-α) has proven to be a clinically effective antiviral and antineoplastic drug for 20 yr *(1)*. Recently, pegylated forms of IFN-α have been commercially produced and have shown superior clinical efficacy to unpegylated IFN-α for reducing HCV viral load with less frequent dosing required for the patient *(2,3)*. The superior clinical efficacy is probably derived from the enhanced serum half-life of the pegylated IFN-α in patients. However, pegylation also reduces the in vitro activity of the core IFN-α protein *(4,5)*. Understanding the structural implications of pegylation on IFN-α activity is critical for biologically characterizing the commercial drug product. In vitro characterization provides a basis for establishing consistency in the manufacturing process for the precursor IFN-α and the final pegylated product. The better characterized the product is, the higher the confidence is for assessing and demonstrating comparability between pegylated products when comparing changes in process of manufacture, site of manufacture, or like-product from different sources.

The process of pegylation introduces several complex factors in the biological characterization of IFN-α. Because the biological activity of IFN-α is, in itself, pleiotrophic *(1)*, a series of biological assays are required to adequately characterize resulting activity postpegylation. These in vitro assays are designed to measure changes in IFN-α activity from a very early receptor-mediated event through the production of interferon-response gene transcripts to subsequent biological endpoints. This chapter defines the methods used to assess receptor-mediated signaling through the JAK/STAT pathway, the interferon-responsive gene (IRG) transcripts that occur as a result of the signaling into the nucleus of the target cell and several subsequent key endpoint assays that measure antiviral activity, antiproliferation, antigen presentation, and cytotoxicity.

From: *Methods in Molecular Biology, vol. 308: Therapeutic Proteins: Methods and Protocols*
Edited by: C. M. Smales and D. C. James © Humana Press Inc., Totowa, NJ

## 2. Materials

### 2.1. STAT Translocation Assay

1. Human Hepatoma (HuH-7) cells were obtained from research sources.
2. HuH-7 culture medium: Dulbecco's modified Eagle's medium (DMEM) containing 10% fetal bovine serum (FBS), 2 m$M$ GlutaMAX-1, 100 U/mL penicillin–streptomycin, and 0.1 m$M$ nonessential amino acids. Other IFN-α responsive cells that attach to plastic and provide a large flat surface area may also be used.
3. 96-Well Packard black view plates (Packard Instruments).
4. Recombinant IFN-α2b and polyethylene glycol (PEG) IFN-α2b.
5. Formaldehyde (3.7% v/v) freshly prepared in Dulbecco's phosphate-buffered saline (DPBS) and prewarmed to 37°C.
6. Triton X-100 (0.5% w/v) prepared in DPBS, which can be stored at 4°C for up to 2 mo.
7. Tween-20 (0.01% w/v) prepared in DPBS, which can be stored at 4°C for up to 2 mo.
8. Anti-STAT1 p84/p91 (clone E-23, Santa Cruz Biotechnology) or a polyclonal anti-STAT2 (clone C-20, Santa Cruz Biotechnology). Prepare primary antibody at a 1:200 dilution in DPBS. All reagent preparations should be performed on the day of the assay unless otherwise indicated. Permeabilization and wash buffers should be brought to room temperature before use. Reagent volumes are typically 200 μL per well except for antibodies, which are added at 50 μL per well.
9. Alexa Fluor® 488 conjugated goat anti-rabbit immunoglobulin G (IgG) highly crossabsorbed (Molecular Probes) and Hoechst 33342 (Molecular Probes) prepared as a mixture in DPBS (1:100/1:2000 respectively). Protect from light.
10. ArrayScan® II High Content Screening System (Cellomics®) with Cytoplasm to Nucleus Translocation Protocol.

### 2.2. Interferon Response Gene
### Reverse Transcriptase-Polymerase Chain Reaction (RT-PCR) Assay in Molt 4 Cells

1. Human T-cell lymphoblastoma (Molt-4) cells American Type Culture Collection (ATCC).
2. Molt-4 culture medium: RPMI 1640 containing 2 m$M$ L-glutamine, 4.5 g/L glucose, 10 m$M$ HEPES, 1.5 g/L sodium bicarbonate, 1.0 m$M$ sodium pyruvate, and 10% FBS. Consult the ATCC catalog for information on how to maintain Molt-4 cell culture.
3. Recombinant human IFN: IFN-α2b and PEG-IFN-α2b.
4. RNeasy Mini Kit for total RNA isolation (Qiagen).
5. Agilent Bioanalyzer (Agilent Technologies).
6. LabChip RNA 6000 microfluidic analysis kit (Agilent Technologies).
7. Heat block.
8. Nuclease free H$_2$O.
9. MultiScribe RT, RNase inhibitor, random hexamers, dNTP mix, MgCl$_2$, and RT reaction buffers (Applied Biosystems).
10. –80°C Storage freezer.
11. 2X PCR Master Mix (Applied Biosystems).
12. TaqMan chemistry-based primer and probe sets for targets.
13. 18s rRNA endogenous control assay-on-demand.
14. ABI SDS 7900HT Sequence Detection System (Applied Biosystems).
15. SDS v2.0 software (Applied Biosystems).
16. 384-Well plates for use with SDS7900HT (Applied Biosystems).
17. Clear optical covers (Applied Biosystems).

## 2.3. Antiviral and Antiproliferation Assays

### 2.3.1. Encephalomyocarditis (EMC) Culture

1. Vero cells (ATCC).
2. Vero cell growth medium: medium 199 containing 100 U/mL penicillin, 100 µg/mL streptomycin, and 10% FBS.
3. Vero cell virus infection medium: medium 199 containing 100 U/mL penicillin, 100 µg/mL streptomycin, and 1% FBS.
4. Hank's balanced salt solution (HBSS).
5. Formaldehyde/crystal violet dye solution is prepared by mixing 37.5 mL for formaldehyde, 375 mL 95% ethanol, 2.0 g crystal violet, 0.75 g NaCl, and $dH_2O$ to 750-mL total volume.
6. EMC virus (ATCC).

### 2.3.2. Antiviral Assay

1. Human lung fibroblast (A549) cells (ATCC). Culture cells according to directions provided by ATCC.
2. Assay medium: DMEM with D-glucose containing 2.7 mg/mL sodium bicarbonate, 20 m$M$ HEPES, 4 m$M$ L-glutamine, 100 U/mL penicillin, 100 µg/mL streptomycin, and 10% FBS.
3. Recombinant IFN-α2b and PEG IFN-α2b.
4. Trypsin (0.25% v/v).
5. DPBS.
6. 0.5% w/v 3-(4,5-Dimethylthiazol-2YL)-2,5-diphenyltetrazolium bromide (MTT) is prepared by dissolving 5 g MTT powder in 1 L DPBS, pH 7.3–7.5. The solution is filter-sterilized (0.45-m$M$ filter) and stored protected from light at 4°C for up to 2 mo.
7. Sodium dodecyl sulfate (SDS) (10% w/v) prepared in $dH_2O$.

### 2.3.3. Antiproliferation Assay

1. Daudi cells (Human Burkitt lymphoma) (ATCC). Culture cells according to directions provided by ATCC.
2. Assay medium: RPMI 1640 containing 2 m$M$ L-glutamine adjusted to contain 1.5 g/L sodium bicarbonate, 4.5 g/L glucose, 10 m$M$ HEPES, 1.0 m$M$ sodium pyruvate, and 10% FBS.
3. Recombinant IFN-α2b and PEG IFN-α2b.
4. Trypsin (0.25% v/v).
5. 0.5% MTT and SDS (10% w/v) prepared in $dH_2O$.

## 2.4. Assays for IFN-α Immune Response

### 2.4.1. Major Histocompatibility Complex (MHC) Class I Expression Assay

1. Molt-4 (ATCC). Molt-4 culture medium: *see* **Subheading 2.2.**, **item 2**. Consult the ATCC catalog for information on how to maintain Molt-4 cell culture.
2. Recombinant human IFN: IFN-α2b and PEG-IFN-α2b.
3. R-Phycoerythrin (R-PE)-conjugated mouse anti-human human leukocyte antigen (HLA)-A, B, C monoclonal antibody (Becton Dickinson). The antibody should be titrated to obtain optimal results, and 20 µL of the titrated antibody volume is used for staining.
4. R-PE-conjugated mouse IgG, isotype control (Becton Dickinson). The amount of isotype control antibody used should be matched to the amount of PE-conjugated mouse anti-

human HLA-A, B, C antibody, and 20 µL of the titrated antibody volume is used for staining.

5. Buffer: 1X PBS containing 1% FBS.
6. Six-Well tissue culture plate.
7. 5 mL Polystyrene round-bottom tube (Becton Dickinson).
8. BDIS FACSCalibur™ equipped with 488-nm argon-ion laser (Becton Dickinson).
9. Beckman GS-6 centrifuge (Beckman).

### 2.4.2. Lymphokine-Activated Killer (LAK)/Natural Killer (NK) Cell Bioassay

1. Human chronic myelogenous leukemia (K562) cells (ATCC).
2. Daudi cells (ATCC). Consult the ATCC catalog for information on how to culture and maintain K562 and Daudi cell cultures.
3. Recombinant human interleukin-2 (rhIL-2).
4. Recombinant human IFN: IFN-α2b and PEG-IFN-α2b.
5. CytoTox96$^R$ Non-Radioactive Cytotoxicity Assay Kit (Promega).
6. Ficoll-Paque Plus.
7. Complete Iscove's modified Eagle's medium (IMDM) medium: IMDM containing L-glutamine, penicillin/streptomycin, and 10% FBS.
8. V-bottom 96-well plate.
9. Flat-bottom 96-well plate.
10. Tissue culture flask.
11. Centrifuge tubes (15 and 50 mL).
12. Multichannel pipet.
13. Beckman GS-6 centrifuge (Beckman).
14. 96-Well plate reader with 490- or 492-nm wavelength.

## 3. Methods

### 3.1. STAT Translocation Assay

IFN-α receptor-mediated signal transduction for antiviral activity involves the JAK/STAT pathway (6). A key step in the pathway is the translocation of the STAT1 homodimeric or STAT1/STAT2 heterodimeric protein complex from the cell cytoplasm to the nucleus. This translocation can be accurately quantitated in near real time using specific fluorescent antibodies and the ArrayScan® II (**Fig. 1**).

### 3.1.1. Cell Culture

1. Plate HuH-7 cells in a 96-well Packard black view plate at a density of 10,000 cells per well in culture medium, and incubate in a humidified incubator at 37°C and 5–6% $CO_2$ for 18–24 h.
2. The cell-plating density and culture conditions should be determined empirically for each different cell line. The optimal density is approx 80–90% confluence.
3. After dispensing cells into plates, allow the cells to attach for 0.5–2 h on a nonvibrating surface (*see* **Notes 1** and **2**).

### 3.1.2. Assay

1. Prepare serial threefold dilutions of IFN-α in complete culture medium.

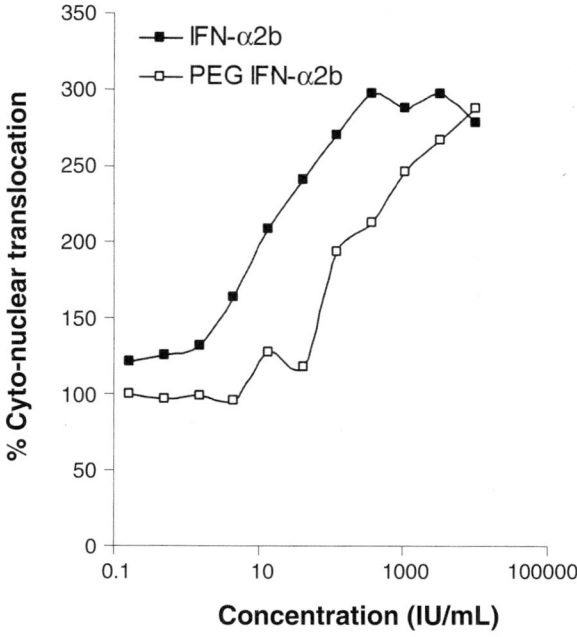

Fig. 1. STAT1 translocation activity of IFN-α2b and pegylated IFN-α2b on HuH-7 cells. HuH-7 cells were incubated for 30 min in the presence of increased concentrations of either IFN-α2b or pegylated IFN-α2b, and the resulting STAT1 translocation was quantitated as the percent of cytonuclear translocation.

2. Incubate the plated cells with IFN-α dilutions at 37°C in 5–6% $CO_2$ for 20–30 min. This time is optimal for HuH-7 cell STAT translocation but should be empirically determined for different cell lines.
3. Quickly aspirate IFN-α containing complete culture medium, and wash the cells under room temperature conditions using DPBS prewarmed to 37°C.
4. Quickly aspirate the DPBS under room temperature conditions and fix the cells for 10 min using 37°C prewarmed 3.7% formaldehyde (*see* **Note 3**).

### 3.1.3. Detection of Intracellular STAT

1. Permeabilize the formaldehyde-fixed cells using a solution of 0.5% Triton X-100/DPBS for 90 s at room temperature.
2. Wash out the solution using DPBS, then incubate the permeabilized cells with the primary antibody for 1 h at room temperature.
3. Wash the cells with a solution of DPBS containing 0.01% Tween-20 for 5–10 min, then wash out the Tween-20 using DPBS.
4. Incubate the cells with an Alexa Fluor 488-conjugated goat anti-rabbit IgG and Hoechst 33342 mixture (1:100/1:2000 dilutions) for 1 h. Keep plates protected from light from this point onward (*see* **Note 4**).
5. Wash the cells with a solution of DPBS containing 0.01% Tween-20 for 5–10 min, then wash out the Tween-20 by washing twice in DPBS.

6. Leave the DPBS in the well with the second wash. Seal the plate with plate sealers.
7. Scan the plates on the ArrayScan II High Content Screening System with a ×10 objective and the XF100 filter using the Cytoplasm to Nucleus Translocation Protocol of the ArrayScan v2.1 software.

### 3.2. IRG RT-PCR Assay in Molt 4 Cells

A well established set of IRG transcripts (mRNA) are known to be upregulated by IFN-α receptor binding and signal transduction *(6)*. The relative quantitation of IRG transcripts can be performed using a two-step RT-PCR fluorogenic TaqMan method using 18s rRNA as the endogenous control.

### 3.2.1. Cell Culture

1. Molt-4 cells are plated into a six-well plate at a density of $3 \times 10^5$ cells/mL in 2 mL culture medium, and incubated in a humidified incubator at 37°C, 5–6% $CO_2$.
2. Before the addition of IFN-α, cells are stimulated for 72 h with the addition of phytohemagglutin (PHA) (1:2000 v/v; *see* **Note 5**).
3. After stimulation, the cells are centrifuged at 250*g*, washed once with DPBS, and then resuspended at $3 \times 10^5$ cells/mL in fresh culture medium.
4. Cells are stimulated with IFN-α2b or pegylated IFN-α2b in the 2–20,000-U/mL range by the addition of 1:10 v/v IFN diluted in culture medium to each well. Serial 0.5 or 1.0 logarithmic dilutions are recommended.
5. Interferon-treated cells are cultured at 37°C, 5–6% $CO_2$ in a humidified incubator, with cells being harvested at time points between 6 and 72 h postinitiation of treatment.
6. Cells are harvested by aspirating the well and centrifuging in a 4°C refrigerated centrifuge at 250*g*, washed once in DPBS, and then lysed.

### 3.2.2. Total RNA Isolation

1. Total RNA is isolated using the RNeasy column-based system (Qiagen) performing a second elution step. DNase treatments are not performed in low-yield conditions, and efforts should be made to design amplicons bridging exon gaps for the purpose of mRNA specificity.
2. Qualitative and quantitative analysis is performed with the RNA LabChip 6000 on the Agilent BioAnalyzer (Agilent Technologies).

### 3.2.3. Reverse Transcription and PCR

1. Reverse transcription reactions are performed in 96-well plates based on 1 μg total RNA in a 100 μL final volume reaction.
2. The typical reaction conditions are as follows: 25°C, 10 min/37°C, 60 min/95°C, 5 min, incorporating final concentrations of 3.125 U/μL MultiScribe RT, 0.4 U/μL RNase inhibitor, 2.5 μ*M* random hexamers, 500 μ*M* dNTP (mix), 5.5 m*M* $MgCl_2$, and 1X RT reaction buffer (Applied Biosystems).
3. The reaction volume is diluted to 1 mL with nuclease-free water and stored at –80°C.
4. Fluorogenic PCR is performed using TaqMan-based chemistry and the ABI 7900HT sequence detection system with a 384-well format. The final reaction volume in each well is 15 μL after incorporating 5 μL template (0.5% diluted RT product).
5. Reactions are run in triplicate, and a single target can be chosen for multiplexed reactions with the 18S rRNA primer-probe set.

6. The targets used for multiplexing are qualified within the laboratory by running an IFN-treated RNA sample standard curve in both single and 18S rRNA-multiplexed reactions over a dilution curve spanning 1024-fold range in concentration (a theoretical 10.0-cycle threshold [Ct] range).
7. 25 ng Total human RNA template (Applied Biosystems) is used for interplate, inter-instrument, and interday control across all plates within an experiment (*see* **Notes 6 and 7**).

### 3.2.4. Data Analysis

1. Data are imported into Microsoft Excel as Ct values, and triplicate reactions are averaged.
2. To normalize the quantity of target mRNA molecules to that of the total RNA, assumed by the understanding that 18S rRNA endogenous levels are practically constant, the Δ-Ct for a given sample is calculated by the equation:

$$[\{Ct_{AVG} (target) - Ct_{AVG} (18s\ rRNA)\}^2]$$

3. To determine the Δ-Δ-Ct or fold change in copy number of a specific target mRNA between samples, the following equation is used:

$$[\{\Delta Ct (untreated) - \Delta Ct (timepointX)\}^2]$$

4. Gene transcript upregulation is indicated by a fold-change value greater than 1.0. Conversely, downregulation is indicated by a fold-change value below 1.0. Almost all triplicate reactions result in Ct standard deviations of below 0.5 (*see* **Note 8**) (**Figs. 2 and 3**).

## 3.3. Antiviral and Antiproliferation Assays

Two important biological consequences of directed IFN-α activation are the resulting antiviral and antiproliferative effects on the target cell. Antiviral assays in a variety of target cells have been reported extensively *(8)*. The assay described in this section measures the amount of extant target cell viability after an experimental infection using EMCV. Antiproliferation assays can be developed in a wide range of cell lines; one of the most commonly used is the Daudi cell line—a B-lymphoblastoma cell line. This assay uses similar endpoint assessment as the antiviral assay, measuring cell oxidative phosphorylation associated with cellular viability and proliferation (**Fig. 4**).

### 3.3.1. Preparation of EMC Virus Stock

1. Vero cells must be expanded into roller bottles to produce virus stock (*see* **Note 9**).
2. From culture flasks, grow Vero cells in growth medium until 75–90% confluent.
3. Remove the medium, rinse the monolayer with HBSS, and add enough trypsin to just cover the monolayer.
4. Gently tilt the flasks until the cells are dislodged, and add fresh growth medium to reseed into increasingly larger culture flasks or roller bottles.
5. To make sufficient stock, six to eight seeded roller bottles are needed.
6. Thaw sufficient EMC virus stock, and dilute 1:100 into Vero cell virus infection medium to prepare 10 mL inoculum for each roller bottle.
7. Remove the cell growth medium from each roller bottle, and rinse the monolayer with HBSS before adding the 10 mL inoculum.
8. Incubate the roller bottle with inoculum at 37°C for 2 h, then add 40 mL fresh Vero cell virus infection medium (without virus) to each bottle and continue to incubate for 48 h.

## PKR mRNA Response

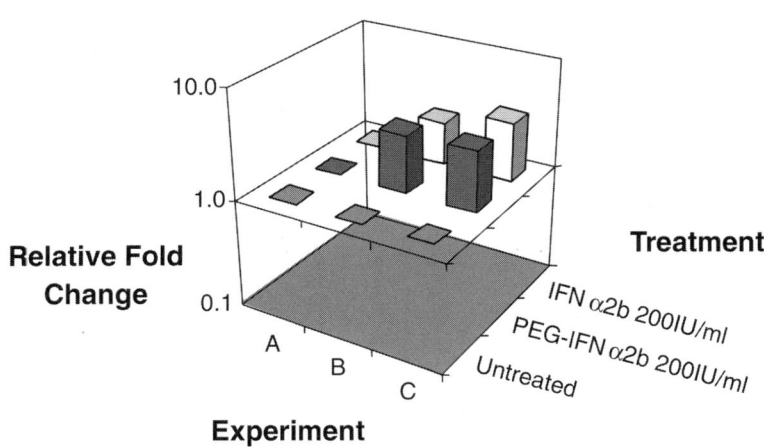

## 2,5 OAS mRNA Response

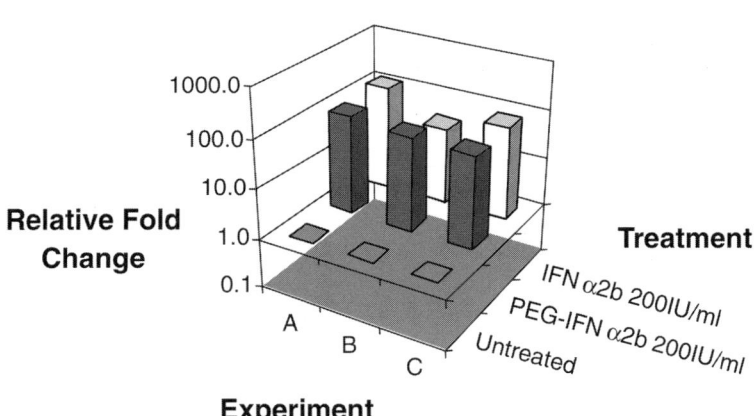

Fig. 2. dsRNA-dependent protein kinase (PKR) and 2'–5' oligoadenylate synthetase (OAS) response to IFN-α2b and PEG IFN-α2b in Molt-4 cells. Relative fold change in mRNA transcript response for two IRGs in three separate experiments.

9. Harvest the cells and virus using a cell scraper, and transfer the contents of the bottles into polypropylene tubes.
10. Rinse the bottles with an additional 20 mL Vero cell virus infection medium and add to the tubes.
11. Lyse the cells in the tubes by performing three freeze/thaw cycles, and centrifuge the lysate at 250$g$, 4°C for 10–15 min.
12. Pool supernatants, and aliquot at 500 μL per tube; store frozen at –80°C.

| Gene | IFN-α2b mean fold | IFN-α2b 95% Conf. | PEG IFN-α2b mean fold | PEG IFN-α2b 95% Conf. |
|------|------|------|------|------|
| 2'-5' OAS | 59.7 ± 1.7 | 29.8–119.4 | 77.4 ± 1.3 | 38.7-154.8 |
| ISG-15 | 33.8 ± 8.8 | 16.9-67.6 | 44.7 ± 8.1 | 22.3-89.4 |
| ISG-54 | 14.5 ± 1.8 | 7.2-29.0 | 19.0 ± 1.6 | 9.5-38.0 |
| PKR | 2.6 ± 1.3 | 1.3-5.2 | 3.3 ± 1.1 | 1.6-6.6 |
| IFI-16 | 2.2 ± 1.5 | 1.1-4.4 | 2.6 ± 1.1 | 1.3-5.2 |

Fig. 3. Performance of IFN-α2b and PEG IFN-α2b-induced IRGs. Mean fold induction (±1 SD) and 95% confidence interval (CI) for the induction of 5 IRGs in Molt-4 cells treated with either IFN-α2b or pegylated IFN-α2b ($n = 4$ experiments). The fold stimulation was normalized to GAPDH mRNA and to untreated Molt-4 cells. The SD is the variance around the mean fold induction for the number of experiments ($n = 4$); the 95% CI is calculated using ± 0.5 Ct around the mean.

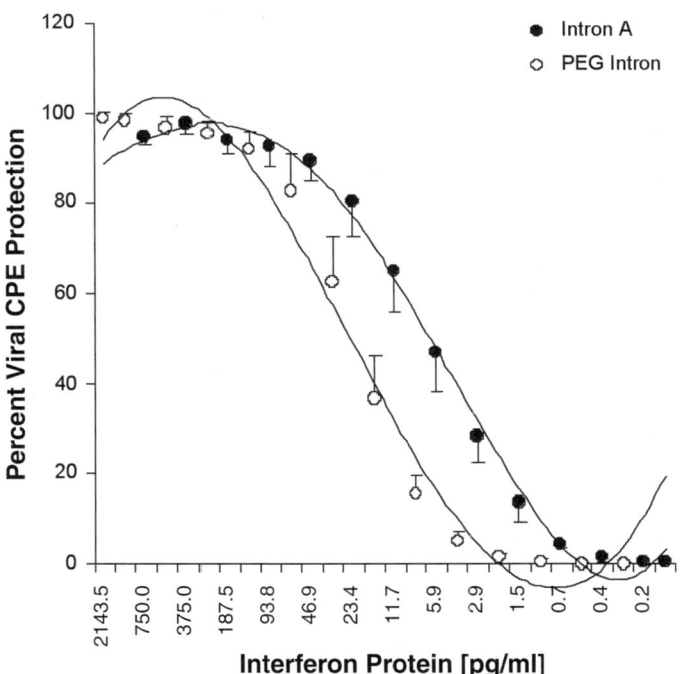

Fig. 4. Antiviral protection response of IFN-α2b and pegylated IFN-α2b. FS-71 cells were cultured in the presence of EMC virus and serial dilutions of IFN-α2b (intron A) or pegylated IFN-α2b (PEG intron). The amount of cytopathic effect (CPE) as determined by MTT was normalized against the maximum CPE for IFN-α2b control run in the same plate.

### 3.3.2. Determination of the Dilution of Virus Stock to Use in Assay

1. Prepare 96-well plates with antiviral assay medium (DMEM) and A549 cells as described in **Subheading 3.3.4.**
2. Prepare a 1:2 serial dilution of the EMC virus stock in assay medium into sterile tubes.
3. Prepare 12 dilutions each with 500-μL final volume.
4. Transfer 50 μL of each virus dilution to duplicate wells in the plates containing the A549 cells, and incubate in a humidified incubator at 37°C, 5% $CO_2$ for 38–42 h.
5. Aspirate the medium from the plate and stain with 50 μL per well of the methanol/crystal violet dye solution for 15 min.
6. Remove the dye from the wells, and rinse with water until all the residual dye is removed.
7. Pat plates on paper towels to remove residual water; air-dry.
8. Determine the highest dilution of the virus that gives complete lysis of the monolayer. This dilution should be used in antiviral assay.

### 3.3.3. Dilution of IFN-α into Test Plates

1. Add 50 μL per well of the assay medium to all wells of a 96-well flat-bottom plate.
2. Add 50 μL of each sample or standard in duplicate to wells in the first column.
3. Serially dilute the samples 1:2 across the plate from the first through the last column.
4. Prepare a matched virus control plate with 50 μL per well of medium (*see* **Note 10**).

### 3.3.4. Cell Culture for Antiviral Assay: Preparation of A549 Cell Suspension

1. Inspect all flasks containing the A549 cells prior to use. A549 cells should be 85–95% confluent with no visible signs of contamination.
2. Aspirate the assay medium, and rinse the cell monolayers with DPBS.
3. Remove the DPBS, and add enough trypsin solution to just cover the bottom of the flasks.
4. Gently tilt the flasks until the monolayer detaches from the surface, then add approx 20–25 mL of assay medium for each 75-cm² flask. For larger flasks, the volumes can be scaled appropriately.
5. Pool the cell suspension from multiple flasks, and centrifuge at 2500g for 5–10 min.
6. Resuspend cells in assay medium, count using a hemacytometer, and adjust the concentration to $1.5 \times 10^5$ cells/mL in growth medium.

### 3.3.5. A549 Antiviral Cell Assay

1. Add 100 μL per well of prepared A549 cell suspension into each well of prepared 96-well plates.
2. Incubate the plates in a humidified incubator at 37°C in 5% $CO_2$ for approx 4 h.
3. To infect the culture, add 50 μL per well of EMC virus diluted in assay medium, and incubate plates 38–42 h or until the monolayer in the virus control plate has been completely disrupted (4+ cytopathic affect).
4. Remove plates from the incubator, and add 50 μL per well of the MTT solution, and then continue incubating plates at 37°C, 5% $CO_2$ for an additional 2.5–4 h.
5. Remove the plates from the incubator, and add 100 μL per well of the 10% SDS solution.
6. Continue incubating overnight in an ambient humidified chamber. Plates are stable at this stage for approx 7 d as long as they are stored in a humidified atmosphere to prevent evaporation.
7. To assess cytopathic protection effect, read the plates using a spectrophotometric plate reader set to 570 nm with a reference wavelength of 630 nm. Quantitation of unknowns is made by comparison to the IFN standard (*see* **Note 11**).

## 3.3.6. Methods for Antiproliferation Assay

### 3.3.6.1. DILUTION OF IFN-A INTO TEST PLATES

1. Add 50 µL per well of antiproliferation assay medium (RPMI) to all wells of a 96-well flat-bottom plate.
2. Add 50 µL of each sample or standard in duplicate to wells in the first column.
3. Serially dilute the samples 1:2 across the plate from the first to the last column.

### 3.3.6.2. CELL CULTURE FOR ANTIPROLIFERATION ASSAY: PREPARATION OF DAUDI CELL SUSPENSION

1. Inspect all flasks prior to use.
2. Cells should be 85–95% confluent with no visible signs of contamination.
3. Aspirate the assay medium, and rinse the back of the flasks to remove any adherent cells.
4. Pool cell suspensions from multiple flasks, and centrifuge at 2500$g$ for 5–10 min.
5. Resuspend cells in assay medium, and count using a hemacytometer.
6. Adjust the concentration to $2.5 \times 10^5$ cells/mL in assay medium.

### 3.3.6.2. DAUDI ANTIPROLIFERATION CELL ASSAY

1. Add 50 µL per well of the Daudi cell suspension, and incubate plates in a humid incubator at 37°C, 5.0% $CO_2$ for 72 h.
2. Add 25 µL per well of the MTT solution, incubate at 37°C, 5.0% $CO_2$ for 4–6 h, and then add 100 µL per well of the 10% SDS solution.
3. Continue incubating overnight at 37°C, 5.0% $CO_2$. Plates are stable at this stage for approx 7 d as long as they are stored in a humidified atmosphere to prevent evaporation.
4. To assess active oxidative phosphorylation associated with cell proliferation, read the plates on a spectrophotometric plate reader set to 570 nm with a reference wavelength of 630 nm.
5. Quantitation of unknowns is made by comparison to the IFN standard.

## 3.4. Assays for IFN-α Immune Response

IFN-α is an immunomodulatory protein that may play an important role in antiviral and antitumor pathogenesis *(1)*. Two key activities associated with IFN-α are the enhancement of antigen presentation through the MHC class I protein complex (**Fig. 5**) and the stimulation of NK and LAK cytotoxic activity *(9–11)* (**Fig. 6** and **Fig. 7**).

### 3.4.1. Molt-4 Culture

1. Culture Molt-4 cells as directed by ATCC.
2. Withdraw Molt-4 cells from culture flasks, count using a hemacytometer, and centrifuge at 250$g$ for 6–10 min.
3. Adjust the concentration to $3 \times 10^5$ cells/mL in fresh growth medium.
4. Add 1.5 mL per well of the prepared Molt-4 cells into each well of six-well plates.
5. Dilute IFN-α and PEG IFN-α in serial twofold into a separate working plate using culture medium.
6. Aliquot 0.5 mL diluted IFN-α into the corresponding wells of the plate containing the Molt-4 cells.
7. Incubate the plate containing the Molt-4 cells with IFN in a humid incubator at 5% $CO_2$, 37°C for 48 h (*see* **Notes 12** and **13**).

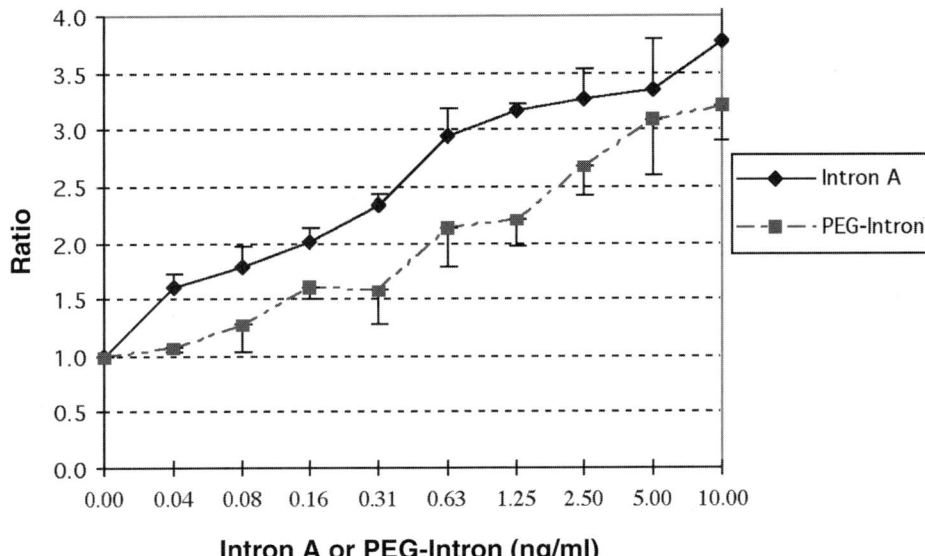

Fig. 5. Molt-4 MHC class I expression. Molt-4 cells were stimulated with IFN-α2b (intron A) or pegylated IFN-α2b (PEG intron) for 48 h, and MHC class I surface expression was detected using fluorescent anti-HLA-A, B, C MAb and flow cytometry.

### 3.4.2. Flow Cytometry Analysis for MHC Class I Expression

1. Remove the plates from the incubator, mix, and transfer the cell suspension into the 5-mL polystyrene tubes.
2. Centrifuge cells at 250*g* for 6–10 min, and resuspend the cell pellet in 50–100 μL PBS containing 1% FBS.
3. Add the anti-HLA-A, B, C antibody or isotype control at the concentration determined by previous optimization experiments, and incubate the cells on the ice in the dark for 30 min.
4. Wash cells once with PBS containing 1% FBS, and resuspend cells into 500 μL PBS containing 1% FBS.
5. Collect and analyze data using a flow cytometer equipped with a 488-nm line laser.
6. Adjust the initial cytometer settings using the isotype control cells, and set a gate around all viable cells.
7. Acquire a minimum of 10,000 cells as list mode file protocols. Data are usually reported as absolute mean fluorescence intensity (MFI) adjusted for the ratio of treated cell MFI-to-untreated cell MFI.

### 3.4.3. LAK/NK Bioassay

#### 3.4.3.1. ISOLATION OF PERIPHERAL BLOOD MONONUCLEAR CELLS (PBMCs)

1. Isolate human PBMCs using Ficoll-Paque Plus gradient centrifugation following the product instruction sheet.
2. Resuspend the PBMCs in IMDM complete medium at a final concentration of $2 \times 10^6$ cells/mL.

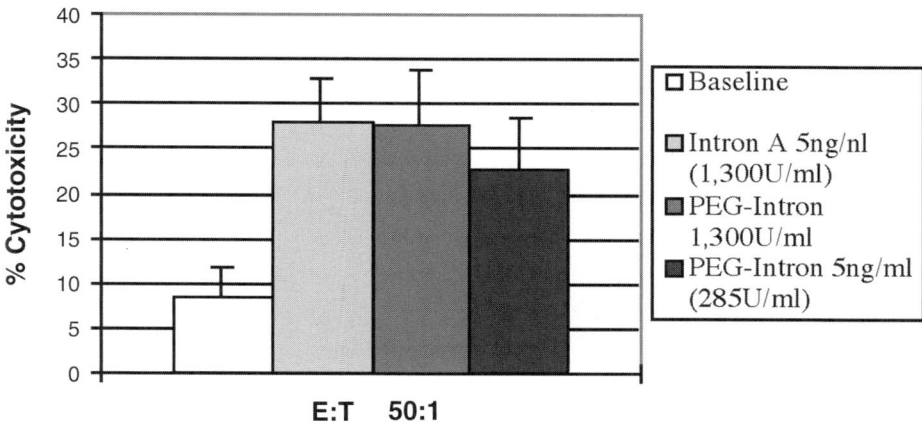

Fig. 6. Effect of intron A and PEG-intron on NK cytotoxicity against K562 cells. PBMCs from six different individuals were incubated with IFN-α2b (intron A) or pegylated IFN-α2b (PEG intron) for 16 h, then assayed at E:T ratios of 50:1. *, $p$ value < 0.05 compared to baseline.

3. Incubate the cells in culture flasks for 1 h in a humidified incubator at 37°C, 5% $CO_2$ to remove adherent cells.
4. Recover nonadherent cells by gently swirling the flask; remove all floating cells with a pipet, and centrifuge cells at 250$g$ for 6–10 min.
5. Discard the supernatant, and resuspend cells in IMDM complete medium.
6. Count the cells, and determine viability by Trypan Blue exclusion. Red blood cells should be lysed by hypotonic shock if heavy contamination is apparent.

### 3.4.3.2. GENERATION OF LAK CELLS AND ENHANCEMENT OF NK CELL CYTOTOXICITY

NK cells manifest cytotoxic activity without prior stimulation by lymphokine and antigen. NK cell cytotoxicity can be tested directly after isolation from peripheral blood. LAK cells require prior activation with IL-2. K562 cells are used as target cells for testing NK cytotoxicity, and Daudi cells are used as target cells for testing LAK cytotoxicity (**Figs. 6** and **7**) (*see* **Notes 14–17**).

### 3.4.3.3. GENERATION OF LAK CELLS

1. Culture PBMCs in IMDM complete medium containing 4 U/mL recombinant human IL-2 (rIL-2), 200 U/mL recombinant human IL-2 (rhIL-2), or 4 U/mL rhIL-2 plus IFN-α for 3 d.
2. Remove the resulting LAK cell suspension by vigorous shaking culture flasks.
3. Complete cell recovery by rinsing flasks with adequate amounts of IMDM complete medium.
4. Centrifuge the cell suspension at 250$g$ for 6–10 min at room temperature, and resuspend at $2 \times 10^7$ cells/mL.

### 3.4.3.4. ENHANCEMENT OF NK CELL CYTOTOXICITY

1. Incubate freshly isolated PBMCs in IMDM complete medium containing IFN-α2b, PEG IFN-α2b, or medium alone for 16 h.

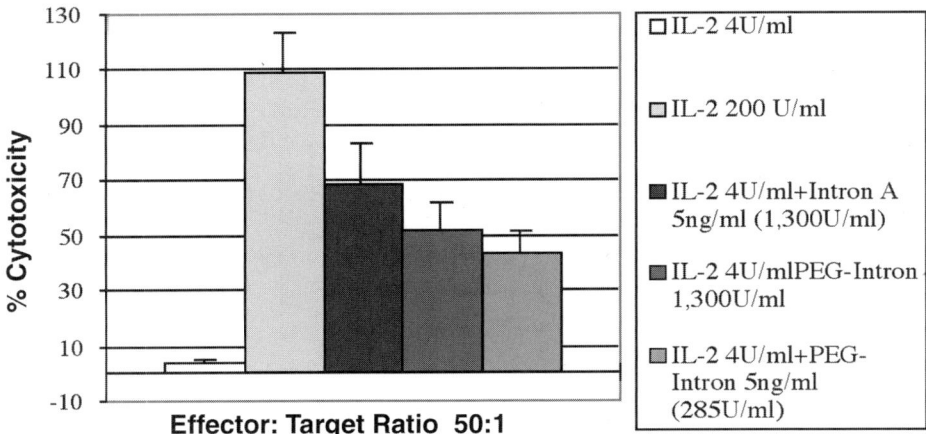

Fig. 7. Effect of intron A and PEG-intron on LAK cytotoxicity against Daudi cells. PBMCs from five different individuals were incubated with IL-2, IL2 + IFN-α2b (intron A) or IL-2 + pegylated IFN-α2b (PEG intron) for 3 d, then assayed at E:T ratio of 50:1.

2.  Remove NK cell suspension by vigorous shaking culture flasks.
3.  Complete cell recovery by rinsing flasks with adequate amounts of IMDM complete medium.
4.  Centrifuge the cell suspension at 250$g$ for 6–10 min at room temperature, and resuspend at $2 \times 10^7$ cells/mL.

### 3.4.3.5. Measurement of NK and LAK Cytotoxicity

1.  Prior to testing, experiments must be performed to optimize the target cell number to ensure an adequate signal-to-noise ratio.
2.  Adjust the effector cells to the appropriate concentration for the desired effector:target (E:T) ratio to achieve optimal signal-to-noise.
3.  Set up the 96-well assay plate following the manufacture's suggestion (CytoTox96 Non-radioactive Cytotoxicity Assay).
4.  Centrifuge the plate at 250$g$ for 4 min to ensure effector and target cells are in contact.
5.  Incubate the assay plate for 4 h in a humidified incubator at 37°C, 5% $CO_2$.
6.  Add 10X lysis solution to Target Cell Maxim LDH Release Control, and incubate for 45 min.
7.  Centrifuge the plate at 250$g$ for 4 min.
8.  Carefully remove 50 μL supernatant to a flat-bottom 96-well plate, add 50 μL substrate mix, and incubate in the dark at room temperature for 30 min.
9.  Add stop solution and read the absorbance on a spectrophotometer at 490 nm using a 630 nm control wavelength within 1 h.

### 3.4.3.6. Calculation of Percentage Cytotoxicity

Cytotoxicity is calculated by the following equation:

$$\% \text{ Cytotoxicity} = \frac{A - B - C}{D - C} \times 100$$

where:

- A = Experimental release: Lactate dehydrogenase (LDH) released into the well containing effector and target cells.
- B = Effector spontaneous release: LDH released from the effector cells cultured in medium alone.
- C = Target spontaneous release: LDH released from the target cells cultured in medium alone.
- D = Target maximum release: LDH released from the target cells incubated with 1.2% Triton.
- E = Culture medium background: For LDH activity contributed by serum corrects the varying amounts of phenol red in the culture medium.

For background absorbance corrections, refer to manufacturer's suggestion (CytoTox96).

## 4. Notes

1. Adherent flat cell lines with a large cytoplasmic surface area will provide the best images for processing.
2. Vibrations from a biological safety cabinet or incubator can cause the cells to settle primarily in the periphery of the well. A sparse cell density in the center of the well will extend the scanning time of the ArrayScan.
3. All reagent preparations should be performed on the day of the assay unless otherwise indicated. Permeabilization and wash buffers should be brought to room temperature before use. Reagent volumes are typically 200 μL per well, except for antibodies, which are added at 50 μL per well.
4. Hoechst 33342 will label DNA and define both individual cells and the nuclear-cytoplasmic boundary for each cell. The algorithm will generate concentric rings from this boundary to define both a nuclear and cytoplasmic region for each cell. The fluorescence intensity of the Alexa488-labeled STAT is measured in both nuclear and cytoplasmic regions for each cell. The mean cytonuclear difference for a user-defined number of cells is reported as a measure of STAT translocation (7).
5. For IFN-α response genes, the kinetics of response needs to be studied between 3- and 72-h post-induction. The optimal kinetics of response is different for each IRG transcript studied. It should also be noted that we work with IFN-α doses in vitro that reflect the typical blood concentrations that have been reported for standard three-weekly IFN-α dosing.
6. The gene transcript of choice for normalization should be carefully studied and specifically designed for the response genes and cell-based system being used. For the IRG transcript response in Molt-4 cells, both glyceraldehyde 3-phosphate dehydrogenase (GADPH) and 18s are suitable for gene transcript normalization.
7. The selection of the primer-probe sets is a key component for this type of assay. We initially developed the assay using our own designed primer-probe sets built from public access databases and scientific literature (5). The Applied Biosystems commercial sets that were used for the results presented in this chapter also performed very well. It is critical when developing or modifying the primer-probe sets to rigorously test against potential artifact.
8. The IRG transcript results reported in Molt-4 cells are correlative with response patterns and response magnitudes that we have observed in typical human PBMC preparations.

9.  For antiviral CPE assays, the choice of the virus and target cell will directly impact the IFN-$\alpha$ dose-response. The assay presented here uses EMC virus infection of A549 cells. Another commonly used virus for these kinds of assays is Vesicular Stomatitis virus. Other commonly used cell lines are Madin-Darby bovine kidney and WISH; fibroblast primary explant cultures are also frequently used, e.g., human foreskin fibroblasts (FS-71; *1,4,8*). For each virus/target cell system, a specific reference standard for the isoform of IFN-$\alpha$ must be used; e.g., if IFN-$\alpha$2b activity is assessed, the World Health Organization standard for IFN-$\alpha$2b should be used for calibration of reference curves.

10. Standard curves must be run in each titration plate to normalize for plate variability. System suitability criteria for the standard curve, such as suitability of fit to a log-log or four-parameter function is used to accept or reject test plates.

11. If the antiviral protection potency is unknown, dose-ranging experiments will be required owing to the limited response range (~1 log) of the typical antiviral plaque assay.

12. Molt-4 cells are most responsive when stimulated the next day after an overnight passaging. There is an optimal range of cell passages for which a single split can be used effectively, after which the signal-to-noise ratio becomes too low. This range must be determined empirically within each laboratory depending on the original source of the Molt-4 cells and past passage history of the cell bank.

13. Many commercial anti-HLA antibodies and fluorescence detection systems are available. It has been our experience that this assay can be used in a two-, three-, and even four-color (fluorophore) combination with other relevant surface antigens or internal cell response signals, as long as proper care is taken with the excitation wavelengths and emission filter set-ups used for the multicolor analyses.

14. Because both assays use human PBMC preparations as the effector cells, the associated variability of these assays is large, even by cellular bioassay standards. Percentages of coefficient of variance can be typically 50–80%. The most effective approach to reduce this variability is to use a consistent pool of known donors. If the pool is limited to males and postmenopausal females who only donate during nonallergy and nonsick periods, we have found that variability can be reduced to 20–40%. A second approach is to use unit-sized packed cell preparations provided by a reliable, known, and previously tested blood center. We have reduced variability to 30–50% using this approach. Typical cell bioassay variability for cell line assays is 20–40%. In both cases, it is highly recommended that all critical comparative studies and samples be assayed using the same preparation and assay set-up day. Because of the variability inherent in these kinds of assays, we recommend using a minimum of three different donors for the final analysis. Because some donors fail to respond at any E:T ratio, additional donors need to be tested. Typically, for an averaged final of $n = 6$ responsive donors, as many as 8–12 donors may need to be tested.

15. The E:T ratio must be empirically determined by each laboratory. We typically use ratios between 10 and 50:1 and always test at two different ratios. Factors affecting this ratio are the quality of the effector cells, quality of the target cells, and the method used for quantitating cytotoxicity.

16. Target cells for both assays need to be healthy to reduce background spontaneous cytotoxicity. Optimization of the cell passage range and preassay split time increase the target cell sensitivity to background response.

17. The method for determining cytotoxicity can be radiochemical (release of $Cr^{51}$) or nonradiochemical, as we have described here. The method used must be considered when adjusting E:T ratios because of the change in sensitivity relative to signal-to-noise.

## References

1. Brassard, D., Grace, M., and Bordens, R. (2002) Interferon-α as an immunotherapeutic protein. *J. Leukoc. Biol.* **71,** 565–581.
2. Glue, P., Tang, J., Rouzier-Panis, R., Raffanel, C., Sabo, R., Gupta, S., et al. (2000) Pegylated interferon alpha 2b: pharmacokinetics, pharmacodynamics, safety, and preliminary efficacy data. *Clin. Pharm. Ther.* **68,** 556–657.
3. Luxon, B., Grace, M., Brassard, D., and Bordens, R. (2002) Pegylated interferons for the treatment of chronic hepatitis C infection. *Clin. Ther.* **24,** 1363–1383.
4. Foser, S., Schacher, A., Weyer, K., Brugger, D., Dietel, E., Marti, S., and Schreitmuller, T. (2003) Isolation, structural characterization, and antiviral activity of positional isomers of monopegylated interferon α-2 (Pegasys). *Protein Expr. Purif.* **30,** 78–87.
5. Grace, M., Youngster, S., Gitlin, G., Sydor, W., Xie, L., Westreich, L., et al. (2001) Structural and biological characterization of pegylated recombinant IFN-α2b. *J. Interferon Cytokine Res.* **21,** 1103–1115.
6. Stark, G., Keer, I., Agarwala, S., Williams, B., Silverman, R., and Schreiber, R. (1998) How cells respond to interferons. *Annu. Rev. Biochem.* **67,** 227–264.
7. Ding, G., Fischer, P., Boltz, R., Schmidt, J., Colaianne, J., Gough, A., et al. (1998) *J. Biol. Chem.* **273,** 28897–28905.
8. Forti, R., Schuffan, S., Davies, H., and Mitchell, W. (1986) Objective antiviral assay of the interferons by computer-assisted data collection and analysis. *Method Enzymol.* **119,** 533–540.
9. Singer, D. and Maguire, J. (1990) Regulation of the expression of class I MHC genes. *Crit. Rev. Immunol.* **10,** 235–257.
10. Ortaldo, J., Phillips, W., Wasserman, K., and Heberman, R. (1980) Effects of metabolic inhibitors on spontaneous and interferon boosted human natural killer cell activity. *J. Immunol.* **125,** 1839–1844.
11. Ortaldo, J., Mason, A., and Overton, R. (1986) Lymphokine-activated killer cells: Analysis of progenitors and effectors. *J. Exp. Med.* **164,** 1193–1198.

# 26

# Characterization of Interferon α2B Pegylated via Carboxyalkylation

## A Case Study

**David C. Wylie, Marcio Voloch, Seoju Lee, Yan-Hui Liu, Collette Cutler, Brittany Larkin, and Susan Cannon-Carlson**

## 1. Introduction

The efficacy of several therapeutic cytokines in vivo is often restricted by their clearance rate, which can be retarded by the covalent attachment of polyethylene glycol (PEG)—a large, highly soluble, nontoxic adduct *(1–5)*. However, pegylation is a double-edged sword. In addition to retarding clearance, it also frequently impairs the biological activity of the modified protein. Consequently, therapeutic biologicals are typically only minimally pegylated. For small cytokines, monopegylation provides a good balance. The most commonly used chemistry involves carboxyalkylation of nucleophilic nitrogens on the target protein. The product is a stable mixture of monopegylated proteins varying in the point of polymer attachment. The relative proportion of each positional isomer is based on protein properties, such as local-charge fields and solvent exposure. However, by modifying the pH of the pegylation reaction, the preferred target side chain can be shifted from ε amines (preferred at pH 10) to α amines and imidazoles (preferred at pH 6.5; *6*) which may have positive effects on biological activity *(7)*.

Pegylated proteins can be difficult to characterize, as the product is usually heterogeneous and often unstable under the harsh conditions used in traditional peptide-mapping analysis. This chapter describes: (1) chemical methods to discern and quantify the distribution of different classes of pegylation linkages in an heterogeneous population of positional isomers; (2) purification methods for resolving positional isomers; (3) a peptide mapping protocol that exploits the large hydrodynamic radius of hydrophilic polymers, such as PEG, to identify the site(s) of pegylation; and (4) a simplified model system for assessing the chemistry of specific peptide sequences for pegylation. To demonstrate how to generate and characterize different compositions of positional

From: *Methods in Molecular Biology, vol. 308: Therapeutic Proteins: Methods and Protocols*
Edited by: C. M. Smales and D. C. James © Humana Press Inc., Totowa, NJ

isomers from the same substrates, the pegylation and chemical analysis of mono-pegylated interferon (IFN) α2B and related synthetic peptides will be used as a case study.

## 2. Materials

1. Human recombinant α interferon (IFN) 2B.
2. Synthetic peptides homologous to selected domains of IFN. The sequences are: Ac-Phe-Gln-Lys-Ala-Glu-Thr-Ile-Pro-NH₂ (referred to as P1, identical to residues 47–54 of IFN) and Ac-Asp-Arg-His-Asp-Phe-Gly-Phe-Pro-Gln-NH₂ (referred to as P2, identical to residues 32–40 of IFN), Asp-Arg-His-Asp-Phe-Gly-Phe-Pro-Gln-Glu-Glu-Phe-Gly-Asn-Gln-Phe-Gln-Lys-NH₂ (referred to as P3).
3. PEG succinimidyl carbonate (SC), 12,000 average molecular weight (SC-PEG-12k) from Shearwater Polymers; *see* **Note 1**.
4. Slide-A-Lyzer (Pierce) dialysis kits (10-kDa cutoff).
5. Neutral pegylation reaction buffer: 50 m*M* sodium phosphate, pH 6.5 (*see* **Note 2**).
6. Basic pegylation reaction buffer: 30 m*M* sodium tetraborate, pH 10.0 (*see* **Note 2**).
7. Quenching solution: 1 *M* glycine.
8. Size exclusion-high-performance liquid chromatography (SE-HPLC) buffer: 100 m*M* sodium phosphate and 150 m*M* NaCl, pH 5.0.
9. Preparative SE column: Superdex 200 HiLoad Size Exclusion Column (2.6 cm × 60 cm, Amersham).
10. Analytical SE column: Superdex 200 HR 10/30 (Amersham).
11. Ion exchange (IE)-HPLC buffer A: 1.8 m*M* citric acid and 3.3 m*M* dibasic sodium phosphate, pH 5.3.
12. IE-HPLC buffer B: 30 m*M* trisodium citrate dihydrate and 70 m*M* sodium phosphate monobasic monohydrate, pH 6.0.
13. Preparative IE HPLC column: SP-5PW (2.15 cm × 15.0 cm, 13-µm particle size; Tosoh Bioscience).
14. Trifluoroacetic acid (TFA), sequencing grade.
15. Digestion buffer: 50 m*M* ammonium bicarbonate, freshly made.
16. Trypsin, TPCK-treated, sequencing grade (Roche Diagnostics).

## 3. Methods

The following methods describe the pegylation (**Subheading 3.1.**), purification (**Subheading 3.2.**), and characterization (**Subheading 3.3.**) of both monopegylated IFN and synthetic peptide homologs of selected IFN sequences. The latter provide both a simpler system for method development as well as homogeneous controls for use in experiments that analyze complex pegylated proteins. Characterization can be classified by quantitative chemical characterization, including acid hydrolysis (**Subheading 3.3.1.**) and cleavage with neutral hydroxylamine (**Subheading 3.3.2.**), which can be performed on both homogeneous and heterogeneous populations, and structural characterization of purified positional isomers by peptide mapping (**Subheading 3.3.3.**) and N-terminal sequence analysis (**Subheading 3.3.3.2.**).

### 3.1. Pegylation of IFN α2B and Related Synthetic Peptides

Cytokines, such as IFN, interact with cell receptor proteins via binding sites that encompass much of their exposed surface. Monopegylation provides the optimal balance between reduction of clearance rate and biological activity. The pegylation reaction should be adjusted to maximize monopegylation while minimizing populations of dipegylated and unmodified protein.

Proteins can be pegylated by various linking chemistries. Shearwater Polymers is a vendor that provides a wide range of linkers that will form stable bonds with nucleophilic nitrogens (*see* **Note 1**). SC-PEG-12k is highly selective for primary amino and imidazole nitrogens and reacts rapidly to form a stable bond.

The most reactive sites in IFN for pegylation by SC-PEG-12k are the imidazole nitrogens of histidine 34, the α amino nitrogen of cysteine 1, and the ε amino nitrogen of lysines 83, 121, 131, and 134. At neutral pH, imidazole and amine carboxylalkylation is strongly favored, and the preferred sites are His34, Cys1, and Lys134. Under basic conditions, pegylation occurs exclusively at ε amino nitrogens. These two reaction conditions are described in **Subheadings 3.1.1.** and **3.1.2.**

### 3.1.1. Reaction at Neutral pH

1. Dialyze IFN against neutral pegylation reaction buffer (1:1000 v/v, o/n), and adjust concentration to 5 mg/mL (260 μ*M*).
2. Reconstitute peptides P1 and P2 in neutral pegylation reaction buffer to a final concentration of 260 μ*M*.
3. In three parallel reactions, add SC-PEG-12k to a final concentration of 520 μ*M* (*see* **Note 3**), and incubate at 22°C for 70 min (*see* **Note 4**).
4. To minimize the formation of multipegylated species, quench each reaction by adding 1 *M* glycine to a concentration of 50 m*M* (*see* **Note 5**).

### 3.1.2. Reaction at Basic pH

1. Dialyze IFN against basic pegylation reaction buffer (1:1000 v/v, o/n), and adjust concentration to 5 mg/mL (260 μ*M*).
2. Reconstitute peptides P1 and P2 in neutral pegylation reaction buffer to a final concentration of 260 μ*M*.
3. In three parallel reactions, add SC-PEG-12k to a final concentration of 260 μ*M* (*see* **Note 3**), and incubate at 22°C for 70 min (*see* **Note 4**).
4. To minimize the formation of multipegylated species, quench each reaction by adding 1 *M* glycine to a concentration of 50 m*M* (*see* **Note 5**).

## 3.2. Purification of Monopegylated IFN Positional Isomers

The products of the IFN reactions are mixes of positional isomers whose composition changes with reaction pH. The desired monopegylated products can be resolved from dipegylated and unmodified IFN by several methods, including IE and SE chromatography. The following section describes resolving monopegylated IFN from other products by SE chromatography. Individual monopegylated positional isomers are purified from this mix by IE chromatography (**Subheading 3.2.3.**).

### 3.2.1. Purification of Monopegylated IFN

1. Dialyze quenched product against SE-HPLC buffer (1:1000 v/v, 5°C, o/n; *see* **Note 6**).
2. Resolve monopegylated IFN by SE chromatography on Superdex 200 HiLoad (2.6 cm × 60 cm) at 0.36 cm/min. Collect 1.33-mL fractions. The monopegylated fraction elutes between 0.47- and 0.52-column volumes.
3. Confirm that the isolated peak contains less than 5% multipegylated and unpegylated IFN by SE-HPLC.

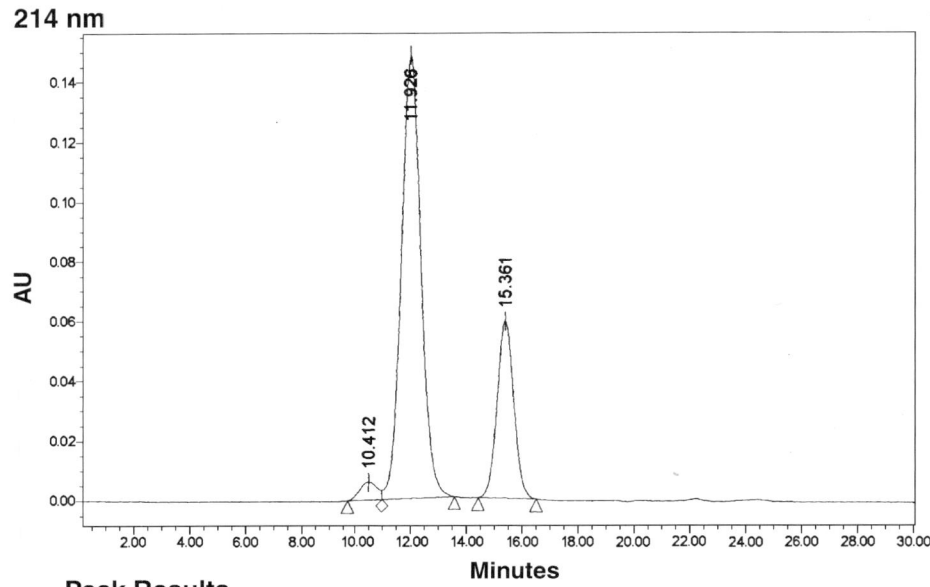

**214 nm**

**Peak Results**

Fig. 1. Representative SE-HPLC of pegylated IFN. Peak assignments based on SDS-PAGE and MALDI-TOF MS are: 10.412, PEG$_2$-IFN; 11.920, PEG-IFN; and 15.361, IFN.

### 3.2.2. Analytical SE-HPLC of Monopegylated IFN

SE-HPLC is performed on a Superdex 200 HR 10/30 (Amersham) column. SE-HPLC buffer is used throughout. Flow is maintained at 0.5 mL/min, and the eluate is monitored at 280 nm. A representative chromatogram is shown in **Fig. 1**.

### 3.2.3. Purification of Individual Monopegylated IFN Positional Isomers

Although the monopegylated IFN purified by SE is a heterogeneous mix of positional isomers, all components have the same net charge. However, because the positional isomers' attachment points differ, the uncharged polymer can mask different surfaces of IFN. This causes differences in surface electrostatic potential that can be exploited by IE-HPLC. These differences are subtle, and complete resolution of all positional isomers is difficult. For some, multiple cycles of IE-HPLC may be required to obtain sufficiently pure material.

Because some linkages formed by these reactions are acid-labile (*see* **Subheading 3.3.1.**), high-resolving separations performed in acid, such as reverse phase (RP)-HPLC, cannot be used.

1. Dialyze monopegylated IFN against IE-HPLC buffer A (1:1000 v/v, 5°C, o/n; *see* **Note 6**).
2. Concentrate to 2.2 mg/mL using a stirred cell (BioMaxx 10 membrane, Millipore, *see* **Note 6**).
3. Apply to the preparative IE-HPLC column equilibrated with IE-HPLC buffer A with a load of 20–65 µg protein/mL resin. Flow is maintained at 6 mL/min (1.65 cm/min) for this and all subsequent steps.

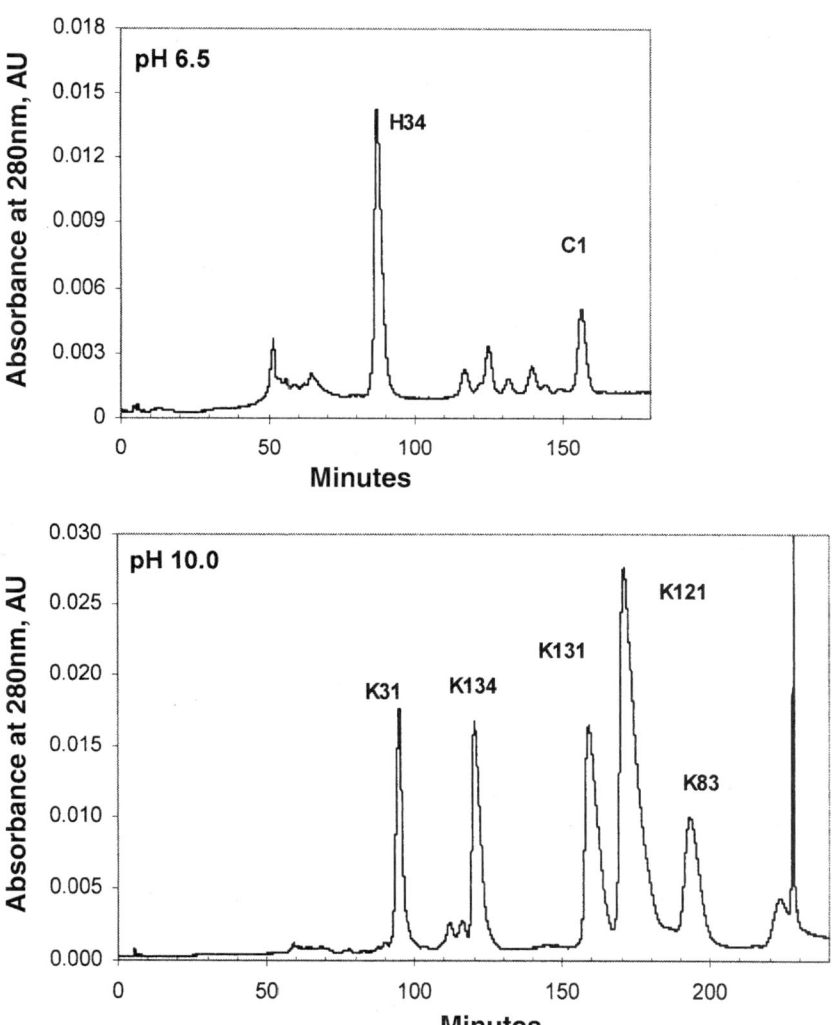

Fig. 2. Resolution of pegylation reaction products by IE-HPLC. Top panel: Products of pH 6.5 (top) and pH 10.0 (bottom) reactions are shown. Peak labels identify positional isomers by PEG-linked residue. Identifications assigned on the basis of peptide mapping. Gradients are described in **Subheading 3.2.3.**

4. Wash with 2 column volumes (cv) of IE-HPLC buffer A.
5. For the pH-6.5 pegylation products, elute with a pH gradient from 0% to 10% buffer B in 100 min, followed by 10–30% B in 80 min. For the pH-10 pegylation products, elute with a pH gradient from 0% to 14% IE-HPLC buffer B in 180 min, followed by 14–20% IE-HPLC buffer B in 40 min. To assure precise collection of peaks, collect fractions manually. Representative chromatograms are shown in **Fig. 2**.

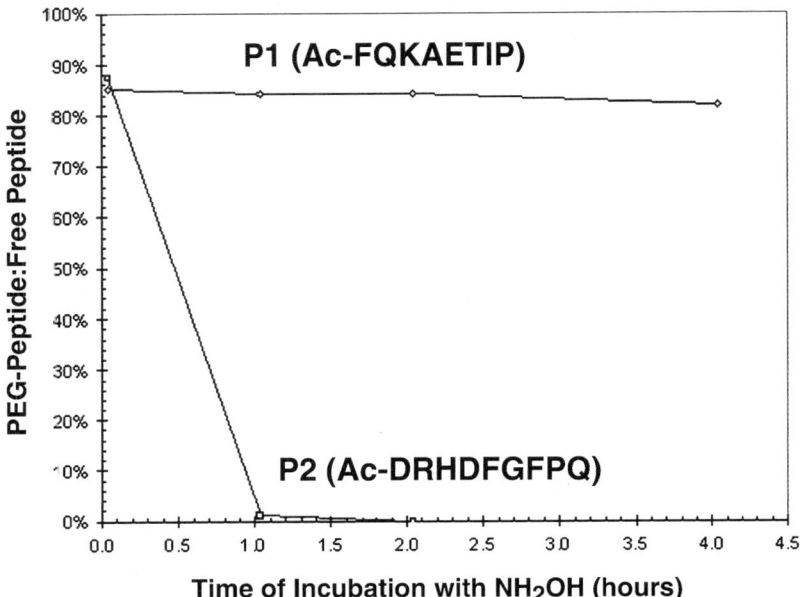

Fig. 3. Hydrolysis of PEG from synthetic PEG-peptides with neutral hydroxylamine. Pegylated synthetic peptides incubated in 400 m*M* hydroxylamine, pH 7.0, were assayed by SE-HPLC at selected times. The ratio of areas underpegylated and free-peptide peaks is plotted against the time of exposure to hydroxylamine.

## 3.3. Characterization of Monopegylated IFN

IFN pegylated at histidine is sensitive to hydrolysis in the presence of either strong acid or neutral hydroxylamine. The latter reagent is highly selective for carboxy-alkylated histidines. In contrast, IFN pegylated at amino groups is resistant to these chemical stresses. Therefore, the extent of depegylation by neutral hydroxylamine provides a quantitative measure of carboxylalkylated histidine residues within a mix of positional isomers.

### 3.3.1. Sensitivity to Acid Hydrolysis

1. Adjust an aliquot (100 µL) of monopegylated IFN to pH 2.1 with TFA and incubate at 22°C for 4 h.
2. Assay both untreated and treated monopegylated IFN by SE-HPLC.

### 3.3.2. Sensitivity to Hydrolysis With Hydroxylamine

The high selectivity of neutral hydroxylamine can be demonstrated with synthetic peptides. Peptide P1 (Ac-Phe-Gln-Lys-Ala-Glu-Thr-Ile-Pro-NH₂) contains lysine but no histidine, whereas the reverse is true for peptide P2 (Ac-Asp-Arg-His-Asp-Phe-Gly-Phe-Pro-Gln-NH₂). Only P2 is depegylated in the presence of neutral hydroxylamine (*see* **Fig. 3**).

1. Dilute monopegylated IFN to a concentration of 0.4 mg/mL into 600 m$M$ hydroxylamine, pH 7.0.
2. At 6 h, dialyze against 40 m$M$ sodium acetate, pH 5.0 (1:1000 v/v, o/n; *see* **Note 6**).
3. Assay both untreated and treated samples by SE-HPLC for monopegylated and unmodified IFN. The ratio of pegylated vs unmodified IFN indicates the degree of depegylation (*see* **Fig. 3**).

## 3.3.3. Peptide Mapping

An optimal peptide-mapping strategy for pegylated proteins should account for two unusual characteristics of these modified substrates. Some linkages, most notably those involving carboxylalkylated histidines, are labile under the conditions used in conventional peptide mapping. Partial hydrolysis can occur during digestion, and complete depegylation occurs when pegylated peptides are resolved by RP-HPLC in dilute TFA. Linkages associated with primary amines of IFN are stable under these stresses. Because of the potential for hydrolysis, peptide mapping is suitable for identification only. The distribution of histidine vs lysine linkage should be determined by the methods described previously (**Subheading 3.3.2.**).

A second unique characteristic of pegylated proteins, which can be exploited, is the unusually large mass of a nonbiological polymer that is linked at discrete point(s) to the protein. By definition, the complete proteolytic digestion of a pure monopegylated positional isomer yields only one pegylated peptide with its extraordinary mass. When unmodified IFN is digested completely with trypsin, the largest resulting peptide has a mass of 3.3 kDa. In contrast, the mass of a pegylated peptide from IFN monopegylated with SC-PEG-12k can be no less than 12 kDa. Consequently, SE-HPLC becomes a highly selective and gentle means to isolate pegylated peptides. This avoids RP-HPLC and the issue of possible depegylation during chromatography.

### 3.3.3.1. DIGESTION OF A PURIFIED MONOPEGYLATED IFN POSITIONAL ISOMER

1. Dialyze a purified positional isomer of monopegylated IFN (100 µg, 1 mg/mL) against digestion buffer (1:1000 v/v, o/n; *see* **Note 6**).
2. Add an aliquot (10% v/v) of trypsin dissolved in 1 m$M$ HCl (1 mg/mL). Incubate overnight at 37°C (*see* **Note 7**)
3. Reduce the resulting peptides with dithiothreitol (to 100 m$M$) and incubate for 1 h at ambient temperature.
4. Resolve pegylated peptide(s) by preparative SE-HPLC on Superdex 200 (1 × 20 cm, 0.5 mL/min). Collect 0.2-mL fractions. A representative chromatogram is shown in **Fig. 4**.
5. Peptides pegylated via carboxylalkylated histidines can be analyzed by matrix-assisted laser desorption/ionization time-of-flight mass spectrometry (MALDI-TOF MS). Owing to the heterogeneous nature of SC-PEG, untreated peptides yield a wide distribution of masses centered at a mass somewhat higher than 12,000 Daltons. Peptides treated with neutral hydroxylamine exhibit the mass of the unmodified peptide, and can be identified either by mass measurement in relation to a synthetic peptide standard or MS/MS (e.g., *see* **Fig. 5**).

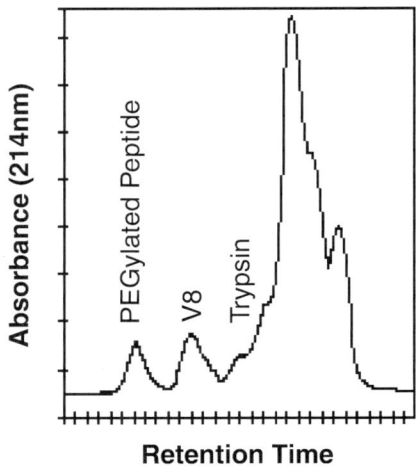

**Retention Time**

Fig. 4. Analytical SE-HPLC of tryptic digestion products. The absorbance (214 nm) of the eluate is plotted against retention time. Labeled peaks were identified by a combination of SDS-PAGE, MALDI-TOF MS, and SE-HPLC of pegylated synthetic peptides.

### 3.3.3.2. N-Terminal Sequence Analysis

The bond between PEG and the peptide is hydrolyzed during Edman degradation. Cycles containing pegylated Lys are characterized by the lack of phenylthiohydantoin (PTH)-Lys and presence of a small peak that elutes near PTH-Leu. Cycles with PTH-His show neither characteristic, but the cycle yields are often low. Unmodified cysteine is destroyed during Edman degradation, and no distinctive peaks associated with pegylated-α-amino-Cys have been observed.

Although N-terminal sequence analysis generally fails to identify the pegylated residue directly, the indirect evidence is usually sufficient. Under the conditions described in **Subheading 3.1.**, only histidines, lysines, and α amines form stable pegylation products, and only pegylated peptides are isolated using the outlined procedures. The presence of a missed tryptic cleavage site, as well as the absence of PTH-Lys and presence of PTH-Leu in cycles where PTH-Lys is expected, together indicate the presence of a pegylated lysine.

To prove linkage at the amino terminus (Cys1), both reduced and nonreduced pegylated peptides should be analyzed. A blocked sequence for reduced N-terminal

Fig. 5. *(opposite page)* MALDI-TOF MS of pegylated and depegylated peptides. Top panel: pegylated peptide obtained from tryptic digestion of the monopegylated IFN His$_{34}$ positional isoform (residues 32–49, Asp-Arg-His-Asp-Phe-Gly-Phe-Pro-Gln-Glu-Glu-Phe-Gly-Asn-Gln-Phe-Gln-Lys). Masses of the depegylated peptide (M + 1, formed during sample preparation with sinapinic acid or CHCA matrix), and both M + 1 and M + 2 ions of the pegylated peptide are listed. Middle panel: same peptide after treatment with neutral hydroxylamine. Bottom panel: spectrum of synthetic peptide P3, which corresponds to residues 32–49 of IFN.

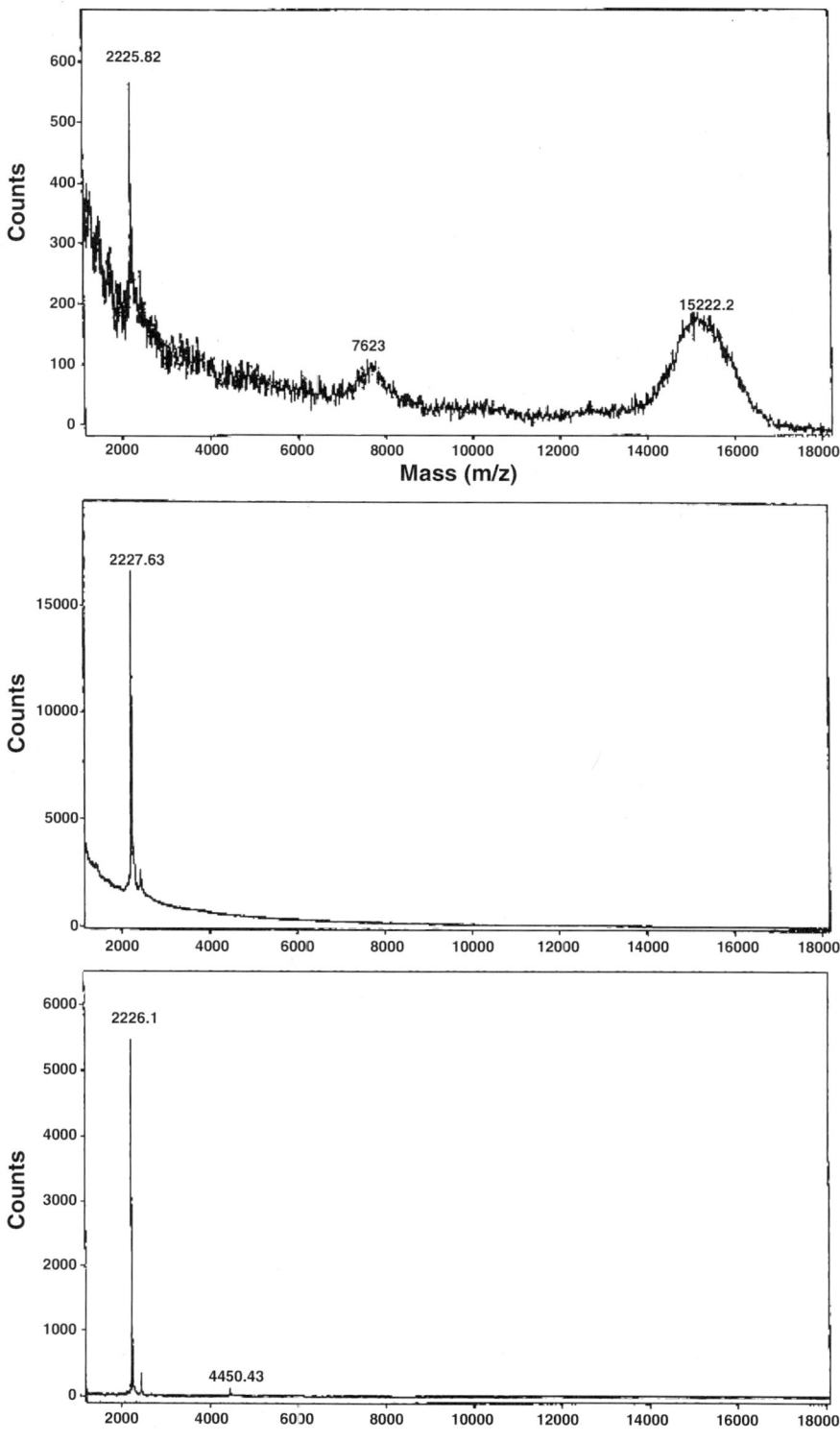

peptide indicates α amino pegylation. If this peptide is not reduced after digestion, then N-terminal sequencing will identify the peptide linked via disulfide bond to Cys1. With tryptic digestion, this peptide spans residues 84–110.

Histidine linkage is the most difficult to establish by N-terminal sequencing. In this case, because the products by Edman degradation of pegylated and unmodified histidine are identical, one must confirm that no other suitable substrate, e.g., lysine, is present in the isolated pegylated peptide. The observations that the peptide molecular weight is too large to represent an unmodified sequence, as well as that no suitable substrates are present other than histidine, represent the limit of proof that can be provided by N-terminal sequencing for this linkage. These observations should be corroborated by hydroxylamine sensitivity.

1. Confirm by SE-HPLC that pegylated peptides collected previously (**Subheading 3.3.3.1., step 4**) are free of undigested protein.
2. Estimate peptide concentration of the isolated pegylated peptide pool. Assume that digestion was complete and recovery across the procedure was 90%. Therefore, peptide concentration is 90% of the substrate mass divided by the pegylated peptide pool volume.
3. Absorb an aliquot (1 nmol) of pegylated peptide onto polyvinylidene fluoride (or equivalent) membrane, and analyze by automated N-terminal sequencing. Results obtained by Agilent sequencers using the Peptide Chemistry kit have given clear results in terms of quantitation.

Identify the sequence and determine cycles, which should yield PTH-Lys. Establish whether PTH-Lys was recovered in the expected amount. If not, determine whether a small peak was eluted near PTH-Leu. A chromatogram showing the products of Edman degradation on a cycle in which pegylated lysine is degraded is shown in **Fig. 6.**

## 4. Notes

1. Shearwater produces a number of linkers that carboxylate primary amines and histidines. Those that form urethane bonds include SC-PEG and PEG benzyltriazole carbonate. Others that form amide bonds include $PEG_2$ succinimide, PEG succinimidyl propionate, and PEG succinimidyl carboxymethylate. The preferred amino acid side chain for all of these reagents can be adjusted by pegylation reaction pH.
2. The pegylation reaction buffer must be devoid of potential reactants, such as amino groups, which eliminates the Tris family of buffers.
3. The reaction rate follows pseudo first-order kinetics and can be effectively controlled by protein and linker concentrations. The solubility of some linkers possessing longer polymer chains (e.g., SC-PEG-30k) can be somewhat limiting.
4. The progress of the pegylation reaction can be monitored by the SE-HPLC assay described in **Subheading 3.2.2.** The reaction rate increases with pH and can be controlled by linker concentration and temperature. It appears to be pseudo-first rate with respect to linker, but this is a simplification. There are several simultaneous reactions:

$$IFN + SC\text{-}PEG \rightarrow PEG\text{-}IFN + N\text{-hydroxysuccinate}$$
$$SC\text{-}PEG + H_2O \rightarrow PEG + N\text{-hydroxysuccinate}$$
$$PEG\text{-}IFN \rightarrow PEG + IFN$$
$$PEG\text{-}IFN + SC\text{-}PEG \rightarrow PEG_2\text{-}IFN + N\text{-hydroxysuccinate}$$

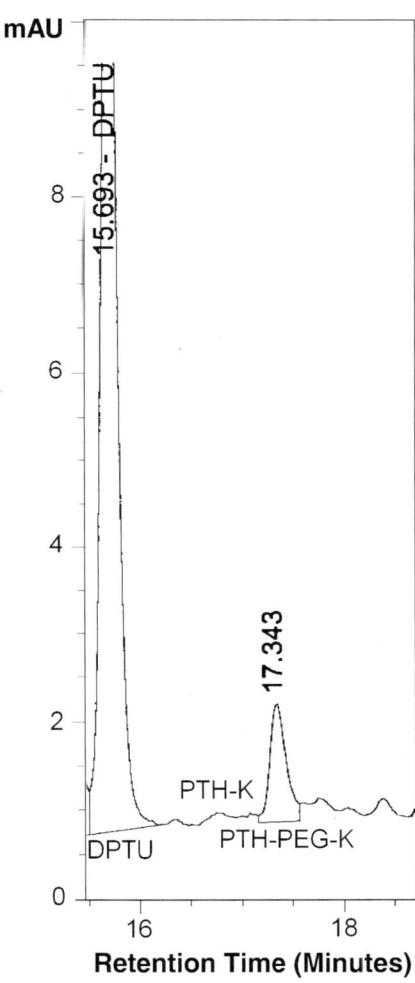

Fig. 6. RP-HPLC of products of Edman degradation (cycle 1). The peptide is Lys-Tyr-Phe-Gln-Arg (residues 121–125 of IFN), pegylated at $Lys_{121}$. Labels indicate the absence of a peak at the expected retention time of PTH-Lys (17.1 min) and the presence of a peak unique (17.3 min) to this cycle and the next (reflects 4% carryover). This peak at 17.3 min is routinely observed in cycles resulting from the degradation of pegylated lysine.

The reaction rate constant for each amine and imidazole side chain is determined by intrinsic reactivity as well as local electrostatic potential and conformation. A typical reaction will have both dipegylated and unmodified protein present at the time of quenching.

5. The amino group of glycine reacts with unbound linker to end the reaction. The products of both reactions are stable in a range of pH 4.5–7.5.

6. Although a dialysis cassette with a molecular weight cut-off of 10,000 Dalton is specified, cassettes with considerably higher molecular weight cutoffs (≤50,000 Dalton) can be substituted. Unlike biological polymers, PEG tends to form a loose rapidly changing

conformation with a considerable hydrodynamic radius. Consequently during dialysis, the PEG polymer behaves as if it is much larger than its actual mass.

7. Some depegylation has been observed during incubation with trypsin. Loss during a mock incubation at 37°C in the absence of trypsin was minimal.

## Acknowledgments

The authors thank Drs. Birenda Pramanik and Urooj Mirza for advice on MS, Drs. Steven Tindall and Russell Wolz for advice on sequence analysis, Roland Mengisen for providing IFN, and Julie Obara and George Soder for sodium dodecyl sulfate polyacrylamide gel electrophoresis (SDS-PAGE) analysis.

## References

1. Inada, Y., Furukawa, M., Sasaki, H., Kodera, Y., Hiroto, M., Nishimura, H., and Matsushima, A. (1995) Biomedical and biotechnological applications of PEG- and PM-modified proteins. *Trends Biotechnol.* **13,** 86–91.

2. Asselin, B. L. (1999) The three asparaginases. Comparative pharmacology and optimal use in childhood leukemia. *Adv. Exp. Med. Biol.* **457,** 621–629.

3. Veronese, F. M., Largajolli, R., Boccu, E., Benassi, C. A., and Schiavon, O. (1985) Surface modification of proteins. Activation of monomethoxy-polyethylene glycols by phenylchloroformates and modification of ribonuclease and superoxide dismutase. *Appl. Biochem. Biotechnol.* **11,** 141–152.

4. Teicher, B. A., Ara, G., Herbst, R., Takeuchi, H., Keyes, S., and Northey, D. (1997) PEG-Hemoglobin: effects on tumor oxygenation and response to chemotherapy. *In Vivo (Attiki)* **11,** 301–311.

5. Kinstler, O. B., Brems, D. N., Lauren, S. L., Paige, A. G., Hamburger, J. B., and Treuheit, M. J. (1996) Characterization and stability of *N*-terminally PEGylated rhG-CSF. *Pharm. Res.* **13,** 996–1002.

6. Wylie, D. C., Voloch, M., Lee, S., Liu, Y.-H., Cannon-Carlson, S., Cutler, C., and Pramanik, B. (2001) *Pharm. Res.* **18,** 1354–1360.

7. Grace, M. J., Lee, S., Bradshaw, S., Liu, M., Lee, S., Chapman, J., et al. (2004) Site of pegylation and PEG molecule size attenuate interferon-alpha anti-viral and anti-proliferative activities through the JAK/STAT signaling pathway. *J. Bio. Chem.* (PMID 15596441).

# 27

## Quantifying Recombinant Proteins and Their Degradation Products Using SDS-PAGE and Scanning Laser Densitometry

### Aaron P. Miles and Allan Saul

## 1. Introduction

Numerous methods are widely used for the quantitation of proteins and their degradation products. Gel electrophoresis, followed by scanning laser densitometry, offers an accurate, sensitive, and reproducible technique that can be used at any stage during the production and purification of recombinant proteins. This technique offers many advantages over others for two main reasons: (1) proteolytic degradation is common among these proteins, and this method is able to accurately quantify degradation in both minute and large amounts; and (2) the common constituents of fermentation broths, refolding tanks, and purification processes do not interfere with the analysis. Analyses of a recombinant malarial protein at two different stages of purity will be used as examples to illustrate this method.

## 2. Materials

1. SDS-PAGE equipment.
2. Coomassie blue-staining solution: 40% methanol, 10% glacial acetic acid, and two PhastGel Blue R tablets (Amersham Biosciences, Piscataway, NJ) per liter.
3. Coomassie blue-destaining solution: 5% methanol and 10% glacial acetic acid.
4. 0.45-µm filtration unit.
5. Concentrated, diafiltered test protein (Pvs25H) fermentation culture supernatant.
6. Purified test protein (Pvs25H) solution.
7. Personal Densitometer SI with ImageQuant software (Molecular Dynamics Inc., Sunnyvale, CA; 2).

## 3. Methods

These methods outline the (1) running of SDS-PAGE gels, followed by visualization with Coomassie blue and (2) scanning and analyzing the gels using laser densitometry and ImageQuant software.

From: *Methods in Molecular Biology, vol. 308: Therapeutic Proteins: Methods and Protocols*
Edited by: C. M. Smales and D. C. James © Humana Press Inc., Totowa, NJ

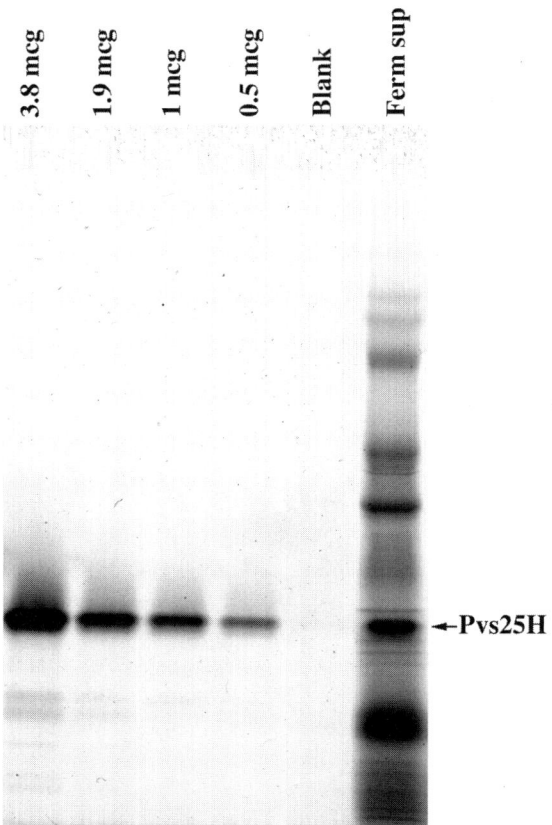

Fig. 1. SDS-PAGE of Pvs25H fermentation culture supernatant with purified Pvs25H standards. Nonreduced 4–20% Tris-glycine SDS-PAGE Coomassie blue-stained gel showing purified Pvs25H standards and concentrated, diafiltered fermentation culture supernatant. Purified Pvs25H was loaded in the indicated amounts for comparison to Pvs25H in the fermentation supernatant.

## 3.1. SDS-PAGE

1. Load the gel with samples as desired, and run according to the manufacturer's instructions.
2. Filter the staining solution using a 0.45-µm filtration unit (*see* **Note 1**).
3. Add staining solution to the gel, bring to a boil in a microwave oven, and agitate for 30 min (*see* **Notes 2** and **3**).
4. Pour off staining solution, add destaining solution, bring to a boil, and agitate for 15 min (*see* **Note 4**).
5. Change destaining solution, bring to a boil again, and continue agitation further for 15 min.
6. Change destaining solution once more, bring to a boil, and continue destaining until gel background is clear or almost clear (~10 min). Take care not to destain the gel too much, as minor protein bands can disappear (*see* **Note 5**).
7. Place the gel in water until scanned. **Figure 1** shows a nonreduced 4–20% Tris-glycine SDS-PAGE gel that contains standards of purified Pvs25H (a recombinant form of the

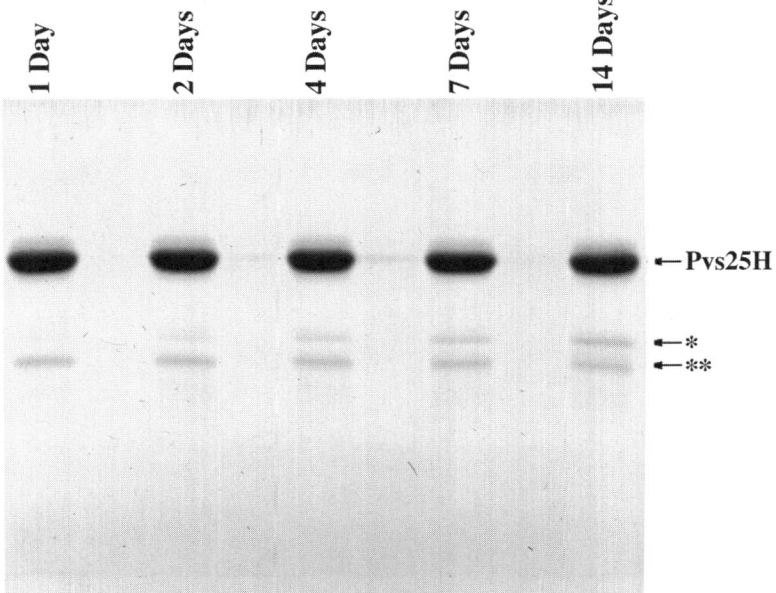

Fig. 2. SDS-PAGE of purified Pvs25H stored at 37°C. Reduced 15% SDS-PAGE Coomassie blue-stained gel showing samples taken at the indicated time points during storage at 37°C. All protein loads are 2 μg, separated by blank lanes. *A minor band accumulating with time whose sequence corresponds to cleavage at a single site within Pvs25H. **A minor band (Pvs25H-derived) present in final purified Pvs25H, not increasing upon storage. This type of gel was chosen to better resolve Pvs25H, which migrates diffusely in a gradient gel (*see* **Fig. 1**).

25-kDa ookinete surface protein from the malaria parasite *Plasmodium vivax*; *1*) and a concentrated, diafiltered fermentation culture supernatant of the same protein. **Figure 2** shows a reduced 15% acrylamide gel loaded with the same purified protein that was subjected to 37°C storage for up to 14 d to monitor stability (*see* **Note 6**). This protein is heavily disulfide-bonded and must be reduced to visualize the degradation that occurs over time. These gels will serve as examples of two applications of scanning laser densitometry: (1) monitoring low-level degradation in a highly purified protein and (2) quantifying yields of a protein recovered in a fermentation broth not yet subject to purification.

## 3.2. Laser Densitometry

1. Scan the gel according to manufacturer's instructions, ensuring that the lanes of the gel run straight from the top to the bottom of the scanning bed.
2. Draw an encompassing line for analysis down each lane of interest, and widen the line appropriately. Include enough length from top to bottom in the analyzed area of each lane and enough width to envelope all visible protein bands. **Figures 3** and **4** illustrate this using the same gels as shown in **Figs. 2** and **1**, respectively.
3. Create a line graph for each line drawn. Examples are shown in **Fig. 5**, showing representations (in optical density units) of purified Pvs25H stored for 1 (A) and 14 (B) d at 37°C, and **Fig. 6**, which illustrates Pvs25H in concentrated, diafiltered fermentation culture

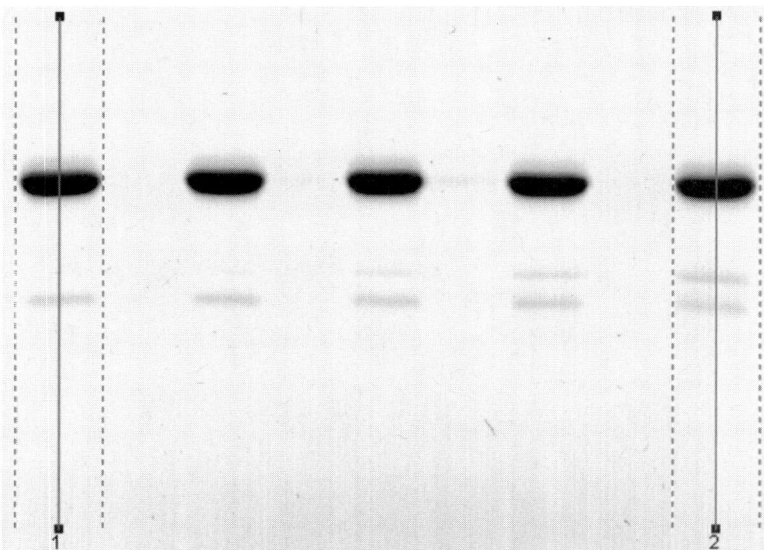

Fig. 3. Gel from **Fig. 2** with lines drawn in ImageQuant for analysis. Shown is the same gel as in **Fig. 2**, with lines drawn down two lanes using ImageQuant. These lines define the length and width of the areas being analyzed. Lengths and widths are adjusted to encompass all protein in each lane. These lines give rise to the line graphs shown in **Fig. 5**.

    supernatant. Note in **Fig. 5B**, the appearance of a minor degradation product over time (peak 2) and its corresponding band in **Fig. 2**.

4. Adjust all relevant settings for the most accurate representation of the baseline signal. Options for this within the ImageQuant software may include "automatic" (the baseline is detected by the software based on manually adjusted settings to appropriately identify peaks), "lowest point" (the most transparent part of the lane is taken as the baseline), and point-by-point manipulation. Other software programs available allow more customized identification and tuning of the baseline and peaks (e.g., PeakFit). It is then usually necessary to redefine the peaks by hand, rather than letting the software do it to fine-tune where each peak begins and ends (*see* **Note 7**). Once the peaks are defined, they are integrated, i.e., the area under each peak is calculated by the software. In this way, one quantifies the proportion of one or several peaks in relation to all others and determines both the total protein in the lane and the percent of each protein of interest.

5. Using the areas generated from the integration of each Pvs25H standard peak, create a graph of area counts vs known protein load using the appropriate software (e.g., Excel, SigmaPlot). Determine the linear range, fit a line for the data points of that range, and obtain an equation for the line. Using this method, the amount of Pvs25H in the fermentation supernatant lane of **Fig. 1** was found to be 1.49 µg. This corresponds to a total production of 500 mg from a 60-L fermentation (*see* **Note 8**). Likewise, the amount of full-length (nondegraded) purified Pvs25H was found to be decreasing with time, whereas a minor degradation product was increasing. The purity of the main band dropped from 95.3% at day 1 to 94.5% by day 14. The degradation rate is thus found to be 0.06%/d at 37°C.

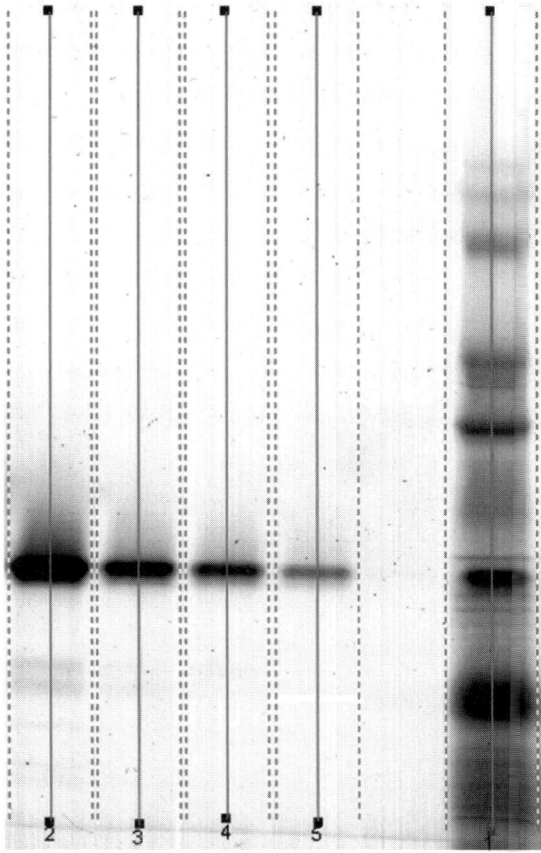

Fig. 4. Gel from **Fig. 1** with lines drawn in ImageQuant for analysis. Shown is the same gel as in **Fig. 1**, with lines drawn down each lane containing protein. The line drawn down the fermentation supernatant lane gives rise to the line graph shown in **Fig 6**.

## 4. Notes

1. It is especially important when using this method to filter the Coomassie blue stain prior to use. This eliminates large particles that otherwise stick to the gel and give rise to high-background readings that interfere with quantitation. Filtered stain leads to much cleaner, more transparent gels. To accelerate the filtering process and to help prevent clogging of the filter, use a conveniently sized filtration unit (e.g., 1 L), and pour the solution in slowly, allowing it to filter through, before pouring more. If too much solution is poured in all at once, large particles quickly block the filter. It is not necessary to use a 0.22-µm filter, which will become clogged more quickly.
2. Microwaving allows staining and destaining to proceed much more quickly than at room temperature, with no loss in sensitivity.
3. Standard staining and destaining times must be adhered to, especially when gel-to-gel comparisons are to be made. Staining for less than 30 min can result in nonuniformity in

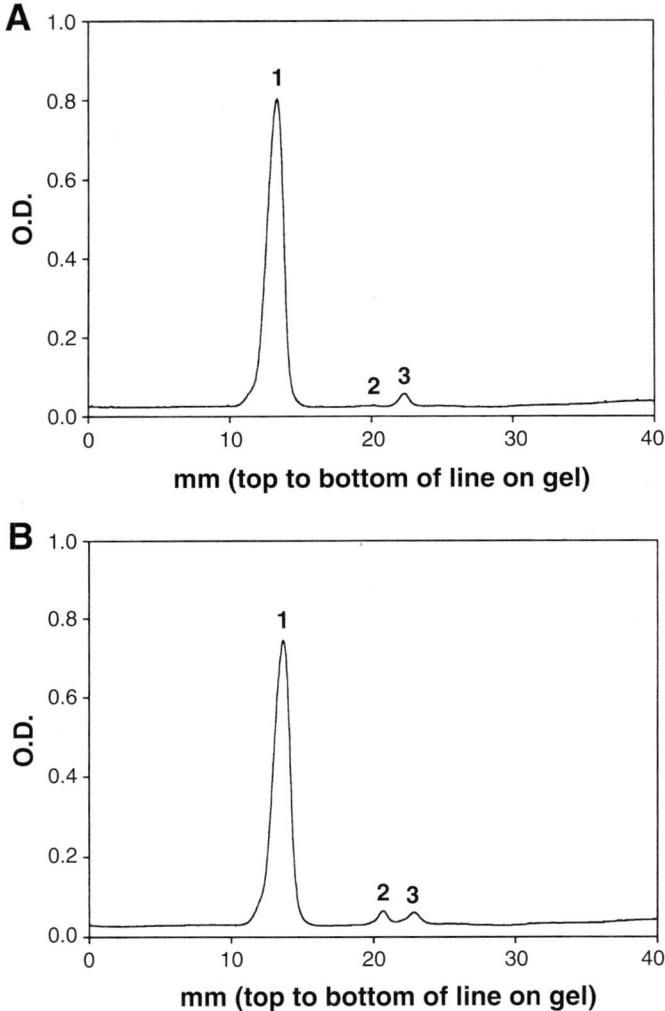

Fig. 5. Line graphs of Pvs25H at 37°C, days 1 and 14. The line graphs for the lines drawn in **Fig. 3** are shown, with (**A**) showing day 1 and (**B**) showing day 14. These graphical representations of the optical density vs millimeters are measured by the scanner at each point from the top to the bottom of the drawn lines. In these cases, the baselines were detected by simply making minor adjustments to the peak detection parameters (baselines not shown). Note the increase in the area of peak 2, a minor degradation product, from day 1 to day 14.

   the binding of the Coomassie blue dye to the proteins, especially in bands containing large amounts of protein. This can, in turn, cause some underestimation in the analysis of those bands.
4. A sponge or paper towel can be added to soak up the stain from the gel, but this is not necessary, especially when microwaving, as the gel will quickly destain.

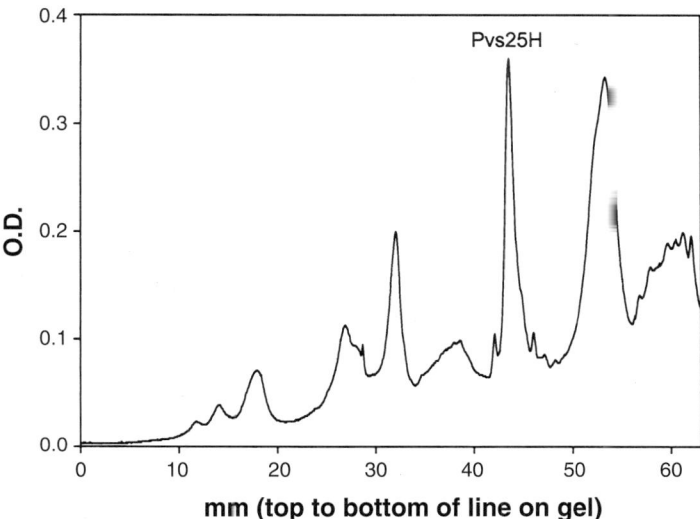

Fig. 6. Line graph of fermentation culture supernatant. The line graph for the line drawn in **Fig. 4** is shown. In this case, the baseline (not shown) was taken as the lowest point (point of lowest optical density; *see* **Note 9**). The peak corresponding to Pvs25H is indicated. The line graphs for the standard lanes are not shown to avoid redundancy.

5. It is unnecessary to destain the gel until the background is completely clear. A small amount of background stain will generally not be detected as a protein peak or peaks by the software and will become part of the baseline. Higher acrylamide concentrations in gradient gels require slightly more time to destain, but with more extended destaining times, protein bands (especially low-molecular-weight bands) can disappear. This will affect the quantitation of yield and purity and should be avoided.

6. After purification of Pvs25H, it was determined that this protein is better resolved on a 15% SDS-PAGE gel than on a 4–20% gradient gel (compare **Fig. 1**, 3.8-µg lane, to **Fig. 2**, 1-d lane). This serves to illustrate an important point to consider when performing densitometry: protein quantitation is more difficult to perform when the protein runs as a diffuse band. Consequently, the error in this technique increases with poorer resolution. Although a good estimate can still be achieved, peak identification can be problematic. If necessary, different gel types should be tested to find the one that best resolves each protein of interest.

7. Focus is often on the proportion of one peak (the protein of interest) in relation to all others. Defining the peaks manually allows the peak of interest to be isolated while, e.g., treating all other peaks as one. A percent yield is then easily calculated.

8. Scanning laser densitometry was found to be especially useful during the production of Pvs25H. This is because Pvs25H is produced as two populations of molecules that migrate well separated by SDS-PAGE (*see* large diffuse band beneath Pvs25H in **Fig. 1**). However, they do copurify during the Ni-NTA capture step. Traditional protein assays (e.g., BCA, Lowry, $A_{280}$) performed at this point would overestimate the yield of the target protein (labeled Pvs25H in **Fig. 1**), because these other methods are incapable of distinguishing between the two forms of Pvs25H that are produced. Densitometry makes

this distinction, with the added benefit that no capture step is required before an estimate of yield is given.

9. When analyzing a sample that is not yet purified, determining the baseline is usually more difficult than when working with purified material. Each sample must be analyzed on a case-by-case basis as appropriate.

## Acknowledgments

The authors wish to thank Anthony W. Stowers, Richard L. Shimp, Vu Nguyen, Jacqueline Glen, Brian Henderson, and Jin Wang for their contributions in producing and purifying the recombinant Pvs25H protein, and Holly McClellan for expert technical assistance.

## References

1. Hisaeda, H., Stowers, A. W., Tsuboi, T., Collins, W. E., Sattabongkot, J. S., Suwanabun, N., et al. (2000) Antibodies to malaria vaccine candidates Pvs25 and Pvs28 completely block the ability of *Plasmodium vivax* to infect mosquitoes. *Infect. Immun.* **68,** 6618–6623.
2. ImageQuant User's Guide Version 5.0. Molecular Dynamics, Inc., 1998.

# 28

## Extraction and Denaturing Gel Electrophoretic Methodology for the Analysis of Yeast Proteins

### Kenneth J. O'Callaghan, Lee J. Byrne, and Mick F. Tuite

### 1. Introduction

Gel electrophoretic analysis of proteins is a well-established method for separating, quantifying, identifying, comparing, and characterizing proteins. This form of analysis consists of a diverse class of methods, with new approaches and improvements being frequently added, including computer software dedicated to the analysis of protein gels. Protein electrophoresis involves four principal steps: protein harvesting, in-gel separation, visualization of separated proteins, and data collection and analysis. Proteins are extremely heterogeneous, and no single method will recover all cellular proteins equally effectively. In its native state, a protein may be soluble, membrane-, or organelle-bound and may show posttranslational and pH-dependent modifications.

Protein species that are difficult to isolate using the general methods described next will require careful selection of extraction buffer, detergent, and other reagents. Several volumes of literature are helpful for troubleshooting difficult proteins (1,2), and there are many published protocols for modifying and improving the harvesting (3), separation (4), visualization (5), and computer analysis (6) steps of protein electrophoresis. It is important to distinguish between approaches that attempt to preserve the native activity of proteins and those that involve some degree of denaturation. Researchers wishing to study protein complexes, polymers, or aggregates will likely attempt to avoid electrophoretic reagents that tend to disrupt protein–protein interactions. For example, semidenaturing detergent-agarose gel electrophoresis (SDD-AGE) was developed recently to visualize intact prion protein polymers in yeast (7). Comparing the behavior of a protein sample between nondenaturing and denaturing gel conditions can also be a useful analytical method (8).

This chapter offers researchers with little or no experience a compilation of integrated protocols to perform a complete denaturing electrophoretic study of either

From: *Methods in Molecular Biology, vol. 308: Therapeutic Proteins: Methods and Protocols*
Edited by: C. M. Smales and D. C. James © Humana Press Inc., Totowa, NJ

native or recombinant protein. Mass spectrometry—the method of choice to confirm protein species identity—is not described here (however, *see* **refs. 9–11**).

## 2. Materials

### 2.1. Protein Extraction From Yeast With Glass Beads in Nondenaturing Buffer for One-Dimensional Sodium Dodecyl Sulfate-Polyacrylamide Gel Electrophoresis (1D SDS-PAGE)

1. YEPD medium: 2% (w/v) glucose, 1% (w/v) bactopeptone, and 1% (w/v) yeast extract.
2. YNB medium: 0.67% (w/v) yeast nitrogen base without amino acids, 2% (w/v) glucose, and amino acids as required for selection by nutritional complementation.
3. Nondenaturing lysis buffer, pH 6.8, contains 25 m*M* Tris-HCl, 150 m*M* NaCl, 10% (v/v) glycerol, 2% (v/v) 100 m*M* phenylmethylsulfonylfluoride (PMSF) stock solution (dissolved in isopropanol), and 1X complete protease inhibitors (Roche). The latter will confer a 1 m*M* final concentration of EDTA (a metalloprotease inhibitor), but additional EDTA (up to 5 m*M* final) may be added if considerable metalloprotease activity is suspected. Complete protease inhibitor can also be used at 2X concentration.
4. Glass beads of approx 0.4–0.5-mm diameter obtained from either BDH or Sigma; Sigma offers beads previously acid-washed.

### 2.2. Bradford Assay for Protein Concentration Determination

1. Coomassie Plus Protein Assay Reagent (Pierce) or other similar ready-to-use Coomassie Blue G-250 reagent.
2. 2 mg/mL Commercial bovine serum albumin (BSA) preparation.
3. Nondenaturing lysis buffer (*see* **Subheading 2.1.**, **item 3**).

### 2.3. 1D SDS-PAGE

1. Complementary gel-caster and electrophoretic tank and power pack.
2. 1 *M* Tris-HCl, pH 8.8.
3. 1 *M* Tris-HCl, pH 6.8.
4. Commercially prepared 30% acrylamide (37.5:1 acrylamide:bisacrylamide).
5. *N,N,N',N'*-tetramethylethylenediamine (TEMED).
6. 20% (w/v) SDS.
7. 10% Ammonium persulfate.
8. Isobutanol.
9. High-quality reverse osmosis (RO) or double-distilled (dd) $H_2O$.
10. Whatman no. 1 filter paper.
11. 5X gel-loading buffer, pH 6.8: 60 m*M* Tris-HCl, 25% (v/v) glycerol, 2% (w/v) SDS, 10% (v/v) 2-mercaptoethanol, and 0.1% (w/v) bromophenol blue.
12. 10X SDS-PAGE running buffer (Tris-glycine-SDS [TGS]): 30.3 g Tris, 144 g glycine, and 10 g SDS; made up to 1000 mL with dd or RO $H_2O$. Dilute to 1X before use.
13. Coomassie Blue gel stain: 0.1% (w/v) Coomassie Brilliant Blue R-250, 45% (v/v) methanol, and 10% (v/v) acetic acid.
14. Destaining Gel: 10% (v/v) methanol and 10% (v/v) acetic acid.

### 2.4. Western Blot Analysis

1. Semidry transfer blotter and power pack.
2. 0.45 μm Hybond C Super nitrocellulose enhanced chemiluminescence (ECL) membrane (Amersham Biosciences).

3. Whatman 3MM paper.
4. Transfer buffer: 2.12 g Tris, 11.26 g glycine, 15% (v/v) methanol, and dd or RO $H_2O$ to 1000 mL.
5. Blocking solution: 5% (w/v) dried milk (e.g., Marvel, Cadbury) in 1X phosphate-buffered saline (PBS).
6. 10X PBS stock: 80 g NaCl, 2 g $KH_2PO_4$, 29 g $Na_2HPO_4 \cdot 12H_2O$, or 11.5 g anhydrous $Na_2HPO_4$. Dilute before use.
7. PA solution: 2% (w/v) dried milk in 1X PBS.
8. Wash solution: 0.05% (v/v) Tween-20 in 1X PBS.
9. SA solution: 1% (w/v) dried milk in 1X PBS.
10. ECL solution: solution 1, pH 8.5: 1 mL 250 m$M$ 3-aminophthalhydrazide (luminol; dissolved in dimethylsulfoxide [DMSO]), 0.44 mL 90 m$M$ coumaric acid (dissolved in DMSO), 10 mL 1 $M$ Tris-HCl with dd or RO $H_2O$ up to 100 mL. Solution 2, pH 8.5: 64 µL 30% $H_2O_2$, 10 mL 1 $M$ Tris-HCl with $H_2O$ up to 100 mL. Combine both solutions 1:1 just before use.
11. Hyperfilm ECL (Amersham Biosciences).
12. Hyperprocessor™ (e.g., Amersham Biosciences) to develop exposed film.
13. Primary antibodies and peroxidase-conjugated secondary antibodies.

## 2.5. Protein Extraction by Beadbeating in Denaturing Buffer for 2D SDS-PAGE

1. YEPD medium (*see* **Subheading 2.1.**, **item 1**).
2. YNB medium (*see* **Subheading 2.1.**, **item 2**).
3. 1.5-mL Screw-cap microfuge tubes.
4. Glass beads (*see* **Subheading 2.1.**, **item 4**).
5. Lysis solution 1: 0.90 g urea, 0.38 g thiourea, 12.6 µL 0.2 $M$ EDTA, 200 µL 10X protease inhibitor solution (*see* **Subheading 2.5.**, **item 6**), and 840 µL sterile dd or RO $H_2O$, which will make 2 mL.
6. 10X protease inhibitor solution: 0.05 g dithiothreitol (DTT), 140 µL 1 $M$ Tris buffer, pH 10.8, two tablets of Complete Mini EDTA-free protease inhibitor cocktail (Roche), 9.8 µL 1 mg/mL pepstatin (dissolved in MeOH), and 330.2 µL dd or RO $H_2O$.
7. Lysis solution 2: 250 µL dd or RO $H_2O$, 0.2 g 3-(3-cholamidopropyl-dimethylammonio)-1-propane sulfonate (CHAPS), 500 µL autoclaved glycerol, and 250 µL 40% Carrier Ampholytes (e.g., BDH Electran Resolyte, cat. no. 44338 2X). This will make 1 mL.
8. Tubeshaker (e.g., Eppendorf Thermomixer Comfort).

## 2.6. The First Dimension: Isoelectric Focusing

1. Rehydration buffer: 9.5 $M$ urea, 2% (w/v) CHAPS, 1% (w/v) DTT, 40 m$M$ Tris, 0.8% (v/v) pharmalyte carriers, and a few grains of bromophenol blue per 25 mL buffer.
2. IEF ceramic strip holders (Amersham Biosciences).
3. 24-cm Immobiline DryStrips pH 4–7 (Amersham Biosciences).
4. Immobiline DryStrip Cover Fluid.
5. Ettan IPGphor IEF unit.
6. 10-mL Disposable pipets approx 30 cm in length.
7. 2-mL Round-bottom microfuge tubes.

## 2.7. The Second Dimension: Separation by Size

1. Equilibration buffer: 6.7 mL 1.5 $M$ Tris-HCl, pH 8.8, 72.1 g urea, 69 mL glycerol, 4.0 g SDS, several grains of bromophenol blue, and dd$H_2O$ made up to 200 mL.

DTT equilibration buffer is 400 mg DTT in 40 mL buffer; iodoacetamide equilibration buffer is 1 g iodoacetamide in 40 mL buffer.

2. Sealing agarose: 0.5% (w/v) agarose in TGS buffer (*see* **Subheading 2.7.**, **item 3**).
3. TGS electrophoresis buffer: 15.1 g Tris base, 72.1 g glycine, 5.0 g SDS, and $H_2O$ made up to 5 L. The pH does not need to be checked.
4. Electrophoresis tank and power supply unit (e.g., Amersham Ettan DALT II system).

## 2.8. Silver Staining of SDS-PAGE Gels

1. Silver fix: 40% (v/v) ethanol and 10% (v/v) glacial acetic acid in $ddH_2O$.
2. Sensitizer: 300 mL ethanol, 2 g sodium thiosulfate, 68 g sodium acetate, and $H_2O$ made up to 1 L.
3. Silver stain: 2.5 g silver nitrate, 0.4 mL formaldehyde, and $H_2O$ made up to 1 L.
4. Developer: 25 g sodium carbonate, 0.2 mL formaldehyde, and $H_2O$ made up to 1 L.
5. Stop solution: 14.6 g disodium-EDTA and $H_2O$ made up to 1 L.

## 2.9. Overview of Gel image Analysis Using Computer Software

1. Recent generation computer with high-data transfer rates and a large (preferably ≥128 MB) random access memory capacity.
2. Internet connectivity.
3. Zip file decompressing program (e.g., WinZip).

## 3. Methods

The following general methods explain how to obtain total protein extracts from *Saccharomyces cerevisiae* and separate them by SDS-PAGE. A method for 1D SDS-PAGE is followed by a method to perform a Western blot, whereas a method for 2D SDS-PAGE is followed by a silver-staining protocol and suggestions for gel analysis using readily available computer software.

## 3.1. Protein Extraction With Glass Beads in Nondenaturing Buffer for 1D SDS-PAGE

1. Produce a starter culture of the yeast strain, growing cells in YEPD medium beyond exponential phase (>1.0 optical density [$OD_{600}$] for selective minimal media). YNB medium may be used as a selective minimal medium for cells requiring nutritional marker complementation, e.g., plasmid transformed cells.
2. Inoculate fresh medium with cells (*see* **Note 1**) from the starter culture, and grow until the $OD_{600}$ of the culture reaches approx 0.5 (*see* **Note 2**).
3. Collect 10 $OD_{600}$ units of cells (*see* **Note 3**) for each sample, centrifuge (1500$g$) for 5 min, wash with sterile $H_2O$, centrifuge, discard aqueous supernatant, and place cell pellet on ice.
4. Thoroughly resuspend the cell pellet in 500 µL cold nondenaturing lysis buffer, and transfer the entire sample to a tapered-end 1.5-mL microfuge tube on ice. Add cold acid-washed (*see* **Note 4**) glass beads (approx 0.5-mm diameter) until their upper edge rests just below the sample meniscus.
5. Close the lid, and vortex for 5 × 30 s at high speed, returning tube to ice for 30 s between each vortexing.
6. Clean the tapered tip of the tube with ethanol, rinse with sterile water, open the lid, and carefully pierce the tube tip with a hot narrow (≤0.5-mm diameter) needle (*see* **Note 5**).

**Table 1**
**Preparation of BSA Standards for a Standard Protein Curve**

| Concentration (μg/mL) | 2 mg/mL BSA (μL) | Lysis buffer (μL) |
|---|---|---|
| 0 | 0 | 100 |
| 100 | 5 | 95 |
| 200 | 10 | 90 |
| 300 | 15 | 85 |
| 400 | 20 | 80 |
| 500 | 25 | 75 |

Quickly place the sample tube inside a new sterile tube, close the lid of the sample tube, and briefly centrifuge (at 4°C) the two tubes together to obtain lysate without glass beads. Keep the lysates cold at all times unless they are being heated for electrophoresis.

7. Do not store protein solutions and lysates at room temperature. Lysates may be finally stored at 4°C (may degrade within hours or days), −20°C with 50% glycerol (stable for months), or −80°C (stable for years; *see* **Note 6**).

## 3.2. Bradford Assay for Protein Concentration Determination

The Bradford assay *(12)* is a colorimetric test in which the absorbance values of protein samples are compared against a standard curve produced with known amounts of a commercial protein solution, such as BSA. The Bradford assay can give false results under certain conditions (*see* **Note 7**).

1. Prepare the materials for a standard curve (**Table 1**) using 2 mg/mL BSA solution (Pierce) and the nondenaturing lysis buffer described previously (*see* **Subheading 2.1.**).
2. Dilute an aliquot of each sample (but not the standards), adding 10 μL lysate to 90 μL nondenaturing buffer. Place 50 μL of each diluted sample and each standard into a labeled cuvet, add 1.5 mL Coomassie reagent, and read the absorbance at $OD_{595}$.
3. Construct a standard curve from the BSA data, and find the points on the curve that match the absorbance values of the samples. The concentration of each sample will be 10 times that indicated on the curve, because the samples were previously diluted 1–9.

## 3.3. 1D SDS-PAGE

SDS-PAGE separates proteins in denaturing conditions in which most protein complexes are dissociated, and polypeptides pass through the gel matrix chiefly as elongate particles. SDS polyacrylamide gels may be made manually or purchased already made (e.g., Novex and Gradipore) as precast units, the latter offering greater reproducibility. SDS polyacrylamide gels usually consist of a long section of resolving (i.e., separating) gel overlaid with a shorter (1.5–3 cm) stacking gel containing the wells into which protein samples are loaded.

### 3.3.1. Preparing the Gel

To make gels, complementary gel-casting and gel-running apparatus are required (e.g., the Bio-Rad Protean system).

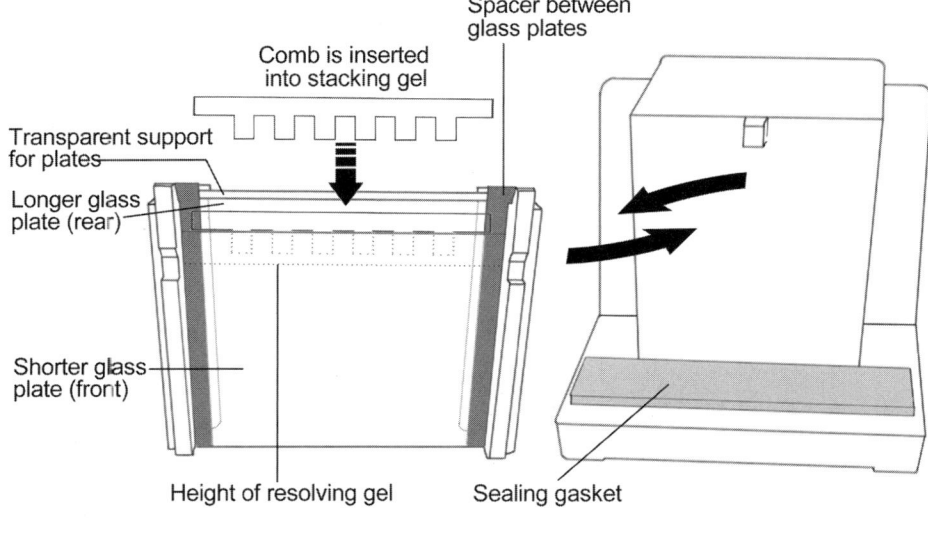

**Gel holder unit**                        **Gel casting unit**

Fig. 1. Polyacrylamide gels are cast between two glass plates secured in a gel-holder unit clipped firmly into a gel-casting frame (above similar to early Bio-Rad Protean system). Spacers determine the gel thickness, and a gasket prevents acrylamide leakage during polymerization. Following polymerization of the stacking gel, the holder unit is unclipped from the casting frame, the comb is removed, and the entire unit is transferred into the electrophoresis tank (not shown). The translucent gels are difficult to see: marking one of the glass plates with an indelible felt-tip pen can help visualize both the lower edge of the stacking gel and the position of the wells for subsequent loading.

1. Assemble the gel-casting unit and gel holder. This consists of a frame into which a removable gel-holder module may be locked. This module holds firmly two glass plates separated by thin (0.75 or 1.0 mm) spacers, where the gel material is sandwiched between the plates (**Fig. 1**). Ensure that the glass plates are clamped securely in the gel holder and that the lower edges of the plates are tightly locked down onto the sealing gasket of the casting unit. Mark the position where the top of the resolving gel will be by drawing a line with an indelible felt-tip pen on one of the glass plates.

2. Make up the acrylamide solution (**Table 2**) for the resolving gel according to the protein sizes for separation (*see* **Note 8**), and carefully dispense the acrylamide through a 5-mL disposable pipet until it reaches the pen mark on the glass plate (*see* **Note 9**).

3. Immediately overlay the resolving gel with a 1-cm layer of $H_2O$-saturated isobutanol (*see* **Note 10**). After about 1 h, the resolving gel will be polymerized.

4. Make the stacking gel solution (**Table 2**).

5. Pour off the isobutanol, and rinse the top of the resolving gel several times with sterile water. Dry any $H_2O$ from the upper inner surfaces of the glass plates by carefully inserting absorbent filter paper; try not to disturb the top of the resolving gel. Pipet an approx 1-cm layer of the stacking gel solution onto the top of the resolving gel.

**Table 2**
**Ingredients for a 10% and 12.5% Polyacrylamide Minigel**

| Reagent | Resolving gel | | Stacking gel |
| | 10% | 12.5% | 4% |
| --- | --- | --- | --- |
| H$_2$O | 1.69 mL | 1.31 mL | 3.05 mL |
| 1 *M* Tris-HCl, pH 8.8 | 1.25 mL | 1.25 mL | – |
| 1 *M* Tris-HCl, pH 6.8 | – | – | 1.25 mL |
| 20% (w/v) SDS | 25.0 µL | 25.0 µL | 50.0 µL |
| 30% (w/v) Acrylamide[a] | 1.5 mL | 1.88 mL | 0.65 mL |
| TEMED[b] | 2.5 µL | 2.5 µL | 5.0 µL |
| 10% (w/v) APS[c] | 25.0 µL | 25.0 µL | 25.0 µL |

[a]37.5:1 ratio of acrylamide:*N*,*N*′-methylene-bisacrylamide.
[b]TEMED, *N*,*N*,*N*′,*N*′-tetramethyl-ethylene-diamine.
[c]APS, ammonium persulfate (store at 4°C for 1 d only).

6. Push the gel comb into the stacking gel solution without trapping air bubbles (*see* **Note 11**). After about 45 min, remove the comb and rinse the wells thoroughly with sterile H$_2$O. The gel is ready to use.

### 3.3.2. Preparing the Sample in Denaturing Loading Buffer

1. Mix the correct volume of protein lysate with 5X gel-loading buffer containing SDS and 2-mercaptoethanol (*see* **Note 12**). About 15–40 µg total protein may be loaded into each well, depending on the dimensions of the well and gel.
2. Denaturing and SDS binding of the protein is only complete after heating. Heat samples (including any molecular weight markers requiring heating) at 100°C for 5–10 min and centrifuge (1100*g*) at room temperature for a few seconds.

### 3.3.3. Loading the Samples and Running the Gel

1. Place the gel holder and gel(s) into a compatible electrophoresis tank (*see* **Note 13**), and fill the upper tank chamber (i.e., interior of the gel-holder unit) completely with 1X SDS-PAGE running buffer (TGS).
2. Load the samples (*see* **Note 14**) with a Hamilton syringe or pipettor equipped with a gel-loading tip.
3. Place the lid onto the electrophoresis tank, connect the wires to a power pack, and set the unit to run at 200 V for about 45 min (*see* **Note 15**). Turn off the power supply when the dye front reaches the bottom of the gel. Remove the gel holder and each individual gel assembly and, using a gel trimming tool, lever open the glass plates and remove the gel (*see* **Note 16**).

### 3.3.4. Coomassie Blue Staining of Gel Proteins

1. Place the gel in Coomassie Blue gel stain at room temperature for 15 min with gentle agitation.
2. Pour off the stain, add destain, and leave at room temperature overnight with gentle agitation (*see* **Note 17**).
3. Obtain an image of the gel, if required, using dedicated equipment or by recording a digital image of the gel using a flatbed digital scanner (**Fig. 2**).

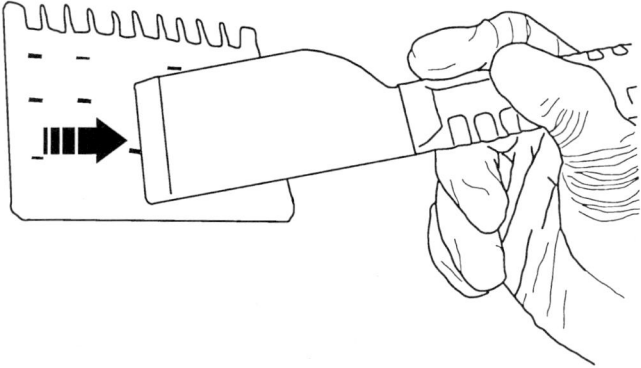

Fig. 2. To obtain a digital image of a stained gel, place the gel with gloved hands onto a flatbed scanner, draw a gel-trimming tool carefully across the upper surface of the gel to remove air bubbles trapped beneath, and scan in transmission mode. Images are best saved in a globally recognized format (e.g., tiff., jpeg.).

## 3.4. Western Blot Analysis

A Western blot is an immuno-method in which all protein bands from a gel are transferred onto a membrane, which is then probed with antibodies for the presence of specific protein species *(13)*. Two identical gels are often electrophoresed for comparison, one being stained with Coomassie and the other blotted.

### 3.4.1. Transferring Gel Proteins to a Nitrocellulose Membrane

1. Cut out one piece of 0.45-µm nitrocellulose membrane (e.g., Hybond C Super, Amersham Biosciences) to the dimensions of the gel, and soak in transfer buffer for 30 min.
2. Cut out six pieces of Whatman 3MM paper, each to the same size as the gel, and soak in transfer buffer.
3. Stack onto the lower plate of the semidry blotting apparatus (anode) three pieces of the 3MM paper, nitrocellulose membrane, gel, and three pieces of 3MM paper (**Fig. 3**). Press down firmly, preferably using a rubber roller, to remove any air bubbles from the gel stack.
4. Attach the top plate (cathode) of the apparatus, connect the wires to the correct polarity sockets of the power pack, and apply 10 V for 1 h (*see* **Note 18**).
5. Carefully peel the upper filter papers and gel back from one corner, keeping them together, and check that the blue dye from the loading buffer has transferred to the membrane (*see* **Note 19**). If transfer has occurred, remove the membrane carefully, cutting one corner diagonally as a visual aid to recall its orientation with respect to the gel.

### 3.4.2. Chemiluminescent Detection of Specific Proteins

1. Gently agitate the membrane for 1 h at room temperature in a clean plastic box or dish containing blocking solution (*see* **Note 20**).
2. Pour off and replace the blocking solution with the primary antibody (*see* **Note 21**) in PA solution. Agitate the membrane gently overnight at 4°C (e.g., in a cold room).
3. Pour off the PA solution, and rinse the membrane for 3 × 15 min in wash solution.

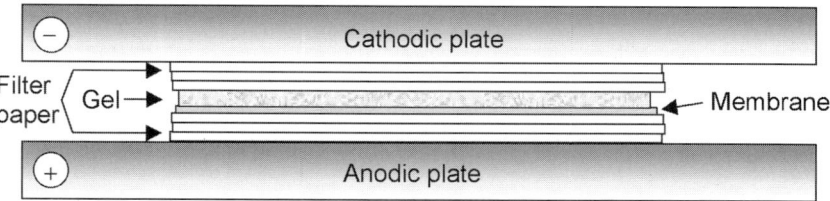

Fig. 3. Assembling a Western Blot in semidry-blotting apparatus. A nitrocellulose membrane sits beneath the gel, and both are sandwiched between filter paper soaked in transfer buffer. An applied current mobilizes proteins downward onto the membrane.

4. Add peroxidase-conjugated secondary antibody (*see* **Note 22**) in SA solution for 1 h at room temperature.
5. Wash the membrane three times as described previously (*see* **Note 23**).
6. Replace the final wash solution with enough ECL solution (*see* **Note 24**) to cover the membrane, incubate for about 1 min, lift the membrane to pour off excess solution, and expose a sheet of Hyperfilm ECL (Amersham) to the membrane (*see* **Note 25**). Place the films into the hyperprocessor. A signal—evident as a dark band on the film—should appear following exposure for 5 s to 5 min, depending on the amount of target protein and antibody titer.

## 3.5. Protein Extraction by Beadbeating in Denaturing Buffer for 2D SDS-PAGE

1. Inoculate 100 mL (*see* **Note 26**) fresh YEPD or YNB medium with cells from a stationary culture (*see* **Note 1**). Incubate the cells at 30°C with 200 rpm, shaking until the culture reaches $OD_{600}$ 0.5–0.6.
2. Centrifuge at 1500$g$ for 2 min at 4°C, and discard the supernatant. Wash the pelleted cells with ice-cold sterile water (*see* **Note 27**), centrifuge as before, and discard supernatant.
3. Prepare autoclaved screw-cap 1.5-mL microfuge tubes, placing 250 mg of acid-washed and autoclaved 0.5-mm diameter glass beads into each tube.
4. Add 160 µL lysis solution 1 (*see* **Note 28**) to the cell pellet, vortex briefly, and transfer entire suspension to a screw-cap tube with beads.
5. Bead-beat for 6 × 30 s at medium speed, returning the tube to ice for 1 min between each beating.
6. Add 40 µL lysis solution 2 to each sample, and vortex briefly. Place tubes into a tubeshaker or similar (e.g., Eppendorf Thermomixer Comfort) and agitate for 6 × 30 s at high speed, with tubes placed on ice for 1 min between each shaking (*see* **Note 29**).
7. Centrifuge at 16,100$g$ for 10 min at 4°C, carefully remove the supernatant above the beads with a pipet, and transfer the lysate to a clean tube.
8. Quantitate the protein concentration for each sample (*see* **Note 7**).
9. Store extracts at –80°C, giving consideration to **Note 6**.

## 3.6. The First Dimension: Isoelectric Focusing (See Note 30)

A charged particle in an electrical field will move toward the electrical pole having opposite charge. The isoelectric point (pI) is the pH value at which a protein (including all its amino acid side chains) has an overall charge of zero. When the pH of a supporting matrix or suspension buffer is below the pI, the protein will be positively charged. When the pH is above the pI, the protein will be negatively charged. We describe here the positional focusing of protein species in a 24-cm DryStrip gel containing an immobilized pH gradient. These commercially produced strips are usually available with immobilized pH 4–7 and pH 3–10 gradients.

1. Add 120 µg sample protein to a sterile 1.5-mL microfuge tube and make up to 450 µL with rehydration buffer.
2. Pipet the 450 µL as a 20-cm line lengthways along the center of a clean 25-cm IPGphor Strip Holder. Carefully remove the transparent cover film from a 24-cm DryStrip (*see* **Note 31**), and place the strip gel downwards onto the sample: avoid trapping air bubbles. Overlay the strip with about 1.5 mL Immobiline DryStrip Cover Fluid, and place the lid onto the Strip Holder.
3. Position the IPGphor Strip Holders onto the baseplate of an Ettan IPGphor IEF unit, placing the ⊕-ends of gels in the ⊕ zone of the plate, and focus the gels by applying a voltage in a stepwise manner (*see* **Note 32**).
4. Record the length of the focusing in volts per hour, remove DryStrips, carefully place each in a labeled 10-mL disposable pipet (*see* **Note 33**), and store at –70°C.

### 3.7. The Second Dimension: Separation by Size

We do not describe the casting of large (approx 26 cm wide) gels for 2D SDS-PAGE, because various forms of equipment for multiple gel casting exist, and the principles are the same as for 1D SDS-PAGE gels. The ingredients have been described previously (**Table 2**) and can be increased proportionally according to the number of gels required (*see* **Note 34**). Isoelectric focusing is performed in the presence of urea, whereas separation by size involves the addition of SDS (*see* **Note 12**). We have routinely separated over 2000 native yeast proteins from one sample using **Subheadings 3.6.** and **3.7.** (**Fig. 5**).

1. Remove prefocused Drystrips from –70°C, and thaw briefly.
2. Remove microfuge tubes from the top (nontapered) ends of pipets, add about 10 mL DTT equilibration buffer to each, replace tubes tightly, lay pipets horizontally and agitate gently on a rocker or shaker for 15 min at room temperature.
3. Remove the microfuge tubes from both ends of the pipet over a sink, and allow the buffer to run off. Add 10 mL iodoacetamide equilibration buffer and agitate gently for 15 min.
4. Remove microfuge tubes from the tapered ends of the pipets, allow buffer to run off and rinse once with dd or RO $H_2O$ (*see* **Note 35**).
5. Lift out the strip with forceps, slide it into the recess above the resolving gel, and carefully push it into place with a spare spacer (*see* **Note 36**). Quickly overlay 2–3 mL sealing agarose onto the strip, and tap the strip onto the top of the resolving gel, checking that no bubbles are trapped between the two gels (**Fig. 4**). Allow the agarose to set.
6. Fill the electrophoresis tank to the specified mark with 1X TGS, place the gels in the tank, and close the lid (*see* **Note 37**).
7. Turn on the power supply, and run the gels for 30 min at 5 W per gel, then for about 7 h at 8 W per gel (or until the dye front reaches the bottom of the gels).
8. Remove the gels from the tank, and fix or stain them immediately.

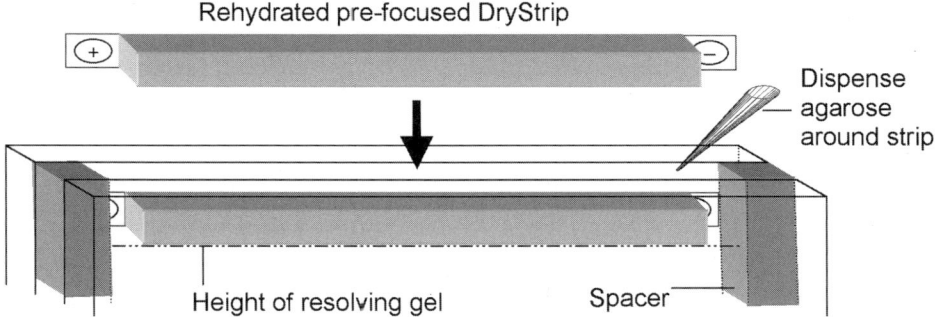

Fig. 4. Inserting a prefocused DryStrip between glass plates containing a resolving gel for 2D separation by molecular mass. Forceps are used to carefully place the strip into the well above the resolving gel. Cooled molten agarose is pipetted onto the strip, covering it, and a spare spacer is used to gently tap the strip downward, dislodging any air bubbles trapped beneath.

## 3.8. Silver Staining of SDS-PAGE Gels

This procedure will work equally well for 1D or 2D SDS-PAGE gels.

1. Place the gel into Silver fix with gentle agitation for at least 1 h (*see* **Note 38**).
2. Replace the fixative with sensitizer and agitate for 30 min (*see* **Note 39**).
3. Replace the sensitizer with dd cr RO $H_2O$, washing the gels for $3 \times 5$ min.
4. Replace the $H_2O$ with silver stain and agitate for about 20 min.
5. Replace the silver stain (*see* **Note 40**) with $H_2O$, washing the gels for $2 \times 1$ min.
6. Replace the $H_2O$ with developer (*see* **Note 41**) for 1 min, pour off and replace with fresh developer (*see* **Note 42**). Observe staining.
7. When the gel is stained sufficiently (*see* **Note 43**), replace the developer with stop solution, and agitate for about 10 min.
8. Replace the solution with $H_2O$, and record an image of each gel (**Fig. 5**).

## 3.9. Obtaining Freely Available Computer Software for Gel Image Analysis

Bioinformatics for 2D protein studies is now well underway with much currently available to assist protein researchers. Several high-quality programs are offered in the public domain. Other programs, e.g., 2D Phoretix (Amersham Biosciences), provide many advanced features, in addition to those found in the programs described next. However, a great deal can be achieved without purchasing a dedicated program. Numerous 2D resources and links are also available through the websites described here.

### 3.9.1. Image J

Image J, made available in the public domain by the US National Institutes of Health (NIH), is a general image analysis program that constantly evolves with new specialized "plug-ins" (*see* **Note 44**), being created by administrators and advanced users. Although Image J does not offer tools for a complete 2D analysis, there are many functions specifically for gel analysis.

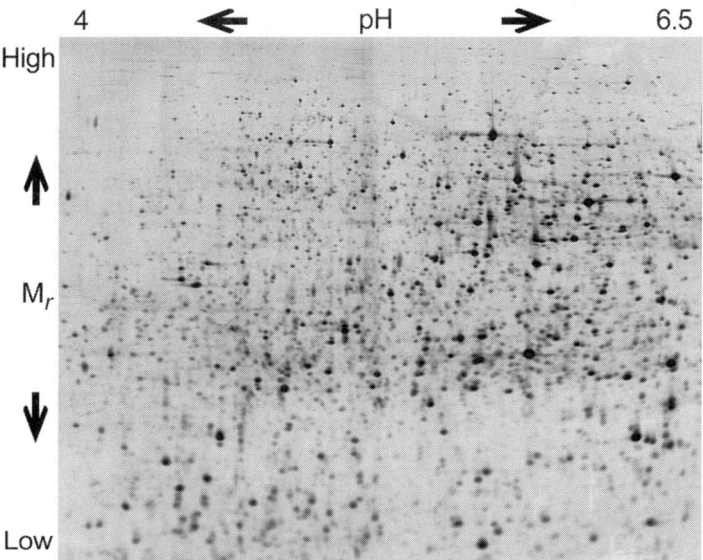

Fig. 5. Silver-stained 26-cm gel showing separation of approx 2000 proteins from *Saccharomyces cerevisiae*. Separation by charge was followed by separation by size, as described in the text.

1. Visit the US National Institutes of Health website for Image J (rsb.info.nih.gov/ij/index.html).
2. Download the latest version available that is compatible with your computer (e.g., Macintosh or PC).
3. Download the examples and documentation provided.
4. Unpack the program and other files using a Zip decompression program.
5. Open your image(s), and utilize any of the gel analysis functions, e.g., lane-finding and quantifying spot volumes.

## 3.9.2. GelScape

GelScape *(14)* is a recently developed, freely available Internet based tool for manipulating, comparing, annotating, and spot marking 1D and 2D gel images. Protein spot intensity may also be quantified. The organization of the webpage is intuitively understandable.

1. Visit the University of Alberta Gelscape download page (www.gelscape.ualberta.ca:8080/htm/index.html).
2. Download the Java Virtual Machine and tutorial files if required.
3. Sign up for a GelScape account.
4. Enter the location of your gel image (gray-scale jpg. or gif.) in the "browse" text box, or click on one of the sample gel images. Alternatively, load more than one image by initiating an image comparison (e.g., between two gels).
5. Click on the command buttons to initiate and perform your analysis.

### 3.9.3. Flicker Web

Flicker Web *(15)* is a Web-based program for comparing and annotating 2D gels and is made available through the US National Cancer Institute (NCI). It is possible to compare different images, even those from different Internet sources. For example, one may input any URL, directing the program to a webpage containing a gel image, and compare that image with any other, including your own. Flicker Web can stretch or shrink images to account for any distortions in one or both gel images. The program allows the user to "flicker" back and forth between any two images, which are exactly superimposed. Because annotated gels may also be viewed, any gel may be compared against a fully annotated 2D gel from a source, such as the Swiss-2D PAGE website (ca.expasy.org/ch2d/), helping to identify protein spots of interest. Moreover, an open-access downloadable form of the Flicker program for use on a stand-alone PC was recently made available and can be obtained as outlined here.

1. Visit the NCI website offering the open-source version of Flicker (open2dprot.source forge.net/Flicker/).
2. Download the open-source program and Java Virtual Machine using the installer available on the download page.
3. Open the program, click on the left image, and open any gel file (jpeg., gif., or tiff.). Click on the right image, and open another gel file. Click on the Flicker radio button, and the comparison will begin. The righthand side of the program window offers various settings for adjusting the image. A helpful manual is available at open2dprot.sourceforge. net//Flicker/indexRefMan.html. This website can also be activated through the Help menu in the program window.

## 4. Notes

1. Overgrown cultures (>0.7 $OD_{600}$) are not generally suitable for protein extraction. Therefore, grow several cultures, each inoculated with a different volume (e.g., 10, 25, 50, or 75 μL) of starter culture, subsequently harvesting cells from the culture with the appropriate $OD_{600}$.
2. For comparative analysis of proteins from different cultures, all cell samples must be harvested at identical growth densities to ensure parity of cell physiology. Many protein species are synthesized at different rates, or not at all, in stationary-phase cells.
3. One $OD_{600}$ unit is the quantity of cells, which, if resuspended in 1.0 mL medium, would give an $OD_{600}$ of 1.0. For example, ×10 $OD_{600}$ units would be obtained from 20 mL cell culture with an $OD_{600}$ of 0.5.
4. To acid-wash beads, place them into a glass flask, add concentrated hydrochloric or nitric acid, and stir several times during a 1-h incubation at room temperature in a fume hood. Pour off the acid, rinse beads exhaustively in tap water, rinse once in $ddH_2O$, autoclave, and dry beads in a drying oven.
5. Pushing the hot needle into a closed tube will cause a momentary pressure increase that discharges the sample lysate out of the hole in the tube.
6. Glycerol (50%) stops lysates from freezing at –20°C; thus, protein can be easily removed for analysis many times without thawing the sample. Frozen protein is susceptible to damage; it cannot be freeze-thawed repeatedly. Although storage at –80°C is desirable, such samples should be stored in small aliquots for once-only removal.

7. The Bradford assay is unstable in the presence of some salts, urea, and a host of other conditions, including many denaturing buffers used to prepare protein lysate. A modified Bradford assay *(16)* can yield better results with some samples. Noninterfering assays for protein determination (e.g., Calbiochem's Non-Interfering Protein Assay Kit, cat. no. 488250), which give a stable result over a wide range of lysate conditions, are designed to offer much greater accuracy and reproducibility.

8. Acrylamide gels (10% and 12.5%) should separate proteins in the approximate ranges of 20–90 kDa and 14–70 kDa, respectively.

9. Try to dispense the acrylamide in one continuous action, making sure no air bubbles are trapped in the gel matrix; tap the plates to shift air bubbles.

10. Pipet sterile-filtered 10% $H_2O$ (v/v) into isobutanol, and shake vigorously. Water that does not enter the isobutanol matrix will fall to the bottom of the vessel. Draw $H_2O$-saturated isobutanol from the upper layer.

11. Some insert the comb at an angle to allow bubbles to escape; bubbles can sometimes be dislodged by tapping the glass plates or moving the comb up and down several times. Do not allow air bubbles to remain directly beneath any well.

12. SDS confers a strong negative charge to proteins, surmounting their relatively insignificant intrinsic charge. Therefore, almost all SDS–protein complexes have similar charge densities and are separated chiefly according to polypeptide size. 2-Mercaptoethanol is a reducing agent employed to reduce disulfide bridges between proteins. These covalent bridges prevent proteins from adopting the random-coil conformation required for size separation. Urea, used in some other protein lysate buffers (often at 5–9 *M*), partially dissociates protein complexes without affecting the charge of individual polypeptides, permitting them to be separated according to their intrinsic charges.

13. The upper chamber (i.e., gel holder) of an SDS-PAGE electrophoresis tank (except in buffer recirculating models) must be sealed. This is achieved by either casting two gels into the gel holder, each forming one of the two sides of the chamber, or by using one gel and one of the dams supplied with the equipment. (A dam serves only to seal the chamber). Before placing the gel holder in the tank, draw a short line aligned exactly with the floor of each well on one of the glass plates to aid the subsequent loading of samples.

14. Carefully place the tip or syringe into the well, using the short line previously drawn opposite the base of the well for guidance. Slowly dispense the sample without expelling air bubbles. If only a few wells are utilized for samples, dispensing 1X loading buffer into the adjacent outer wells may help prevent edge effects during electrophoresis. Make a record of the name of each sample and which well it occupied.

15. A minigel (~10 × 10 cm) may be run for approx 45 min at 200 V, whereas large gels will take longer. Gels of 20-cm length or more may be run overnight (at ~50 V for 16 h or more). Some researchers prefer to use relatively low voltages, because the excess heat that can accompany high voltages can cause gel distortion.

16. Gels (especially those of 1.0-mm thickness) can be removed from glass plates by gripping them at one end with gloved hands and lifting, or they may be floated off into liquid ($H_2O$) in a dish. Take care not to tear the gel.

17. Minigels may be stained and destained conveniently in the plastic boxes in which molecular biology enzymes are often supplied. Glass "casserole" dishes may also be used. Agitate gently by placing the container on a flatbed rocker, Bellydancer (Stovall, Life Science Inc.), or similar device.

18. In a Bio-Rad Transblot Semi-dry Transfer Cell, a gel of about 10 × 10 cm will transfer satisfactorily in about 1 h at 10 V, with the current likely not exceeding 0.2 Å; two gels of

size 10 × 10 cm should be transferred with a current not exceeding 0.4 Å, and so on. Excessive current produces overheating and poor sample transfer.

19. The transfer of blue-loading buffer dye from the gel to the nitrocellulose membrane is a good indication that efficient transfer of proteins has taken place. If there is no transfer of dye, this likely indicates a transfer problem. If only some of the dye has transferred, carefully reset the gel in place on the membrane, and reapply 10 V until proper transfer has occurred.

20. Obtain a box slightly larger than the membrane; 25 mL solution should adequately cover the membrane.

21. Primary antibodies (raised against your target protein) are often used at a dilution of 1:1000 (i.e., add 10 µL antibody to 10 mL PA solution). Antibodies are costly; try to use a container in which 10 mL (minigel) or 25 mL (large gel) solution will cover the membrane.

22. Use secondary antibodies raised against immunoglobulins from the animal in which the primary antibody was produced (often rabbit) and that are conjugated to peroxidase—the substrate for the development of a detectable ECL signal. For example, if rabbits were used to produce the primary antibody against your target protein, use peroxidase-conjugated anti-rabbit immunoglobulin as a secondary antibody.

23. After these three washes (**Subheading 3.4.2., step 5**) perform the remainder of the protocol in the dark room.

24. The luminol and coumaric acid (in DMSO) used in ECL solution (*see* **Subheading 2.4., item 9**) should be stored at –20°C in 1-mL aliquots. Solutions 1 and 2 may be stored in the dark at 4°C for weeks, but should not be combined until just before use.

25. Prepare a rectangular piece of clingfilm (e.g., Saran wrap) with edges about 1 cm larger than the membrane. Place the membrane onto the nonopening leaf of a film developer cassette, and carefully overlay the clingfilm onto the membrane, avoiding bubbles and folds. Some researchers stick the edges of the clingfilm down with masking tape onto the base of the cassette, because repeated opening of the cassette often displaces the membrane. Turn off lights (other than the darkroom safelight), remove a piece of hyperfilm ECL from its safety box, cut a sheet of film to the size of the film cassette, place the film over the membrane, and close and lock the cassette. Expose different sheets of film for 30 s, 1 min, 2 min, and 5 min, respectively, to obtain a useful set of blots for each gel.

26. One replicate requires 100 mL medium. Grow a larger culture if cells are treated to different experimental conditions. For example, for a pairwise comparison, grow 200 mL cells, and divide equally into two flasks before applying experimental and control treatments.

27. If you have divided the 100-mL culture between two 50-mL Falcon tubes, the two cell pellets can be combined into one tube during this step.

28. Although the EDTA and protease inhibitor solutions for solution 1 may be prepared in advance, solution 1 should be made up just before use. For solution 2, the CHAPS should be dissolved in the $H_2O$ before adding the other reagents. Lysis solution 2 and protease inhibitor solution may be stored in aliquots at –20°C for several months.

29. If a cooled tubeshaker is used, tubes may be shaken for 3 min continuously at 4°C.

30. **Subheadings 3.6.** and **3.7.** assume the reader has access to dedicated 2D SDS-PAGE equipment. Researchers who want to adapt existing laboratory equipment for 2D SDS-PAGE experiments should consult other sources (*17*).

31. DryStrips should be placed onto samples very soon after pipetting. It is helpful to fold and bend back a few millimeters of the cover film from the ⊕-ends of all the DryStrips to be used for a particular experiment before loading the samples, because this will help rapid subsequent processing of the strips.

32. The IPGphor should be operated with an upper limit of 50 µA per strip. Gel strips (24 cm) can be run (20°C) at 30 V for 12 h (gel rehydration phase), 500 V for 1 h, 1000 V for 1 h, and 8000 V until the strips have been focused for about 100,000 V/h in total.

33. For each strip, obtain two 2-mL capacity round-bottomed microfuge tubes (or similar), and cut off the caps. Heat a scalpel blade, and cut off neatly the top (nontapered) end of the 10-mL pipets (one for each strip) at about the 12 mL (+2 mL) mark. Using forceps and avoiding damage to the gel area, place each strip into a pipet, then place a 2-mL tube tightly over each end, and store at −70°C.

34. Precast gels can also be used. The main difference from 1D SDS-PAGE gels is that there is no stacking gel. Instead, focused strips are secured in molten agarose above the resolving gel as described in the main text.

35. This can be done by shooting $H_2O$ from a plastic dispenser aimed into the top of the pipet.

36. Holding the spacer perpendicular to the upper edge of the glass plates, carefully push down onto the plastic backing of the DryStrip. Gently tap the strip down to level it and dislodge any bubbles beneath.

37. The Amersham Ettan DALT II separation tank contains parallel horizontal guides through which the glass plates of the gels are pushed. Lubricating the guides with 1X TGS will help the gels slide in.

38. Use a plastic-lidded box just larger than the gel, and agitate on a wide-platform shaker on slow setting.

39. Use two pieces of magnetic strip (Kent County Supplies, UK), each one-gel width in length, to immobilize large gels while pouring off liquids. Place the tray so that it partially overhangs the benchtop. Drop one magnet onto the upper margin of one end of the gel, and attach the second magnet underneath the tray in a position directly beneath the first. The two strips should magnetize through the plastic, clamping one end of the gel and permitting liquid to be poured off quickly without tearing the gel.

40. Silver stain waste must be discarded as hazardous waste, not poured down the sink. Discoloration of sink surfaces following any accidental spillage of silver can be removed with nitric acid.

41. The developer and stop solutions must be made before they are required. Both solutions include reagents that are difficult to dissolve and require stirring for about 10 min with a magnetic flea.

42. Initially, the developer solution precipitates any unbound silver stain present in the staining box, clouding the solution.

43. Development of the stain is continuously progressive and *must* be halted before overstaining ruins the gel. Agitate gels in the developer solution by rocking the box in your hands. Spots should become evident after a few minutes. Vigilantly observe the background areas of the gel away from developing spots. Stop the development when background staining is accompanied by little further spot definition. Act promptly, noting that it will take 30–60 s to pour off the developer fluid and add the stop solution.

44. A "plug-in" is a small piece of software that adds a specific feature or function to a larger program.

## Acknowledgments

We are indebted to Angela Dunn for her useful suggestions on some of the described procedures. Work in the authors' laboratory is funded by the Biotechnology and Biological Sciences Research Council, United Kingdom.

## References

1. Bollag, D. M. and Edelstein, S. J. *Protein Methods*. Wiley-Liss, New York, 1991.
2. Ramsby, M. L. and Makowski, G. S. Differential detergent fractionation of eukaryotic cells, in *2-D Proteome Analysis Protocols* (Link, A. J., ed.), Humana Press, Totowa, New Jersey, 1999, pp. 53–66.
3. Qoronfleh, M. W., Benton, B., Ignacio, R., and Kaboord, B. (2003) Selective enrichment of membrane proteins by partition phase separation for proteomic studies. *J. Biomed. Biotechnol.* **4**, 249–255.
4. Guttman, A. (2003) Gel and polymer-solution mediated separation of biopolymers by capillary electrophoresis. *J. Chromatogr. Sci.* **41**, 449–459.
5. Unlu, M., Morgan, M. E., and Minden, J. S. (1997) Difference gel electrophoresis: a single gel method for detecting changes in protein extracts. *Electrophoresis* **18**, 2071–2077.
6. Swiss Institute of Bioinformatics. Tools and software packages on Expert Protein Analysis System (ExPASy) Proteomics Server. Available at ca.expasy.org.
7. Kryndushkin, D. S., Alexandrov, I. M., Ter-Avanesyan, M. D., and Kushnirov, V. V. (2003) Yeast [*PSI+*] prion aggregates are formed by small Sup35 polymers fragmented by Hsp104. *J. Biol. Chem.* **278**, 49636–49643.
8. Manabe, T., Yamaguchi, N., Mukai, J., Hamada, O., and Tani, O. (2003) Detection of protein-protein interactions and a group of immunoglobulin G-associated minor proteins in human plasma by nondenaturing and denaturing two-dimensional gel electrophoresis. *Proteomics* **3**, 832–846.
9. Zhu, H., Bilgin, M., and Snyder, M. (2003) Proteomics. *Annu. Rev. Biochem.* **72**, 783–812.
10. Binz, P. A., Hochstrasser, D. F., and Appel, R. D. (2003) Mass spectrometry-based proteomics: Current status and potential use in clinical chemistry. *Clin. Chem. Lab. Med.* **41**, 1540–1551.
11. Nørregaard Jensen, O., Wilm, M., Shevchenko, A., and Mann, M. Sample preparation methods for mass spectrometric peptide mapping directly from 2-DE gels, in *2-D Proteome Analysis Protocols* (Link, A. J., ed.), Humana Press, Totowa, New Jersey, 1999, pp. 53–66.
12. Bradford, M. M. (1976) A rapid and sensitive method for the quantitation of microgram quantities of protein utilizing the principle of protein-dye binding. *Anal. Biochem.* **72**, 248–254.
13. Towbin, H., Staehelin, T., and Gordon, J. (1979) Electrophoretic transfer of proteins from polyacrylamide gels to nitrocellulose sheets: Procedure and some applications. *Proc. Natl. Acad. Sci. USA* **76**, 4350–4354.
14. Young, N., Chang, Z., and Wishart, D. S. (2004) GelScape: a web-based server for interactively annotating, manipulating, comparing and archiving 1D and 2D gel images. *Bioinformatics* **20**, 976–978.
15. Lemkin, P. F. and Thornwall, G. (1999) Flicker image comparison of 2-D gel images for putative protein identification using the 2DWG meta-database. *Mol. Biotechnol.* **12**, 159–172.
16. Ramagli, L. S. Quantifying protein in 2-D PAGE solubilization buffers, in *2-D Proteome Analysis Protocols* (Link, A. J., ed.), Humana Press, Totowa, New Jersey, 1999, pp. 99–103.
17. Sarmiento, M. High resolution, 2-D protein electrophoresis using nondedicted equipment, in *2-D Proteome Analysis Protocols* (Link, A. J., ed.), Humana Press, Totowa, New Jersey, 1999, pp. 133–145.

# 29

## Characterization of Therapeutic Proteins by Membrane and In-Gel Tryptic Digestion

### C. Mark Smales, Rosalyn J. Marchant, and Michèle F. Underhill

### 1. Introduction

The fate and effectiveness of a therapeutic protein product within a patient is, in part at least, determined by potential product variation that leads to differential activity and the half-life of the protein reagents in vivo, and these variations can also contribute towards antigenicity. To assure therapeutic product quality and authenticity, it is therefore necessary to have in place a (or several) method(s) to characterize the possible variation in therapeutic protein products. This is required to determine whether variation in the detailed molecular composition of complex biopharmaceuticals is controlled during bioprocessing. One traditional approach to therapeutic protein characterization involves the proteolytic digestion of the protein of interest in solution followed by high-performance liquid chromatography (HPLC) analysis of the resulting peptide mixture on a reverse-phase column (e.g., see **refs. 1,2**). The peptides are then usually collected either manually or by a fraction collector, concentrated, and analyzed by either mass spectrometry or N-terminal Edman sequencing. However, such an approach requires relatively large amounts of protein (up to milligrams) and is time-consuming.

Although many new methods for therapeutic protein characterization are now available, peptide fingerprinting still remains one of the most commonly utilized methods both industrially and in academic research. With the increased sensitivity of mass spectrometry techniques, it is now possible to undertake such analysis on microgram amounts of protein. Furthermore, digestion and processing can be undertaken on solid-phase supports (e.g., polyvinylidene fluoride [PVDF] membrane) or in gel slices excised from polyacrylamide gels (*3–5*). The use of solid-phase PVDF 96-well plates lends itself to the automation of the digestion procedure and can be linked to glycosylation analysis (*6*). Automated technology is now available for high-throughput of both plate form and in-gel digestion (or combinations of both), which can also be linked to automated mass spectrometry analysis. Even without automation,

From: *Methods in Molecular Biology, vol. 308: Therapeutic Proteins: Methods and Protocols*
Edited by: C. M. Smales and D. C. James © Humana Press Inc., Totowa, NJ

the 96-well format allows the rapid turnaround of a large number of samples, allowing the monitoring of a therapeutic protein throughout production and packaging. This chapter describes methods for both 96-well PVDF and in-gel tryptic digestion for the characterization of therapeutic proteins.

## 2. Materials

### 2.1. Equipment

1. Pipet tips.
2. Eppendorf tubes.
3. Centrifugal vacuum evaporator.
4. PVDF 96-Well microtiter plates.
5. Vacuum manifold for 96-well plate format (*see* **Note 1**).

### 2.2. Reagents

1. Milli-Q water or equivalent.
2. Methanol.
3. Acetonitrile.
4. Ammonium bicarbonate.
5. Silver destaining solution: 1:1 solution of potassium ferricyanide (30 m$M$) and sodium thiosulfate (100 m$M$).
6. Coomassie-destaining reagent: 50% (v/v) acetonitrile and 0.2 $M$ ammonium bicarbonate.
7. Reducing reagent: 10 m$M$ dithiothreitol and 5 m$M$ ammonium bicarbonate.
8. Iodoacetamide solution: 55 m$M$ iodoacetamide and 5 m$M$ ammonium bicarbonate.
9. Trypsin solution: sequencing grade trypsin (e.g., Promega) at 20 ng/µL in 25 m$M$ ammonium bicarbonate solution.
10. Trifluoroacetic acid (TFA).
11. Reducing buffer: 8 $M$ urea, 360 m$M$ Tris-HCl, and 3.2 m$M$ EDTA, pH 8.6.
12. Iodoacetic acid.
13. Dithiothreitol.
14. Polyvinylpyrrolidone 360.

## 3. Methods

### 3.1. In-Gel Tryptic Digestion

1. Separate the therapeutic protein of interest by either one-dimensional (1D) or two-dimensional (2D) sodium dodecyl sulfate-polyacrylamide gel electrophoresis analysis and detect spots by visualization with appropriate staining (*see* **Note 2**).
2. Excise protein band (1D gels) or spots (2D gels) using a cut-off pipet tip, and place into an Eppendorf tube that has been washed with methanol.
3. For silver-stained gels, destain by covering the gel piece with silver-destaining solution for 60 min at room temperature with agitation. For Coomassie-stained gels, destain by covering the gel piece with Coomassie-destaining reagent, then incubate at room temperature for 60 min with agitation (*see* **Notes 3** and **4**).
4. Pipet off the destaining solution, then wash the gel piece twice with 50 µL Milli-Q water (*see* **Note 5**).
5. Wash three times with 50% (v/v) acetonitrile and 25 m$M$ ammonium bicarbonate solution for 10 min, then dry in a centrifugal vacuum evaporator for 30 min.

6. Add sufficient reducing reagent to cover the gel piece, and incubate at 50°C for 45 min.
7. Pipet off reducing reagent, then add sufficient iodoacetamide solution to cover the gel piece, and incubate in the dark at room temperature for 1 h.
8. Remove iodoacetamide solution, and wash the gel piece twice with 50% (v/v) acetonitrile and 25 mM ammonium bicarbonate solution. Dehydrate in a centrifugal vacuum evaporator as previously described.
9. Rehydrate each gel piece in enough trypsin solution to just cover the gel piece, and incubate at 4°C for 1 h.
10. Remove excess trypsin solution, then cover gel pieces with 25 mM ammonium bicarbonate solution.
11. Incubate overnight at 37°C (*see* **Note 6**).
12. Centrifuge tubes to spin down liquid, then remove supernatant (*see* **Note 7**).
13. Extract peptides from the gel using 20 µL 0.1% (v/v) TFA solution followed by further extraction using 20 µL 50% (v/v) acetonitrile and 5% (v/v) TFA solution (*see* **Note 8**).
14. Combine all supernatant extracts, and evaporate to dryness in a centrifugal vacuum evaporator.
15. Store peptide extract at –80°C or resuspend for HPLC or mass spectrometry analysis (*see* **Note 9**).

## 3.2. Digestion of Therapeutic Proteins Using a PVDF 96-Well Plate Format

1. Fill the outside wells of the PVDF membrane 96-well plate with 150 µL Milli-Q water (*see* **Note 10**).
2. Wash the remaining PVDF membrane wells with 100 µL methanol, 300 µL Milli-Q water, followed by 100 µL reducing buffer. Between each step, draw the solution through the membrane using the vacuum manifold (*see* **Note 11**).
3. Resuspend the protein sample of interest in 100 µL reducing buffer. Alternatively, if the protein of interest is already in solution, dilute 20-fold in reducing buffer.
4. Apply protein solution(s) to the PVDF membrane wells that had previously been washed with methanol and so forth as appropriate, and draw the protein solution through the membrane using the vacuum manifold (*see* **Note 12**).
5. Wash the wells with 500 µL reducing buffer (*see* **Note 13**).
6. Reduce the immobilized protein by adding 50 µL reducing buffer containing 100 mM dithiothreitol, then incubate the plate at 37°C for 1 h.
7. Draw the reducing agent off under vacuum, then add 50 µL reducing buffer containing 100 mM iodoacetic acid. Incubate in the dark at room temperature for 30 min.
8. Draw iodoacetic acid solution off, then block the membrane by incubation at room temperature with a 1% (w/v) polyvinylpyrrolidone 360 solution (in $H_2O$, 150 µL) for 1 h (*see* **Notes 14** and **15**) to prevent adsorption of trypsin to the membrane.
9. Draw the blocking solution off by applying vacuum, then wash the membrane several times with 50 mM ammonium bicarbonate solution.
10. Add 50 µL trypsin solution, then incubate overnight at 37°C.
11. Pipet off the trypsin solution, and place in an Eppendorf tube.
12. Wash the well/membrane with 50% acetonitrile containing 2% TFA (20 µL), then combine this wash with the solution from **step 11**, which consists of a peptide mixture.
13. Concentrate down the peptide mixture using a centrifugal vacuum evaporator until only 1–10 µL solution remains.
14. The peptide solution may now be either spotted directly onto a matrix-assisted laser desorption ionization time-of-flight (MALDI-TOF) mass spectrometer plate or injected onto

a reverse-phase C18 HPLC column for analysis. The HPLC column may also be connected to an electrospray mass spectrometer for on line analysis of the individual peptides as they elute (*see* **Note 16**).

## 4. Notes

1. Although it is possible to buy a vacuum manifold specifically for 96-well plates, you can build your own for a fraction of the cost. We make our own out of a tissue culture flask (use a T175-sized flask). Simply cut a portion out of the top of the flask using a hot knife, which is slightly smaller than the size of the 96-well plate. The plate can then sit over this area. Adapt a cap with a piece of tubing that can be attached to a vacuum cap, and seal around the vent of the flask. Place the 96-well plate on top of the flask, and attach the tubing to the vacuum. You now have a very cheap and efficient vacuum drainage system. Be careful not to apply too high a vacuum or the flask will split. The liquid should drain through and collect rapidly in the bottom of the tissue culture flask.

2. 1D and 2D polyacrylamide gel electrophoresis are described elsewhere in this book. However, a good description of 2D electrophoresis that works for most therapeutic proteins is presented in **ref. 6**. If staining with silver stain before undertaking in-gel digestion, it is imperative that glutaraldehyde is omitted in the sensitizing step.

3. Destaining is usually undertaken until all color has been removed from the gel piece, which may require repetition of this step. However, recent reports suggest that destaining is not required to obtain reproducible and acceptable proteolytic digestion (*4*).

4. Destaining of silver-stained gels is often easier to undertake using a 1% solution of hydrogen peroxide ($H_2O_2$). Destaining should be complete within 5–10 min when using $H_2O_2$.

5. The destaining solution should now be colored and the gel piece clear. If not, repeat the destaining step.

6. Incubation can be achieved in a shorter period of time. Incubation for 4 h is usually sufficient to achieve acceptable digestion. Prolonged incubation should be avoided.

7. After overnight incubation at 37°C, most of the liquid will be in the cap of the tube because of evaporation and condensation. It is therefore necessary to spin this down before removing the supernatant.

8. This extraction procedure should extract most peptides out of the gel. It is possible to keep each extraction separate and analyze them separately. Using TFA does suppress ionization during mass spectrometric analysis, particularly when using electrospray mass spectrometry.

9. Peptides can be stored at –80°C for prolonged periods of time. However, for particular samples, it may be difficult to resolubilize peptides if concentrated to dryness. If this is the case, concentrate down to a few microliters, and store in a liquid form.

10. Filling the outside wells with water helps prevent evaporation and drying out of the sample wells during prolonged incubation periods.

11. Aqueous buffers cannot be drawn through the PVDF membrane until treated with methanol.

12. If you have a large volume of protein, this can be applied to the membrane in several steps, drawing the liquid off under vacuum after each application. This is an excellent and quick way of concentrating the protein of interest.

13. This will require passing several lots of reducing buffer through the membrane under vacuum, as 500 µL does not fit into the well.

14. It is necessary to block the membrane to prevent the trypsin immobilizing on the membrane once it is introduced.

15. At this stage, *N*-glycan analysis can be undertaken on the immobilized protein as described in **ref. 6**.
16. MALDI-TOF mass spectrometry analysis will give a peptide mass fingerprint. However, by using an HPLC connected to an electrospray mass spectrometer, it is possible to obtain both a peptide chromatogram and a mass measurement for each resolved peak. This can improve your chances of observing any changes in your tryptic map owing to differential posttranslational modifications or protein modifications during bioprocessing.

## References

1. Smales, C. M., Moore, C. H., and Blackwell, L. F. (1999) Characterization of lysozyme-estrone glucuronide conjugates. The effect of the coupling reagent on the substitution level and sites of acylation. *Bioconjug. Chem.* **10,** 693–700.
2. Smales, C. M., Pepper, D. S., and James, D. C. (2000) Protein modification during antiviral heat bioprocessing. *Biotechnol. Bioeng.* **67,** 177–188.
3. Deutzmann, R. (2004) Structural characterization of proteins and peptides. *Methods Mol. Med.* **94,** 269–297.
4. Terry, D. E., Umstot, E., and Desiderio, D. M. (2004) Optimized sample-processing time and peptide recovery for the mass spectrometric analysis of protein digests. *J. Am. Soc. Mass Spectrom.* **15,** 784–794.
5. Courchesne, P. L., Luethy, R., and Patterson, S. D. (1997) Comparison of in-gel and on-membrane digestion methods at low to sub-pmol level for subsequent peptide and fragment-ion mass analysis using matrix-assisted laser-desorption/ionization mass spectrometry. *Electrophoresis* **18,** 369–381.
6. Smales, C. M., Pepper, D. S., and James, D. C. (2002) Protein modification during antiviral heat-treatment bioprocessing of factor VIII concentrates, factor IX concentrates, and model proteins in the presence of sucrose. *Biotechnol. Bioeng.* **77,** 37–48.

# 30

# Oligosaccharide Release and MALDI-TOF MS Analysis of N-Linked Carbohydrate Structures from Glycoproteins

Rodney G. Keck, John B. Briggs, and Andrew J. S. Jones

## 1. Introduction

With the growth in the number of recombinant proteins being developed for use as human therapeutics, there also exists a corresponding need for new and highly sensitive analytical techniques to characterize these new pharmaceuticals. Many new assays were developed to monitor and characterize the heterogeneity that naturally exists in these complex molecules *(1,2)*. For glycoprotein-based therapeutics, one major source of heterogeneity arises from the oligosaccharides present on the protein *(3)*. This oligosaccharide heterogeneity is a natural outcome of the glycoprotein synthesis pathways in the host cells, such as Chinese hamster ovary cells (CHO; *4*) or baby hamster kidney cells (BHK; *5*). For some glycoproteins, e.g., erythropoietin *(6)* and rituximab *(7)*, the distribution of oligosaccharides can affect the in vitro biological activity of the molecule. In these cases, it is imperative to have analytical methods in place that measure the oligosaccharide distribution during the drug development phase, as well as after regulatory approval for the market. Oligosaccharide profiling of glycoproteins has become a common tool in both academic laboratories, as well as in the commercial environment. Most techniques employ enzymatic digestion to release the N-linked oligosaccharides; however, there are various methods used for detection of the released oligosaccharides. These include analysis of fluorescently labeled oligosaccharides by capillary electrophoresis *(8–10)* or high-performance liquid chromatography (HPLC; *11,12*), along with analyses of the unlabeled oligosaccharides by high-performance anion-exchange chromatography (HPAEC; *13*) and matrix-associated laser desorption ionization time-of-flight mass spectrometry (MALDI-TOF MS; *14–16*).

In our work, MALDI-TOF MS has become the method of choice for detection and profiling of unlabeled neutral or sialylated oligosaccharides. This chapter describes a high-throughput method for the release and MALDI-TOF MS analysis of N-linked oligosaccharides. This method permits the routine preparation and analysis of N-linked oligosaccharides from microgram amounts of protein samples. Results of these

From: *Methods in Molecular Biology, vol. 308: Therapeutic Proteins: Methods and Protocols*
Edited by: C. M. Smales and D. C. James © Humana Press Inc., Totowa, NJ

analyses can then be used to assess lot-to-lot comparability or characterize a new glycoprotein. Integration of the resulting MALDI-TOF mass spectra provides quantitative information about the relative oligosaccharide distribution of a particular sample or group of samples from which decisions about, e.g., cell culture or recovery processes, are made.

## 2. Materials

1. Millipore 96-well Multiscreen-IP plate (part no. MAIP N45 10) with 0.45-μm polyvinylidene fluoride (PVDF) membrane well bottoms (Millipore, Billerica, MA).
2. MultiScreen Vacuum Manifold (Millipore).
3. Reduction and alkylation (RCM) buffer: 6 $M$ guanidine hydrochloride, 360 m$M$ Tris and 2 m$M$ ETDA, pH 8.3.
4. Dithiothreitol (DTT).
5. Iodoacetic acid (IAA).
6. HPLC-grade or milli-Q water (18 Me$\Omega$).
7. HPLC-grade methanol.
8. Reagent-grade 1 $M$ NaOH.
9. Reagent-grade glacial acetic acid.
10. 1% Polyvinylpyrrolidone-360 (PVP-360) in water.
11. Peptide-N-glycosidase F (PNGase F; glycerol-free), formulated in Tris-acetate (New England BioLabs, Beverly, MA). Store at –20°C.
12. 10 m$M$ Tris acetate, pH 8.3.
13. Compact reaction columns, compact reaction column filters, and luer-lock lids (USB order nos. US13929USB, US13913, and US13928, respectively; United States Biochemical Corporation, Cleveland, OH).
14. 60-mL Luer-lock syringe.
15. AG 50W-X8 cation-exchange resin, hydrogen form, 100–200 mesh (Bio-Rad, Hercules, CA).
16. 2,5-Dihydroxybenzoic Acid (DHB).
17. 5-Methoxysalicylic Acid (5-MSA).
18. 2,4,6-Trihydroxyacetophenone (THAP).
19. Ammonium citrate dibasic.
20. MALDI-TOF MS capable of delayed extraction in the linear mode with an acceleration voltage of 20 kV.
21. Small glass vacuum desiccator.
22. Edwards E2M1.5 vacuum pump.
23. Oligosaccharide standards (Prozyme, San Leandro, CA or QA-Bio, San Mateo, CA).

   a. Neutral standards: Man 5 and NA4 (see **Table 1** for structures).
   b. Acidic standards: A1 and A3.

24. Caesar Workstation 7.2 integration software (SciBridge Software, Trabuco Canyon, CA).
25. Microsoft Excel software (Microsoft, Bellevue, WA).

## 3. Methods

### 3.1. PVDF Plate Preparation

1. Add 250 μL water to each well of the Multiscreen-IP plate that will not be loaded with a sample (see **Note 1**).
2. To wet the PVDF membrane, add 100 μL HPLC-grade methanol to each well that is designated to receive a sample. Place the plate onto the multiscreen vacuum apparatus,

**Table 1**
**Neutral Oligosaccharide Structures, Nomenclature, and Masses**

| Neutral oligosaccharide structures | Name | Mass Daltons $(M + Na^+)$ |
|---|---|---|
| Man<br>Man   Man   Man–GlcNAc–GlcNAc<br>Man | Man-5 | 1258.1 |
| GlcNAc – Man<br>   Man–GlcNAc–GlcNAc<br>GlcNAc – Man | NGA2, 2000 | 1340.2 |
|       Fuc<br>Gal – GlcNAc – Man    &#124;<br>   Man–GlcNAc–GlcNAc<br>Gal – GlcNAc – Man | NA2F, 2120 | 1810.6 |
| Gal – GlcNAc – Man<br>Gal – GlcNAc – Man   Man<br>Gal – GlcNAc – Man   Man–GlcNAc–GlcNAc<br>Gal – GlcNAc – Man   Man | NA4, 4040 | 2395.2 |
|       Fuc<br>GlcNAc – Man   &#124;<br>   Man–GlcNAc–GlcNAc<br>GlcNAc – Man | NGA2F, 2100 | 1486.4 |
|       Fuc<br>GlcNAc – Man   &#124;<br>   Man–GlcNAc–GlcNAc   + K<br>GlcNAc – Man | $2100 + K^+$<br>Potassium adduct of 2100 | 1502.4<br>($K^+$ replaces $Na^+$) |
| GlcNAc – Man<br>Gal – {   Man–GlcNAc–GlcNAc<br>GlcNAc – Man | 2010 | 1502.4 |
|       Fuc<br>GlcNAc – Man   &#124;<br>Gal – {   Man–GlcNAc–GlcNAc<br>GlcNAc – Man | 2110 | 1648.5 |
| Gal – GlcNAc – Man<br>   Man–GlcNAc–GlcNAc<br>Gal – GlcNAc – Man | 2020 | 1664.5 |

**Table 1** *(continued)*

| Neutral oligosaccharide structures | Name | Mass Daltons (M + Na+) |
|---|---|---|
| Fuc \| <br>GlcNAc – Man ╲ <br>GlcNAc ╲ Man–GlcNAc–GlcNAc <br>Man ╱ <br>GlcNAc ╱ | 3100 | 1689.6 |
| Gal – { GlcNAc – Man ╲ <br>GlcNAc ╲ Man–GlcNAc–GlcNAc (Fuc) <br>Man ╱ <br>GlcNAc ╱ } | 3110 | 1851.7 |
| 2 Gal – { GlcNAc – Man ╲ <br>GlcNAc ╲ Man–GlcNAc–GlcNAc (Fuc) <br>Man ╱ <br>GlcNAc ╱ } | 3120 | 2013.8 |
| Fuc \| <br>Gal – GlcNAc – Man ╲ <br>Gal – GlcNAc ╲ Man–GlcNAc–GlcNAc <br>Man ╱ <br>Gal – GlcNAc ╱ | 3130 | 2175.9 |

and apply gentle vacuum to pull the methanol through the PVDF membrane. Rinse each sample well three times with 250 µL water, pulling the water through the membrane after each rinse with gentle vacuum.

### 3.2. Reduction and Alkylation (see Note 2)

1. Rinse each sample well twice with 50 µL RCM buffer, pulling the buffer through the membrane after each rinse with gentle vacuum. Add 50 µL RCM buffer to each sample well. Add 50 µg glycoprotein sample to the appropriate well, and pull the solution through the membrane with gentle vacuum. The maximum volume of a well is approx 250 µL, so if the sample volume exceeds 200 µL, this process should be repeated, starting from adding 50 µL RCM buffer until all the volume of the sample has been applied to the PVDF membrane.
2. After the samples are applied, rinse each sample well twice with 50 µL RCM buffer, pulling each rinse through the PVDF membrane with gentle vacuum. Wipe the top and bottom of plate with a clean Kimwipe (*see* **Note 3**).
3. Freshly prepare a 0.1 *M* DTT solution in the RCM buffer. Add 50 µL DTT solution to each sample well. Place the covered plate in a 37°C incubator for 1 h. Place an open vessel of water in the incubator to humidify the atmosphere and prevent evaporative loss of liquid from the sample wells.
4. After the reduction step, the DTT solution is removed by placing the plate onto the vacuum apparatus and applying gentle vacuum to pull the solution through the membrane. Rinse each sample well three times with 250 µL water, pulling the water through the membrane

following each rinse with gentle vacuum. Wipe the top and bottom of plate with a clean Kimwipe (*see* **Note 3**).

5. Freshly prepare a 1 *M* IAA solution in 1 *M* NaOH. Dilute this stock solution 1:10 with RCM buffer to obtain a working solution of 0.1 *M* IAA.

6. Add 50 μL 0.1 *M* IAA solution to each sample well. Place the covered plate in the dark for 30 min at ambient temperature.

7. After the alkylation step, remove IAA by placing the plate onto the vacuum apparatus. Apply a gentle vacuum to the plate to pull the solution through the membrane. Each sample well should be rinsed three times with 250 μL water, pulling the water through the membrane after each rinse with gentle vacuum. Use a clean Kimwipe to wipe the top and bottom of plate (*see* **Note 3**).

### 3.3. Blocking the PVDF Membrane

1. Add 100 μL 1% aqueous solution of PVP-360 to each sample well. Incubate the covered plate for 30 min at ambient temperature. This step blocks any remaining protein-binding sites on the PVDF and prevents adsorptive losses of the PNGase F enzyme.

2. After blocking, remove PVP-360 by applying a gentle vacuum to the plate and pulling the solution through the membrane. Rinse each sample well three times with 250 μL water, pulling the water through the membrane after each rinse with gentle vacuum. Wipe the top and bottom of plate with a clean Kimwipe (*see* **Note 3**).

### 3.4. PNGase F Digestion

1. Prepare a 20 IUB milliunit/mL PNGase F solution in 10 m*M* Tris acetate buffer, pH 8.3 (*see* **Note 4**).

2. Add 25 μL PNGase F solution to each sample well. Place the covered plate in the 37°C humidified incubator for 3 h.

3. After digestion, transfer each sample into a labeled 500-μL microcentrifuge tube (*see* **Note 5**).

### 3.5. Acidification of Released Oligosaccharides

1. Prepare a 1.5 *M* acetic acid solution in water.

2. Add 2.5 μL acetic acid solution to each microcentrifuge tube (150 m*M* final acetic acid concentration), and mix the sample by pipet action. Incubate for 2 h at ambient temperature (*see* **Note 6**).

### 3.6. Removal of Cations From Released Oligosaccharides

1. Assemble the compact reaction columns by firmly inserting the filter into the column using the plunger provided with the columns. One column is needed for each sample.

2. Pipet 700–800 μL cation-exchange resin into a compact reaction column (*see* **Note 7**). Rinse each column with approx 5 mL water using a 60-mL syringe equipped with the luer-lock lid adapter. After rinsing, expel as much of the water in the columns as possible using the 60-mL syringe filled with air. Place each of the columns in a 1.5-mL microcentrifuge tube, and spin the columns in a microcentrifuge for 1–2 min at 16,000*g* to remove all residual water from the cation-exchange resin.

3. Place each column in a labeled 1.5-mL microcentrifuge tube, and add 3–5 μL sample to the top of the cation-exchange resin. Spin the column/microcentrifuge tube combination in a microcentrifuge using the pulse spin button to slowly move the sample down the cation-exchange column into the microcentrifuge tube (*see* **Note 8**).

### 3.7. MALDI-TOF MS Matrix Preparation

1. Super DHB (sDHB) positive mode matrix.

   a. Prepare 1 m$M$ NaCl solution in 25% aqueous ethanol (solvent A).
   b. Prepare 4 mg/mL DHB solution in solvent A.
   c. Prepare 0.2 mg/mL 5-MSA solution in solvent A.
   d. Combine equal amounts of the DHB and 5-MSA matrices to form the sDHB matrix (*see* **Note 9**).

2. THAP negative mode matrix.

   a. Prepare a solution consisting of 13.33 m$M$ ammonium citrate/acetonitrile, 75/25 (solvent B).
   b. Prepare a 2 mg/mL THAP solution in solvent B (*see* **Note 9**).

### 3.8. Preparation of Sample Spots

1. Positive mode analysis for neutral oligosaccharides. Spot a 0.5-µL aliquot of each sample and the neutral standard mix onto the stainless steel MALDI plate, then apply a 0.5-µL aliquot of the sDHB matrix on top of each sample spot. Place the spotted stainless steel plate into a small vacuum desiccator, and apply vacuum to quickly dry all the spots.
2. Negative mode analysis for sialylated oligosaccharides. Spot a 0.5-µL aliquot of each sample and the acidic standard mix onto the stainless steel MALDI plate, then apply a 0.5-µL aliquot of the THAP matrix on top of each sample spot. Place the spotted stainless steel plate into a small vacuum desiccator, and apply vacuum to quickly dry all the spots (*see* **Note 10**).

### 3.9. MALDI-TOF MS Analysis of Released Oligosaccharides (see Note 11)

1. Neutral oligosaccharides are analyzed in the positive mode and are detected as the sodium adducts. First, acquire spectra from the neutral oligosaccharide calibration mixture. Smooth the neutral oligosaccharide spectra using the Savitsky-Golay 19-point algorithm, then calibrate the instrument using the two-point external calibration method (*see* **Note 12**). After the instrument is calibrated, analyze the unknown samples in the positive mode.

   a. Instrument settings: Positive mode, linear configuration, and accelerating voltage set at 20 kV; delayed extraction on and set at 60 ns; grid voltage at 93.2%; and guide wire voltage at 0.05%. The laser power setting must be empirically determined for each instrument. For neutral oligosaccharides, apply enough laser power to obtain good signal intensities, along with an acceptable signal-to-noise ratio. Too much laser power broadens the peaks. Also, desialylation of sialylated oligosaccharides can alter neutral glycan distribution. Currently, we are using a laser power setting of 1600 for neutral oligosaccharides, but this value can change as the laser ages or after installation of a new laser (*see* **Note 13**).
   b. To ensure consistent results between analysts and across assays, spectra are acquired in the summing mode for 240 laser shots for each sample in the positive mode (*see* **Note 14**).

2. Sialylated oligosaccharides are analyzed in the negative mode. Acquire spectra from the acidic oligosaccharide calibration mixture, then smooth the acidic oligosaccharide spectra using the Savitsky-Golay 19-point algorithm. Use the smoothed spectra to calibrate the instrument using the two-point external calibration method (*see* **Note 12**). Following calibration, analyze the unknown samples in the negative mode.

   a. Instrument settings: Negative mode, linear configuration, and accelerating voltage set at 20 kV; delayed extraction on and set at 140 ns; grid voltage at 92.2%; and guide wire voltage at 0.05%. Again, the laser power setting for sialylated oligosaccharides must be empirically determined for each instrument (*see* **Note 15**). Sialylated oligosaccharides are much more sensitive to laser power than are the neutral oligosaccharides because desialyation can occur at higher laser powers. Currently, we are using a laser power setting of 1900 for sialylated oligosaccharides. Yet, this setting varies based on laser age or the installation of a new laser (*see* **Note 13**).

   b. For consistent results between analysts and across assays, spectra are obtained in the summing mode for 256 laser shots for each sample in the negative mode.

3. Samples can also be acquired in an automated mode if this feature is available on the MALDI-TOF MS instrument.

4. Smooth the acquired spectra using the Savitsky-Golay 19-point algorithm, and print a copy of the labeled spectra.

## 3.10. Interpretation of MALDI-TOF MS Spectra

1. When analyzing the MALDI-TOF MS spectra of released oligosaccharides, it is important to be aware of factors that can affect the quality and interpretation of the spectral data. Regarding data quality, two attributes are generally considered to be of primary importance in assessing the value of a spectrum: resolution and signal-to-noise ratio (*see* **Note 16**). However, when the purpose of acquiring the spectrum is to determine the oligosaccharide distribution, spectral resolution is not usually considered a limiting factor. Actually, the act of processing a spectrum for integration smoothes the spectrum, thereby removing the resolution of isotopic forms of the oligosaccharide. Alternatively, the signal-to-noise ratio is a factor that can affect the quality of the data obtained when assessing the oligosaccharide distribution (*see* **Note 17**).

2. When interpreting MALDI-TOF MS spectra, the analyst must be aware that the spectra contain peaks that result not only from the sodium oligosaccharide adducts of expected structures but from oligosaccharides associated with either other components present in the sample or present in the matrices used in ionization. As previously discussed (*see* **Note 8**), in positive-mode spectra, a common ion observed is from an oligosaccharide related to a potassium ion, in addition to a sodium ion. Another typically observed ion in the positive mode is an ion that results from the association of a sialylated oligosaccharide with one or more sodium ions present in the sDHB matrix. Generally, a series of peaks whose associated masses differ from one another by 22 Daltons is observed (**Fig. 1**). In most cases, there are $(n + 1)$ ions in the series, where $n$ is the number of sialic acid residues present on the sialylated oligosaccharide.

3. In the negative mode, a commonly observed ion is one that results from the association of a neutral oligosaccharide with the citrate present in the THAP matrix. This ion is found in the negative mode at a mass that is equal to the ion observed in the positive mode plus 168 Daltons. The difference in mass is due to the variation between the mass of the oligosaccharide related to a sodium ion in the positive mode and the mass of the oligosaccharide associated with a citrate ion in the negative mode. For example, in **Fig. 2**, the citrate adduct of the 2120 neutral oligosaccharide is the peak at 1977 Daltons (marked with an asterisk). Fortunately, when present, this ion tends to be at relatively low intensity.

4. After acquiring acceptable spectra (*see* **Note 19**) from all samples, convert the unsmoothed .ms spectra into .prn or .txt files, which contain *x*-axis values (mass information) and *y*-axis values (relative abundance) from each of the sample spectra. The .prn or .txt files are used by the integration program to obtain peak area values.

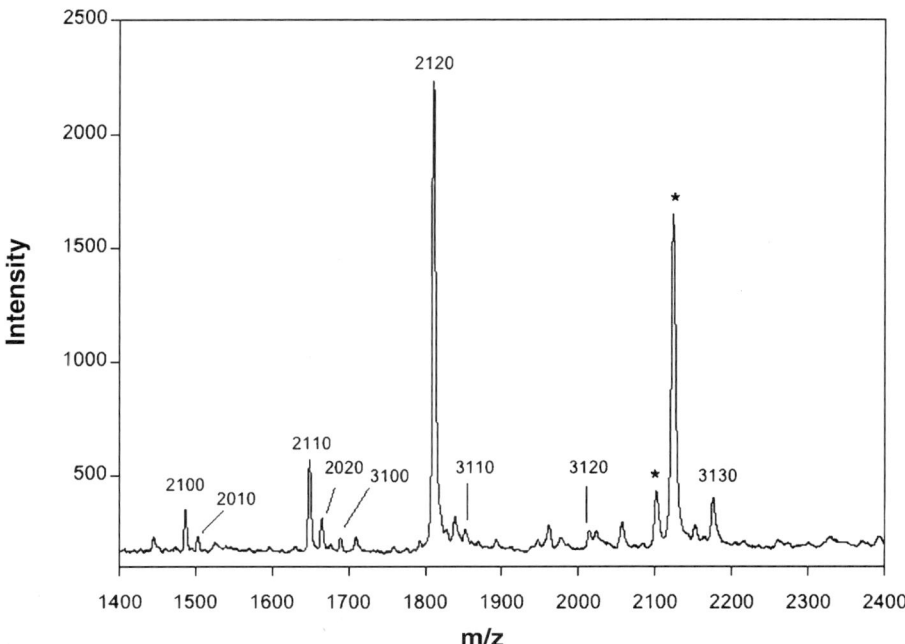

Fig. 1. Positive-ion mode spectrum of neutral oligosaccharides. Peaks identified as sialylated oligosaccharides associated with sodium are labeled with an asterisk (*). For an explanation of oligosaccharide nomenclature used in the **Figs. 1–3**, *see* **Tables 1** and **2** (*see* **Note 18**).

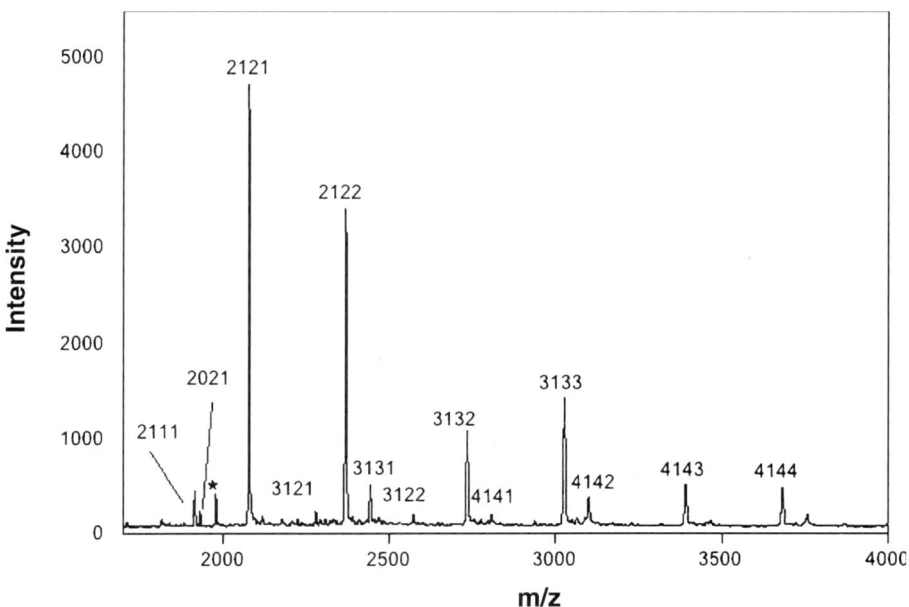

Fig. 2. Negative-ion mode spectrum of sialylated oligosaccharides. Peak identified as neutral oligosaccharides associated with citrate are labeled with an asterisk (*).

**Table 2**
**Sialylated Oligosaccharides Structures, Nomenclature, and Masses**

| Acidic oligosaccharide structures | Name | Mass Daltons (M – H⁻) |
|---|---|---|
| NeuAc – { Gal – GlcNAc – Man \ / Gal – GlcNAc – Man } Man–GlcNAc–GlcNAc | A1, 2021 | 1931.7 |
| NeuAc – { Gal – GlcNAc – Man \ / Gal – GlcNAc – Man } Man–GlcNAc–GlcNAc (Fuc) | A1F, 2121 | 2077.9 |
| NeuAc – Gal – GlcNAc – Man \ / NeuAc – Gal – GlcNAc – Man — Man–GlcNAc–GlcNAc | A2, 2022 | 2223.0 |
| NeuAc – Gal – GlcNAc – Man \ / NeuAc – Gal – GlcNAc – Man — Man–GlcNAc–GlcNAc (Fuc) | A2F, 2122 | 2369.2 |
| NeuAc – Gal – GlcNAc – Man \ NeuAc – Gal – GlcNAc \ Man / Man–GlcNAc–GlcNAc NeuAc – Gal – GlcNAc / | A3, 3033 | 2879.6 |
| NeuAc – Gal – { GlcNAc – Man \ / GlcNAc – Man } Man–GlcNAc–GlcNAc (Fuc) | 2111 | 1915.8 |
| GlcNAc – Man \ NeuAc – { Gal – GlcNAc \ Man / Man–GlcNAc–GlcNAc Gal – GlcNAc / } | 3121 | 2281.1 |

389

**Table 2** *(continued)*

| Acidic oligosaccharide structures | Name | Mass Daltons (M − H⁻) |
|---|---|---|

| Acidic oligosaccharide structures | Name | Mass Daltons (M − H⁻) |
|---|---|---|
| NeuAc − { Gal – GlcNAc – Man \ / Gal – GlcNAc \ / Gal – GlcNAc / Man } Man–GlcNAc–GlcNAc | 3131 | 2443.2 |
| 2 NeuAc − { Gal – GlcNAc – Man \ / Gal – GlcNAc \ / Gal – GlcNAc / Man } Man–GlcNAc–GlcNAc, Fuc | 3122 | 2572.3 |
| 2 NeuAc − { Gal – GlcNAc – Man \ / Gal – GlcNAc \ / Gal – GlcNAc / Man } Man–GlcNAc–GlcNAc, Fuc | 3132 | 2734.5 |
| NeuAc − { Gal – GlcNAc – Man \ Man / Gal – GlcNAc – Man / Gal – GlcNAc – Man \ Man / Gal – GlcNAc – Man } Man–GlcNAc–GlcNAc, Fuc | 4141 | 2808.6 |
| NeuAc – Gal – GlcNAc – Man \ / NeuAc – Gal – GlcNAc \ Man / NeuAc – Gal – GlcNAc / Man–GlcNAc–GlcNAc, Fuc | 3133 | 3025.8 |
| 2 NeuAc − { Gal – GlcNAc – Man \ Man / Gal – GlcNAc – Man / Gal – GlcNAc – Man \ Man / Gal – GlcNAc – Man } Man–GlcNAc–GlcNAc, Fuc | 4142 | 3099.8 |
| 3 NeuAc − { Gal – GlcNAc – Man \ Man / Gal – GlcNAc – Man / Gal – GlcNAc – Man \ Man / Gal – GlcNAc – Man } Man–GlcNAc–GlcNAc, Fuc | 4143 | 3391.1 |
| NeuAc – Gal – GlcNAc – Man \ Man / NeuAc – Gal – GlcNAc – Man / NeuAc – Gal – GlcNAc – Man \ Man / NeuAc – Gal – GlcNAc – Man } Man–GlcNAc–GlcNAc, Fuc | 4144 | 3682.4 |

5. Open the .prn or .txt files with the Caesar Workstation software (*see* **Note 20**). The .prn or .txt files are then smoothed using the Savitsky-Golay 19-point algorithm. The smoothed spectra are integrated into the files using acceptable integration parameters. Transfer the integration values to an Excel spreadsheet to calculate relative area percent values for each sample.

## 4. Notes

1. Water will not flow through the PVDF membrane unless the membrane is first wetted with methanol. Therefore, water placed in the unused wells will remain throughout the procedure. The unused wells are filled with water to minimize evaporative losses during the 37°C-incubation steps and to ensure adequate vacuum is applied to those wells containing samples.
2. The routine method employs denaturation of the sample by reduction and alkylation to maximize the accessibility of the protein–glycan bond to the glycanase enzyme to ensure efficient and complete removal of the oligosaccharides. This step might not always be required, but reproducible complete oligosaccharide removal should be demonstrated before reduction/alkylation is routinely omitted.
3. It is very important to make sure that no liquid is adhering to the outside of the wells to prevent siphoning of the liquids from the well during the incubation steps.
4. We specified IUB units because every supplier of PNGase F seems to have a different PNGase F activity unit definition, but most include a conversion factor to obtain IUB unit equivalents. For the New England BioLabs' PNGase F, 1 IUB milliunit equals 65 NEB units, whereas 1 Prozyme unit equals 1 IUB unit.
5. To facilitate the transfer of all the released oligosaccharides, it helps to tip the plate slightly to gather the liquid at one edge of the well. The oligosaccharide solution recovered from each well can be stored at –70°C overnight.
6. Acidification of the samples converts the oligosaccharides from the glycosylamine form into an oligosaccharide with a hydroxyl group on the reducing terminus. If this step is omitted, the oligosaccharides will be retained on the cation-exchange column, and no oligosaccharide peaks will be observed in the MALDI-TOF MS spectra. Because the acidified samples can be stored at –70°C for at least 1 mo, this a convenient stopping point.
7. The cation-exchange resin must be prewashed with copious amounts of Mill-Q water to remove impurities and colored material. After washing, the cation-exchange resin can be stored at 4°C in 50-mL Falcon tubes under water. To use, remove from the refrigerator, and gently invert the tube until the resin is once again dispersed in the water. Then, use a 1-mL Pipetman to transfer 700–800 µL to each of the compact reaction columns.
8. Removal of potassium ions from the released oligosaccharide samples is one of the most important steps of the entire procedure. The presence of potassium ions in a sample can introduce ambiguity into the analysis. For example, the potassium adduct of NGA2F has the same mass as NA2. The success of potassium removal by the cation-exchange step can vary for unexplained reasons. For instance, within the same set of samples, one sample might show a fairly high level of potassium adducts, whereas the remaining samples show very low and acceptable levels.
9. Both the sDHB and THAP matrices solutions can be stored at –70°C and can be reused.
10. The THAP matrix/sample spots must be rehydrated before analysis, which can be accomplished by breathing over the spots.
11. One limitation of this method is the inability to link the relative oligosaccharide distributions obtained in the positive mode with those acquired from the same sample in the

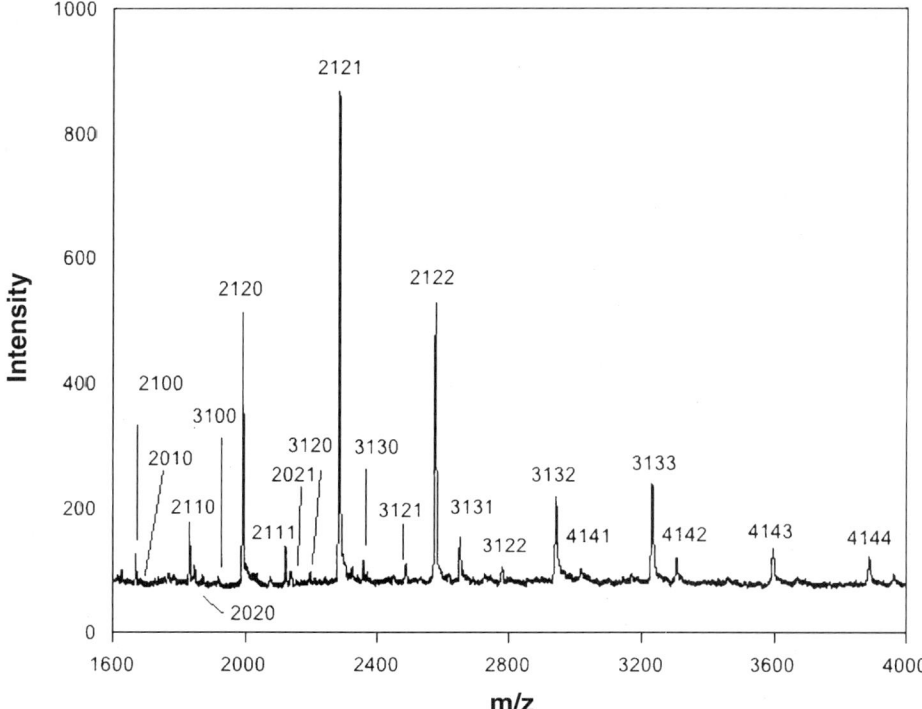

Fig. 3. Negative-ion mode spectrum of ANSA-labeled oligosaccharides. The oligosaccharide mixture used in the labeling procedure was the same one used to generate the positive- and negative-mode spectra (**Figs. 1** and **2**).

negative mode. We have developed a method in which the released oligosaccharides are derivatized with a singularly charged fluorescent reagent, 5-amino-2-napthalenesulfonic acid (ANSA). After derivatization, all of the oligosaccharides (both neutral and sialylated) are negatively charged and can be analyzed in the negative mode using the THAP matrix. **Figure 3** shows the neutral and sialylated oligosaccharides released from TNKase in a single negative-mode spectrum. More information on this method is provided in Chapter 31.

12. Neutral and acidic oligosaccharide standard solutions were constructed from commercially available neutral and sialylated oligosaccharides, respectively. The target concentration was 20 pmol of each oligosaccharide/µL. We prepared a large pool of each oligosaccharide standard mixture in water, and these pools are stored at –70°C. Portions of each oligosaccharide mixture was subaliquoted into 500-µL microcentrifuge tubes. Typically 10 µL is placed in each tube; these working aliquots are also stored at –70°C. The neutral pool consists of Man 5, NGA2, NA2F, and NA4, whereas the acidic standard contains A1, A1F, A2, A2F, and A3. For calibration purposes, we use Man 5 and NA4 as positive-mode calibrants, and for the negative mode, we use A1 and A3.

13. We periodically measure the power output of the laser using a Moleotron single-channel joulemeter over a large range of laser settings to verify that the power output is linear in the range we commonly use. This is especially important after installation of a new laser.

14. The sDHB matrix has a limited lifetime in the MALDI-TOF source and will completely evaporate from the plate after a short period of time. We assume it has a lifetime of 30 min once the plate is inserted into the source. Therefore, we generally do not spot more than 16 samples at a time. THAP is very stable in the source, and we find that our sample number is not limited.

15. We use our acidic standard to determine the optimal laser power setting for the sialylated oligosaccharides. We acquire spectra from the acidic standard with increasing laser power until we begin to see desialylation of the A2F standard, which can be observed as an increase in the A1F signal. The laser power is then decreased to a level that results in efficient signal intensity without any desialylation. This becomes our working laser power for the sialylated oligosaccharides.

16. A factor that can profoundly limit the ability to improve the spectral signal-to-noise ratio is the dynamic range of the instrument. Such limitations imposed by the dynamic range of the mass spectrometer are especially pronounced when analyzing a mixture containing one or more dominant forms of oligosaccharide and a number of oligosaccharides present in relatively small abundances. In this case, attempts to increase the signal-to-noise ratio of the small peaks in the spectrum could be restricted because the signal of the large peaks in the spectrum could saturate the detector if the signal is increased too much.

17. One element that can decrease the spectral signal-to-noise ratio, thereby adversely affecting the determination of the oligosaccharide distribution, is oligosaccharide heterogeneity. This decrease in the signal-to-noise ratio stems from the fact that with increasing heterogeneity, there is a decrease in the concentration of the individual oligosaccharides that constitute the oligosaccharide mixture. Thus, if the oligosaccharide release resulted in a mixture containing 20 different oligosaccharides in roughly equal amounts, the overall signal obtained would be 1/20 of the signal that would be acquired if a single type of oligosaccharide was present. The effect of glycan heterogeneity can be particularly pronounced in negative-mode spectra, because greater heterogeneity exists in sialylated oligosaccharides. This increased heterogeneity is the result of the additional variation introduced by the variability in the degree of sialylation.

There are several ways to compensate for a reduced signal-to-noise ratio. One is to increase the number of acquisitions that are summed to obtain the final spectrum. It is recommended that over 200 acquisitions be summed to obtain the final spectrum. However, summing more than the 200 provides diminishing returns on spectral signal-to-noise. Another way to compensate for a low signal-to-noise ratio is to increase the laser power used in the acquisition. However, as previously stated, if the laser power used in the acquisition is too high, peak broadening and desialylation can result. In some cases, compensation is possible by concentrating the oligosaccharide mixture. This can be accomplished by evaporating the sample to dryness, then reconstituting the oligosaccharides in a smaller volume. Because only 3–5 μL are required for the cation-exchange step, and the digest volume is typically around 25 μL, a fivefold concentration is easily achievable. If further signal enhancement is required, the glycoprotein of interest can be added to more than one well, and the released glycans from each well can be combined, evaporated, and reconstituted to obtain a concentrated oligosaccharide solution. However, concentration of the oligosaccharide mixture does not always lead to better spectral quality, as any interfering substances present in the mixture are also concentrated.

18. The code employed in these figures was described in Papac et al. (*1*). The four-digit ABCD code is based on the known structures of the common mammalian N-linked oligosaccharides, with the core structure (GlcNAc-GlcNAc-Man$_3$) designated as "0000."

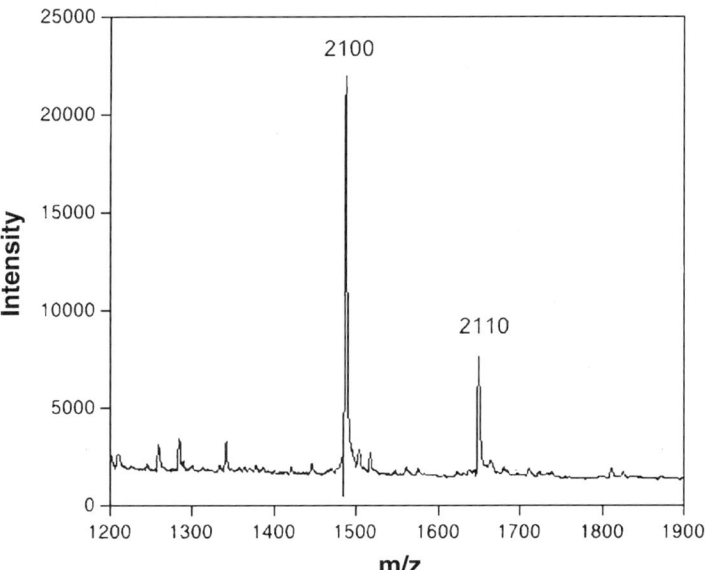

Fig. 4. Positive-mode MALDI-TOF MS analysis of oligosaccharides released from a recombinant monoclonal antibody. This spectrum meets our acceptance criteria with low levels of potassium adducts and sufficient signal strength to allow for quantitation of minor oligosaccharide species.

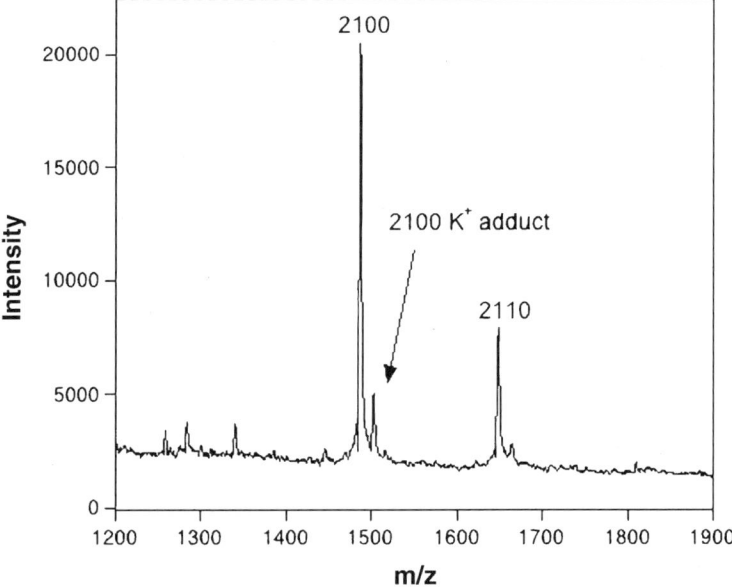

Fig. 5. Positive-mode MALDI-TOF MS analysis of oligosaccharides released from a recombinant monoclonal antibody. This spectrum does not meet our acceptance criteria, because the level of the 2100 potassium adduct is too high, despite that the signal strength is sufficient for quantitation of the minor oligosaccharide species.

The first digit represents the number of additional GlcNAc outside the core structure and is generally the number of antennae in the glycan. The second digit denotes the presence or absence of a fucose in the core structure. The third digit denotes the number of Gal present, and the fourth digit represents the number of sialic acids present. This code is simply a shorthand description of the monosaccharide composition and incorporates structural inferences from the restricted structures commonly synthesized by mammalian cells.

19. Mass spectral acceptance criteria should be defined to ensure consistency of results between analysts and across assays. Currently, our acceptance criteria are based on visual inspection of the spectra to verify that sufficient signal intensity was achieved. (We strive for a peak signal intensity between 15,000 and 30,000.) We also verify that the amount of NGA2F potassium adduct is low or nonexistent. **Figure 4** shows a typical spectrum that meets these acceptance criteria. whereas the spectrum in **Fig. 5** shows an unacceptable amount of the NGA2F potassium adducts. We are in the process of developing less subjective criteria. Although we have not chosen the attribute, such criterion might depend on the signal-to-noise ratio of a predefined oligosaccharide.

20. Open the .prn or .txt files with the Caesar Workstation software using the XY file type in the ASCII import dialog box. If the .prn or .txt file has any header lines, enter the number of header lines in the Header rows dialog box.

## Acknowledgments

The authors thank Moira Kelly for the neutral mode spectra of the release oligosaccharides from a recombinant monoclonal antibody.

## References

1. Papac, D. I., Briggs, J. B., Chin, E. T., and Jones, A. J. S. (1998) A high-throughput microscale method to release N-linked oligosaccharides from glycoproteins for matrix-assisted laser desorption/ionization time-of-flight mass spectrometric analysis. *Glycobiology* **8**, 445–454.

2. Harris, R. J. (1995) Processing of C-terminal lysine and arginine residues of proteins isolated from mammalian cell culture. *J. Chrom.* **705**, 129–134.

3. Rudd, P. M., Leatherbarrow, R. J., Rademacher, T. W., and Dwek, R. A. (1991) Diversification of the IgG molecule by oligosaccharides. *Molecular Immuno.* **28**, 121369–121378.

4. Inoue, N., Watanabe, T., Kutsukake, T., Saitoh, H., Tsumura, H., Arai, H., and Takeuchi, M. (1999) Asn-linked sugar chain structures of recombinant human thrombopoietin produced in Chinese hamster ovary cells. *Glycoconj. J.* **16**, 707–718.

5. Hoffman, R. C., Andersen, H., Walker, K., Krakover, J. D., Patel, S., Stamm, M. R., and Osborn, S. (1996) Peptide, disulfide, and glycoylation mapping of recombinant human thrombopoietin from ser1 to arg246. *Biochemistry* **35**, 14849–14861.

6. Takeuchi, M., Inoue, N., Strickland, T. W., Kubota, M., Wada, M., Shimizu, R., et al. (1989) Relationship between sugar chain structure and biological activity of recombinant human erythropoietin produced in Chinese hamster ovary cells. *Proc. Natl. Acad. Sci.* **86**, 7819–7822.

7. Gazzano-Santoro, H., Ralph, P., Ryskamp, T. C., Chen, A. B., and Mukku, V. R. (1997) A non-radioactive complement-dependent cytotoxicity assay for anti-CD20 monoclonal antibody. *J. Immunol. Methods* **202**, 163–171.

8. Ma, S. and Nashabeh, W. (1999) Carbohydrate analysis of a chimeric recombinant monoclonal antibody by capillary electrophoresis with laser-induced fluorescence detection. *Anal. Chem.* **71**, 5185–5192.

9. Jackson, P. (1990) The use of polyacrylamide-gel electrophoresis for the high-resolution separation of reducing saccharides labelled with the fluorophore 8-aminonaphthalene-1,3,6-trisulphonic acid. Detection of picomolar quantities by an imaging system based on a cooled charge-coupled device. *Biochem. J.* **270,** 705–713.

10. Hase, S., Hara, S., and Matsushima, Y. (1979) Tagging of sugars with a fluorescent compound, 2-aminopyridine. *J. Biochem.* (Tokyo) **85,** 217–220.

11. Lines, A. C. (1996) High-performance liquid chromatographic mapping of the oligosaccharides released from the humanized immunoglobulin CAMPATH™1H. *J. Pharm. Biomed. Analysis.* **14,** 601–608.

12. Takahasi, N. (1996) Three-dimensional mapping of N-linked oligosaccharides using anion-exchange. hydrophobic and hydrophilic interaction modes of high-performance liquid chromatography. *J. Chrom.* **720,** 217–225.

13. Spellman, M. W., Basa, L. J., Leonard, C. K., Chakel, J. A., O'Connor, J. V., Wilson, S., and van Halbeek, H. (1989) Carbohydrate structure of human tissue plasminogen activator expressed in Chinese hamster ovary cells. *J. Biol. Chem.* **264,** 14100–14111.

14. Karas, M., Ehring, H., Nordhoff, E., Stahl, B., Strupat, K., Hillenkamp, F., et al. (1993) Matrix-assisted laser desorption/ionization mass spectrometry with additives to 2,5-dihydroxybenzoic acid. *Org. Mass Spectrom.* **28,** 1476–1481.

15. Strupat, K., Karas, M., and Hillenkamp, F. (1991) 2,5-Dihydroxybenzoic acid: a new matrix for laser desorption-ionization mass spectrometry. *Int. J. Mass Spectrom. Ion Proc.* **111,** 89–102.

16. Harvey, D. J. (1993) Quantitative aspects of matrix-assisted laser desorption mass spectrometry of complex oligosaccharides. *Rapid Commun. Mass Spectrom.* **7,** 614–619.

# 31

## Capillary Electrophoresis of Carbohydrates Derivatized With Fluorophoric Compounds

**Stacey Ma, Wendy Lau, Rodney G. Keck, John B. Briggs, Andrew J. S. Jones, Kathy Moorhouse, and Wassim Nashabeh**

## 1. Introduction

Carbohydrates are known to play a key role in the therapeutic use of recombinant proteins *(1)*. The fundamental understanding of the biological roles of the carbohydrate moieties demands high-performance analytical tools to accomplish both the structural characterization and routine analysis of carbohydrates. Among the various analytical techniques developed currently, nuclear magnetic resonance and mass spectrometry are indispensable tools for the structural elucidation of carbohydrates *(2–4)*.

Alternatively, the routine profiling and quantitative analysis of carbohydrates are accomplished mostly by chromatographic and planar electrophoretic techniques. Among them, the most widely used techniques are high-pH anion-exchange chromatography with pulsed amperometric detection *(5,6)*, liquid chromatography with pre- or post-column derivatization schemes *(7–9)*, and fluorophore-assisted carbohydrate electrophoresis *(10,11)*.

More recently, capillary electrophoresis (CE) has emerged as a powerful tool in carbohydrate analysis *(12)*. CE is simply the instrumental version of planar electrophoresis, which uses detection systems adapted from high-performance liquid chromatography (HPLC). Consequently, many electrolyte systems that were originally developed for traditional electrophoresis (as well as precolumn derivatization schemes, which afforded sensitive detection of carbohydrates separated by HPLC) have been readily adapted for CE analysis coupled with on-column laser-induced fluorescence (LIF) detection. To date, this is the most sensitive detection method employed by CE, with a limit of detection in the low-nanomolar range. The development of LIF detection also triggered the development of several fluorophoric-tagging compounds with superior fluorogenic properties, including 8-aminonaphthalene-1,3,6-trisulfonic acid (ANTS; *13*), aminonaphthalene sulfonic acids *(14)*, and 8-aminopyrene-1,3,6-trisulfonic

From: *Methods in Molecular Biology, vol. 308: Therapeutic Proteins: Methods and Protocols*
Edited by: C. M. Smales and D. C. James © Humana Press Inc., Totowa, NJ

acid (APTS; *15–17*). The advantages of CE over the traditional methods, including enhanced separation efficiencies and shorter analysis time, are extensively documented in the literature *(18–20)*.

This chapter describes three applications for the carbohydrate analysis by CE using two different fluorophoric compounds for derivatization and the materials and methods required to perform the analyses are described. The first application is the analysis of N-linked oligosaccharides on a recombinant immunoglobulin IgG$_1$ (IgG$_1$) monoclonal antibody after derivatization with APTS. The second is the quantification of terminal monosaccharides after derivatization with APTS. The third application is the analysis of N-linked oligosaccharides from a glycoprotein derivatized with 5-amino-2-naphthalenesulfonic acid (ANSA).

## 2. Materials

### 2.1. Analysis of N-Linked Oligosaccharides Derivatized With APTS

The following materials are for profiling and determination of the relative distribution of N-linked oligosaccharides on recombinant monoclonal antibodies released by peptide *N*-glycosidase F (PNGase F) and derivatized with APTS:

1. Enzyme: PNGase F from commercial sources, including New England Biolabs (Beverly, MA), Calbiochem (La Jolla, CA) and Prozyme (San Leandro, CA). Storage and expiration date per manufacturer's recommendation.
2. PNGase F digestion buffer: 20 m*M* sodium phosphate, 50 m*M* EDTA, and 0.02% (w/w) sodium azide, pH 7.5. (Reagent is stable for up to 1 mo at 2–8°C.)
3. Deionized or distilled water.
4. Internal standard (IS): 50 µg/mL maltoheptaose.
5. Fluorophoric compound: 19 m*M* APTS in 15% (v/v) acetic acid. (Reagent is stable for 2 wk at ≤10°C in the dark.)
6. Reducing agent: 1 *M* sodium cyanoborohydride (NaBH$_3$CN) in tetrahydrofuran (THF). (Reagent is stable for 3 mo at ambient temperature.)
7. Separation buffer: Carbohydrate Separation Gel Buffer (Beckman Coulter, Fullerton, CA). Storage and expiration date per manufacturer's recommendation.

### 2.2. Quantification of Terminal Monosaccharides Derivatized With APTS

These materials are for quantification of the terminal monosaccharide, e.g., *N*-acetyl glucosamine (GlcNAc), on a glycoprotein released by specific enzyme and derivatized with APTS:

1. Enzyme: β-*N*-acetylhexosaminidase derived from Jack Bean (Canavalia ensiformis) from commercial sources. Storage and expiration date per manufacturer's recommendation.
2. Enzyme formulation buffer: 20 m*M* sodium citrate, 250 mg/mL bovine serum albumin, and 0.1% (w/w) sodium azide, pH 6.0, adjusted with concentrated phosphoric acid. (Reagent is stable for 2 mo at 2–8°C.)
3. Digestion buffer: 100 m*M* sodium citrate, pH 5.0, adjusted with concentrated phosphoric acid. (Reagent is stable for 2 mo at 2–8°C.)
4. Deionized or distilled water.
5. IS: maltose at a concentration appropriate for the amount of glycoprotein being digested.
6. Monosaccharide standard: *N*-acetyl-D-glucosamine.
7. Ice-cold ethanol, 200 proof.

8. Fluorophoric compound: 19 m$M$ APTS in 15% (v/v) acetic acid. (Reagent is stable for 2 wk at ≤–10°C in the dark.)
9. Reducing agent: 1 $M$ NaBH$_3$CN in THF. (Reagent is stable for 3 mo at ambient temperature.)
10. Separation buffer: Carbohydrate Separation Gel Buffer (Beckman Coulter). Storage and expiration date per manufacturer's instructions.

### 2.3. Analysis of N-Linked Oligosaccharides Derivatized With ANSA

These materials are for profiling and determination of the relative distribution of N-linked oligosaccharides on glycoproteins released by PNGase F and derivatized with ANSA.

1. Enzyme: PNGase F (glycerol-free; New England Biolabs). Storage and expiration date per manufacturer's recommendation.
2. Enzyme buffer: 10 m$M$ Tris-acetate, pH 8.3. (Reagent is stable for up to 1 mo at 2–8°C.)
3. Deionized or distilled water.
4. Fluorophoric compound stock solution: 200 m$M$ ANSA in dimethylsulfoxide (DMSO; *see* **Note 1**). Prepare fresh.
5. Glacial acetic acid.
6. Reducing agent: 1 $M$ NaBH$_3$CN in THF. (Reagent is stable for 3 mo at ambient temperature.)
7. Acetonitrile.
8. Buffer A: 96% acetonitrile and 4% water.
9. Buffer B: 70% acetonitrile and 30% water.
10. Separation buffer: Carbohydrate Separation Gel Buffer (Beckman Coulter). Storage and expiration date per manufacturer's directions.

### 2.4. Equipment

1. Microcon-30 concentrator or PD-10 columns for buffer exchange purposes.
2. Water bath at 37°C, 55°C, and 75°C.
3. Heating block at 95–100°C.
4. SpecCN SPE column (ANSYS Technologies) for purification of ANSA-derivatized oligosaccharides.
5. Centrifugal vacuum evaporator.
6. CE system equipped with LIF detection systems (Beckman Coulter or Bio-Rad Laboratories, Hercules, CA).
7. Fused silica capillary with a neutral hydrophilic coating at the inner wall. The capillary dimensions are 50 µm inner diameter (ID), 375 µm outer diameter (OD), and varying lengths from 30 to 60 cm.

## 3. Methods

### 3.1. Relative Distribution of N-Linked Oligosaccharides Derivatized With APTS

A general method for the analysis of N-linked oligosaccharides on recombinant IgG$_1$ monoclonal antibody is described here. The individual oligosaccharides can be profiled, and changes in the extent or nature of the glycosylation can be quantified based on the relative distribution of the oligosaccharides. The method involves three steps:

1. Enzymatic release of the oligosaccharides from the antibody with PNGase F.
2. Derivatization of the released oligosaccharides with APTS.
3. Analysis of the APTS-oligosaccharide conjugates by CE with LIF detection.

### 3.1.1. Enzymatic Release of N-Linked Oligosaccharides From a Recombinant IgG₁ Monoclonal Antibody

1. Typically 300 µg or 2 nmol antibody sample is required for the analysis (*see* **Note 2**).
2. Buffer exchange the antibody sample using a Microcon-30 concentrator into 40 µL PNGase F digestion buffer (**Subheading 2.1., item 2**).
3. Add 10 µL (10 mIU or as specified by the manufacturer) PNGase F to the sample, and incubate for approx 15 ± 2 h at 37°C in a water bath.
4. Add 50 µL IS (maltoheptaose) solution to each sample. If needed, substitute the internal standard with water.
5. Precipitate the deglycosylated protein by heating at 95°C for 5 min, followed by centrifugation at 10,000g for 10 min. After heating, the solution may appear cloudy.
6. Transfer 80 µL supernatant containing the oligosaccharides to a clean 1.5-mL microcentrifuge tube. Be careful not to disturb the precipitate.
7. Dry the supernatant in a centrifugal vacuum evaporator at low heat for 50–90 min until a translucent oligosaccharide pellet is obtained. At this point, the sample may be stored frozen at ≤ –10°C for up to 1 wk, or proceed to the derivatization step.

### 3.1.2. Derivatization of Oligosaccharides With APTS

1. Add 15 µL APTS (**Subheading 2.1., item 5**) to the dried oligosaccharide pellet. Mix well until pellet is dissolved.
2. Add 5 µL NaBH₃CN (**Subheading 2.1., item 6**). Mix well by vortexing, and pulse centrifuge briefly.
3. Incubate samples at 55°C for 2 h (*see* **Note 3**).
4. Quench the reaction by adding 500 µL water to samples (or approx 25-fold dilution). Mix well by vortexing and pulse centrifuge briefly. At this point, the sample may be stored at room temperature up to 16 h or frozen at ≤ –10°C up to 1 wk prior to CE analysis.

### 3.1.3. CE Analysis

1. Transfer an aliquot of the sample (typically 40–200 µL) to a sampling vial for CE analysis.
2. For CE analysis of APTS-oligosaccharide conjugates, the CE system is equipped with a 3-mW Argon-Ion laser with excitation wavelength at 488 nm and an emission bandpass filter at 520 ± 20 nm.
3. All capillaries are of 50-µm ID and 20-cm effective length (inlet to detector).
4. Set the capillary temperature at 20 ± 2°C.
5. Set the polarity from negative (at the inlet) to positive (at the detector end).
6. Each CE analysis run includes four steps.

   a. Rinse the capillary with the separation buffer for 1 min at a minimum pressure of 20 psi.
   b. Inject the samples hydrodynamically for 4 psi × seconds (i.e., 8 s at 0.5 psi).
   c. Wait for 0.1 min with both ends of the capillary dipped in vials filled with water after injection to minimize sample cross-contamination.
   d. Perform separation under constant voltage mode at an electric field of 740 V/cm. The run time is 4.5 min.

7. A typical profile of the CE analysis of N-linked oligosaccharides is shown in **Figs. 1** (*see* **Note 4**).
8. The relative distribution of the N-linked oligosaccharides is determined based on the corrected peak areas (CPA) of the individual peaks (*see* **Note 5**).

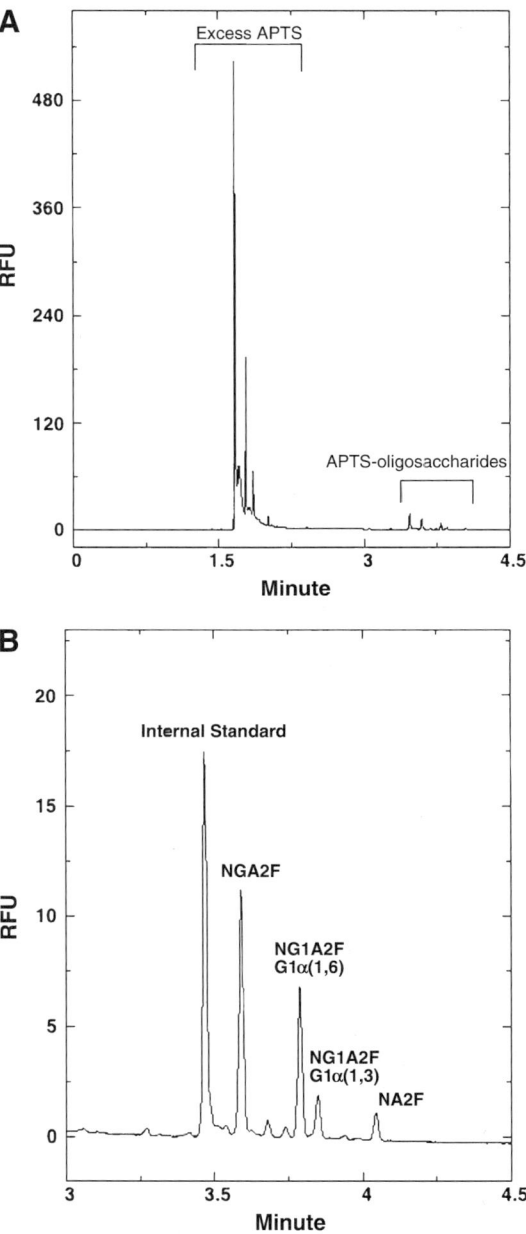

Fig. 1. (**A**) Typical CE profile obtained for the N-linked oligosaccharides analysis of a recombinant IgG$_1$ monoclonal antibody. (**B**) Expanded-scale view showing the APTS-oligosaccharides of interest. Internal standard refers to maltoheptaose. NGA2F refers to N-linked asialo-, agalacto-, biantennary, and core-substituted with fucose structure. NG1A2F refers to N-linked asialo-, mono-galactosylated biantennary, and core-substituted with fucose structure. The G1α(1,6) and G1α(1,3) isomers refer to the structures with the single-terminal galactose residue on the mannose α(1,6) and mannose α(1,3) arms, respectively. NA2F refers to N-linked asialo-, galactosylated biantennary, and core-substituted with fucose structure.

9. At the end of analysis and before storing the capillary, purge the capillary with purified water for 1 min, followed with air or nitrogen for 3 min. At this point, the capillary can be either stored in the instrument or at room temperature in the cartridge cassette.

## 3.2. Quantification of Terminal Monosaccharides Derivatized With APTS

A general method for the quantification of specific terminal monosaccharides of interest on a glycoprotein is described here. The amount of terminal monosaccharides, e.g., GlcNAc can be quantified and reported as moles of GlcNAc per mole of protein. The method involves four steps.

1. Determination of the protein concentration.
2. Enzymatic release of the terminal GlcNAc residues from the antibody with β-*N*-acetyl-hexosaminidase.
3. Derivatization of the released GlcNAc residues with APTS.
4. Quantification of the APTS-GlcNAc conjugates by CE with LIF detection and determination of the terminal GlcNAc amount on a molar ratio basis relative to the protein amount using a four-point GlcNAc standard curve.

### 3.2.1. Protein Concentration Determination

1. Typically we use 5 mg glycoprotein for accurate analysis (*see* **Note 6**).
2. Buffer exchange the glycoprotein into the enzyme digestion buffer using PD-10 columns. Add 1 mL sample onto the PD-10 columns, and discard eluent. Add 1.5 mL digestion buffer onto the column, and discard eluent. Then, add 2 mL digestion buffer onto the column, and collect eluent.
3. Transfer 500 µL buffer-exchanged sample to a clean tube for protein concentration determination. Determine the protein concentration of the buffer-exchanged sample based on absorbance at $A_{max}$ with corresponding extinction coefficient (*see* **Note 7**).
4. Reserve the remaining protein for GlcNAc quantification.

### 3.2.2. Enzymatic Release of Terminal GlcNAc Residues From a Glycoprotein

1. Transfer 100 µL remaining buffer-exchanged sample to each of the following tubes: one tube for the glycoprotein sample (if required, prepare samples in duplicate or triplicate to improve assay precision) and four tubes for the GlcNAc standards that will be used to produce the standard curve (*see* **Note 8**).
2. Add 20 µL β-*N*-acetylhexosaminidase or (20 U enzyme/mg glycoprotein or as specified by the manufacturers) to the glycoprotein sample. Add 20 µL enzyme formulation buffer (**Subheading 2.2.**, **item 2**), instead of the enzyme, to the four GlcNAc standards (*see* **Note 8**).
3. Incubate all the samples for approx 18 ± 1 h at 37°C in a water bath.
4. Pulse centrifuge all the samples after incubation. Add 50 µL IS (maltose) solution and 50 µL water to the glycoprotein sample. For the four GlcNAc standards, add 50 µL maltose solution and 50 µL GlcNAc standards representing an appropriate concentration range (*see* **Note 9** and **Subheading 2.2.**, **items 5** and **6**).
5. Heat the digest at 95–100°C for 5 min in a dry-heating block. Pulse centrifuge, and add 170 µL (or equal volume) cold ethanol (**Subheading 2.2.**, **item 7**) to precipitate the protein. Mix well and centrifuge at 10,000*g* for 10 min.
6. Transfer 250 µL supernatant to a new tube. Be careful not to disturb the precipitate.
7. Dry the supernatant in a centrifugal vacuum evaporator at approx 65°C (or high-drying rate) for about 1.5–2 h to a translucent or crystalline pellet. At this point, the sample may be stored frozen at ≤ –10°C for up to 1 wk, or proceed to the derivatization step.

### 3.2.3. Derivatization of Monosaccharide With APTS

1. Add 15 µL APTS (**Subheading 2.2.**, **item 8**) to the dried oligosaccharide pellet. Mix well until the pellet is dissolved.
2. Add 5 µL NaBH$_3$CN (**Subheading 2.2.**, **item 9**). Mix well by vortexing, and pulse centrifuge briefly.
3. Incubate the samples at 75°C for 1.5–2 h.
4. Quench the reaction by adding 1 mL purified water (or approx 50-fold dilution). Mix well by vortexing, and pulse centrifuge briefly. At this point, the sample may be stored frozen at ≤ –10°C up to 2 wk prior to CE analysis.

### 3.2.4. CE Analysis

Repeat **steps 1–5** of **Subheading 3.1.3.**

1. Each CE analysis run includes four steps:
   a. Rinse the capillary with the separation buffer for 1 min at a minimum pressure of 20 psi.
   b. Inject the samples hydrodynamically for 2 psi × seconds (i.e., 4 s at 0.5 psi).
   c. Wait for 0.1 min with both ends of the capillary dipped in vials filled with water after injection to minimize sample cross-contamination.
   d. Perform separation under constant voltage mode at an electric field of 648 V/cm. The run time is 3 min.
2. A typical profile of the CE analysis of terminal GlcNAc is shown in **Fig. 2**.
3. Calculate the CPA of the GlcNAc and the IS (maltose) peak (*see* **Note 5**).
4. Determine the relative CPA of GlcNAc by dividing its CPA with the CPA of the IS. Generate a four-point standard curve by plotting the relative CPA of GlcNAc/IS vs the corresponding GlcNAc concentration (**Fig. 3**). Use a linear curve fit and the correlation coefficient, *R*, should be greater than or equal to 0.99.
5. Based on the standard curve slope and intercept, determine the molar amount of GlcNAc in the glycoprotein digest.
6. Determine the molar amount of glycoprotein in the digest by multiplying the volume used for the digest (in this case, 100 µL) and the protein concentration previously determined (**Subheading 3.2.1.**).
7. Determine the GlcNAc:protein ratio by dividing the number of moles of GlcNAc by the number of moles of glycoprotein (*see* **Subheading 3.2.4.**, **steps 5** and **6**).
8. At the end of analysis and before storing the capillary, purge the capillary with purified water for 1 min, followed with air or nitrogen for 3 min. At this point, the capillary can be either stored in the instrument or at room temperature in the cartridge cassette.

## 3.3. Relative Distribution of N-Linked Oligosaccharides Derivatized With ANSA

Although APTS derivatization offers high resolution and rapid analysis for the neutral oligosaccharides as described in **Subheading 3.1.**, the singly charged ANSA facilitates the resolution improvement for sialylated oligosaccharides, because only a single charge is added to the ANSA-oligosaccharide conjugates compared to the three charges when APTS is conjugated to the oligosaccharides. Therefore, a general method for the analysis of N-linked oligosaccharides on glycoproteins after derivatization with ANSA is described here. The individual oligosaccharides can be profiled, and changes

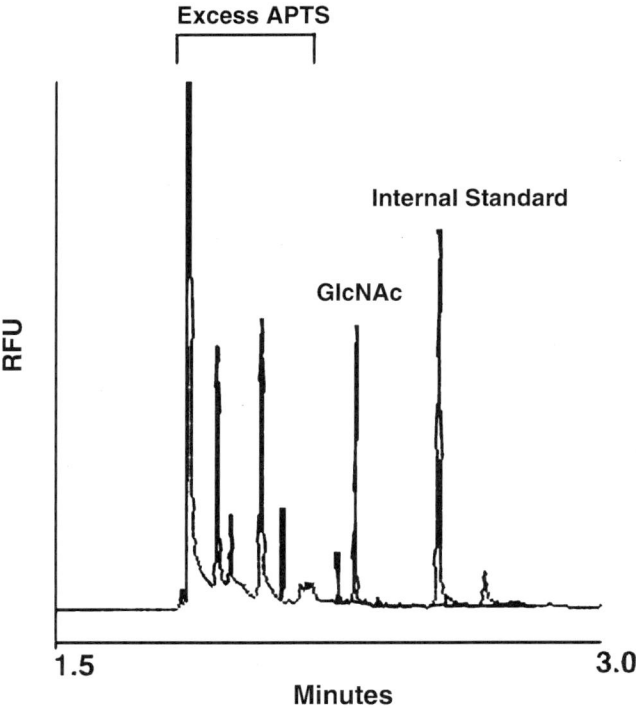

Fig. 2. Typical CE profile obtained for the terminal monosaccharides (GlcNAc) analysis of a glycoprotein. Internal standard refers to maltose.

in the extent or nature of the glycosylation can be quantified depending on the relative distribution of the ANSA-oligosaccharides. The method involves four steps.

1. Enzymatic release of the oligosaccharides from the glycoprotein with PNGase F.
2. Derivatization of the released oligosaccharides with ANSA.
3. Purification of derivatized oligosaccharides to remove excess ANSA reagent using a solid-phase extraction column.
4. Analysis of the ANSA-oligosaccharide conjugates by CE with LIF detection.

### 3.3.1. Enzymatic Release of N-Linked Oligosaccharides From a Glycoprotein

1. The detailed procedure is described in Chapter 30. Follow **Subheadings 3.1.–3.4.** (*see* **Note 10**).
2. After digestion, combine the solution containing oligosaccharides from four separate wells for the subsequent analysis (*see* **Note 11**).
3. Dry the supernatant containing the oligosaccharides in a centrifugal vacuum evaporator for about 1–2 h to a translucent or crystalline pellet. At this point, the sample may be stored frozen at ≤ –70°C overnight, or proceed to the derivatization step.

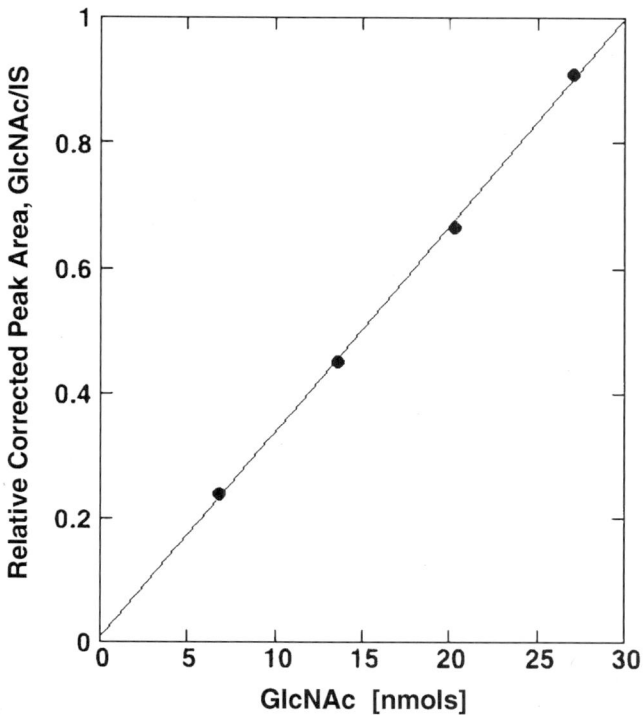

Fig. 3. A typical four-point standard curve of GlcNAc derivatized with APTS.

### 3.3.2. Derivatization of Oligosaccharides With ANSA

1. Add 75 µL glacial acetic acid to 225 µL 0.2 *M* ANSA solution to yield an ANSA solution of 150 m*M* ANSA in 25% acetic acid (*see* **Note 12** and **Subheading 2.3.**, **items 4** and **5**). Mix well by vortexing.
2. Add 5 µL water to the dried oligosaccharides pellet. Mix well until the pellet is dissolved.
3. Add 8 µL 150 m*M* ANSA solution. Mix well by vortexing, and pulse centrifuge briefly.
4. Add 8 µL NaBH$_3$CN (**Subheading 2.3.**, **item 6**). Mix well by vortexing, and pulse centrifuge briefly.
5. Incubate the samples at 37°C overnight.

### 3.3.3. Purification of ANSA-Derivatized Oligosaccharides

1. Equilibrate the SpecCN SPE columns by adding 1 mL acetonitrile to each column, followed by 1 mL water and 2 mL acetonitrile.
2. Add 200 µL acetonitrile to each sample. Vortex.
3. Transfer the ANSA-oligosaccharide sample (**Subheading 3.3.2.**, **step 5**) to the SpecCN column, and incubate at room temperature for about 15 min.
4. Wash the column with 1 mL acetonitrile, and follow with four additions of 1.5 mL buffer A (**Subheading 2.3.**, **item 8**).
5. Add 1.5 mL buffer B (**Subheading 2.3.**, **item 9**) to the SPE column to elute the ANSA-oligosaccharides with decreasing concentration of acetonitrile, and collect the eluate.

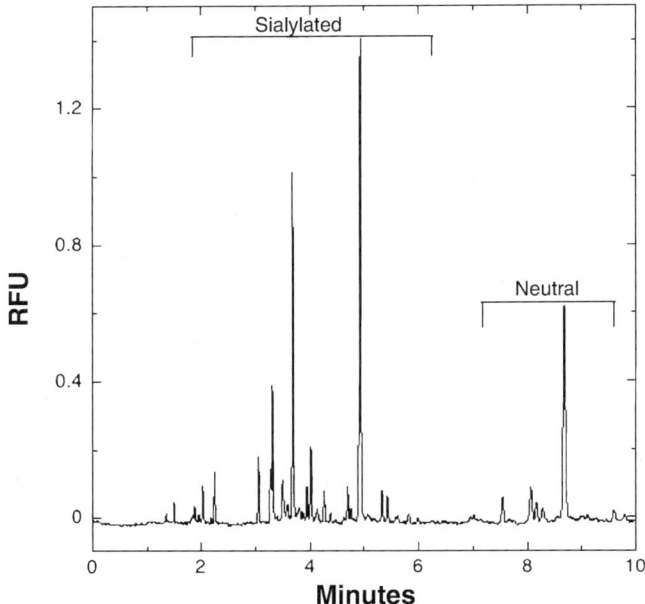

Fig. 4. Typical CE profile obtained for the analysis of N-linked oligosaccharides derivatized with ANSA.

6. Concentrate the pooled eluates containing the ANSA-oligosaccharides in a centrifugal vacuum evaporator for about 1 h to a translucent or crystalline pellet. Then, the sample may be stored frozen at –70°C for up to 1 mo, or proceed to the CE analysis.

### 3.3.4. CE Analysis

1. Add approx 50 µL water to the ANSA-oligosaccharides pellet. Mix well until the pellet is dissolved.
2. Transfer about 40 µL ANSA-oligosaccharide solution to a sampling vial prior to CE analysis.
3. For CE analysis of ANSA-oligosaccharide conjugates, the CE system is equipped with a 3-mW He-Cd laser with excitation wavelength at 325 nm and an emission bandpass filter at 460 ± 40 nm.
4. All capillaries are of 50-µm ID and 20 cm effective length (inlet to detector).
5. Set the capillary temperature at 20 ± 2°C.
6. Set the polarity from negative (at the inlet) to positive (at the detector end).
7. Each CE analysis run includes four steps:
   a. Rinse the capillary with the separation buffer for 1 min at a minimum pressure of 20 psi.
   b. Inject the samples hydrodynamically for 2.5 psi × seconds (i.e., 5 s at 0.5 psi).
   c. Wait for 0.1 min with two ends of the capillary dipped into a set of vials filled with water after injection to minimize sample cross-contamination.
   d. Perform separation under constant voltage mode at an electric field of 968 V/cm.
8. A typical profile of CE analysis of ANSA-oligosaccharides is shown in **Fig. 4**.

9. The relative distribution of the N-linked oligosaccharides is determined based on the CPA of the individual peaks (*see* **Note 5**).

10. At the end of analysis and before storing the capillary, purge the capillary with purified water for 1 min, followed with air or nitrogen for 3 min. The capillary can be stored in the instrument or at room temperature in the cartridge cassette.

## 4. Notes

1. To facilitate the dissolution of ANSA in DMSO, incubate the solution at 65°C for 3 h prior to use.

2. At least 50 µg $IgG_1$ is recommended for the analysis described here. Greater than 99% digestion is typically achieved using these conditions.

3. The derivatization yield is not 100% under these conditions. However, the absolute yield was found to have no impact on the relative distribution of the oligosaccharides *(19)*.

4. The two positional isomers of NG1A2F are separated at baseline, and the characterization was previously reported *(19)*.

5. The CPA is defined as the peak area of a given peak divided by its migration time. This correction is needed to adjust for the inherent bias in CE with on-column detection, because each peak migrates through the detection window at different speeds. For example, the peak is broader if it migrates pass the detector at a lower speed; hence, the apparent peak area is larger.

6. The protein amount required for this assay depends on the amount of terminal monosaccharide of interest present on the protein. The amount estimated here is recommended for the monosaccharide-to-protein ratio in the range from 1 to 5 mol/mol. Clearly, the accuracy of the protein concentration will affect the accuracy of the mol/mol value observed.

7. Protein concentration can be determined by other techniques, such as amino acid analysis. If ultraviolet absorbance measurement is performed, it is important to measure the $A_{max}$ between 0.5 and 1.0 absorbance units and correct for the light scattering by subtracting the $A_{320}$ reading from $A_{max}$.

8. The effect of sample matrixes on the derivatization yield was documented in the literature *(19)*. To minimize the yield difference between the sample and GlcNAc standard curve, it is crucial to mimic the protein sample matrix in the GlcNAc standards.

9. It is essential to use an internal standard in a quantitative assay to improve assay precision by correcting the inherent variability from sample to sample observed during the APTS derivatization and injection in CE. The IS solution should target an appropriate concentration so that the peak area of the IS is similar to that of the GlcNAc peak in the sample. The GlcNAc standard curve should target a range to ensure the sample concentration is in the middle of the standard curve.

10. In principle, this procedure is similar to that for the monoclonal antibody with the addition of reduction and alkylation steps before PNGase F digest. This may be required for some glycoproteins to achieve complete digestion. On the contrary, for the $IgG_1$ antibodies (as previously described in **Subheading 3.1.**), complete digestion is achieved even when the intact antibody is digested with PNGase F.

11. Typically 50 µg glycoprotein is required for the analysis. The actual amount of glycoprotein required for profiling and relative distribution determination will depend on the size of the glycoprotein, degree of glycosylation, and the extent of glycosylation heterogeneity. In general, 50 µg glycoprotein is a good starting point.

12. The optimal acetic acid concentration ranges from 15% to 25%, in which the derivatization yield remains constant, and no desialylation was observed.

# References

1.  Varki, A. (1993) Biological roles of oligosaccharides: all the theories are correct. *Glycobiology* **3,** 97–130.
2.  Lee, K. B., Loganathan, D., Merchant, Z. M., and Linhardt, R. (1990) Carbohydrate analysis of glycoproteins. *Appl. Biochem. Biotechnol.* **23,** 53–80.
3.  Chaplin, M. F. and Kennedy, J. F. *Carbohydrate Analysis: A Practical Approach,* 2nd ed., IRL Press, Oxford, UK, 1994.
4.  Papac, D. I., Briggs, J. B., Chin, E. T., and Jones, A. J. S. (1998) A high-throughput microscale method to release N-linked oligosaccharides from glycoproteins for matrix-assisted laser desorption/ionization time-of-flight mass spectrometric analysis. *Glycobiology* **8,** 5445–5454.
5.  Townsend, R. R., Hardy, M. R., and Lee, Y. C. (1989) Separation of oligosaccharides using high-performance anion-exchange chromatography with pulsed amperometric detection. *Methods Enzymol.* **179,** 65–76.
6.  Basa, L. and Spellman, M. W. (1990) Analysis of glycoprotein-derived oligosaccharides by high-pH anion-exchange chromatography. *J. Chromatogr.* **499,** 205–220.
7.  Hase, S. (1996) Pre-column derivatization for chromatographic and electrophoretic analyses of carbohydrates. *J. Chromatogr.* **720,** 173–182.
8.  Hase, S. Pre- and post-column detection-oriented derivatization techniques in HPLC of carbohydrates, in *Carbohydrate Analysis: High Performance Liquid Chromatography and Capillary Electrophoresis* (El Rassi, Z., ed.), Elsevier, Amsterdam, NL, 1995, pp. 551–575.
9.  Honda, S. and Suzuki, S. (1984) Common conditions for high-performance liquid chromatographic microdetermination of aldoses, hexosamines and sialic acid in glycoproteins. *Anal. Biochem.* **142,** 167–174.
10. Jackson, P. (1990) The use of polyacrylamide-gel electrophoresis for the high resolution separation of reducing sugars labeled with the fluorophore 8-amino-naphthalene-1,3,6-trisulfonic acid. *Biochem. J.* **270,** 705–713.
11. Starr, C., Masada, R. I., Hague, C., Skop, E., and Klock, J. (1996) Fluorophore-assisted-carbohydrate-electrophoresis, FACE® in the separation, analysis and sequencing of carbohydrates. *J. Chromatogr. A* **720,** 295–321.
12. El Rassi, Z. and Mechref, Y. (1996) Recent advances in capillary electrophoresis of carbohydrates. *Electrophoresis* **17,** 275–301.
13. Chiesa, C. and Horvath, Cs. (1993) Capillary zone electrophoresis of malto-oligosaccharides derivatized with 8-aminonaphthalene-1,3,6-trisulfonic acid. *J. Chromatogr.* **645,** 337–352.
14. Chiesa, C. and O'Neill, R. A. (1994) Capillary zone electrophoresis of oligosaccharides derivatized with various aminonaphthalene sulfonic acids. *Electrophoresis* **15,** 1132–1140.
15. Evangelista, R. A., Liu, M. S., and Chen, F. A. (1995) Characterization of aminopyrene trisulfonate derivatized sugars by capillary electrophoresis with laser-induced fluorescence detection. *Anal. Chem.* **67,** 2239–2245.
16. Chen, F. A. and Evangelista, R. A. (1995) Analysis of mono- and oligosaccharide isomers derivatized with aminopyrene trisulfonate by capillary electrophoresis with laser-induced fluorescence. *Anal. Biochem.* **230,** 273–280.
17. Guttman, A. and Pritchett, T. (1995) Capillary gel electrophoresis separation of high-mannose type oligosaccharides derivatized by aminopyrene trisulfonic acid. *Electrophoresis* **16,** 1906–1911.

18. Chiesa, C., O'Neill, R. A., Horvath, Cs, and Oefner, P. J. Analysis of glycoproteins, oligo- and monosaccharides, in *Capillary Electrophoresis in Analytical Biotechnology* (Righetti, P. G., ed.), CRC Press, Boca Raton, FL, 1996, pp. 277–430.

19. Ma, S. and Nashabeh, W. (1999) Carbohydrate analysis of a chimeric recombinant monoclonal antibody by capillary electrophoresis with laser-induced fluorescence detection. *Anal. Chem.* **71,** 5185–5192.

20. Ma, S. and Nashabeh, W. Analysis of protein therapeutics by capillary electrophoresis, in *CE in Biotechnology: Practical Applications for Protein and Peptide Analyses* (Chen, A., Nashabeh, W., and Wehr, T., eds.), Chromatographia, Wiesbaden, Weinheimer, 2001, pp. S-75–S-89.

# 32

# High-Throughput LC/MS Methodology for $\alpha(1\rightarrow3)$Gal Determination of Recombinant Monoclonal Antibodies

## A Case Study

**Lihua Huang and Charles E. Mitchell**

## 1. Introduction

Antibodies are the first line of defense of the adaptive immune response and are found in blood plasma and extracellular fluids. Many monoclonal antibodies (MAbs) have been marketed and are being tested in clinical trials as therapeutics in immune and inflammatory diseases, oncology, and other therapeutic areas *(1,2)*. Development of these therapeutic protein–based drugs is a multistep process typically involving the selection of an appropriate production cell line, evaluation of a library of monoclonal clones, and optimization of bioreactor growth conditions. Each step requires a careful evaluation of the quantity and quality of the antibodies produced. Of particular interest are the carbohydrate chains attached to all immunoglobulin G (IgG) heavy chains. One such structure is Galactose (Gal)$\alpha(1\rightarrow3)$Gal$\beta(1\rightarrow4)$GlcNAc, which is an epitope of one of the natural human antibodies *(3)*. Traditionally, only titer is considered as a criterion for clone selection of therapeutic proteins for development. For certain cell lines, aside from the clone titer, it may also be necessary to consider oligosaccharide structures as a criterion of clone selection.

Traditional methods *(4–7)* for carbohydrate structural characterization require pure material and have long cycle times. Thus, they are unsuitable to support the molecular biology stage for clone selection. This chapter discusses a high-throughput method using mass spectrometry (MS) to estimate the relative percentage of oligosaccharide containing $\alpha(1\rightarrow3)$-linked Gal residues for the evaluation of the relatively large numbers of samples generated in clone selection and bioreactor optimization.

From: *Methods in Molecular Biology, vol. 308: Therapeutic Proteins: Methods and Protocols*
Edited by: C. M. Smales and D. C. James © Humana Press Inc., Totowa, NJ

## 2. Materials

1. Mobile phase A: 0.15% formic acid aqueous solution.
2. Mobile phase B: 0.12% formic acid acetonitrile.
3. 50 mg/mL Dithiothreitol aqueous solution (reducing reagent).
4. A manifold system (Millipore).
5. 96-Well Multiscreen plate with low-protein absorption filter membrane (Millipore).
6. 96-Well plates.
7. 3 $M$ Tris-HCl buffer, pH 8.0.
8. 1X Dulbecco's phosphate-buffered saline (PBS; Invitrogen Corporation).
9. α-Glycosidase kit (from green coffee bean; Prozyme).
10. Recombinant protein A Sepharose Fast Flow (Amersham Pharmacia Biotechnology AB).
11. High-performance liquid chromatography (HPLC) system (Waters Alliance Model 2695).
12. Reversed-phase Vydac C4 2. 1 × 50-mm column (Vydac).
13. Quadrupole time-of flight (Q-tOF) Micro Mass spectrometer (Micromass).

## 3. Methods

The methods described below outline the (1) capture of MAb, (2) liquid chromatography (LC)/MS analysis of captured MAb, and (3) mass spectral data analysis.

### 3.1. Capture of MAb

To quantitatively determine the relative percentage of oligosaccharides containing α-Gal residues, it is necessary to obtain a good-quality mass spectrum. In the early stages of fermentation development, crude cell culture normally contains many other proteins, bovine antibodies, DNA, lipids, detergents, and salts. These impurities will decrease mass spectral quality obtained if the crude cell culture is directly analyzed by LC/MS. It is a good practice to have a simple purification step before LC/MS analysis. Protein A strongly binds the Fc region of human IgG antibodies at neutral-to-alkaline pH to form a protein A–IgG complex. By adjusting the pH to 3 or less, the IgG is released from the complex. LC/MS analysis for the captured antibody does not require any further sample preparation and only needs approx 1 μg quantity of antibody material for a quality mass spectrum. Use of a 96-well plate with immobilized protein A resin is a simple and rapid purification method, which can handle high-throughput samples, and the captured material obtained by this method is sufficient for LC/MS analysis.

### 3.1.1. Immunoprecipitation of MAb

1. Transfer 1–2 μL immobilized resin with protein A into each well of a 96-well plate (*see* **Note 1**), to which crude cell culture will be added.
2. Add 100–300 μL crude cell culture containing the MAb of interest into each well for capture.
3. Seal the plate with a cover sheet.
4. Incubate the plate at ambient temperature for about 2 h with gentle shaking to let the protein A bind to the antibody.
5. Put the plate on a vacuum manifold, and filter to remove cell culture (*see* **Note 2**).
6. Wash the resin by adding approx 100 μL PBS buffer into each well, then apply vacuum to the manifold system to remove the solution. Repeat this step one more time.

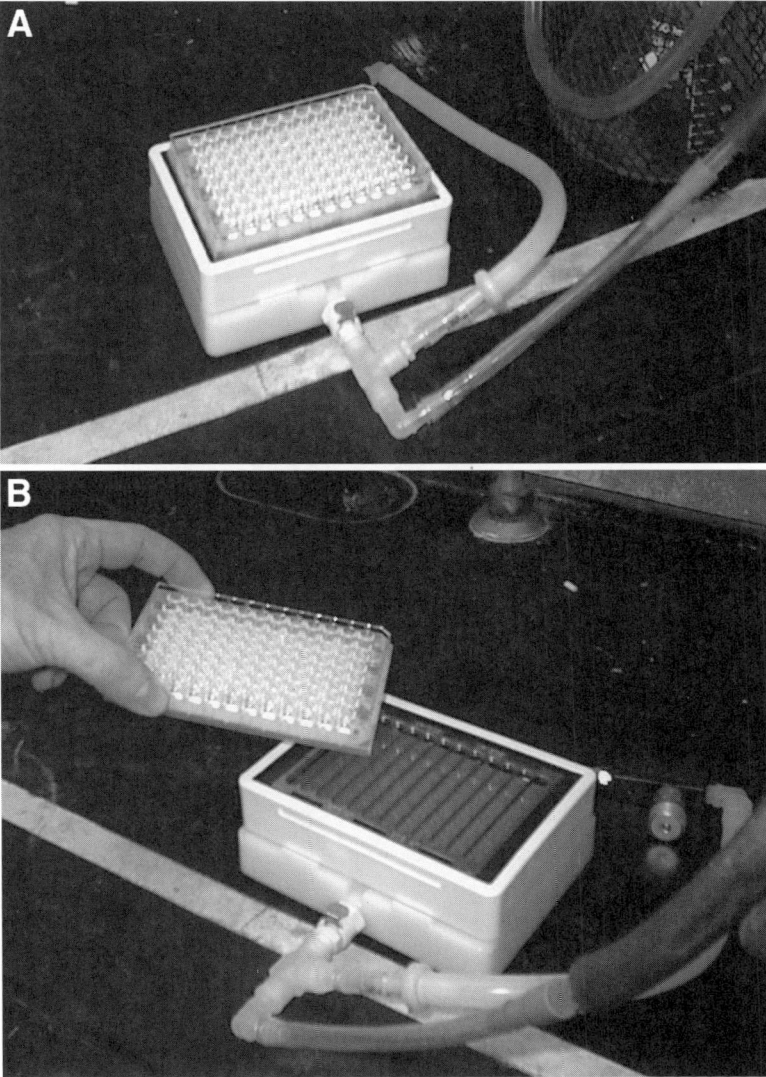

Fig. 1. The manifold with Multiscreen plate (**A**) and 96-well collection plate (**B**).

### 3.1.2. Release of MAb

1. Add approx 50 µL 0.15% formic acid aqueous solution into each sample well (*see* **Note 3**).
2. Incubate the plate for 5 min to release the antibody from the immobilized protein A resin.
3. Place a 96-well plate inside the manifold system (*see* **Fig. 1**). This plate is used to collect the filtrate, which contains the antibody of interest.
4. Apply vacuum, and collect the filtrate (*see* **Note 4**).
5. Transfer the solution into HPLC vials (*see* **Note 5**).

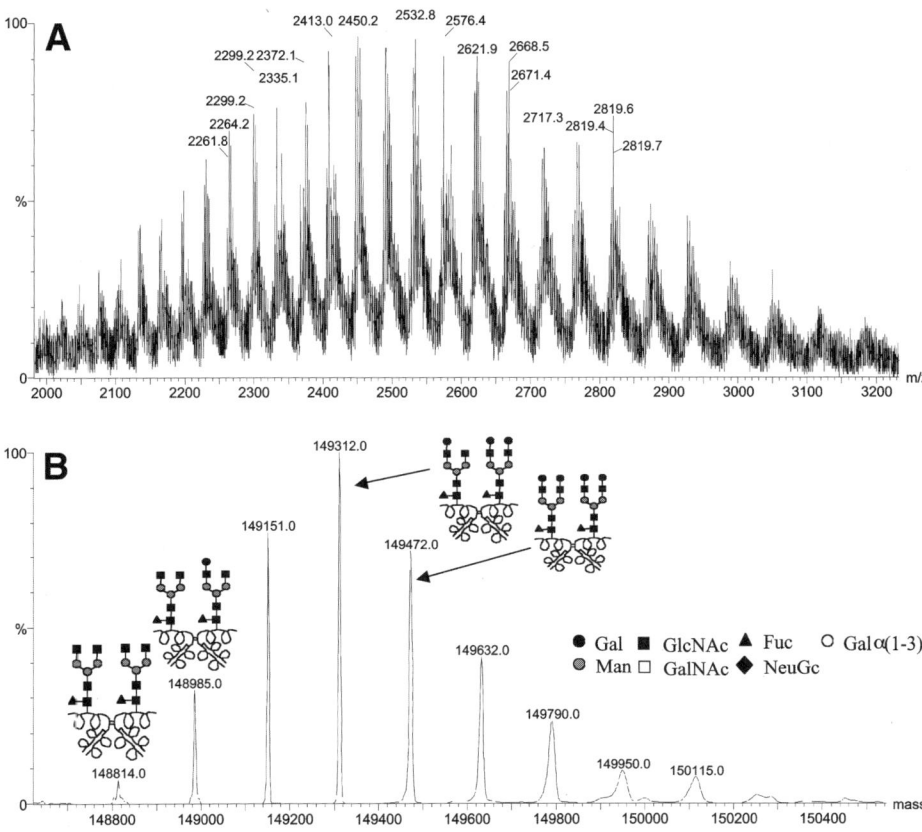

Fig. 2. Mass spectrum (**A**) and deconvoluted mass spectrum (**B**) of intact monoclonal IgG antibody.

## 3.2. LC/MS Analysis

IgG antibody has two heavy chains and two light chains. The Fc region of each heavy chain contains one *N*-linked glycosylation site, which has several important functions *(8,9)*. Direct LC/MS analysis (**Fig. 2**) for an intact antibody shows product-quality information, such as the existence of a half antibody, which cannot be acquired by LC/MS analysis for the reduced antibody. However, oligosaccharide analysis from the mass spectrum of an intact antibody is much more complex, and some oligosaccharide structures cannot be assigned. In addition, the sensitivity of LC/MS analysis for an intact antibody is less than half of that for the reduced antibody, which is very important for low-titer clone samples. Hence, it is appropriate to analyze reduced antibodies by LC/MS to obtain information on oligosaccharide structures.

### 3.2.1. Partial Reduction of IgG Antibody

1. Add 10 µL of 3 *M* Tris buffer, pH 8.0, into each antibody vial to adjust the pH to 8 or more. Check the pH with pH paper. Add more 3 *M* Tris, pH 8, if necessary.

2. Add 1 μL 50 mg/mL DTT solution into each vial, and incubate the vials at ambient temperature for approx 5 min (*see* **Note 6**) to partially reduce the antibody into the heavy and light chains.

### 3.2.2. HPLC Conditions

1. Wash a Vydac C4 2.1 × 50-mm HPLC column with mobile phase B at 0.5 mL/min for 5 min, then 20% B for 10 min at 0.5 mL/min (*see* **Note 7**).
2. Inject approx 20 μL (containing 1–10 μg antibody; *see* **Note 8**) of sample solution onto the HPLC column.
3. Elute the sample using the following gradient (*see* **Note 9**).

| Time (min) | Flow rate (mL/min) | Mobile phase A (%) | Phase B (%) |
|---|---|---|---|
| 0 | 0.5 | 80 | 20 |
| 2 | 0.5 | 50 | 50 |
| 2.1 | 0.2 | 50 | 50 |
| 7 | 0.2 | 10 | 90 |
| 9 | 0.2 | 10 | 90 |
| 10 | 0.5 | 80 | 20 |
| 14 | 0.5 | 80 | 20 |

### 3.2.3. Mass Spectrometer Conditions

1. Connect HPLC tubing to a switch valve, then to a Micromass Q-TOF mass spectrometer.
2. Set a switch valve program to let the HPLC stream go directly to the mass spectrometer between 2 and 9 min, and go to waste for the remaining times.
3. Set the mass spectrometer conditions as shown (*see* **Note 10**).

Electrospray mode: Positive
Capillary: 3000 V
Mass scan range: 600–1990
Scan time (min): 2–9
Sample cone: 35 V
Source temperature: 105°C
Collision energy: 10 eV
Cone gas flow: 50 L/h
Desolvation temperature: 250°C
Desolvation gas flow: 300 L/h

## 3.3. LC/MS Data Analysis

The procedure for LC/MS data analysis is to (1) obtain an average mass spectrum of the heavy chain from a total ion chromatogram (TIC; *see* **Fig. 3**); (2) deconvolute an average mass spectrum to obtain a zero-charged mass spectrum of the heavy chain; (3) assign each mass to glycoform (oligosaccharide); and (4) integrate mass peak area, and calculate relative percentage of each oligosaccharide and oligosaccharides with α(1→3)-linked Gal residues.

### 3.3.1. Mass Spectrum of the Heavy Chain

1. Open a TIC (**Fig. 3**) with "Chromatogram" function on MassLynx software.
2. Obtain an average mass spectrum that covers the whole heavy-chain peak.

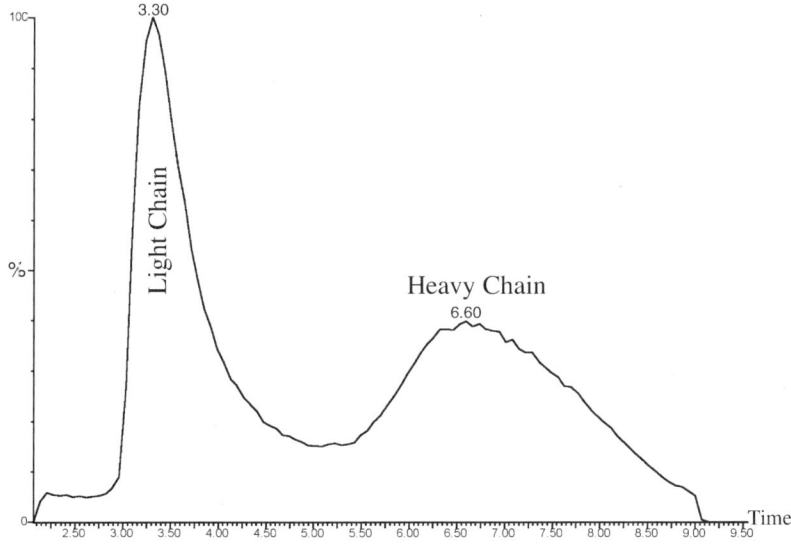

Fig. 3. TIC of LC/MS analysis for partially reduced MAb.

### 3.3.2. Deconvoluted Mass Spectrum of the Heavy Chain

1. Subtract the baseline of the average mass spectrum of the heavy chain if necessary.
2. Deconvolute the average mass spectrum with MaxEnt1 function on MassLynx to give a zero-charged mass spectrum of the heavy chain.

### 3.3.3. Mass Peak Assignment (***Fig. 4***)

For the first time, oligosaccharide containing α-Gal residues can be assigned by comparing the LC/MS analysis for the sample treated with and without α-galactosidase. α-Galactosidase only cleaves the terminal α(1→3, 4, or 6)-linked Gal residue *(10)*. In fact, α-galactosyltransferase only transfers α-Gal residues to Gal (acceptor) with β linkage *(11)*. Hence, if the oligosaccharide is biantennary but has three or four Gal residues based on the mass obtained, it most likely contains one or two α-linked Gal residues. Once the oligosaccharides with α-Gal residues are identified, they can be assigned based on the masses obtained.

1. Add 10 µL 250 m*M* sodium phosphate buffer, pH 5.0, into 40 µL captured (**Subheading 3.1.1.**, **steps 1–6** and **Subheading 3.1.2.**, **steps 1–5**).
2. Add 4 µL α-galactosidase solution into the sample vial.
3. Incubate the solution at 37°C for 4 h.
4. Repeat **Subheading 3.2.1.**, **steps 1–2** to **Subheading 3.3.2.**, **steps 1–2**.

### 3.3.4. Peak Integration and Relative Percentage Calculation

After obtaining a deconvoluted mass spectrum for partially reduced MAb, the area of each mass peak can be obtained by selecting the "Integration" function on MassLynx

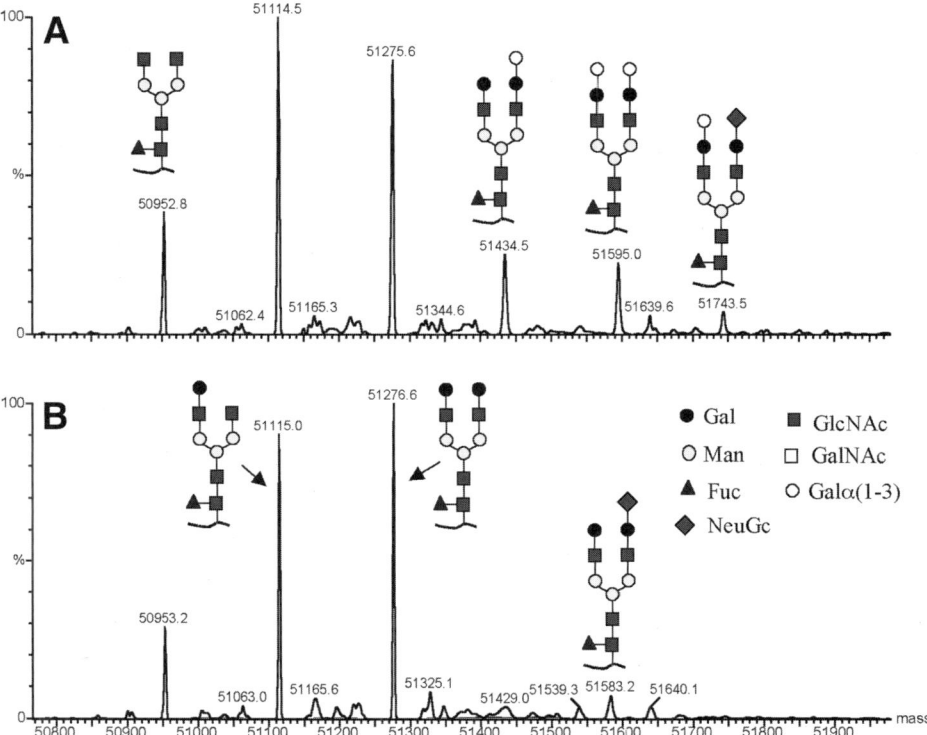

Fig. 4. Deconvoluted mass spectra of partially reduced heavy chain of MAb before (**A**) and after (**B**) treatment with α-galactosidase.

software. The relative percentage of each oligosaccharide is calculated according to **Eq. 1**. The percentage of oligosaccharides with α-linked Gal residue is equal to the sum of all oligosaccharides with α-linked Gal residues (*see* **Note 11**).

$$\text{Relative percent } (\%) = \frac{\text{Peak area of an oligosaccharide mass}}{\sum \text{Peak area of each oligosaccharide mass}} \times 100\% \qquad (1)$$

## 4. Notes

1. If only a few samples are to be analyzed, Eppendorf tubes can be used instead of 96-well plates.
2. Some crude cell culture samples may contain many particles that will block the filter membrane. Centrifugation may be needed for these samples. To rapidly filter crude cell culture material, cut the cover seal over used sample wells, and leave the cover seal for empty wells.
3. Other buffers, such as 0.1 *M* citrate buffer, pH 3.0, are also used for releasing antibody from the immobilized protein A.

4.  If the volume for releasing solution (0.15% formic acid in water) is too small, it is very difficult to drop filtrates into the wells of the collecting plate, and the droplets may hang onto the back of the plate. Before taking away the sample plate, tap the plate to allow droplets to fall into the wells.
5.  If an HPLC is equipped with a 96-well plate autosampler, this step could be omitted.
6.  The intermolecular disulfide bonds between the heavy and light or heavy and heavy chains are easily reduced under high pH. The incubation time is not necessary. However, intramolecular disulfide bonds are normally not reduced under these conditions. Partially reduced MAb will slowly form half antibody or molecules with two heavy chains and one light chain under partially reduced condition after 1 d at 4°C.
7.  If running one blank before running samples, the wash step is not necessary.
8.  The amount of sample injected for LC/MS analysis depends on instrument sensitivity, column size, and HPLC mobile system. For very low concentrations of sample, 1- or 0.3-mm diameter columns can be used. Without changing any other conditions, 1/4–1/10 of the material is needed when using these small-diameter columns.
9.  Several factors affect the sensitivity of the mass spectrometer and the quality of the recorded spectral quality. The acidic modulator strongly affects the sensitivity of the mass spectrometer. Trifluoroacetic acid (TFA) is a good ion-paring reagent. Using TFA as the acidic modulator normally gives nice peak shape, but TFA is not good for MS. The negative counter ion of TFA will suppress mass signals. Alternatively, using formic acid as an acidic modulator increases the sensitivity about 5–10 times than TFA for MS analysis of antibody. However, formic acid normally gives unsymmetrical peaks and low resolution compared with TFA. In addition, the formic acid/acetonitrile/$H_2O$ may have sample carryover problems. It may be necessary to add a zig-zag washing step (with TFA mobile-phase system) after sample gradient, or add a short-washing run between the sample runs.
10. Different instruments require different conditions. It is important to optimize the conditions to get the best quality of mass spectrum. Among them, the sample cone (or declustering potential for Sciex's instrument) needs to be carefully adjusted. Because glycosidic linkages are more labile compared with peptide-amide bonds, a large-sample cone voltage may cause fragmentation of glycosylation. However, a sample-cone voltage that is too low decreases the sensitivity of the mass spectrometer.
11. Two oligosaccharides (*see* below) have isobaric mass. Oligosaccharide 1 contains one α-linked Gal, and oligosaccharide 2 does not contain any α-linked Gal residue. But oligosaccharide 1 typically only accounts for a small percent of the isobaric mass.

Galα(1→3)Galβ(1→4) { GlcNAcβ(1→2)Manα(1→6) \ ... Fucα(1→6) \ Manβ(1→4)GlcNAcβ(1→4)GlcNAcβ(1→N)-R    1
GlcNAcβ(1→2)Manα(1→3) /

Galβ(1→4)GlcNAcβ(1→2)Manα(1→6) \ ... Fucα(1→6) \ Manβ(1→4)GlcNAcβ(1→4)GlcNAcβ(1→N)-R    2
Galβ(1→4)GlcNAcβ(1→2)Manα(1→3) /

## Acknowledgments

The authors would like to thank Drs. Amareth Lim, Bryan Harmon, and Bruce Meiklejohn for their critical reading and comments.

# References

1. Andreakos, E., Taylor, P. C., and Feldmann, M. (2002) Monoclonal antibodies in immune and inflammatory diseases. *Curr. Opin. Biotech.* **13**, 615–620.
2. Trikha, M., Yan, L., and Nakada, M. T. (2002) Monoclonal antibodies as therapeutics in oncology. *Curr. Opin. Biotech.* **13**, 609–614.
3. Galili, U., Buehler, J., Shohet, S. B., and Macher, B. A. (1987) The human natural anti-Gal IgG III. The subtlety of immune tolerance in man as demonstrated by crossreactivity between natural anti-Gal and anti-B antibodies. *J. Exp. Med.* **165**, 693–704.
4. Davies, M. J. and Hounsell, E. F. HPLC and HPAEC of oligosaccharides and glycopeptides, in *Methods in Molecular Biology, vol 76: Glycoanalysis Protocols* (Hounsell, E. F., ed.), Humana, Totowa, NJ, 1998, pp. 79–100.
5. Papac, D. I., Wong, A., and Jones, A. J. S. (1996) Analysis of acidic oligosaccharides and glycopeptides by matrix-assisted laser desorption/ionization time-of-flight mass spectrometry. *Anal. Chem.* **68**, 3215–3223.
6. Hardy, M. R. Glycan labeling with the fluorophores 2-aminobenzamide and anthranilic acid, in *Techniques in Glycobiology* (Townsend, R. R. and Hotchkiss, A. T., ed.), Marcel Dekker, Inc., New York, NY, 1997, pp. 359–375.
7. Honda, S. and Kakehi, K. (1996) Analysis of glycoproteins, glycopeptides and glycoprotein-derived oligosaccharides by high-performance capillary electrophoresis. *J. Chromatogr. A.* **720**, 377–393.
8. Lund, J., Takahashi, N., Pound, J. D., Goodall, M., and Jefferis, R. (1996) Multiple interactions of IgG with its core oligosaccharide can modulate recognition by complement and human Fcγ receptor I and influence the synthesis of its oligosaccharide chains. *J. Immunol.* **157**, 4963–4969.
9. Jefferis, R., Lund, J., and Pound, J. D. (1998) IgG-Fc-mediated effector functions: molecular definition of interaction sites for effector ligands and the role of glycosylation. *Immunol. Rev.* **163**, 59–76.
10. Zhu, A. and Goldstein, J. (1994) Cloning and functional expression of a cDNA encoding coffee bean α-galactosidase. *Gene* **140**, 227–231.
11. Aubert, M., Crotte, C., Bernard, J.-P., Lombardo, D., Sadoulet, M.-O., and Mas, E. (2003) Decrease of human pancreatic cancer cell tumorigenicity by α1,3galactosyltransferase gene transfer. *Int. J. Cancer* **107**, 910–918.

# 33

## Carbohydrate Structural Characterization of Fas Ligand Inhibitory Protein

### Lihua Huang, Charles E. Mitchell, Lei Yu, P. Clayton Gough, and Alice Riggin

### 1. Introduction

Decoy receptor 3 (DcR3) is a novel member of the tumor necrosis factor (TNF) receptor superfamily that plays an important role in regulating apoptosis in normal physiology *(1–5)*. The engineered (Arg218Gln) soluble form of DcR3 was termed Fas ligand inhibitory protein (FLINT; LY498919) in an effort to create a molecule that was less susceptible to proteolysis *(5)*. FLINT, expressed in Chinese ovary (CHO) cells, is a therapeutic glycoprotein with 271 amino acid residues, 10 disulfide bonds, 1 potential *N*-glycosylation site, and 2 potential *O*-glycosylation sites. Oligosaccharide structures strongly influence FLINT pharmacokentics *(6)*. Hence, it is essential to characterize carbohydrate structures of FLINT in routine testing to monitor batch-to-batch consistency. There are many ways to characterize the carbohydrate structures of a glycoprotein. In this chapter, we will describe routine carbohydrate characterization of FLINT to monitor its batch-to-batch consistency. The methods include (1) neutral and amino monosaccharide composition analysis; (2) sialic acid content assay; and (3) oligosaccharide profiling by weak-anion exchange (WAX) high performance liquid chromatography (HPLC) with fluorescence detection, and additional glycosylation characterization, including site mapping and oligosaccharide structural elucidation.

### 2. Materials

#### 2.1. Monosaccharide Composition Analysis

1. 4 *M* Trifluoroacetic acid (TFA) aqueous solutions.
2. Fucose (Fuc; Fluka).
3. Galactose (Gal; Fluka).
4. Glucose (Glc; Fluka).
5. Mannose (Man; Fluka).
6. Galactosamine HCl (GalNH$_2$·HCl; Fluka).

From: *Methods in Molecular Biology, vol. 308: Therapeutic Proteins: Methods and Protocols*
Edited by: C. M. Smales and D. C. James © Humana Press Inc., Totowa, NJ

7. Glucosamine HCl (GlcNH$_2$·HCl; Fluka).
8. Biocompatible ternary HPLC system with pulsed amperometric detector (PAD; Dionex).
9. Mobile phase A: 200 m*M* sodium hydroxide solution.
10. CarboPac PA10 column and amino trap guard column (Dionex).
11. Centrifugation vacuum system, e.g., Savant Speed-Vac.

## *2.2. Sialic Acid Content Assay*

1. Neuraminidase (*Arthrobacter ureafaciens*; Roche Diagnostics).
2. 150 m*M* Sodium acetate (Mallinckrodt) buffer, pH 5.2.
3. *N*-acetylneuraminic acid (NeuAc; Roche Diagnostics).
4. Mobile phase B: 96 m*M* sodium hydroxide and 100 m*M* sodium acetate aqueous solution.
5. CarboPac PA1 and guard columns (Dionex).

## *2.3. Oligosaccharide Profiling by WAX-HPLC With Fluorescence Detection*

1. Protein: Peptide-*N*-glycosidase F (PNGase F; Prozyme).
2. 50 mg/mL (DL)-dithiothreitol (DTT) solution.
3. 100 mg/mL Iodoacetic acid solution.
4. 10 m*M* ammonium bicarbonate buffer.
5. 3 *M* Tris-HCl buffer, pH 8.0.
6. Urea, ACS reagent grade.
7. 2-Aminobenzamide (2-AB) fluorescent-labeling kit (Prozyme)
8. 10% Acetic acid solution.
9. P2-BioGel (Bio-Rad): Reconstituted or swell the P2-BioGel to 50% (w/v) in purified water.
10. Mobile phase C: 20% acetonitrile (v/v) in reagent-grade water.
11. Mobile phase D: 250 m*M* ammonium acetate, pH 4.5, in 20% acetonitrile aqueous solution.
12. Gradient HPLC system equipped with a fluorescence detector (Agilent).
13. WAX column (Glycosep C, 4.6 × 150 mm or Vydac WAX 0.75 × 5 CM 301VHP575P).

## *2.4. Glycosylation Site Mapping*

1. Trypsin-TPCK treated (Worthington Biochemical Corporation).
2. 7 *M* Guanidine and 0.5 *M* Tris buffer, pH 8.0.
3. Endoproteinase Asp-N (Sigma).
4. *O*-Glycanase (Prozyme).
5. Mobile phase E: 0.15% formic acid in H$_2$O.
6. Mobile phase F: 0.12% formic acid in acetonitrile.
7. Capillary HPLC system (Agilent).
8. Micromass quadrupole time-of-flight (Q-TOF) Micro (Micromass).

## *2.5. Oligosaccharide Structural Characterization*

1. β-Galactosidase (sp).
2. β-*N*-acetylglucosaminidase (sp).
3. α-Mannosidase (Jack Bean).

## 3. Methods

The methods described below outline (1) neutral and amino monosaccharide composition analysis; (2) sialic acid content assay; (3) oligosaccharide profiling; (4) glycosylation site mapping; and (5) oligosaccharide structure elucidation.

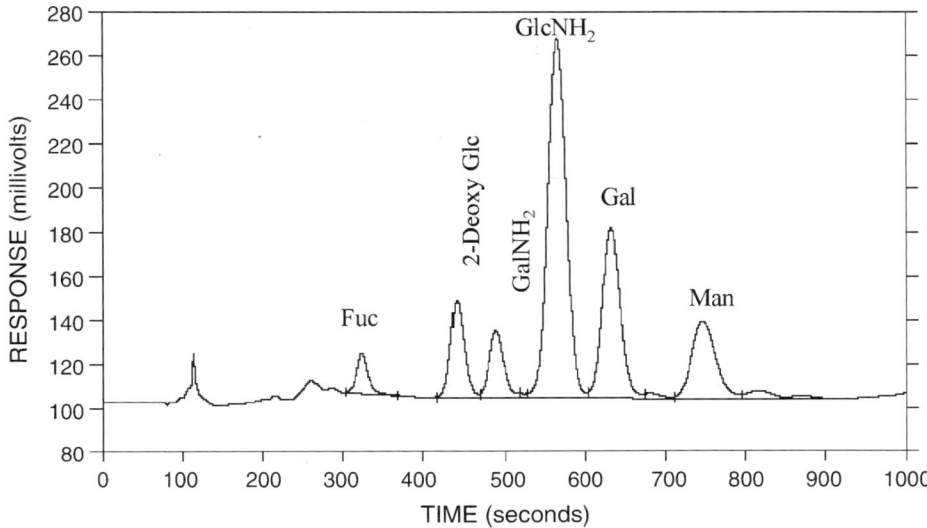

Fig. 1. HPAEC/PAD of FLINT neutral and amino monosaccharides separated on a CarboPA10 column with amino-trap column.

## 3.1. Neutral and Amino Monosaccharide Composition Analysis (see Fig. 1)

Neutral (Gal, Man, and Fuc) and amino (GlcNH$_2$ and GalNH$_2$) monosaccharides are released under acidic hydrolysis conditions. Under these conditions, sialic acid is completely destroyed. A different method (see **Subheading 3.2.**) is required to analyze the sialic acid content of FLINT.

### 3.1.1. Acid Hydrolysis

1. Transfer 200 µg FLINT into a clear glass vial with screw cap, then adjust it with reagent water to 1 mg/mL (see **Note 1**).
2. Add 200 µL of 4 *M* TFA solution into the vial, and incubate it at 110°C for 4 h (see **Note 2**).
3. Lyophilize the sample to dryness under a centrifugation vacuum system.
4. Prepare an internal standard solution at 7.5 µg/mL 2-deoxyglucose.
5. Add 400 µL internal standard solution into the sample vial.
6. Prepare the following standard solutions with the internal standard solution (see **Note 3**).

| Monosaccharide | Concentration (µg/mL) |
| --- | --- |
| Fuc | 7.0 |
| Gal | 15.0 |
| Glc | 3.0 |
| Man | 20.0 |
| GalNH$_2$ · HCl | 5.0 |
| GlcNH$_2$ · HCl | 40.0 |

7. Dilute each standard solution with the internal standard solution to 1:7, 1:3, 1:1, 3:1, and 1:0.

### 3.1.2. High-Performance Anion-Exchange Chromatography (HPAEC)/PAD Analysis

1. Equilibrate a CarboPac PA10 column/amino trap column with 11% mobile phase A and 89% reagent water at a flow rate of 1 mL/min.
2. Inject 20 µL of reagent water (blank), the standard or sample solutions.
3. Elute the neutral and amino monosaccharides with the following gradient at flow rate of 1 mL/min.

| Time (min) | % Mobile phase A | % Reagent water |
|---|---|---|
| 0 | 11 | 89 |
| 20 | 11 | 89 |
| 20.1 | 100 | 0 |
| 30 | 100 | 0 |
| 30.1 | 11 | 89 |
| 45 | 11 | 89 |

4. Detect monosaccharides (PAD and gold electrode) with the following pulse potential waveform (7).

| Step | Time (s) | Potential (V) | Integration |
|---|---|---|---|
| 0 | 0.00 | +0.10 | |
| 1 | 0.20 | +0.10 | Begin |
| 2 | 0.40 | +0.10 | End |
| 3 | 0.41 | −2.00 | |
| 4 | 0.42 | −2.00 | |
| 5 | 0.43 | +0.60 | |
| 6 | 0.44 | −0.10 | |
| 7 | 0.50 | −0.10 | |

5. Quantitate the neutral and amino monosaccharides according to the standard curves obtained after calibration with the internal standard.

## 3.2. Sialic Acid Content (Fig. 2)

### 3.2.1. Sialic Acid Release

1. Prepare 1 mg/mL FLINT solution, and transfer 100 µL solution into a 1.5-mL polypropylene microcentrifuge tube with a screw cap.
2. Add 20 µL 150 m$M$ sodium acetate buffer, pH 5.2, 10 µL reagent water, and 20 µL neuraminidase solution (1 U/mL) into the tube (see **Note 4**).
3. Mix the solution, and incubate at 37°C for 1–3 h (see **Note 5**).
4. Heat at 95–100°C for 5 min.
5. Add 100 µL reagent water into the tube, centrifuge to pellet possible precipitate, and transfer supernatant into HPLC vial.
6. Prepare five standards (NeuAc) in 7.5 m$M$ sodium acetate buffer, pH 5.2, from 2 to 10 µg/mL (see **Note 6**).

### 3.2.2. HPAEC-PAD Analysis

1. Equilibrate a CarboPac PA1 column/a guard column with mobile phase B at a flow rate of 1 mL/min.
2. Inject 25 µL reagent water (blank), standard, or sample solution.

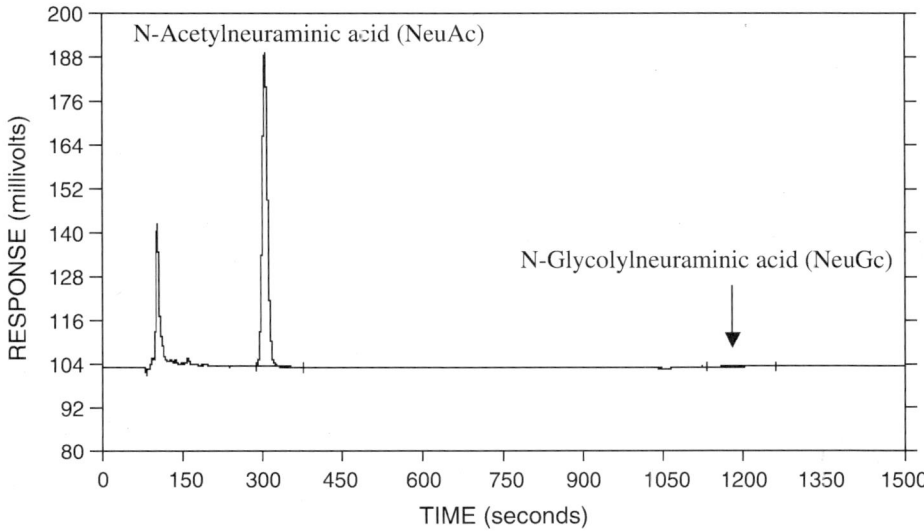

Fig. 2. HPAEC/PAD of FLINT sialic acid separated on a CarboPA1 column with a guard column. NeuGc level (<1%) is very low for recombinant FLINT in CHO.

3. Elute the sample using an isocratic gradient at a flow rate of 1 mL/min for 40 min.
4. Detect sialic acid (PAD and gold electrode) with the pulse potential waveform shown in **Subheading 3.1.2., step 4**.
5. Quantitate the NeuAc according to the standard curve.

### 3.3. Oligosaccharide Profiling by WAX-HPLC/Fluorescence (Fig. 3)

### 3.3.1. Oligosaccharide Release (see **Note 7**)

1. Transfer 30 µg FLINT solution (<300 µL) into a microcentrifuge tube.
2. Weigh approx 50 mg urea, and add to the sample tube.
3. Add 50 µL 3 *M* Tris-HCl buffer, pH 8.0, and 3 µL 50 mg/mL DTT solution into the sample tube.
4. Mix and incubate at 37°C for approx 15 min.
5. Add 5 µL 100 mg/mL iodoacetoamide solution into the tube.
6. Mix and incubate the solution at ambient temperature in the dark for approx 15 min.
7. Transfer the sample solution onto a 1-mL polypropylene disposable spin column, which was packed with P6 resin and equilibrated with 10 m*M* ammonium bicarbonate buffer.
8. Elute the protein with 300 µL 10 m*M* ammonium bicarbonate buffer, and collect the 300-µL effluent.
9. Add 1 µL PNGase F to the collected fraction, and incubate the solution at 37°C for 2–18 h.
10. Add 3 µL 10% acetic acid solution into sample solution to precipitate the deglycosylated protein, centrifuge to pellet the precipitate, and transfer supernatant into a clear vial (*see* **Note 8**).

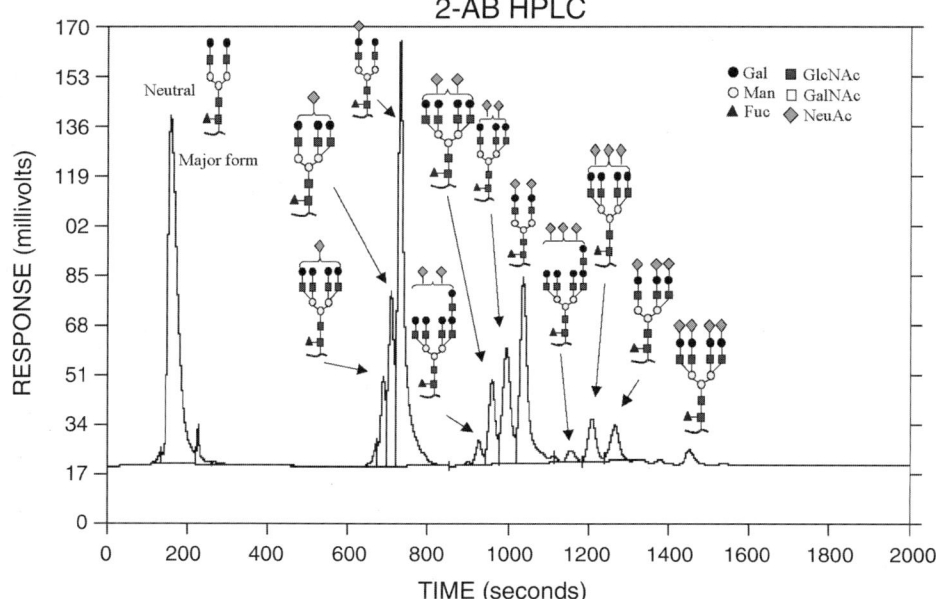

Fig. 3. Fluorescence chromatogram of WAX-HPLC for 2-AB-labeled oligosaccharides of FLINT. Peak assignment was based on capillary HPLC/MS and neuraminidase (*a.u.*) digestion for the collected each fraction.

### 3.3.2. Oligosaccharide Labeling and Removal of Excess-Labeling Reagent

1. Evaporate 300 μL supernatant solution under a vacuum centrifugation system at ambient temperature until dry.
2. Prepare 2-AB label kit according to the manufacturing instructions (*see* **Note 9**).
3. Add 2–5 μL mixture of labeling reagents into the dried oligosaccharide vial, mix completely, and incubate the solution at 65°C for 1.5–2.5 h.
4. Add 50 μL reagent water to the sample vial, and load the solution onto an 1-mL spin column, which was packed with 0.5-mL P-2 BioGel resin and washed with reagent water, and discard the effluent (*see* **Note 10**).
5. Wash the sample vial with 50 μL reagent water, load the solution onto the column, and discard the effluent.
6. Elute the column with 250 μL reagent water, and collect the effluent in an Eppendorf tube or HPLC sample vial.

### 3.3.3. WAX-HPLC With Fluorescence Detection

1. Equilibrate a WAX column, Glycosep C, or Vydac with mobile phase C.
2. Inject 25 μL water (blank) or the sample solution.
3. Elute oligosaccharides using the following gradient at flow rate of 0.4 mL/min.

| Time (min) | % Mobile phase C | % Mobile phase D |
|:---:|:---:|:---:|
| 0 | 100 | 0 |
| 2 | 100 | 0 |
| 32 | 55 | 45 |
| 34 | 10 | 90 |
| 40 | 10 | 90 |
| 41 | 100 | 0 |
| 55 | 100 | 0 |

4. Set a fluorescence detector with excitation wavelength at 330 nm and emission wavelength at 420 nm.

## 3.4. Glycosylation Site Mapping

FLINT contains one potential *N*-glycosylation site, Asn144. From monosaccharide composition analysis of FLINT, the molecule contains approx 0.6 mol GalNAc/mol protein. The result indicates that FLINT possibly contains one or more partially *O*-glycosylation sites. A carbohydrate marker ion scan *(8,9)* of enzymatic digests of FLINT by liquid chromatography/mass spectrometry (LC/MS) was applied to map-glycosylated peptides. *O*-Glycosylated peptides were detected by LC/MS analysis of the PNGase F-treated digests of FLINT. *O*-glycosylation sites were determined based on the peptide sequence found or by N-terminal sequencing analysis for *O*-glycopeptides.

### 3.4.1. Trypsin Digestion of FLINT

1. Transfer 50 µg FLINT solution into a 500-µL Eppendorf vial, and lyophilize the solution to dryness under vacuum centrifugation system (Speed-Vac).
2. Add 4.5 µL 7 *M* guanidine, 0.5 *M* Tris buffer, pH 8.0, and 0.5 µL 50 mg/mL DTT solution into the vial.
3. Vortex the vial, and incubate at 37°C for 30 min.
4. Add 9 µL 10 m*M* ammonium bicarbonate buffer and 1 µL 100 mg/mL iodoacetamide solution into the vial.
5. Mix and incubate the solution at ambient temperature in the dark for 10 min.
6. Dilute the solution with 80 µL 10 m*M* ammonium bicarbonate buffer, and add 2 µL 1.0 mg/mL TPCK-treated trypsin solution into the vial.
7. Incubate the solution at 37°C for 2 h, and heat the solution at 100°C to destroy the remaining trypsin activity.
8. Transfer 60 µL tryptic digest into another vial for treatment with Asp-N.
9. Dilute the remaining solution with 10 m*M* NH₄HCO₃ buffer into 200 µL final volume.

### 3.4.2. Treatment of PNGase F for the Tryptic Digest

1. Transfer 50 µL tryptic digest in a micro-HPLC vial.
2. Add 0.5 µL PNGase F solution into the vial, and incubate it at 37°C for 2 h.

### 3.4.3. Identification of Glycopeptides by LC/MS (see **Note 11** and **Fig. 4**)

1. Inject 5 µL tryptic digests (prior to or after treatment of PNGase F) of FLINT onto a C8 reverse-phase HPLC column (1 × 50 mm) equilibrated in mobile-phase E and coupled to the ionization probe of a Q-TOF Micromass spectrometer.
2. Elute the (glyco)peptides at 20 µL/min with a linearly increasing gradient of mobile-phase F from 0% to 15% in 15 min, then to 29% in 21 min.

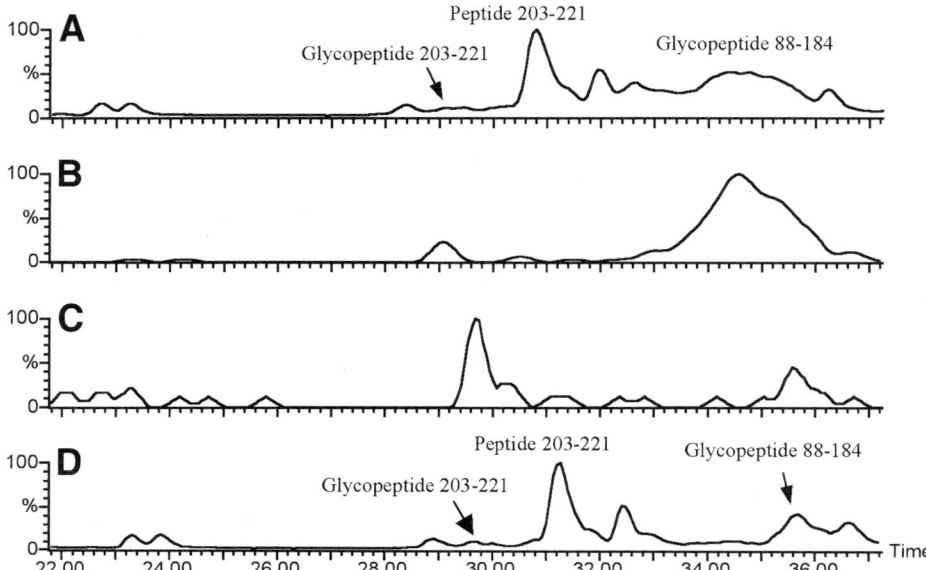

Fig. 4. HPLC/MS of the tryptic digest of FLINT. (**A,D**) Total ion chromatograms (TICs);
(**B,C**) XICs at 204.

3. Record two positive-ion spectra with the conditions for the first at a sample-cone voltage
   of 70 V and a scanning mass/charge range of 100–1000; for the second, record at a sample
   cone of 25 V and a scanning mass/charge range of 300–2000.
4. Reconstruct extracted ion chromatograms of carbohydrate marker (oxonium) ions:
   204 for HexNAc; 366 for HexHexNAc; 292 for NeuAc; and 274 for dehydro-NeuAc
   from the first-ion spectra (*see* **Note 12**).
5. Reconstruct average mass spectra of glycopeptides that were obtained from the second-
   ion spectra.

From the extracted-ion chromatogram (XIC; **Fig. 4B**) at 204, two peaks around
29 and 35 min possibly contain glycopeptides. Mass spectral data showed the peptides
for those two peaks are peptide 203–221 and peptide 88–184. After treatment with
PNGase F, the XIC chromatogram (**Fig. 4C**) at 204 indicated that the two peaks
remaining at retention time around 29 and 36 min may contain glycopeptides. A peak
near 36 min still contains peptide 88–184. These results show that peptide 88–184
contains both *N*- (Asn144) and *O*-glycosylation sites. To separate *N*- and *O*-glyco-
sylation sites for site determination, the tryptic digest of FLINT was further treated
with endoproteinase Asp-N.

### 3.4.3. Asp-N Treatment of the Tryptic Digest

1. Add 20 µL 0.02 mg/mL Asp-N solution into 60 µL tryptic digest.
2. Mix and incubate the solution at 37°C for 2 h.

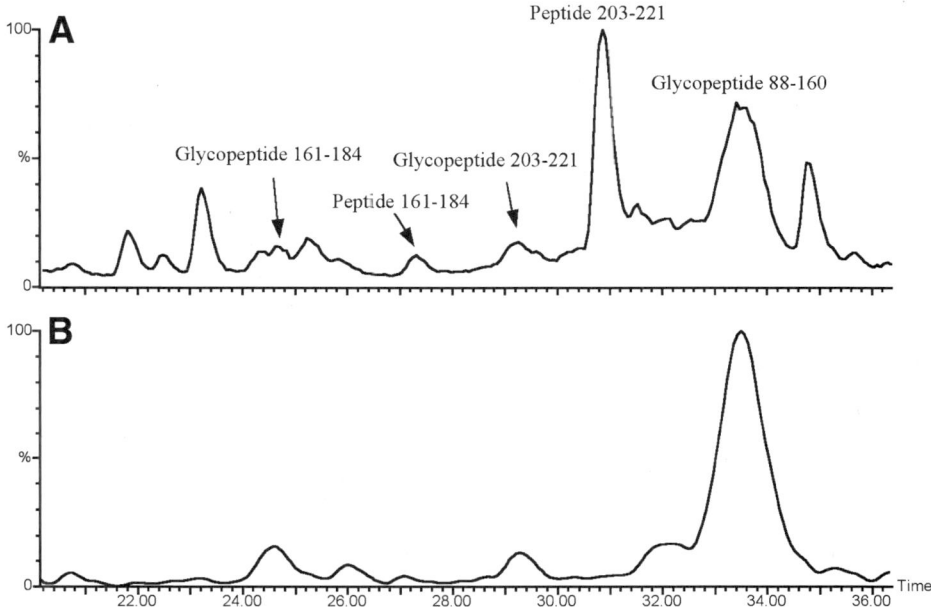

Fig. 5. HPLC/MS of the Asp-N-treated tryptic digest of FLINT. (**A**) TIC; (**B**) XIC at 204.

### 3.4.4. Treatments of PNGase F for Asp-N-Treated Tryptic Digest

1. Transfer 10 µL of the above digest, and dilute it with 20 µL 10 m*M* ammonium bicarbonate buffer.
2. Add 0.5 µL PNGase F into it, and incubate it at 37°C for 2 h.

### 3.4.5. Identification of Glycopeptides by LC/MS (see **Note 13** and **Fig. 5**)

Use the conditions in **Subheading 3.4.3.** to analyze the Asp-N-treated tryptic digest before and after treatments with PNGase F.

## 3.5. Oligosaccharide Structures

N-Linked oligosaccharide structures of recombinant glycoprotein, expressed in CHO cells, are well-documented *(10–13)*. Hence, the major glycan structures of FLINT could be elucidated based on their masses obtained via a combination of LC/MS analysis and simple sequential exoglycosidase digestions.

### 3.5.1. Treatments of O-Glycanase for the Asp-N-Treated Tryptic Digest

1. Dilute 40 µL digest with 100 µL reagent water and 40 µL of 250 m*M* sodium phosphate buffer, pH 6.0.
2. Add 4 µL of 10 U/mL neuraminidase (*a.u.*) solution into the digest (sample 1), and transfer 30 µL solution into a micro-HPLC vial.
3. Add 2 µL *O*-glycanase solution into the vial, and incubate it at 37°C overnight (12–18 h).

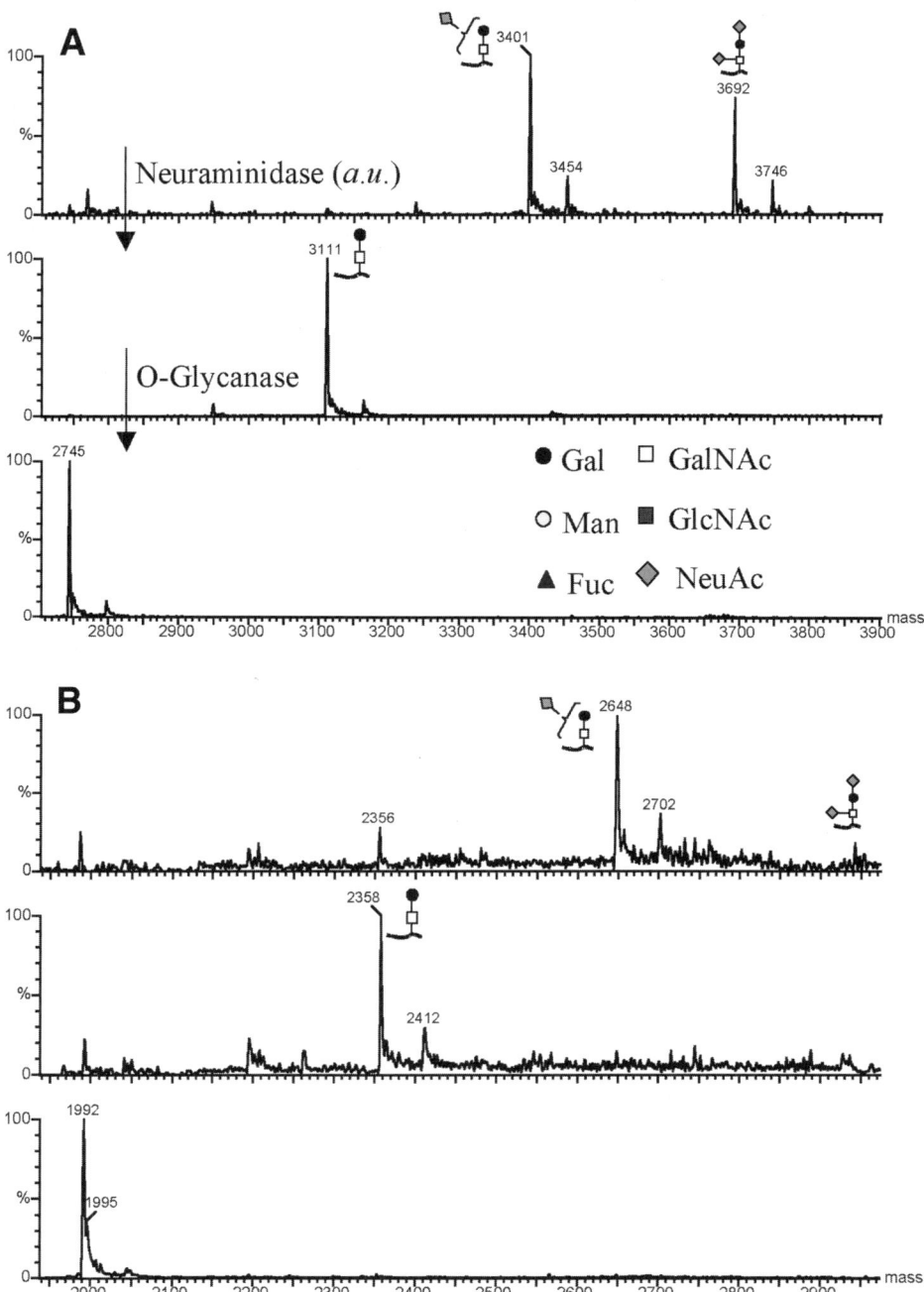

Fig. 6. Mass spectra of the glycopeptides 161–184 (**A**) and 203–221 (**B**) before and after treatment with neuraminidase (*a.u.*) and *O*-glycanase.

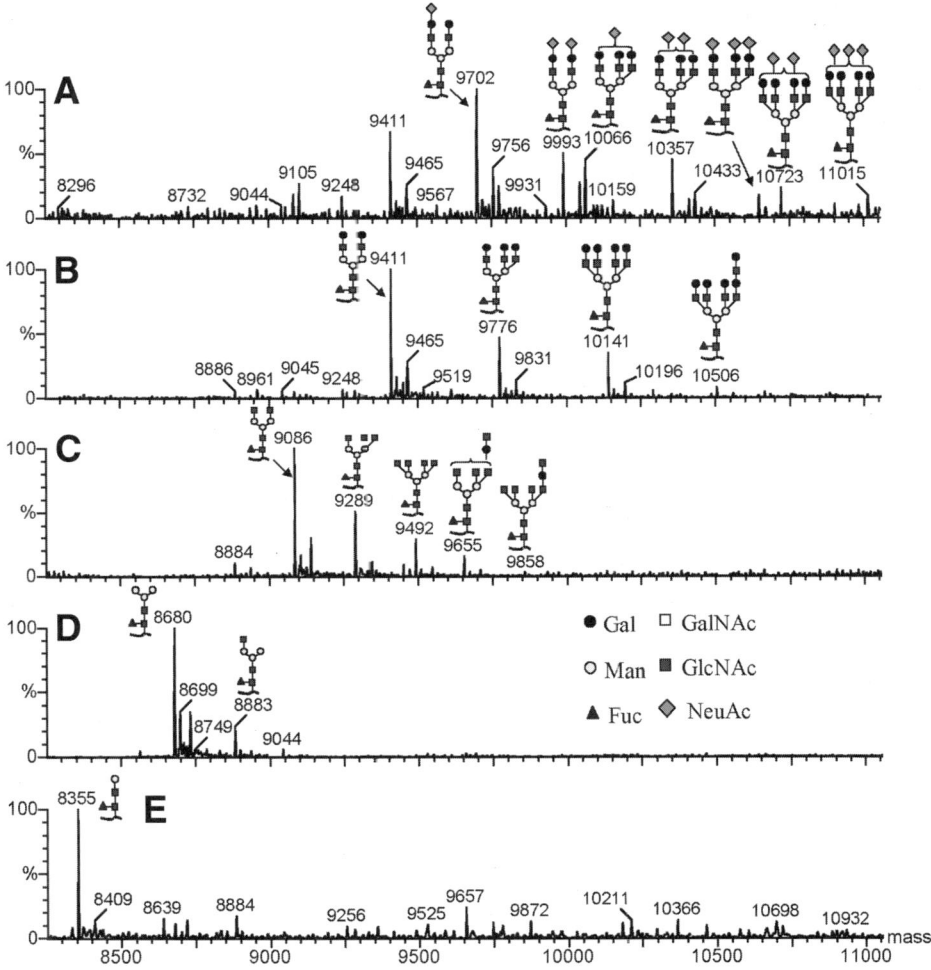

Fig. 7. Mass spectra of the glycopeptide 88–160 for sequential exoglycosidase digestions.

## 3.5.2. Identification of Glycopeptides by LC/MS (see **Note 14** and **Fig. 6**)

1. Inject 5 μL of each sample onto a C8 reverse-phase HPLC column (1 × 50 mm) equilibrated in mobile-phase E and coupled to the ionization probe of a Q-TOF Micromass spectrometer.
2. Elute the (glyco)peptides at 20 μL/min with a linearly increasing gradient of mobile-phase F from 0% to 15% in 15 min, then to 29% in 21 min.
3. Record a positive-ion spectrum with the conditions at sample cone of 25 V, collision energy of 5 eV, and scanning mass/charge range of 300–2000.

## 3.5.3. Sequential Treatment of Exoglycosidases

1. Transfer 90 μL sample 1 into a micro-HPLC vial, and add 5 μL β-galactosidase (*s.p.*) (sample 2).

2. Transfer 60 μL sample 2 into a micro-HPLC vial, and add 5 μL β-acetylglucosaminidase (*s.p.*) (sample 3).
3. Transfer 30 μL sample 2 into a micro-HPLC vial, and add 3 μL α-mannosidase (*j.b.*) and 3 μL 5X buffer into it (sample 4).
4. Incubate samples 1–4 at 37°C overnight (12–18 h).

### 3.5.4. Identification of Glycopeptides by LC/MS (see **Note 15** and **Fig. 7**)

Use the conditions described in **Subheading 3.5.2.** to analyze samples 1–4.

## 4. Notes

1. The amount of glycoprotein required for hydrolysis will depend on the sensitivity of the electrochemical detector and monosaccharide composition. For quantitative purposes, the minimum amount of glycoprotein is at least 100 times more than the detection limit or 10 times more than the quantitation limit for the lowest percentage of monosaccharide in a glycoprotein. For example, if the detection limit for each monosaccharide is 10 pmol, and GalNAc is the lowest percent of monosaccharide for FLINT, less than one residue per FLINT (approx 30 kDa) molecule, the minimum amount for one injection is more than 100 pmol, i.e., greater than 3 μg FLINT. Hence, for multiple injections, over 30 μg FLINT is needed for quantitative determination of monosaccharide composition.
2. 2 or 4 *M* HCl has frequently been used for monosaccharide hydrolysis. The best hydrolysis conditions are determined based on a series of experiments for monosaccharide recovery yield.
3. The sensitivity of the electrochemical detector is decreased during a long-batch run, especially for an old pulse-potential waveform. Using an internal standard can remedy the sensitivity problem of the electrochemical detector and can also help monosaccharide peak assignments.
4. Hydrolysis under 0.1 *N* HCl at 70°C for 1 h to release sialic acid from glycoproteins is another commonly used method. However, for FLINT, neuraminidase (*a.u.*) digestion provided better quantitative data.
5. The optimal hydrolysis time is determined from the experiment design. For FLINT, 1-h incubation is enough to completely release sialic acid residues.
6. Using an internal standard (e.g., 2,3-dehydro-3-deoxyneuraminic acid) will give more consistent data (*see* **Note 3**).
7. For complete oligosaccharide release, it is necessary to denature the glycoprotein before adding PNGase F. There are multiple ways to denature protein. One of the most common methods is to boil the glycoprotein in 0.5% sodium dodecyl sulfate (SDS) and 5% β-mercaptoethanol, then add 10% nonionic detergent, such as *n*-octyl glucoside or nonidet P-40, to the mixture to neutralize SDS before treatment of PNGase F.
8. For general glycoproteins, pH adjustments may not precipitate deglycosylated proteins. Microcon 10 concentrator (Millipore) or extraction with 75% cold ethanol can be used to separate protein and oligosaccharide components.
9. 2-AB labeling reagent containing 70% DMSO, 30% glacial acetic acid with 0.25 *M* 2-AB, and 0.1 *M* NaCNBH$_3$ can be self-prepared.
10. A hydrophilic separation disk (supplied with labeling kit) could be used to remove excess 2-AB dye, but the method described in this chapter is more robust and simpler based on our experience. According to our studies, the results from both separation methods give similar results.

11. When using formic acid as an acid modulator for HPLC separation, mass signals increase approx 10-fold compared with 0.1% TFA as an acid modulator. However, carbohydrate moieties of glycopeptide or glycoprotein are very easy to fragment when using formic acid. To obtain a mass spectrum of intact glycopeptide or glycoprotein, the parameters of the mass spectrometer should be adjusted.

12. Peptides may also be fragmented on higher sample-cone voltage to give the ions, one of which may be very close to a carbohydrate-marker ion. Hence, using one carbohydrate-marker ion to reconstruct XIC may not be enough to tell which peak contains glycopeptide. It can be resolved by checking the mass spectrum of each peak indicated by the marker ion XIC or by using several marker ions to reconstruct XICs.

13. When the Asp-N-treated tryptic digest was analyzed by LC/MS, three peaks were found to contain three glycopeptides (*see* **Fig. 5**). A peak around 34 min contains *N*-glyco-peptide 88–160 (*see* **Fig. 7**). Peaks near 25 min and 29 min contain *O*-glycopeptides. *O*-glycopeptide 203–221 contains only one Thr216 residue and no Ser residue; obviously, it is an *O*-glycosylation site. *O*-glycopeptide 161–184 contains multiple Ser and Thr residues. *O*-glycosylation site Thr174 was determined by N-terminal sequencing of the collected *O*-glycopeptide. Aglycosylated peptides 161–184 and 203–221 were found in peaks around 28 and 32 min. Thus, both *O*-glycosylation sites are partially occupied.

14. *O*-Glycanase only cleaves O-linked glycan with a structure of Gal$\beta(1\rightarrow3)$ GalNAc$\alpha(1\rightarrow O)$ *(14)*. *O*-glycanase digestion removed *O*-glycans from both *O*-glycopeptides after desialylation. The results demonstrate that both glycopeptides contain *O*-glyco-sylation, and the major *O*-glycan is a mono- and disialylated core class 1 structure (Gal$\beta(1\rightarrow3)$GalNAc$\alpha(1\rightarrow O)$).

15. When the glycopeptide 88–160 of FLINT was treated with neuraminidase (*a.u.*), the major masses (**Fig. 7B**) obtained matches asialo di-, tri-, and tetra-antennary oligosaccharides. Further treatment of $\beta$-galactosidase (*s.p.*), which cleaves $\beta(1\rightarrow4)$-linked Gal, removed almost all Gal residues (**Fig. 7C**), indicating that FLINT N-linked oligosaccharides contain $\beta(1\rightarrow4)$-linked Gal residues. After sequential treatment of neuraminidase (*a.u.*), $\beta$-*N*-acetylglucosaminidase (*s.p.*), and $\alpha$-mannosidase (*j.b.*), the mass (**Fig. 7E**) obtained for glycopeptide 88–160 is 8355 Daltons, in which glycan matches a tet-rasaccharide of Man$\beta(1\rightarrow4)$GlcNAc$\beta(1\rightarrow4)$[Fuc$\alpha(1\rightarrow6)$]GlcNAc$\beta(1\rightarrow N)$. These results confirmed N-linked oligosaccharide assignments (shown in **Fig. 7A**) and are consistent with those obtained by WAX-HPLC analysis for 2-AB-labeled oligosaccharides of FLINT (**Fig. 3**).

## Acknowledgments

The authors would like to thank Drs. Ralph Riggin and Bruce Meiklejohn for their critical reading and comments.

## References

1. Kramer, P. H., Behrmann, I., Dhein, J., and Debatin, K. M. (1994) Regulation of apoptosis in the immune system. *Curr. Opin. Immunol.* **64,** 279–289.

2. Yu, K.-Y., Kwon, B., Ni, J., Zhai, Y., Ebner, R., and Kwon, B. S. (1999) A newly identi-fied member of tumor necrosis factor receptor superfamily (TR6) suppresses LIGHT-mediated apoptosis. *J. Biol. Chem.* **274,** 13733–13736.

3. Zhang, J., Salcedo, T. W., Wan, X., Ullrich, S., Hu, B., Gregorio, T., et al. (2001) Modu-lation of T-cell responses to alloantigens by TR6/DcR3. *J. Clin. Invest.* **107,** 1459–1468.

4. Wroblewski, V. J., Witcher, D. R., Becker, G. W., Davis, K. A., Dou, S., Micanovic, R., et al. (2003) Decoy receptor 3 (DcR3) is proteolytically processed to a metabolic fragment having differential activities against Fas ligand and LIGHT. *Biochem. Pharmacol.* **65,** 657–677.

5. Wroblewski, V. J., McCloud, C., Davis, K. A., Manetta, J., Micanovic, R., and Witcher, D. R. (2003) Pharmacokinetics, metabolic stability, and subcutaneous bioavailability of a genetically engineered analog of DcR3, FLINT [DcR3(R218Q)], in cynomolgus monkeys and mice. *Drug Metab. Disp.* **31,** 502–507.

6. Jenkins, N., Witcher, D. R., and Wroblewski, V. J. (2002) Enhancing sialic acid content of FLINT (Fas ligand inhibitory protein) R218Q with improved protease-resistance by having additional glycosylation sites engineered into the native FLINT sequence. *WO 2002060949.* 42 pp.

7. Rocklin, R. D., Clarke, A. P., and Weitzhandler, M. (1998) Improved long-term reproducibility for pulsed amperometric detection of carbohydrates via a new quadruple-potential waveform. *Anal. Chem.* **70,** 1496–1501.

8. Carr, S. A., Huddleston, M. J., and Bean, M. F. (1993) Selective identification and differentiation of *N*- and *O*-linked oligosaccharides in glycoproteins by liquid chromatography-mass spectrometry. *Protein Sci.* **2,** 183–196.

9. Huddleston, M. J., Bean, M. F., and Carr, S. A. (1993) Collisional fragmentation of glycopeptides by electrospray ionization LC/MS and LC/MS/MS: methods for selective detection of glycopeptides in progress digests. *Anal. Chem.* **65,** 877–884.

10. Spellman, M. W., Leonard, C. K., Basa, L. J., Gelineo, I., and van Halbeek, H. (1991) Carbohydrate structures of recombinant soluble human CD4 expressed in Chinese hamster ovary cells. *Biochemistry* **30,** 2395–2406.

11. Spellman, M. W., Basa, L. J., Leonard, C. K., et al. (1989) Carbohydrate structures of human tissue plasminogen activator expressed in Chinese hamster ovary cells. *J. Biol. Chem.* **264,** 14,100–14,111.

12. Watson, E., Bhide, A., and van Halbeek, H. (1994) Structure determination of the intact major sialylated oligosaccharide chains of recombinant human erythropoietin expressed in Chinese hamster ovary cells. *Glycobiology* **4,** 227–237.

13. Amoresano, A., Siciliano, R., Orru, S., Napoleoni, R., Altarocca, V., De Luca, E., et al. (1996) Structural characterisation of human recombinant glycohormones follitropin, lutropin and choriogonadotropin expressed in Chinese hamster ovary cells. *Eur. J. Biochem.* **242,** 608–618.

14. Fan, J. Q., Yamamoto, K., Kumagai, H., and Tochikura., T. (1990) Induction and efficient purification of endo-alpha-N-acetyl-D-galactosaminidase from *Alcaligenes* sp. *Agric. Biol. Chem.* **54,** 233–234.

# 34

## Top-Down Characterization of Protein Pharmaceuticals by Liquid Chromatography/Mass Spectrometry

*Application to Recombinant Factor IX Comparability— A Case Study*

**Jason C. Rouse, Joseph E. McClellan, Himakshi K. Patel, Michael A. Jankowski, and Thomas J. Porter**

## 1. Introduction

Recombinant protein pharmaceuticals have revolutionized the treatment of a variety of medical ailments, including cancer, autoimmune diseases, and hemostatic disorders. Proteins manufactured with eukaryotic expression systems may be complex and heterogeneous because of posttranslational modifications (PTMs) and differential proteolytic processing. At one time, detailed characterization and definition of the protein structure were difficult, and the manufacturing process defined the product. If process changes were made, clinical trials were required to demonstrate product equivalence prior to regulatory agency acceptance of the product manufactured by the modified process. To adopt new manufacturing processes in a timely manner, the biopharmaceutical industry and regulatory agencies have worked together over the last few years to develop new guidance documents based on knowledge gained from industry experience in the manufacture and clinical testing of protein pharmaceuticals *(1,2)*. Manufacturers of protein pharmaceuticals consistently strive to deliver the highest quality product in a cost-efficient manner. This can be accomplished through optimization of the production process and incorporation of new technologies to enhance product purity and yield. Process improvements may include a change of the host cell line, enhancement of the cell culture medium or cell culture management, or modifications to the purification process. In some cases, an additional manufacturing site is brought online to augment production capacity.

Steady advancements in analytical techniques and methodology now allow a protein pharmaceutical to be considered a "defined biologic" (formerly known as "well-characterized protein"). This designation provides the opportunity to demonstrate the

From: *Methods in Molecular Biology, vol. 308: Therapeutic Proteins: Methods and Protocols*
Edited by: C. M. Smales and D. C. James © Humana Press Inc., Totowa, NJ

"comparability" of the product manufactured by a new process to that produced with the original manufacturing process. Establishing structural and functional comparability by a protocol that includes predefined acceptance criteria can minimize the need to conduct extensive preclinical or clinical testing. Successful demonstration of comparability can therefore improve product development timelines or accelerate the implementation of changes to a commercial manufacturing process.

Mass spectrometry (MS) has emerged as an integral component of comparability programs because it is a rapid, highly informative analytical technique capable of detecting structural changes in peptides and proteins with high selectivity and sensitivity. MS is especially powerful in protein structural analysis when interfaced with reversed-phase high-performance liquid chromatography (RP-HPLC). The structure of recombinant proteins is traditionally assessed through RP-HPLC peptide mapping, which typically involves disulfide bond reduction, cysteine alkylation, and enzymatic digestion, followed by RP-HPLC profiling of the component peptides. The established combination of peptide mapping with MS, which links peptide RP-HPLC elution time with molecular weight and structure, provides verification of the intended amino acid sequence, definition of the N and C termini, and site-specific detection and characterization of the PTMs or degradation products (**Table 1**).

Top-down characterization *(3)* is a newer MS approach for protein identification and characterization that involves accurate mass determination of intact proteins, protein subunits, or large proteolytic fragments with the option of gas-phase ion fragmentation of the intact species. This approach is complementary to peptide mapping with MS (**Note 1**) and, when gas-phase ion fragmentation is employed, the structural endpoints are very similar to in-solution proteolysis. However, in contrast to peptide mapping, the top-down approach provides 100% amino acid sequence coverage, because an accurate mass determination for a particular protein isoform (or related gas-phase fragment ion) is a direct link to its amino acid composition, N and C termini, and PTMs (**Table 1**). Additionally, the top-down strategy has the potential to reduce the number of upstream sample handling and separation steps. If top-down analysis is performed in the liquid chromatography (LC)/MS mode, the simple HPLC profiles, as compared to that of a peptide map, allow protein recovery determination and efficient data analysis. Because intact proteins are more resistant than peptides to the collisional activation processes that occur during MS analysis, labile PTMs (e.g., sulfate and carboxylate on tyrosine-*O*-sulfate and γ-carboxyglutamate, respectively) are largely retained. This phenomenon results in a representative distribution of the protein isoform heterogeneity in the mass spectrum.

Thorough protein characterization by the top-down strategy ideally requires a mass analyzer with high-resolving power and accurate mass-measurement capabilities for unambiguous mass determinations of whole proteins or related gas-phase fragment ions *(6)*. With high-resolution tandem mass spectrometry (MS/MS) and specialized fragmentation techniques, initial top-down characterization strategies focused on the generation of extensive, systematic gas-phase fragmentation from intact proteins. This provided almost complete amino acid sequence in several studies for protein identification and site-specific PTM characterization (**Note 2**). Alternatively, the top-down

**Table 1**
**Characterization Methods for Recombinant Glycoproteins**

| Analyte | Method | Information obtained | Considerations |
|---|---|---|---|
| Component peptides | Reduction, alkylation, proteolytic digestion, C18 RP-HPLC/MS (mass) and nanoESI-QTOF MS/MS (sequence) | Verification of amino acid sequence<br>Definition of N and C termini<br>Site-specific detection and identification of PTM and degradation products | ≤100% Sequence coverage<br>Complicated HPLC profiles with proteolysis artifacts<br>Peptides with PTM may be lost, have low recoveries, or altered distributions |
| Component *N*- and *O*-glycans | Release: PNGase F (*N*-linked) or hydrazinolysis (*N*- and *O*-linked) or β-elimination (*O*-linked)<br>Profiling: underivatized glycans by HPAEC-PED or derivatization at reducing end: normal phase HPLC with fluorescence detection<br>Mass analysis: MALDI-TOF MS | Elucidation of glycan structures and isomers | Allows for direct analysis of glycans<br>Requires extensive sample manipulation |
| Intact protein isoforms | C4 or phenyl RP-HPLC/ESI-QTOF MS and nanoESI-QTOF MS | Accurate mass<br>Sequence composition<br>N- and C-terminal processing<br>PTM composition<br><br>Isoform distribution<br>Extent of heterogeneity<br>New species<br>Correlation to peptide map, *N*-glycan profile, and other essays | 100% Sequence coverage<br>Simple HPLC profiles<br>Retention of labile PTM (i.e., sulfation)<br>Efficient data analysis<br>Determination of protein recovery<br>Reduced sample handling |

PNGase F, Peptide-*N*-glycosidase F; HPAEC-PED, high-performance anion-exchange chromatography-pulsed electrochemical detection (PED).

**Table 2**
**Mass Analyzer Comparison[a]**

| MS instrument | Ionization mode and mass analyzer | Resolving power (m/Δm) | Mass accuracy External Calibration (Daltons [%]) |
|---|---|---|---|
| Bruker Reflex | MALDI-TOF | 900 (Insulin)[b] | ±0.6 (0.01) |
| | | 600 (Trypsinogen)[c] | ±3.0 (0.01) |
| Waters Micromass Platform II | ESI-Quadrupole | 450 (Insulin) | ±0.6 (0.01) |
| | | 450 (Trypsinogen) | ±2.4 (0.01) |
| Waters Micromass Q-Tof-2 | ESI-QTOF (V-Optics) | 10,500 (Insulin) | ±0.1 (0.002) |
| | | 2300 (Trypsinogen) | ±0.5 (0.002) |
| Waters Micromass Q-Tof API-US | ESI-QTOF (W-Optics) | 21,000 (Insulin) | ±0.06 (0.001) |
| | | 2450 (Trypsinogen) | ±0.2 (0.001) |

[a]Used at Wyeth BioPharma.
[b]Insulin = 5729.601 Daltons (bovine).
[c]Trypsinogen = 23,981.0 Daltons (bovine).

approach has successfully been implemented on stock and customized mass analyzers with modest resolving power using MS, MS/MS, and the standard collision-induced dissociation (CID) technique, which induces cleavage at the weaker backbone amide bonds (**Note 3**). Many advances in MS during the last 5 yr have centered on improving the performance of conventional mass analyzers (**Table 2**). The latest hybrid quadrupole time-of-flight (QTOF) mass analyzers (**Fig. 1**) feature resolving powers above 10,000 (m/Δm) with sensitivity in the low-femtomole range, an extended mass range, and stability for accurate mass measurements *(24,25)*. These capabilities make this instrument effective for top-down characterization of intact proteins, especially for the exact mass determinations of intact protein isoforms with high sensitivity (**Note 6**).

---

Fig. 1. *(opposite page)* Schematic of the Waters Micromass Q-Tof-2 mass spectrometer. This instrument was used for all RP-HPLC/MS analyses of activated rFIX samples. FIX activation products in the eluent from the HPLC (flowing 0.100–0.133 mL/min) are desolvated and ionized via the Z-Spray ESI source. The gas-phase analyte ions are accelerated into the QTOF mass analyzer by the potential difference between the ion source block and the extractor cone. All ions are then focused by the hexapole ion bridge. Ions ranging from *m/z* 800 to 4000 are selected with the quadrupole mass filter (**Note 4**). These ions then enter the collision cell at the appropriate axial translational energy of 5 or 10 eV. Through cooling collisions with argon gas at approx $6 \times 10^{-5}$ mbar, the ion beam becomes more monoenergetic and collimated prior to orthogonal acceleration into the TOF analyzer. The orthogonal acceleration ion optics (**Note 5**) successively accelerate segments of the quadrupole ion beam into the TOF analyzer for mass measurement every 88 μs (for this work), which is the necessary time for detection of all ions from *m/z* 800 to 4000. In the reflectron TOF mass analyzer, the *m/z* measurement is based on the time it takes an accelerated ion to traverse the total distance from the pusher region through the reflectron to the MCP detector.

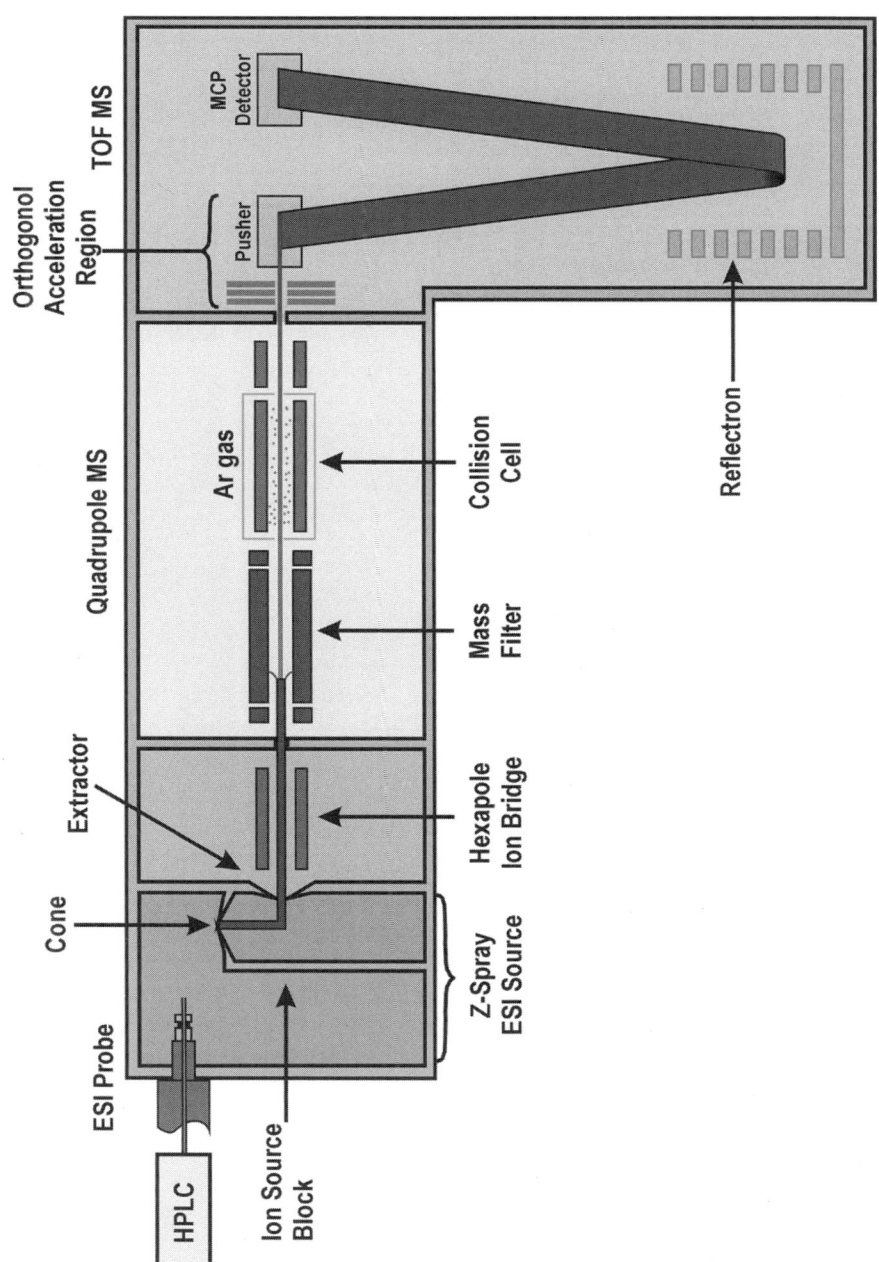

Fig. 1.

439

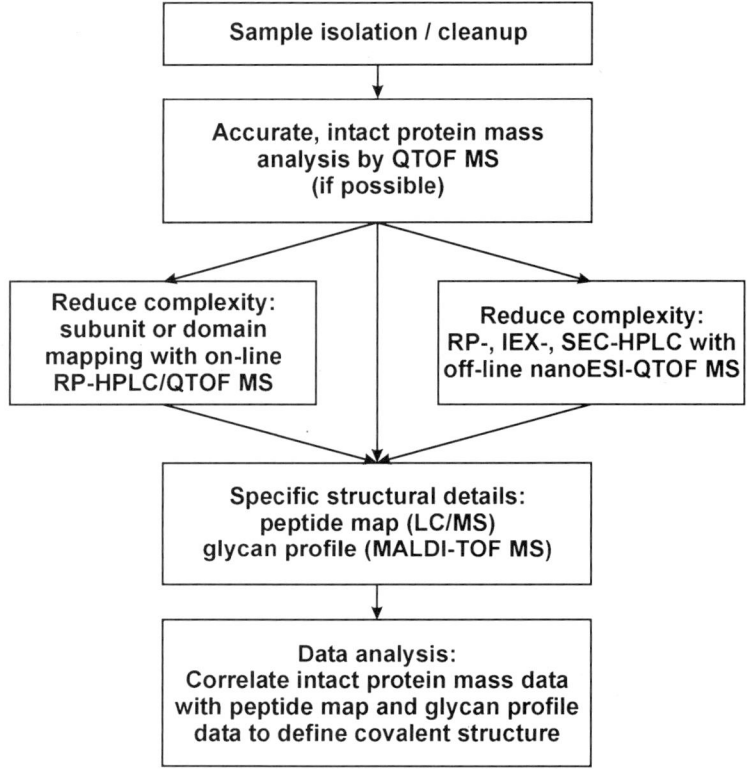

Fig. 2. Our top-down characterization strategy for recombinant proteins.

In our experience, the QTOF mass analyzer, when combined with the powerful maximum entropy (MaxEnt-1) mass deconvolution software *(28,29)*, has provided the unique capability to baseline resolve the mass signals of multiple, variably processed, glycosylated isoforms of recombinant glycoproteins with mass accuracies better than 0.005% (**Note 7**). The overall isoform profiles of proteins, ranging in mass up to 150 kDa, were found to correlate with structural information from other assays. Thus, our top-down characterization strategy (**Fig. 2**), which involves the QTOF mass analyzer and MaxEnt-1 software, is built upon accurate mass determinations and protein isoform mass profiles for full-length glycoproteins (if possible), protein subunits, or large proteolytic fragments as an efficient method to survey the global structural heterogeneity. However, instead of fragmentation of intact protein isoforms in the QTOF mass analyzer, our strategy depends on the previously established structural framework, as determined by the MS-characterized peptide map and *N*- and *O*-glycan profiles, to provide the specific structural details that complement the intact mass information (**Table 1**).

Our version of top-down protein characterization has become an important factor in our structural studies and comparability assessments of product candidates and com-

**Domains**

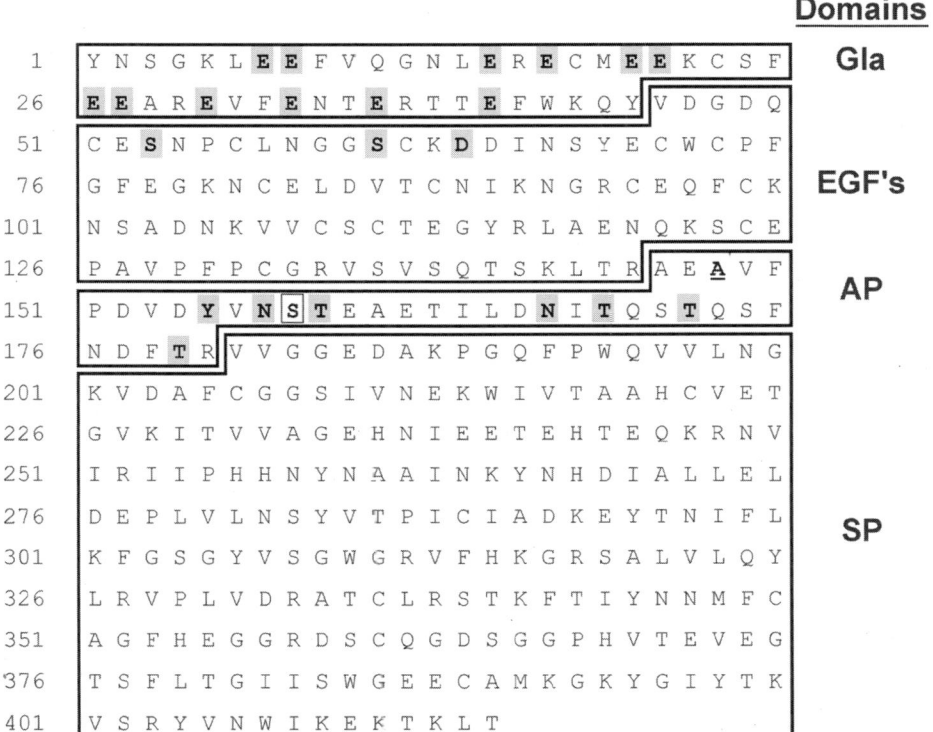

Fig. 3. Sequence and domain structure of FIX. Sites of PTMs are highlighted in shaded boxes. The 12 γ-carboxyglutamates are indicated in the Gla domain. The epidermal growth factor (EGF) domains contain two unique *O*-glycans at Ser[53] and Ser[61] and one β-hydroxy-aspartate at Asp[64]. The activation peptide (AP) contains tyrosine-*O*-sulfate at Tyr[155], two N-linked glycans at Asn[157] and Asn[167], and O-linked glycans at Thr[159], Thr[169], Thr[172], and Thr[179] (**Notes 12** and **13**). Phosphorylation at Ser[158] (open box) is found in plasma-derived FIX but not in rFIX. rFIX is the Ala[148] (underlined) isotype. SP = serine protease domain.

mercial products during evaluation of new reference materials, process scale-up, process improvements, and implementation of new manufacturing facilities. The precise mass determination and isoform profiles acquired by RP-HPLC/electrospray ionization (ESI) QTOF MS, in conjunction with comprehensive product release assay data, have supported the structural comparability and consistency in the samples under evaluation (**Note 8**). The side-by-side analysis of protein samples with LC/MS minimizes sample handling and provides data sets that are optimal for comparative purposes. We have used this top-down characterization approach in comparability studies with protein pharmaceuticals, such as recombinant human bone morphogenetic protein-2 (*30*) and the clotting factor, B-domain deleted factor VIII (BDDrFVIII) (*31,32*; **Note 9**). Additionally, this technique has been extremely useful in the characterization of recombinant monoclonal antibodies (mAbs) during product development (*30,33*; **Note 10**).

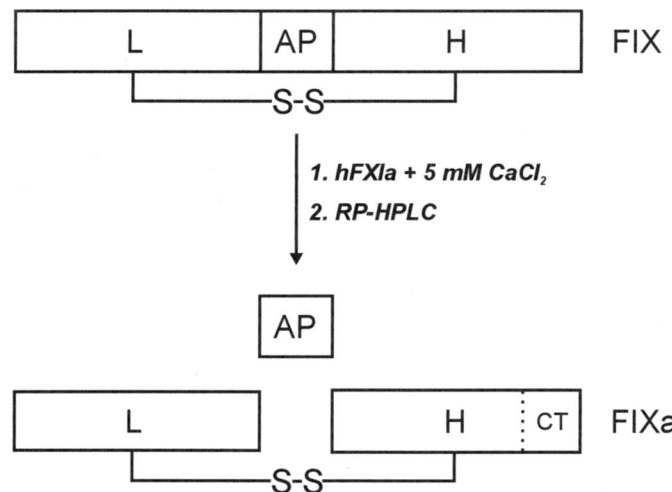

Fig. 4. Activation of FIX with FXIa. L, light chain; AP, activation peptide; H, heavy chain; CT, C-terminal peptide. FXIa cleaves FIX after Arg[145] and Arg[180] to generate FIXa—the enzymatically active product of activation. Under the conditions employed in this work, low-level cleavages were observed in the CT region of the HC, generating CT peptides (**Notes 14** and **15**).

This chapter presents the application of our top-down characterization approach for assessing structural comparability of a well-characterized glycoprotein, recombinant human factor IX (rFIX; *34*), produced at two manufacturing sites (**Note 11**). rFIX is a multidomain protein that contains numerous PTMs (**Fig. 3**). Specific proteolytic cleavage of rFIX at two peptide bonds was employed to liberate the activation peptide (AP; **Fig. 4**). Chromatographic separation of the rFIX fragments permitted acquisition of mass spectra of reduced complexity when compared to the mass spectrum of intact rFIX. This procedure improved the signal-to-noise ratio and resolution of the rFIX isoforms, which made accurate mass assignments possible for the full range of species present. The experimental methods employed for the top-down characterization of rFIX, including data evaluation and MS acceptance criteria for the determination of structural comparability, are presented here.

## 2. Materials

### 2.1. Equipment

1. HPLC System: Alliance 2695 equipped with a 2487 UV/Vis detector (Waters Corporation, Milford, MA).
2. Chromatography column: Symmetry C4 RP-HPLC column (3.9 × 150 mm, 5 μm, part no. 186000286, Waters Corporation).
3. HPLC flow splitter: Low pressure micro-splitter valve (P-451, Upchurch Scientific, Oak Harbor, WA).

4. Syringe pump: Pump 22 (Harvard Apparatus, South Natick, MA).
5. Mass Spectrometer: Q-Tof-2 (Micromass MS Technologies, Waters Corporation, Milford, MA), a hybrid QTOF mass analyzer equipped with a Z-spray ion source for ESI (**Fig. 1**).

## 2.2. Software

1. MassLynx 3.5 for NT (Micromass MS Technologies, Waters Corporation) is the Q-Tof-2 instrument control, data acquisition, calibration, and mass spectrum processing software.
2. Probabilistic maximum entropy analysis (MaxEnt-1 module, Micromass MS Technologies, Waters Corporation) is used to deconvolute the multiple-charged mass data (typical for an ESI mass spectrum) into a zero-charge mass spectrum. This software enables straightforward elucidation of molecular weight values for the different protein isoforms without prior knowledge of the protein species present in the sample.
3. Protein Analysis Worksheet (PAWS), version 2000.06.08 for Windows 95/98/NT/2000 (Genomic Solutions, Ann Arbor, MI), is used for the determination of theoretical mass values (monoisotopic and average mass) for intact proteins and subunits based on the predicted amino acid sequences and PTMs.

## 2.3. Reagents

1. rFIX test materials: multiple batches of rFIX manufactured at sites 1 and 2 were chosen for analysis. Only batches meeting the quality-control release test specifications were eligible for top-down MS analysis and other characterization tests. Representative data obtained for one rFIX batch from each manufacturing site are presented in this chapter.
2. Enzyme: human factor XIa (hFXIa; HCXIA-0160, Haematologic Technologies, Inc.).
3. HPLC-grade solvents: Water (Purelab Plus UV, US Filter, type PL5112 02), trifluoroacetic acid (TFA, 1-mL vials, cat. no. 28904, Pierce), and acetonitrile (HPLC grade, cat. no. AX0145-1, EMD Chemicals, Inc.).
4. Chromatography mobile phases: 0.1% (v/v) TFA in water (mobile-phase A) and 0.1% (v/v) TFA in 95% acetonitrile, 5% water (mobile-phase B).
5. Q-Tof-2 mass axis calibrant: sodium iodide (NaI, Sigma) at 2 mg/mL in 50:50 2-propanol:water (v/v) is prepared daily.
6. Insulin solution: bovine insulin (Sigma) at 2 pmol/$\mu$L in 50:50 acetonitrile:water (v/v) with 2% (v/v) formic acid is prepared daily.
7. Trypsinogen solution: bovine trypsinogen (Sigma) at 2 pmol/$\mu$L in 50:50 acetonitrile:water with 2% (v/v) formic acid is prepared immediately prior to each analysis.
8. Formulation buffer: 10 m$M$ L-histidine, 260 m$M$ glycine, 1% (w/v) sucrose, and 0.005% (w/v) polysorbate 80, pH 6.8.
9. NaCl and $CaCl_2$.

# 3. Methods
## 3.1. Activation of rFIX (Fig. 4)

1. Activation of rFIX with hFXIa is performed in a formulation buffer supplemented with 0.1 $M$ NaCl and 5 m$M$ $CaCl_2$ at an enzyme-to-substrate ratio of 1:100 (w/w) at 37°C for 1 h and 15 min (*35,36*; **Note 16**). The final concentration of rFIX is 2.2 mg/mL.
2. After incubation, snap-freeze the activation reaction in liquid $N_2$, and store at −80°C until LC/MS analysis.

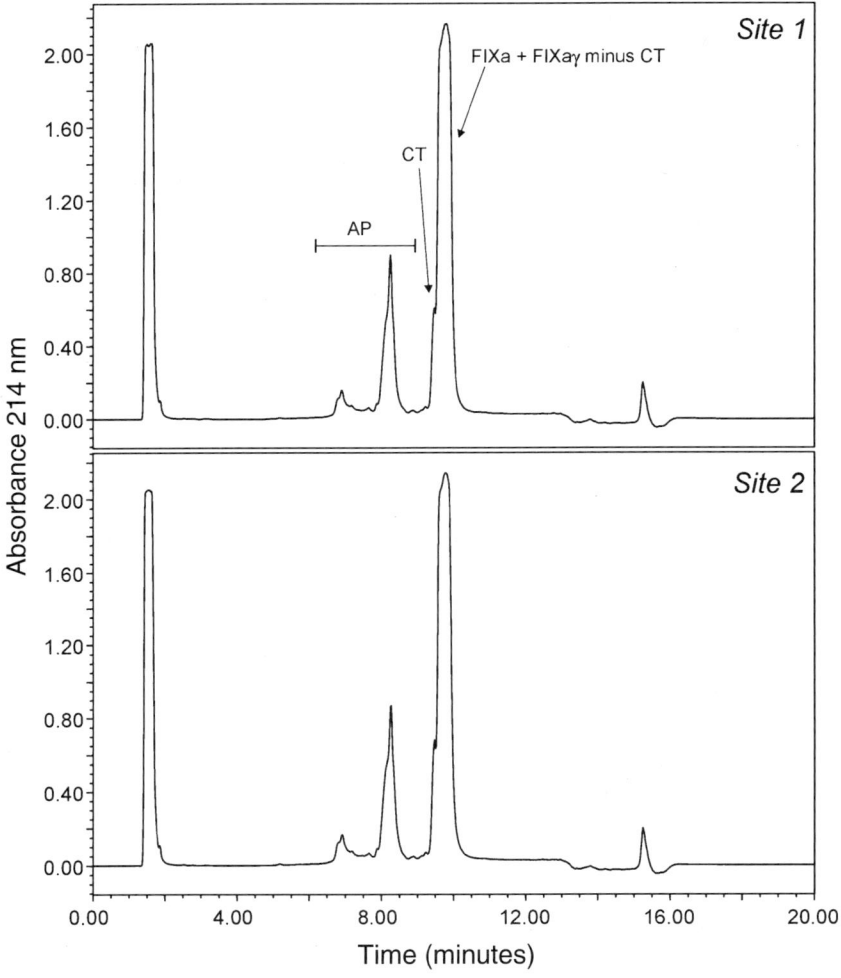

Fig. 5. RP-HPLC separation of the rFIX activation products (**Notes 14** and **17**).

**Table 3**
**RP-HPLC Gradient**

| Time (min) | %A | %B |
| --- | --- | --- |
| Initial | 90 | 10 |
| 1 | 90 | 10 |
| 10 | 40 | 60 |
| 10.1 | 0 | 100 |
| 12 | 0 | 100 |
| 12.1 | 90 | 10 |
| 25 | 90 | 10 |

### 3.2. Chromatography of AP and FIXa (Fig. 5)

1. The activation reaction is thawed and injected (approx 250 µg, 114 µL) onto a Symmetry C4 RP-HPLC column and eluted at a flow rate of 1.0 mL/min with the gradient shown in **Table 3** (*see* **Fig. 5** and **Note 15**).
2. The effluent (1 mL/min) from the HPLC system is split postcolumn so that 0.100–0.133 mL/min is directed into the ESI probe of the Q-Tof-2 through 0.005 in. inner diameter (ID) peek tubing. The remaining effluent is directed to waste.

### 3.3. Mass Analysis of AP and rFIXa

The set up of a multiuser Q-Tof-2 for precise protein mass determinations by LC/MS is a 2-d process. Day 1 involves cleaning and configuring the instrument, setting the acquisition parameters, and assessment of instrument performance. At the end of d 1, the instrument temperature is allowed to equilibrate overnight with all voltages on and all gases flowing. Activities on d 2 involve mass calibration of the instrument, activation of temperature compensation mode, and optimization of ESI parameters for protein LC/MS. At the end of d 2, the ion source block and desolvation gas temperatures are increased for LC/MS, and the instrument temperature is allowed to equilibrate overnight. On d 3, the instrument is ready for LC/MS data acquisition.

1. Q-Tof-2 instrument preparations (d 1). All experiments utilize the standard ESI probe and the Z-spray ion source. Flush the metal capillary in the ESI probe with 50:50 acetonitrile:water (v/v), then rinse the exterior of the ESI probe tip with 100% methanol. Close the isolation valve, and remove the external sample cone, cone gas nozzle, and grounding baffle. To ensure full sensitivity, clean these parts in accordance with the manufacturer's instructions, which includes a scrubbing step with 10% (v/v) formic acid in water and sequential ultrasonication steps in 10% (v/v) formic acid in water, 50:50 methanol:acetone (v/v), and 100% 2-propanol. After this cleaning procedure, dry the external sample cone, cone gas nozzle, and grounding baffle with high-purity nitrogen gas, and reinstall all components. Reboot the host and embedded computers with the manufacturer's shutdown and start-up procedures, and reinitialize the MassLynx software before the acquisition of multiple back-to-back LC/MS runs. All ion source, quadrupole, and TOF parameters should match the predetermined historical values. Set the capillary voltage on the ESI probe to 3000 V. Maintain argon gas at a pressure of approx $6 \times 10^{-5}$ mbar in the collision cell region. Adjust the nitrogen flow rates for the desolvation gas (drying gas) to 400 L/h and cone gas (curtain gas) to 50 L/h. Set the ion source block temperature to 80°C and desolvation gas temperature to 120°C. Set the in/out position of the ESI probe to the furthest distance from the sample cone to minimize ion source contamination. Infuse insulin, trypsinogen, and NaI solutions with a syringe pump through 0.005 in ID PEEK tubing at 5 µL/min for performance testing and calibration. Use the MassLynx "sample list" interface to acquire all MS data. Optimize the left/right position of the ESI probe to between 1 and 4 mm from the sample cone for the best signal. Adjust the valve that controls the nitrogen gas for ESI nebulization (at the probe tip) for the best signal. Optimal signal is found between 1/4 and 1/2 turn from the closed position when solutions are flowing at 5 µL/min.
2. Q-Tof-2 acquisition parameters (d 1). Activate the manual pusher, which determines the duty cycle of the TOF in sampling the quadrupole ion beam. Set the pusher unit to pulse every 88 µs to enable spectral acquisition out to *m/z* 4000. Set the MS profile parameters

to transport all ions between *m/z* 800 and 4000 through the quadrupole mass filter and hexapole collision cell with optimal sensitivity (**Note 18**). This is achieved with the following settings: dwell time at *m/z* 1000 is 20%; ramp time from *m/z* 1000–2000 is 60%; and dwell time at *m/z* 2000 is 20%, where the indicated percentage is the percent of total scan time. (The dwell and ramp values are indicated as "mass," not *m/z*, in MassLynx.) Utilize a total scan time of 5 s for LC/MS analyses of intact glycoproteins; accumulate signal for 4.9 s, and set the interscan delay to 0.1 s for data upload (**Note 19**).

3. Q-Tof-2 performance assessments (d 1). Assess mass analyzer resolving power with the insulin solution using a cone voltage (CV) of 37 V and a collision energy (CE) of 10 eV. Evaluate the resolving powers for the multiple-charged species at *m/z* 956.4, 1147.5, and 1434.4. Determine the resolving power with the peak-width definition, $m/\Delta m$, where m is the *m/z* value of a prominent isotope peak, and $\Delta m$ is the peak width (as *m/z*) at half height (FWHM). The resolving power calculated for each insulin species should exceed 10,000 and match historical values. Optimize the left/right ESI probe position so that the counts for the most abundant insulin ion (*m/z* 1147.5) are between 400 and 450 counts per 5 s scan. It is important to keep the counts for the most intense ion below 450 counts per 5 s scan to avoid signal saturation of the time-to-digital converter (TDC) and prevent an apparent loss in resolving power. Ascertain the instrument sensitivity for proteins with the trypsinogen solution using a CV of 50 V and CE of 10 eV. Optimize the left/right ESI probe position so that the counts for the multiple-charged trypsinogen species at *m/z* 2181 are as intense as possible, and evaluate the number of counts per scan (assuming 5 s total scan time). Accept counts above 250 (minimum); typical values range between 350 and 500 after ESI probe position and nebulizer gas optimization. Ensure that the trypsinogen signals and total ion current (TIC) trace are stable for each successive scan. After performing all cleaning, set-up, and assessment procedures, and prior to mass axis calibration, allow the instrument temperature to equilibrate overnight with all voltages on and all gases flowing.

4. Q-Tof-2 external calibration and temperature compensation set-up (d 2). With the instrument temperature at equilibrium, ensure that the dynamic calibration (DXC) temperature compensation circuitry is not activated. The DXC temperature compensation adjusts (in real time) mass axis calibration drift during LC/MS caused by changes in instrument and room temperatures. Activate analog channel 4, which connects the TDC to the temperature sensor on the TOF mass analyzer. The temperature sensor monitors the thermal expansion and contraction of the flight tube. Infuse the trypsinogen solution, and acquire data for 5–10 min before calibration to warm up the acquisition electronics. After moving the ESI probe far to the right of the sample cone to prevent contamination with NaI, infuse the NaI calibration solution. Acquire NaI mass data with a CV of 37 V, a CE of 10 eV, and the same instrument and scan settings as described in **Subheadings 3.3.1.** and **3.3.2.** for 3 min. Keep the most intense NaI-related peaks (e.g., *m/z* 1971.6) below 450 counts (assuming 5 s total scan time) to avoid signal saturation of the TDC acquisition unit but above 400 counts to obtain optimal peak shapes and sensitivity for the species above *m/z* 3000. Combine (sum together) all NaI mass spectra acquired over the 3 min, smooth the combined spectra (three-channel window, one smooth, and Savitzky–Golay algorithm), and center the data (centroid method at the 80% level by area). Use MassLynx to derive the calibration equation with a fourth order polynomial curve fit. Activate the DXC temperature compensation circuitry immediately after finalizing the calibration parameters in the software. The DXC constant is determined by the manufacturer's installation engineer, and it typically ranges from 70 to 100 ppm/°C. Repeat the analysis of the trypsino-

gen solution to reoptimize the ESI probe position (move closer to the sample cone) and nebulization gas flow. Combine the trypsinogen mass spectra acquired over 1 min, smooth (14-channel window, 1 smooth, and Savitzky-Golay algorithm), and center the data (centroid method at the 80% level by area). Check the mass accuracy of the $m/z$ 2181 peak; the measured value should be within 0.002% of the theoretical value of 2181.089. Raise the ion source block temperature to 115°C and desolvation gas temperature to 275°C. These temperatures are optimized for ESI of HPLC effluent flow rates between 0.100 and 0.150 mL/min. Allow the instrument temperature to equilibrate overnight before LC/MS analysis with all voltages on and all gases flowing.

5. LC/MS mass spectral acquisition (d 3). Prior to LC/MS, the instrument temperature must be at equilibrium. Connect the analog output signal from the 2487 UV/Vis detector to analog input 1 on the Q-Tof-2 to provide UV data complementary to the TIC trace. Connect the HPLC outlet tubing to the flow splitter, and set the HPLC to flow at 1 mL/min at initial solvent conditions (**Table 3**). All tubing is 0.005 in ID PEEK. Adjust the split flow rate from the flow splitter to the ESI probe to be between 0.100 and 0.150 mL/min; the remaining 0.85–0.9 mL/min should pass to waste or, if desired, to a fraction collector. Do not connect the flow splitter outlet tubing to the ESI probe at this time. Once the flow splitter is adjusted, maintain the flow from the HPLC at 1 mL/min; otherwise, the flow splitter will have to be readjusted. If necessary, readjust the valve controlling the nitrogen gas for ESI nebulization. The optimal signal is found between one-half and one turn from the closed position for flow rates 10 μL/min or more for the trypsinogen solution. Upon starting a LC/MS acquisition, infuse the trypsinogen solution at 10 μL/min from the syringe pump into the ESI probe for 2 min before connecting the flow splitter outlet tubing to the ESI probe. The effluent from the flow splitter should not be connected to the mass spectrometer until after the sample flow-through has passed through the outlet tubing. After this occurs, connect the flow splitter outlet tubing to the ESI probe, and acquire LC/MS data. After all chromatographic peaks have eluted, disconnect the flow splitter outlet tubing from the ESI probe, reconnect the syringe pump outlet tubing, and infuse the trypsinogen solution again (10 μL/min for 2 min) prior to stopping the LC/MS data acquisition. Acquisition of trypsinogen at the beginning and end of each LC/MS data file provides instant evaluation of run-to-run instrument sensitivity and determination of mass accuracy variation and drift for all experiments (**Note 20**). LC/MS experiments can be performed for several days without much deterioration in mass accuracy and sensitivity. Day-to-day performance is tracked with freshly prepared trypsinogen solutions.

6. LC/MS spectral processing. Each chromatographic peak contains multiple mass spectra. Combine all mass spectra across a chromatographic peak using MassLynx. Ultimately, this mass spectrum is submitted to the MaxEnt-1 module for mass deconvolution. If there is visible chemical background under the mass spectral signals, execute background subtraction in MassLynx in advance of mass deconvolution (**Note 21**). The specific background subtraction parameters for activated rFIX species are shown in **Table 4**. No mass spectral smoothing is performed because it is incompatible with MaxEnt-1.

7. MaxEnt-1 mass deconvolution. Conversion of the multiple-charged data to zero-charge data provided for the straightforward identification of the C-terminal (CT), FIXa, and AP isoforms in each chromatographic peak. Use the "uniform Gaussian" damage model in MaxEnt-1. Select the input $m/z$ range in MassLynx; choose all charge states of the protein(s) by default. For intact proteins, mass accuracy can be increased by limiting the input $m/z$ range to the 10 to 12 most abundant charge states that have well-defined peak shapes. Measure the average peak width at half the height for an abundant, fully resolved,

**Table 4**
**Background Subtraction and MaxEnt-1 Mass Deconvolution Parameters**
**for the Mass Spectra of the Activated rFIX Species**

| Procedure/rFIX species | AP[a] | FIXa[b] | CT[c] |
|---|---|---|---|
| Background subtraction | Not required | Required | Not required |
| Polynomial order | | 50 | |
| Below curve % | | 1 | |
| MaxEnt-1 mass deconvolution | | | |
| Input *m/z* range | 1700–3000 | 1039–3814 | 1000–2200 |
| Output mass range | 9000–14,000 | 43,000–46,000 | 9000–12,000 |
| Resolution (Dalton/channel) | 1 | 1 | 1 |
| Peak width at half height (Dalton) | 1.45 | 0.80 | 1.50 |
| Minimum intensity ratios (left%/right%) | 33/33 | 66/66 | 33/33 |

[a]*See* **Fig. 7**.
[b]See **Fig. 8**.
[c]See **Table 7**.

multiple-charged ion that is centrally located in the selected input *m/z* range. Use the default 33%/33% as the "left/right minimum intensity ratio." This limits the consideration of adjacent peaks in the same series based on their relative heights; 33% implies that the adjacent peak to the left or right of any particular peak in the multiple-charged envelope is at least 33% as intense. Produce a "survey" zero-charge mass spectrum with MaxEnt-1 at a resolution of 10.0 Daltons per channel and a wide output mass range between 5000 and 160,000 Daltons to establish the presence of all protein components in the chromatographic peak. Stop MaxEnt-1 after several iterations. If necessary, the harmonic artifacts, which are peaks at half and twice the mass of each protein component, may be reduced by raising the left/right minimum intensity ratio to approx 66%/66%. Repeat the MaxEnt-1 survey mass deconvolution until optimized. Use 33%/33% for polypeptides of lower mass, particularly if they are glycosylated, because the abundances of the charged ions are not likely to have the ideal protein-charge envelope with gradual intensity changes. Produce the final zero-charge mass spectrum with MaxEnt-1 for each specific protein component found in the survey at a resolution of 1.0 Daltons per channel with an appropriate narrower output mass range. Set the output mass range to include approx 3000 to 5000 Daltons on either side of the approximate survey-derived mass for each large protein and to include approx 300 to 500 Daltons on either side of the approximate survey-derived mass for smaller protein subunits. The specific MaxEnt-1 parameters used for activated FIX species are shown in **Table 4**. Unless otherwise noted, use the same parameters determined from the survey deconvolution. Allow MaxEnt-1 to iterate to convergence. Center the resulting peaks with the centroid method at the 80% level to obtain masses accurate to the tenths place. Ensure that the MaxEnt-1 deconvolution is a precise representation of the data; the isoform peaks and their ratios in the zero-charge mass spectrum should match or closely resemble an average of those in the multiple-charged data series. Higher mass accuracy and more exact zero-charge profiles are obtained by the judicious use or absence (if possible) of background subtraction (**Note 21**). It is important to use the identical input *m/z* range and MaxEnt-1 deconvolution parameters (i.e., output mass range, resolution, peak width, minimum intensity ratio, and 80% centroid) for each protein species when processing LC/MS data for comparability.

8. LC/MS analysis for rFIX comparability. Optimize the CV and CE parameters for LC/MS of the AP and FIXa chromatographic species to ensure full retention of labile PTMs and, in severe cases, to prevent peptide-bond fragmentation (**Note 22**). Test a range of CV values at a CE of 5 and 10 eV with repeat injections of activated rFIX. Choose the optimal CV and CE parameter combinations for each FIXa species based on the balance between signal strength and minimal analysis artifacts (e.g., the loss of N-linked glycan antennal arms or sialic acid residues). For AP, a reduced CV of 20 V and CE of 5 eV were required to minimize N-linked glycan trimming caused by energetic collisions (*see* **Fig. 6** and **Note 23**). For FIXa, a CV of 37 V and a CE of 10 eV was required to achieve strong signal strengths (**Fig. 6**). To assess comparability of AP, all batches of activated rFIX are injected for LC/MS analysis with a CV of 20 V and CE of 5 eV. For FIXa and CT, all batches of activated rFIX are reinjected for LC/MS analysis with a CV of 37 V and CE of 10 eV. Perform all LC/MS experiments side-by-side in 1 d, if possible, to minimize retention time and mass drift.

## 3.4. Evaluation of Mass Data for Comparability

1. Overall comparability assessment. Our comparability programs include product release testing, animal pharmacokinetic studies, forced protein degradation, biophysical analyses, nonroutine HPLC assays, and top-down characterization with MS. A final review of comparability includes the analysis of results from all of these tests.

2. Comparability acceptance criteria for top-down characterization. Because MS analysis is not routinely performed as part of product release testing, an extensive batch history on which to base quantitative and statistical acceptance criteria was not available. However, criteria were set according to the known instrument specifications and qualitative isoform content. Typically, our first approach is to take a bird's-eye view of the mass envelopes generated for each protein sample, then compare sections of the overall mass profiles. If there are new species detected in the new material, these species and their potential impact are evaluated further in the context of the results from other testing. Frequently, a change in the proportion of existing isoforms is acceptable if it is found to be inconsequential in other tests, such as bioactivity or pharmacokinetics. Large differences in the levels of existing species would require further investigation into the manufacturing process parameters. Specific acceptance criteria for this study were:

- The RP-HPLC chromatograms of rFIX activation products generated from site 2 rFIX must be comparable to those generated from site 1 rFIX with no new species detected.
- No new species should be detected in the mass spectra of site 2 material when compared with those of site 1 material.
- Masses of the isoforms found in rFIX from manufacturing site 2 must agree with the masses of the analogous isoforms found in rFIX from site 1 within a specified mass (mass = $x$ Daltons, where $x = 1$ for AP; 2 for FIXa, and 1 for the CT peptides).
- The measured masses must agree with the theoretical masses according to instrument specifications.
- Analogous mass spectra of activation products from sites 1 and 2 must contain similar peak distributions.

3. Isoform selection. There is a potential for "data overload" because a large number of isoform signals are generated from top-down MS analysis of complex glycoproteins. To assess comparability, we selected important and representative isoforms across the

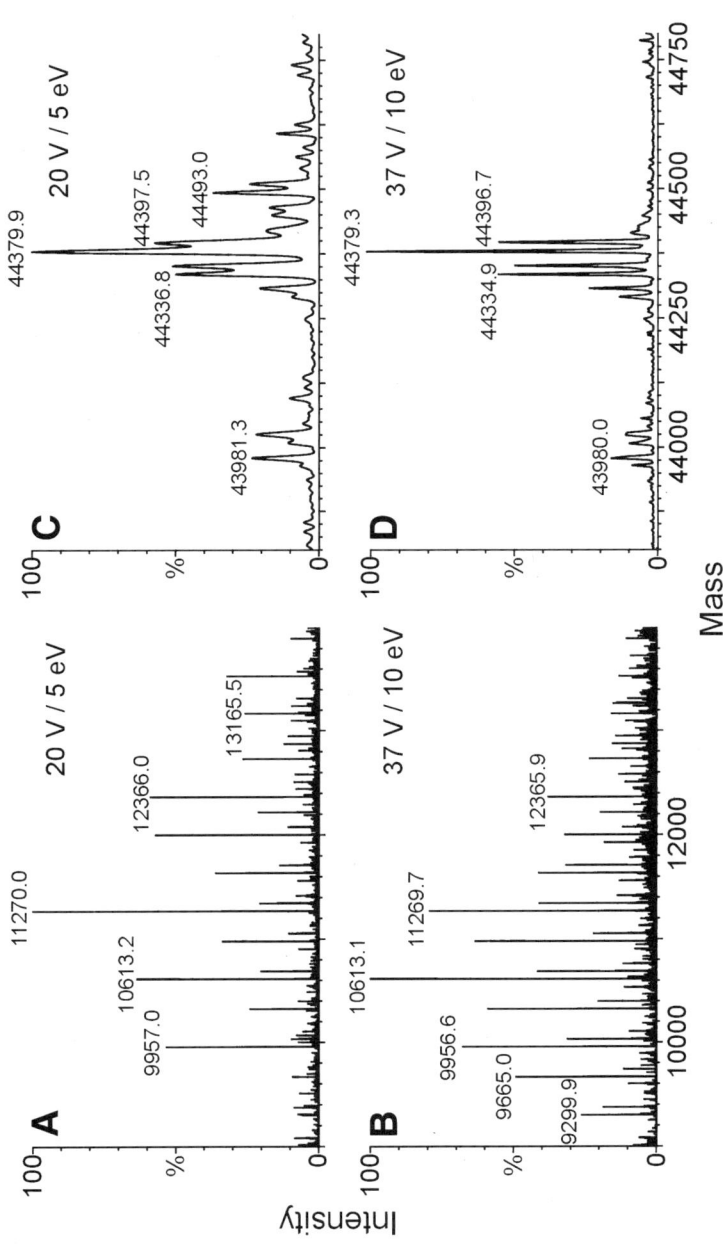

Fig. 6. Optimization and effect of the CV and CE mass spectrometric parameters in the RP-HPLC/ESI-QTOF MS analysis of AP and FIXa (**Notes 22** and **23**). Shown are the resultant zero-charge mass spectra from (**A**) AP region acquired with a CV of 20 V and a CE of 5 eV; (**B**) AP region acquired with a CV of 37 V and a CE of 10 eV; (**C**) FIXa region acquired with a CV of 20 V and a CE of 5 eV; and (**D**) FIXa region acquired with a CV of 37 V and a CE of 10 eV. *See* **Tables 5** and **6** for information relating to the identity of the peaks in the mass spectra. For the AP and FIXa regions, the zero-charge mass spectra were deconvoluted with the same MaxEnt-1 parameters as reported in **Figs. 7** and **8**, respectively.

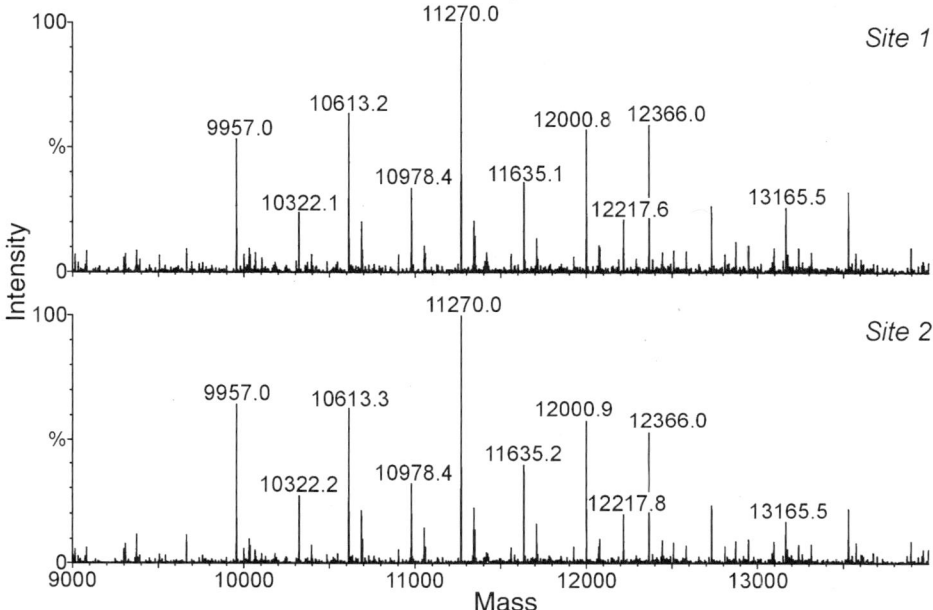

Fig. 7. Analysis of AP from activated rFIX by RP-HPLC/ESI-QTOF MS. Zero-charge mass spectra of the AP region (**Fig. 5**) that compares rFIX material from sites 1 and 2. *See* **Table 5** for information relating to the identity of the peaks in the mass spectra. The AP region was summed and then deconvoluted to obtain a zero-charge mass spectrum.

mass range. Certainly, specific comparisons of selected mass signals in the absence of an overall evaluation of the mass spectra and in the absence of a screen for previously uncharacterized new species can give misleading results. Identical mass signals can be pulled out of spectra that are otherwise very different, and this should be avoided.
4. rFIX comparability results. The RP-HPLC chromatographic profiles of hFXIa-activated rFIX batches manufactured at site 2 were comparable to those of hFXIa-activated rFIX batches manufactured at site 1 (**Fig. 5**). All minor and major peaks were present in the chromatograms, and no new peaks were detected in chromatograms of activated rFIX from site 2.

Zero-charge mass spectra from the AP region of each respective RP-HPLC chromatogram (**Fig. 5**) are presented in **Fig. 7**. The peak distribution in the mass spectrum of the AP from site 2 rFIX batches was similar to that of the AP from site 1 rFIX, and no new peaks were observed. The major peaks in each mass spectrum represent the microhetereogeneity found in the *N*-glycans at the two fully occupied *N*-glycosylation sites. The theoretical and observed masses for selected isoforms are presented in **Table 5**. The observed differences in AP glycoform masses between the site 2 and site 1 batches for analogous species were 1 Dalton or less. Additionally, the measured masses for each AP glycoform when compared to their respective theoretical masses showed mass

**Table 5**
**Masses of AP Glycoforms Observed via RP-HPLC/ESI-QTOF MS**

| AP identity[a] | Theoretical mass (Dalton)[b] | Site 1 (Dalton)[c] | Site 2 (Dalton) |
|---|---|---|---|
| 3A/3SA/1F; 3A/3SA/1F | 9956.7 | 9957.0 | 9957.0 |
| 3A/3SA/1F; 3A/3SA/1F + 1 R | 10322.0 | 10322.1 | 10322.2 |
| 3A/3SA/1F; 4A/4SA/1F | 10613.3 | 10613.2 | 10613.3 |
| 3A/3SA/1F; 4A/4SA/1F + 1 R | 10978.6 | 10978.4 | 10978.4 |
| 4A/4SA/1F; 4A/4SA/1F | 11269.9 | 11270.0 | 11270.0 |
| 4A/4SA/1F; 4A/4SA/1F + 1 R | 11635.2 | 11635.1 | 11635.2 |
| 4A/4SA/1F; 4A/4SA/1F + 2 R | 12000.6 | 12000.8 | 12000.9 |
| 4A/4SA/1F; 4A/4SA/1F + 1 O[d] | 12217.7 | 12217.6 | 12217.8 |
| 4A/4SA/1F; 4A/4SA/1F + 3 R | 12365.9 | 12366.0 | 12366.0 |
| 4A/4SA/1F; 4A/4SA/1F + 2 O[d] | 13165.6 | 13165.5 | 13165.5 |

[a]Abbreviations: A, antennae; SA, sialic acid (N-acetyl neuraminic acid); F, fucose; R, poly-N-acetyl lactosamine (GalGlcNAc); O, O-linked glycan.

[b]Theoretical masses were calculated for the selected AP glycoforms using PAWS and are reported as average masses.

[c]Observed masses were calculated using MaxEnt-1 as described in the text.

[d]O-glycosylation is found at Thr[159], Thr[169], and Thr[172] with partial occupancy and is composed of Gal-GalNAc (i.e., core 1) with one and two sialic acids (N-acetyl neuraminic acid). For the AP region denoted in **Fig. 5**, the minor peak (averages 20% from integration of AP region) represents AP with both N- and O-glycosylation, whereas the major peak represents AP with N-glycosylation only.

differences of 0.005% or less, consistent with the specifications of the Q-Tof-2 mass spectrometer.

Zero-charge mass spectra from the FIXa region of each RP-HPLC chromatogram (**Fig. 5**) are presented in **Fig. 8**. The peak distribution in the mass spectra of the FIXa from site 2 rFIX batches were similar to those of the FIXa from site 1 rFIX, and no new peaks were observed. The major peaks in each mass spectrum represent FIXa with 11 and 12 4-carboxyglutamic acid (Gla) residues and partial β-hydroxylation at Asp[64]. The theoretical and observed masses for selected isoforms are presented in **Table 6**. The observed mass differences between site 1 and 2 batches for analogous FIXa isoforms were 2 Daltons or less. Additionally, the measured masses for each FIXa isoform in comparison to their respective theoretical masses showed mass differences of 0.005% or less, similar to the specifications of the Q-Tof-2 mass spectrometer.

The peak distribution in the mass spectra of the CT peptides derived from site 2 rFIX were like those of the CT peptides derived from site 1 rFIX (data not shown); no new peaks were observed. The theoretical and observed masses of the CT peptides from representative rFIX batches from sites 1 and 2 are presented in **Table 7**. The mass difference between the masses of CT peptides and their respective theoretical masses was less than 1 Dalton. The measured masses for each CT peptide vs their respective theoretical masses showed mass differences of 0.005% or less, which is consistent with the measurements of the Q-Tof-2 mass spectrometer.

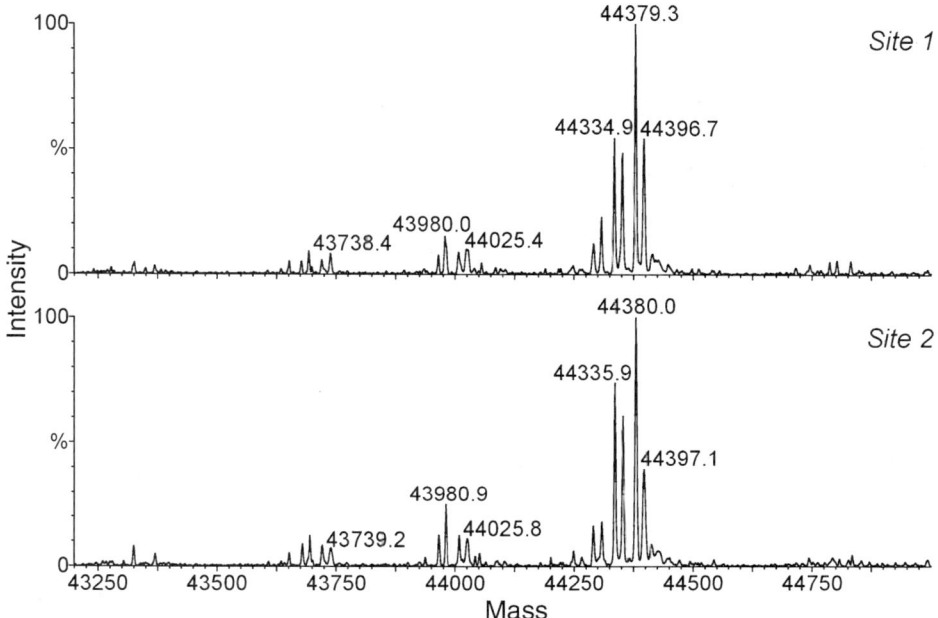

Fig. 8. Analysis of FIXa by RP-HPLC/ESI-QTOF MS. Zero-charge mass spectral profiles of the FIXa peak (**Fig. 5**) comparing rFIX material from sites 1 and 2. *See* **Table 6** for information relating to the identity of the mass spectral peaks. The FIXa region was summed, background subtracted, and then deconvoluted to obtain a zero-charge mass spectrum.

## Table 6
### Masses of FIXa Isoforms Observed via RP-HPLC/ESI-QTOF MS

| FIXa identity[a] | Theoretical mass (Dalton)[b] | Site 1 (Dalton)[c] | Site 2 (Dalton) |
|---|---|---|---|
| $L^{1-145}$, $H^{181-415}$, 12 Gla, β-OH[d,e,g] | 43739.5 | 43738.4 | 43739.2 |
| $L^{1-142}$, $H^{181-415}$, 11 Gla, β-OH[d,f,g] | 43981.7 | 43980.0 | 43980.9 |
| $L^{1-142}$, $H^{181-415}$, 12 Gla, β-OH[d,f,g] | 44025.7 | 44025.4 | 44025.8 |
| $L^{1-145}$, $H^{181-415}$, 11 Gla[d,f] | 44336.1 | 44334.9 | 44335.9 |
| $L^{1-145}$, $H^{181-415}$, 12 Gla[d,f] | 44380.1 | 44379.3 | 44380.0 |
| $L^{1-145}$, $H^{181-415}$, 12 Gla, β-OH[d,f,g] | 44396.1 | 44396.7 | 44397.1 |

[a]Abbreviations: L, light chain; H, heavy chain; Gla, 4-carboxyglutamic acid; β-OH, β-hydroxylation.

[b]Theoretical masses were calculated for the selected isoforms with disulfide bonds included using PAWS and are reported as average masses.

[c]Observed masses were calculated using MaxEnt-1 as described in the text.

[d]The O-linked glycan at $Ser^{53}$ is Glc(Xyl)$_2$.

[e]The O-linked glycan at $Ser^{61}$ is Fuc.

[f]The O-linked glycan at $Ser^{61}$ is Fuc-GlcNAc-Gal-NeuAc.

[g]β-hydroxylation is present at $Asp^{64}$.

**Table 7**
**Masses of CT Peptides Observed via RP-HPLC/ESI-QTOF MS**

| CT peptide identity | Theoretical mass (Dalton)[a] | Site 1 (Dalton)[b] | Site 2 (Dalton) |
|---|---|---|---|
| Val[328]-Thr[415] | 9770.1 | 9769.9 | 9770.4 |
| Val[328]-Thr[415c] | 9788.1 | 9787.5 | 9787.7 |
| Ser[319]-Thr[415] | 10814.4 | 10814.4 | 10814.7 |

[a]Theoretical masses were calculated for the CT peptides with disulfide bonds included using PAWS and are reported as average masses.

[b]Observed masses were calculated using MaxEnt-1 as described in the text.

[c]The peptide bond, Arg[238]-Ser[339] for this CT peptide is presumed to be hydrolyzed based on the observed 18-Dalton mass increase. However, a disulfide bond between Cys[336] and Cys[350] keeps this CT peptide intact as Val[328]-Thr[415].

Based on top-down characterization data, the primary structure and PTMs of rFIX manufactured at sites 1 and 2 were judged comparable. The mass spectra of analogous rFIX fragments were similar to each other with no new species detected, and all masses met the acceptance criteria. The LC/MS data agree with the results of forced degradation testing, biophysical analyses, bioactivity assays, and the nonclinical pharmacokinetic study. These results indicate that there is no significant impact on rFIX structure and function when this protein pharmaceutical is manufactured at the new manufacturing facility.

## 4. Notes

1. Recently, top-down characterization and peptide mapping with MS have been employed in parallel experiments to comprehensively characterize complex mixtures of proteins *(4,5)*. Peptide mapping and sequencing with MS was shown to be more effective than the top-down approach in providing identifications for the mixture of unknown proteins owing to the increased dynamic range of peptide mapping-based techniques. However, the sequence information from the top-down techniques helped to confirm the assignments of the major protein species detected by peptide mapping. Accurate mass determinations from the top-down approach revealed the PTMs in the mass profile and allowed full-length species to be distinguished from N or C-terminally truncated isoforms. Both reports emphasize that the precise mass measurements from the top-down approach provided additional attributes that strengthened the protein identifications and helped define the intact protein structures more thoroughly.

2. Detailed structural information for intact proteins was initially accomplished with Fourier transform-ion cyclotron resonance (FT-ICR) MS/MS *(7)*. To date, extensive contiguous amino acid sequence and PTM information for several intact proteins up to 45 kDa *(8–14)* have been demonstrated using FT-ICR with activated ion—electron capture dissociation (AI-ECD)—an exclusive technique for FT-ICR instruments *(15)*. The FT-ICR mass analyzer features very high-resolving powers that exceed $10^5$ and mass accuracies less than 0.001% for peptides and less than 0.005% for proteins *(7)*. The AI-ECD technique induces cleavages at NH–αCHR bonds along the protein backbone in an indiscriminate manner and breaks the intramolecular noncovalent bonds between two fragment

ions to maximize the number of cleaved sites *(15,16)*. For proteins larger than 70 kDa, limited proteolysis was performed to generate smaller polypeptide fragments, ranging in size from 5 to 48 kDa, which were more amenable to gas-phase sequencing with FT-ICR MS/MS *(17)*.

3. There are several reports specifying the use of a modified quadrupole ion-trap mass analyzer that is capable of systematically converting the unresolved multiple-charged protein fragment ions to all single-charged ions by gas-phase ion/ion proton transfer reactions *(18)*. This provides straightforward spectral interpretation in the absence of high resolution, resulting from the uniformity of the charge states *(19)*. Abundant sequence information *(20,21)* and site-specific PTM characterization *(22,23)* for whole proteins up to 26 kDa have been demonstrated with this customized system.

4. In the analysis of large polypeptides, only the ions between *m/z* 800 and 4000 are allowed to pass through the first-quadrupole mass filter, whereas ions less than *m/z* 800 are destabilized and filtered out using the MS profile settings.

5. QTOF mass analyzers have increased sensitivity when compared with other instrument types because of their inherently high-duty cycles. Recent improvements in the orthogonal acceleration ion optics for Q-Tof-2 mass analyzers combined with improved ion beam spatial and kinetic energy distributions have increased resolving powers to above 10,000 $(m/\Delta m)$ for ESI-generated ions without compromising sensitivity. In the "pusher" unit of a Q-Tof-2, a voltage gradient is applied across the ion beam through separate "pusher" and "puller" pulses in an attempt to minimize turnaround time and improve resolution.

6. Fragmentation of intact proteins on a QTOF with CID has resulted in limited stretches of amino acid sequence information, which is not adequate for thorough characterization. However, these "sequence tags" were extremely valuable for the identification of unknown proteins and their attributes when combined with the accurate mass measurement *(26,27)*.

7. Depending on the glycoprotein, these mass signals will coalesce fully or partially when analyzed by matrix-assisted laser desorption/ionization time-of-flight (MALDI-TOF) MS.

8. Although top-down characterization provides the means for rapid analysis of new species and changes in the isoform profile, a comparability assessment is supported only after the results from all routine testing and nonroutine studies are obtained and evaluated together.

9. BDDrFVIII, although considerably less complex than full-length rFVIII, still poses an analytical challenge because it has a molecular weight of 170 kDa (80-kDa light chain and 90-kDa heavy chain) and modifications, including tyrosine sulfation, *N*- and *O*-linked glycans, and N- and C-terminal processing *(31)*. The intact light chain was analyzed by online RP-HPLC/ESI-QTOF MS, but the intact heavy chain was too complex and exceeded the capabilities of the QTOF. Proteolytic cleavage with thrombin into 6, 43, 50, and 73 kDa domains followed by online RP-HPLC/ESI-QTOF MS was required for analysis of the entire molecule and permitted a comparability evaluation of materials produced by two manufacturing processes *(32)*.

10. For certain mAbs, minor N-terminal processing and low-level modifications were difficult to identify with the initial peptide-map characterization because of small or coeluting peaks in a complex profile. However, these modifications were readily apparent in the mass profile from the complementary top-down approach. mAbs, before and after disulfide bond reduction, were analyzed off-line by nanoelectrospray ionization interfaced to the QTOF mass analyzer. This allowed valid mass determination for the intact mAb tetramer and the light- and heavy-chain subunits, respectively. Modifications detected on the intact mAb tetramer were easily localized on the light or heavy chain. The intact, as

well as the reduced and alkylated, mAbs were also analyzed online by RP-HPLC/ESI-QTOF MS to obtain complementary mass information from chromatographically resolved species. The accurate mass data verified structural composition, whereas the isoform profiles revealed the *N*-glycosylation patterns in the Fc domain of the heavy chain, C- and N-terminal heterogeneity of both the light and heavy chain, and other PTMs *(30,33)*.

11. rFIX manufactured at the two manufacturing facilities was also compared by product release testing, forced degradation studies, biophysical analyses, bioactivity testing, and nonclinical pharmacokinetics.

12. Site-specific assignments for the glycans on the AP were previously determined by subdigestion of the RP-HPLC-fractionated, de-*N*-glycosylated AP with endoproteinase Glu-C, followed by N-terminal sequencing and MALDI-TOF MS postsource decay sequencing.

13. The low-abundance AP species containing the *O*-glycan at $Thr^{179}$ was not observed after FXIa activation. This is presumably a result of steric hindrance of the $Arg^{180}$-$Val^{181}$ cleavage site. This form of the AP was found after Lys-C cleavage of rFIX.

14. Activation of rFIX with hFXIa predominantly generates AP ($Ala^{146}$-$Arg^{180}$) and FIXa by cleavage of the $Arg^{145}$-$Ala^{146}$ and $Arg^{180}$-$Val^{181}$ bonds (**Fig. 4**). FIXa consists of L chain ($Tyr^1$-$Arg^{145}$) linked to H chain ($Val^{181}$-$Thr^{415}$) by one disulfide bond. The heterogeneous AP elutes across two peaks, and the FIXa elutes in the major peak (**Fig. 5**).

15. Along with the major cleavages, there are low-level cleavages observed in the HC at the $Arg^{318}$-$Ser^{319}$, $Arg^{327}$-$Val^{328}$, and $Arg^{338}$-$Ser^{339}$ peptide bonds, generating a species referred to as FIXaγ. FIXaγ consists of the light chain linked to HCγ by one disulfide bond. HCγ is $Val^{181}$-$Arg^{318}$, $Val^{181}$-$Arg^{327}$, or $Val^{181}$-$Arg^{338}$. The CT peptides remain bound noncovalently in nondenaturing conditions. Under the denaturing conditions of RP-HPLC, the FIXaγ/CT peptide complex is dissociated. The CT peptides, $Ser^{319}$-$Thr^{415}$, $Val^{328}$-$Thr^{415}$, and $Ser^{339}$-$Thr^{415}$, elute as a shoulder on the main peak. The N-terminal portion of FIXaγ (FIXaγ minus the CT peptides) coelutes with FIXa.

16. To control for any potential experimental variables that might impact the results, all activation reactions should be carried out with the same lot of hFXIa side-by-side in the same experiment.

17. The early eluting AP peak contains N- and O-linked glycans; the later eluting AP peak contains N-linked glycans with low levels of O-linked glycans. In this work, the mass spectra were summed across the entire AP region of the chromatogram. Representative chromatograms of the activation products of one batch from each site are shown in **Fig. 5**.

18. The MS profile provides control of the RF voltages on the multipole devices for transporting a specific *m/z* range of ions to the TOF analyzer. The *m/z* range between 800 and 4000 is the typical region where the multiple-charged ions for intact proteins are detected on the Q-Tof-2. There are sharp (0.75) and gradual (2) cut-off values associated with the low- and high-mass values, respectively, which are important when setting up the MS profile. When the low-mass value is set to 1000, all ions above *m/z* 750 are transmitted, and all ions below *m/z* 750 are filtered out. For the high-mass value, 2000 is required for optimal sensitivity to *m/z* 4000. By not transmitting all of the ions below *m/z* 750, the sensitivity is further maximized for the higher *m/z* protein ions.

19. More specifically, one mass spectrum is uploaded to the computer hard drive every 4.9 s. This mass spectrum represents the signal accumulation in the TDC from the TOF across *m/z* 800 to 4000 for all orthogonally accelerated ion packets during this period. Longer total scan times of 5 s minimize the repetitive 0.1-s interscan delay, which helps maxi-

mize the signal. Faster total scan times of 2 s (1.9-s signal accumulation and 0.1-s interscan delay) are employed if peptides or both peptides and proteins are anticipated in an LC/MS experiment, because the chromatographic peak widths for peptides are typically more narrow than those of proteins. The identical total scan time that is utilized in LC/MS is also used for instrument calibration. Instrument testing with trypsinogen and insulin is always performed with total scan times of 5 s.

20. Embedded trypsinogen masses in an LC/MS data file also allow the use of the MassLynx "lock mass" feature for correcting mass accuracy drift. Do not use NaI at high-ion source block temperatures, because it will quickly contaminate the RF hexapole and reduce sensitivity.

21. The chemical background is usually minimal in LC/MS, and visible chemical noise may represent either incomplete protein desolvation, small-molecule and metal-ion adducts of the protein, overlapping charge species, or unresolved protein isoforms. Use a "polynomial order" of 0 or 1 with a "below curve %" (i.e., the percentage of data points below the zero line) of 35 or 40 to remove level and constant background from the mass spectrum. Use higher polynomial order values, such as 10, 20, 50, 75, and 99 on a mass spectrum with more complex patterns of chemical noise. Choose the "make graph of fitted polynomial" option to view the straight line or curve fit overlaid with the mass spectral data. The straight line or curve should attempt to mirror the mass spectral data and noise pattern. It should be positioned below the mass spectral data trace but not exceed the root-mean-square of the noise.

22. CV is analyte-dependent, and it represents the ion acceleration potential between the ion source block and extractor cone. The region between the ion source block and the extractor cone in the Q-Tof-2 is turbulent, as the two apertures bridge pressures between approx 760 and approx $10^{-5}$ to $10^{-4}$ Torr. respectively. High-CV potentials are usually required for sensitive protein analysis, but these can induce in-source artifacts for proteins with labile PTMs and Asp-Pro peptide bonds. CE represents the axial translational energies of ions that enter the collision cell. The collision cell region contains argon gas at approx $6 \times 10^{-5}$ Torr (*see* **Fig. 1**). For high resolution and accurate mass analysis, CEs of 5 to 10 eV are utilized to collisionally cool ions against argon. At a CE of 10 eV, fragmentation artifacts for sialylated glycopeptides and proteins with labile PTMs can occur from increased collisional activation events.

23. For both CV and CE, careful selection of both parameters is required to ensure mass spectrometric artifacts from energetic collisions are kept to a minimum. The AP glycoform profile in **Fig. 6**, panel B is skewed as a result of the increased energies of collision caused by the higher voltages employed. New peaks representing the loss of sialic acid residues, entire antennal arms, and O-linked glycans were observed. A CV of 20 V and a CE of 5 eV (**Fig. 6**, panel A) was used in acquiring AP spectra, because the sensitivity was maintained and the in-source and collision cell fragmentation events were minimized. The glycoform profile at a CV of 20 V and a CE of 5 eV suited well with the *N*-glycan profile data obtained with high-performance anion-exchange chromatography-PED. Unfortunately, a CV of 20 V and CE of 5 eV was not ideal for the higher molecular weight FIXa species (**Fig. 6**, panel C); a CV of 37 V and CE of 10 eV is required for proper focusing and optimal detection as shown by the increased resolving power for FIXa in **Fig. 6**, panel D. Because the ion energetics were different for AP and FIXa, each batch of activated rFIX was analyzed twice by RP-HPLC/ESI-QTOF MS using the parameter sets, 20 V CV and 5 eV CE and 37 V CV and 10 eV CE, for the AP and FIXa, respectively.

## Acknowledgments

The authors sincerely thank Drs. Marta Czupryn, Mark Hardy, Lisa Marzilli, and Hubert Scoble at Wyeth BioPharma for their support and critical review of the manuscript. The authors are grateful for the technical support that our colleagues at Waters and Micromass MS Technologies have provided over the years.

## References

1. Food and Drug Administration. (2003) Guidance for industry, comparability protocols-protein drug products and biological products-chemistry, manufacturing, and controls information, draft guidance. Available at www.fda.gov/cber/guidelines.htm.
2. ICH. (2003) ICH Q5E, Comparability of biotechnological/biological products, note for guidance on biotechnological/biological products subject to changes in their manufacturing process (CPMP/ICH/5721/03). Available at www.hc-sc.gc.ca/hpfb-dgpsa/tpd-dpt/q5e_step2_notice_e.html.
3. Kelleher, N. L., Lin, H. Y., Valaskovic, G. A., Aaserud, D. J., Fridriksson, E. K., and McLafferty, F. W. (1999) Top down versus bottom up protein characterization by tandem high-resolution mass spectrometry. *J. Am. Chem. Soc.* **121,** 806–812.
4. VerBerkmoes, N. C., Bundy, J. L., Hauser, L., Asano, K. G., Razumovskaya, J., Larimer, F., et al. (2002) Integrating "top-down" and "bottom-up" mass spectrometric approaches for proteomic analysis of Shewanella oneidensis. *J. Proteome Res.* **1,** 239–252.
5. Nemeth-Cawley, J. F., Tangarone, B. S., and Rouse, J. C. (2003) "Top Down" characterization is a complementary technique to peptide sequencing for identifying protein species in complex mixtures. *J. Proteome Res.* **2,** 495–505.
6. Balogh, M. P. (2004) Debating resolution and mass accuracy. *LCGC North America* **22,** 118–130.
7. McLafferty, F. W. (2001) Tandem mass spectrometric analysis of complex biological mixtures. *Int. J. Mass Spectrom.* **212,** 81–87.
8. Ge, Y., Lawhorn, B. G., El Naggar, M., Strauss, E., Park, J. H., Begley, T. P., and McLafferty, F. W. (2002) Top down characterization of larger proteins (45 kDa) by electron capture dissociation mass spectrometry. *J. Am. Chem. Soc.* **124,** 672–678.
9. Sze, S. K., Ge, Y., Oh, H., and McLafferty, F. W. (2002) Top-down mass spectrometry of a 29-kDa protein for characterization of any posttranslational modification to within one residue. *Proc. Natl. Acad. Sci. USA* **99,** 1774–1779.
10. Sze, S. K., Ge, Y., Oh, H., and McLafferty, F. W. (2003) Plasma electron capture dissociation for the characterization of large proteins by top down mass spectrometry. *Anal. Chem.* **75,** 1599–1603.
11. Shi, S. D., Hemling, M. E., Carr, S. A., Horn, D. M., Lindh, I., and McLafferty, F. W. (2001) Phosphopeptide/phosphoprotein mapping by electron capture dissociation mass spectrometry. *Anal. Chem.* **73,** 19–22.
12. Ge, Y., El Naggar, M., Sze, S. K., Oh, H. B., Begley, T. P., McLafferty, F. W., et al. (2003) Top down characterization of secreted proteins from Mycobacterium tuberculosis by electron capture dissociation mass spectrometry. *J. Am. Soc. Mass. Spectrom.* **14,** 253–261.
13. Ge, Y., Lawhorn, B. G., El Naggar, M., Sze, S. K., Begley, T. P., and McLafferty, F. W. (2003) Detection of four oxidation sites in viral prolyl-4-hydroxylase by top-down mass spectrometry. *Protein Sci.* **12,** 2320–2326.
14. Pesavento, J. J., Kim, Y.-B., Taylor, G. K., and Kelleher, N. L. (2004) Shotgun annotation of histone modifications: a new approach for streamlined characterization of proteins by top down mass spectrometry. *J. Am. Chem. Soc.* **126,** 3386–3387.

15. Horn, D. M., Ge, Y., and McLafferty, F. W. (2000) Activated ion electron capture dissociation for mass spectral sequencing of larger (42 kDa) proteins. *Anal. Chem.* **72**, 4778–4784.
16. Zubarev, R. A. (2003) Reactions of polypeptide ions with electrons in the gas phase. *Mass Spectrom. Rev.* **22**, 57–77.
17. Forbes, A. J., Mazur, M. T., Patel, H. M., Walsh, C. T., and Kelleher, N. L. (2001) Toward efficient analysis of >70 kDa proteins with 100% sequence coverage. *Proteomics* **1**, 927–933.
18. Reid, G. E. and McLuckey, S. A. (2002) "Top down" protein characterization via tandem mass spectrometry. *J. Mass Spectrom.* **37**, 663–675.
19. Stephenson, J. L., McLuckey, S. A., Reid, G. E., Wells, J. M., and Bundy, J. L. (2002) Ion/ion chemistry as a top-down approach for protein analysis. *Curr. Opin. Biotechnol.* **13**, 57–64.
20. Hogan, J. M. and McLuckey, S. A. (2003) Charge state dependent collision-induced dissociation of native and reduced porcine elastase. *J. Mass Spectrom.* **38**, 245–256.
21. Amunugama, R., Hogan, J. M., Newton, K. A., and McLuckey, S. A. (2004) Whole protein dissociation in a quadrupole ion trap: identification of an a priori unknown modified protein. *Anal. Chem.* **76**, 720–727.
22. Reid, G. E., Stephenson, J. L., Jr., and McLuckey, S. A. (2002) Tandem mass spectrometry of ribonuclease A and B: *N*-linked glycosylation site analysis of whole protein ions. *Anal. Chem.* **74**, 577–583.
23. Hogan, J. M., Pitteri, S. J., and McLuckey, S. A. (2003) Phosphorylation site identification via ion trap tandem mass spectrometry of whole protein and peptide ions: bovine alpha-crystallin A chain. *Anal. Chem.* **75**, 6509–6516.
24. Blackburn, R. K. and Moseley, M. A., III. (1999) Quadrupole time-of-flight mass spectrometry: a powerful new tool for protein identification and characterization. *Am. Pharm. Rev.* **2**, 49–59.
25. Morris, H. R., Paxton, T., Dell, A., Langhorne, J., Berg, M., Bordoli, R. S., et al. (1996) High sensitivity collisionally-activated decomposition tandem mass spectrometry on a novel quadrupole/orthogonal-acceleration time-of-flight mass spectrometer. *Rapid Commun. Mass Spectrom.* **10**, 889–896.
26. Nemeth-Cawley, J. F. and Rouse, J. C. (2002) Identification and sequencing analysis of intact proteins via collision-induced dissociation and quadrupole time-of-flight mass spectrometry. *J. Mass Spectrom.* **37**, 270–282.
27. Thevis, M., Ogorzalek Loo, R. R., and Loo, J. A. (2003) Mass spectrometric characterization of transferrins and their fragments derived by reduction of disulfide bonds. *J. Am. Soc. Mass Spectrom.* **14**, 635–647.
28. Ferrige, A. G., Seddon, M. J., Green, B. N., Jarvis, S. A., and Skilling, J. (1992) Disentangling electrospray spectra with maximum entropy. *Rapid Comm. Mass Spectrom.* **6**, 707–711.
29. Green, B. N., Hutton, T., and Vinogradov, S. N. Analysis of complex protein and glycoprotein mixtures by electrospray ionization mass spectrometry with maximum entropy processing, in *Protein and Peptide Analysis by Mass Spectrometry* (Chapman, J. R., ed.), Humana, Totowa, NJ, 1996, pp. 279–294.
30. Rouse, J. C., Marzilli, L. A., Johnson, K. A., McClellan, J. E., and Czupryn, M. J. "Top down" glycoprotein characterization by high resolution mass spectrometry and its application to biopharmaceutical development. *Well Characterized Biopharmaceuticals 2004: 8th Symposium on the Interface of Regulatory and Analytical Sciences for Biotechnology Health Products,* Washington DC, 2004.

31. Sandberg, H., Almstedt, A., Brandt, J., Castro, V. M., Gray, E., Holmquist, L., et al. (2001) Structural and functional characterization of B-domain deleted recombinant factor VIII. *Semin. Hematol.* **38(2 Suppl. 4),** 4–12.

32. Czupryn, M. Current analytical methodologies: application to comparability assessment of recombinant clotting factors. PDA/IABS Conference, State of the art analytical methods for the characterisation of biological therapeutic products and assessment of comparability, Bethesda, MD, 2003.

33. Mehndiratta, P., Grunder, B. C., Marzilli, L. A., McClellan, J. E., Tangarone, B. S., Porter, T. J., and Rouse, J. C. LC/MS characterization of recombinant monoclonal antibodies at the pre-development stage. Sixth International Symposium on Mass Spectrometry in the Health and Life Sciences: Molecular and Cellular Proteomics, San Francisco, CA, 2003.

34. Bond, M., Jankowski, M., Patel, H., Karnik, S., Strang, A., Xu, B., et al. (1998) Biochemical characterization of recombinant factor IX. *Semin. Hematol.* **35(2 Suppl. 2),** 11–17.

35. DiScipio, R. G., Kurachi, K., and Davie, E. W. (1978) Activation of human factor IX (Christmas factor). *J Clin. Invest.* **61,** 1528–1538.

36. Fujikawa, K., Legaz, M. E., Kato, H., and Davie, E. W. (1974) The mechanism of activation of bovine factor IX (Christmas factor) by bovine factor XIa (activated plasma thromboplastin antecedent). *Biochemistry* **13,** 4508–4516.

# 35

## Sample Preparation Procedures for High-Resolution Nuclear Magnetic Resonance Studies of Aqueous and Stabilized Solutions of Therapeutic Peptides

### Mark J. Howard

## 1. Introduction

This chapter covers the preparation of a purified nuclear magnetic resonance (NMR) peptide sample for analysis considering purity, pI, oxidation status, and solubility of the sample, along with the choice and preparation of the NMR sample tube. The NMR analysis of therapeutic peptides is an area that is still underdeveloped, but it offers both great promise and utility in drug screening and design (1–3). Ligand-target NMR-screening methods, such as saturation transfer difference NMR (STD-NMR; 4) rely on correct NMR assignments of potential ligands to relate binding proximity to ligand-target structural arrangement. Furthermore, both nascent and stabilized therapeutic peptide structures from aqueous-based media can be related to binding efficacy and can enhance the design of future therapeutics via *in silico* analysis of data that is, in turn, reliant on correct NMR assignment. Accurate assignment and structural analysis of such peptides is ultimately obtained from high-quality NMR data that is achieved from meticulous sample preparation. It is feasible to run the necessary NMR experiments for the structural elucidation of a single high-quality 20-mer peptide sample in 12–18 h and to assign and obtain a structure from the data in 1–5 d. This is in contrast to the processes required for determining protein structures (5). The structural study of therapeutic peptides is long overdue as a routine analytical step and holds potential for the future. It is anticipated that the procedures described here will encourage research groups and scientists to prepare their peptides of interest for such structural work and present them to the high-resolution NMR facility in their organization for data acquisition and analysis.

## 2. Materials

### 2.1. Equipment

1. PC with Internet access.
2. High-performance liquid chromatography (HPLC) or fast protein liquid chromatography (FPLC) for peptide purification (if necessary).

From: *Methods in Molecular Biology, vol 308: Therapeutic Proteins: Methods and Protocols*
Edited by: C. M. Smales and D. C. James © Humana Press Inc., Totowa, NJ

3. 5-mm Outer diameter (od) NMR sample tubes: for a standard volume of 500–600 µL, use Norell 509-UP or Wilmad® Emperor 535-PP7 tube. For small volumes of 250–350 µL, use Shigemi BMS-005 or BMS-005T (*see* **Note 1**).
4. NMR tube labels.
5. NMR Tube Washer (not essential, but useful) and suction/vacuum unit or pump to operate.
6. Extended tip (7 or 9 in.) NMR Pasteur pipets to retrieve and add samples to NMR tubes.
7. pH meter.
8. Microfuge.
9. Microcentrifuge tubes: 0.5 or 1.5 mL.
10. Polypropylene centrifuge tubes: 15 mL with screw-cap closure.
11. Laboratory film.
12. Freeze-dryer.

## 2.2. Reagents

1. Peptide of interest, HPLC-purified and freeze-dried (typically between 0.5 and 2 mg).
2. Stock solutions of 1.0 $M$ $Na_2HPO_4$, 1.0 $M$ $NaH_2PO_4$, and 1.0 $M$ NaCl to make standard NMR sample buffer (phosphate-buffered saline: 25 m$M$ phosphate and 100 m$M$ NaCl) with the pH set within the phosphate buffer range of 5.8–8.0.
3. 0.1 $M$ NaOH and 0.1 $M$ HCl stock solutions.
4. High-purity (e.g., Milli-Q) water.
5. Deuterium oxide, at least 99% atom $^2$H, NMR-grade.
6. Trifluoroethanol (TFE)-$d_3$, at least 98% atom $^2$H, NMR-grade.
7. Dithiothreitol (DTT)-$d_{10}$, at least 98% atom $^2$H (if needed).
8. Sodium azide (if needed).
9. Concentrated nitric acid.
10. Liquid nitrogen (for freeze-drying).

## 3. Methods

### 3.1. Sample Considerations

The peptide should primarily be pure. If you have synthesized the peptide within your laboratory, either have it HPLC-purified or use a reverse-phase chromatography column with a FPLC system to separate undesirable components before freeze-drying. In the latter preparatory case, beware of residual buffer salts that will redissolve when the peptide is dissolved in the NMR buffer system.

Before reaching for reagents and equipment, obtain the peptide properties prior to proceeding. Obtained such information using the ProtParam tool found at the ExPASy Molecular Biology Server website (http://ca.expasy.org/tools/protparam.html). Enter the single-letter code of your peptide sequence into ProtParam and press the "Compute Parameters" button to obtain the results. In addition to displaying peptide composition, information can be obtained regarding amino acid composition, molecular weight, and estimates of the pI and extinction coefficients (if your peptide contains tryptophan, tyrosine, or cysteine residues). Check your peptide sequence to ensure that it contains amino acids that will support its solubility in water. The peptide must be soluble enough to create approx 1 m$M$ peptide solution in the NMR buffer. Peptides with high proportions of hydrophobic amino acids are unlikely to do this, and their solubility in water will be at micromolar or even lower concentrations at saturation.

### 3.2. Buffer Preparation and Considerations

With your datasheet from ExPASy (*see* **Subheading 3.1.**), plan your ideal NMR buffer system. Unless you have a specific pH at which you wish to study your peptide, use a pH at least 0.5 U away from the estimated pI for the peptide (*see* **Note 2**). Unless your peptide is sensitive to phosphate (*see* **Note 3**), prepare the buffer as follows.

1. Use a 15-mL polypropylene conical centrifuge tube (e.g., Falcon) to prepare your NMR buffer.
2. If you are keeping the sample for some time or concerned about microbial growth within the sample, add sodium azide to the NMR buffer. First, make 10 mL of a 7.7 m$M$ stock of sodium azide by dissolving 5 mg sodium azide in 10 mL high-purity water. Second, aliquot 65 μL sodium azide stock into the base of the 15-mL polypropylene conical centrifuge tube to be used for the buffer.
3. Make up 10 mL NMR buffer system using the phosphate buffer stock solutions to make 25 m$M$ [Na$_2$HPO$_4$/NaH$_2$PO$_4$] and 100 m$M$ NaCl. To prepare 10 mL of a pH 6.4 NMR buffer system, add 64 μL 1.0 $M$ Na$_2$HPO$_4$ stock, 186 μL 1.0 $M$ NaH$_2$PO$_4$ stock, and 1 mL 1.0 $M$ NaCl stock. Make up the volume to 10 mL using high-purity water.
4. Add DTT to your NMR buffer system if your peptide contains one or more cysteine residues. DTT (5 mg) in 10 mL buffer will provide 3.2 m$M$ DTT in your NMR buffer stock. You should add excess DTT in comparison to your desired final peptide concentration in the sample (*see* **Note 4**).
5. Check the NMR buffer pH with a pH meter, and bring the buffer to the appropriate pH by adding small aliquots of NaOH or HCl stock solutions.
6. Freeze this buffer at –20°C for later use if necessary. This buffer can be stored for 6 mo at –20°C without any problems. The 15-mL polypropylene conical centrifuge tube (e.g., Falcon) with the cap sealed using laboratory film makes for a useful storage vessel.
7. If your sample requires degassed buffer, follow the procedure in **Note 5** before adding to the solid peptide.

### 3.3. NMR Tubes

Find the highest-quality tubes that associate best with the NMR system you are using. Your first option should always be to use the tubes and tube preferences (i.e., sample volume and/or depth of liquid) recommended by your facility manager or NMR technician. Included in **Subheading 3.3.1.** are the standard volumes used at Kent for a Varian UnityINOVA 600 MHz NMR spectrometer with a 2002-generation triple-resonance Varian probe.

### 3.3.1. NMR Tube Types

NMR tubes come in several forms, but for peptide NMR work, use either a standard NMR tube as shown in **Fig. 1A** or a microtube as shown in **Fig. 1B,C**. A 600-μL sample volume is required in a standard NMR tube. Standard tubes of choice for 600 MHz NMR are either Norell-509UP or Wilmad 535-PP7 that are washed as described in **Subheading 3.3.2.** before use (*see* **Note 6**). You require 1.2 mg of a 2.0-kDa peptide to create a 600-μL NMR sample of 1 m$M$ concentration. If you have a NMR microtube, such as a Shigemi BMS-005, which requires 330 μL for the sample volume, only 0.66 mg 2-kDa peptide is needed to create the same 1-m$M$ sample. Here is the benefit of microtubes: less sample, less peptide preparatory time, and less cost.

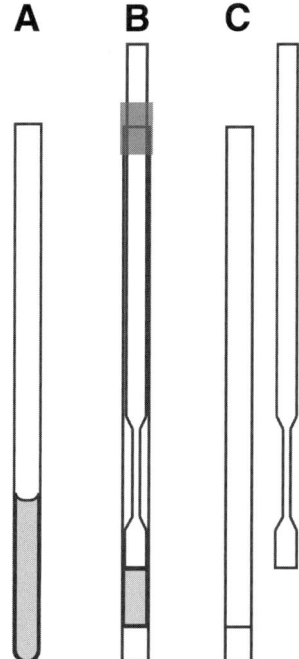

Fig. 1. Standard (**A**) and micro (**B**)-NMR tubes are shown with sample volumes in gray. (**C**) The microtube and plunger assembly. The microtube plunger is held in place using laboratory film wrapped around the plunger and tube, which is shown as the dark gray band at the top of the tube in (B).

Your NMR system volumes for both standard and microtubes may vary from these values, and it is generally found that older-generation NMR probes can tolerate smaller volumes, such as 500 μL for a standard tube and 250 μL for a Shigemi tube.

### 3.3.2. Cleaning NMR Tubes

Clean NMR tubes are important because the only NMR signals observed should arise from your sample, not from tube contaminants. Also, such contaminants may render your peptide useless and only fit for waste. Wash all tubes (even new tubes received directly from the supplier) by following this procedure.

1. Soak your tubes in concentrated nitric acid for 1 h.
2. Carefully remove the tubes from the acid, and rinse well with high-purity water. Be rigorous in rinsing the tube, which is best achieved using a NMR tube washer that draws solvent from a reservoir to rinse the inside of the tube, as shown in **Fig. 2**.
3. If using a tube washer, rinse each tube with 500 mL high-purity water using this device (*see* **Note 7**). If a NMR tube washer is not used, take extreme care to ensure all acid is removed from the tube by rigorous manual rinsing.

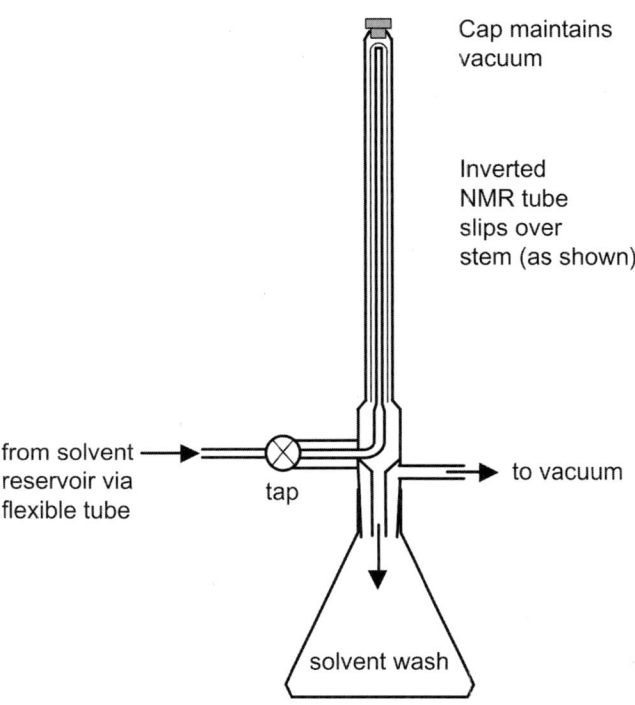

Fig. 2. Example of a NMR tube washer that uses an external reservoir of solvent to flush the tube using suction (*see* **Note 7**).

4. Tubes can be dried inverted at room temperature by applying a vacuum or using dry nitrogen or argon-bottled gas to blow air into an inverted tube. Resist placing NMR tubes in an oven to dry, as this warps the camber and destroys all the costly craftsmanship.

## 3.4. Peptide NMR Sample Preparation

### 3.4.1. Standard Sample Preparation

As mentioned previously, NMR analysis of a 2.0-kDa peptide will require between 0.66 and 1.2 mg peptide depending on which NMR tube is used. However, if there is sufficient "well-behaved" peptide (i.e., very soluble and not prone to aggregation), create a 2.0–2.5 m$M$ peptide concentration in your NMR sample. This requires between 1.65 and 3.00 mg of a 2.0-kDa peptide for a 2.5 m$M$ peptide concentration (depending on the tube of choice). The additional peptide would enable you to obtain high-quality NMR data in a shorter time period and achieve better water suppression in your NMR spectra.

To prepare your sample for analysis, place your desired amount of freeze-dried purified peptide in a 1.5-mL microcentrifuge tube, and follow the procedure based on the type of tube used. *See also* **Notes 5** and **8** for samples requiring inert atmospheres.

### 3.4.1.1. Standard NMR Tubes

1. Add 540 µL NMR sample buffer to the peptide. (Note that this is 90% (v/v) of the total required volume of 600 µL; *see* **Note 9**.)
2. Add 60 µL deuterium oxide. (Note that this is 10% (v/v) of the total required volume of 600 µL.) The deuterium oxide is added as an external lock standard for the NMR and should be present between 5% and 10% (v/v) of the sample volume (*see* **Note 9**).
3. If your recommended final volume is lower than 600 µL, adjust the volumes required accordingly (e.g., a 500 µL final volume will require 450 µL NMR sample buffer and 50 µL deuterium oxide).

### 3.4.1.2. Micro-NMR Tubes

1. Add 300 µL NMR sample buffer to the peptide. (Note that this is 90% (v/v) of the total required volume of 330 µL; *see* **Note 9**.)
2. Add 30 µL deuterium oxide. (Note that this is 10% (v/v) of the total required volume of 330 µL.) The deuterium oxide is added for the same reason as outlined in **Subheading 3.4.1.1.**, **step 2** for a standard NMR tube; *see* **Note 9**.
3. If your recommended final volume is lower than 300 µL, adjust the volumes required accordingly, e.g., a 250 µL final volume will require 225 µL NMR sample buffer and 25 µL deuterium oxide for 10% (v/v).

### 3.4.1.3. Preparing and Placing the Solubulized Peptide Sample in a NMR Tube

1. Place the cap on the microcentrifuge tube, and agitate the tube by flicking it with your finger. The use of vortex mixers is not advised, as flocculation and aggregation can be observed as a result of the severity of this mixing method.
2. Once mixed, spin the tube for 1 min in a microfuge at 10,000*g* to remove any debris.
3. Use an extended tip NMR Pasteur pipet to draw up the peptide solution, and transfer to the NMR tube of choice, taking care not to disturb any debris that has been pelleted by centrifugation.
4. For a standard NMR tube: cap the tube, seal with laboratory film, and affix label (*see* **Note 8**).
5. For a micro-NMR tube: place the plunger section into the tube, slide it down into the top of the solution, and remove the air bubble. Pull up plunger section to create the largest possible volume of liquid devoid of air, and seal plunger in place with laboratory film to create a microtube with the sample, as shown in **Fig. 1B**. Affix label (*see* **Note 8**).
6. Once a sample tube has been prepared, it can be stored for a short time at 4°C until the NMR spectrometer is ready to run all planned experiments.

## 3.4.2. Introducing TFE

TFE is usually added in amounts between 10% and 50% (v/v) and can be introduced to the NMR sample buffer before using the combined solution to dissolve the peptide, or it can be added to a sample that has previously been prepared for NMR (i.e., a peptide already dissolved in the NMR buffer solution). Preparing the buffer prior to dissolving the peptide only requires the correct volume ratios (v/v) of TFE, with the NMR sample buffer to be mixed before the combined buffer being used to dissolve the sample as described earlier. In contrast, adding up to 50% (v/v) TFE to a previously prepared peptide solution can create unwanted interactions, including aggregation or structural breakdown, particularly at the interface between the TFE and buffer. If TFE is added to a sample that already exists, follow these steps.

1. Remove sample from the NMR tube with an extended-tip NMR Pasteur pipet, and place sample in a microcentrifuge tube.
2. Draw up your required amount of TFE: between 10% and 50% (v/v) depending on your experiment.
3. Add the TFE slowly (in 10-µL aliquots), taking time to mildly agitate the tube between additions. Watch the contents of the microcentrifuge tube carefully, and check if any precipitation occurs.
4. Once addition is complete, allow the sample to stabilize inside the microcentrifuge tube for 30–60 min.
5. Spin the microcentrifuge tube for 1 min in a microfuge at 13,000 rpm to remove any debris.
6. Use an extended-tip NMR Pasteur pipet to draw up the peptide solution, and transfer to the NMR tube of choice, being careful not to disturb any debris that has been pelleted by centrifugation.
7. For a standard NMR tube: cap the tube, seal with laboratory film, and affix label (*see* **Note 8**).
8. For a micro-NMR tube, place the plunger section into the tube, slide it down into the top of the solution, and remove the air bubble. Pull up plunger section to create the largest possible volume of liquid without air, and seal plunger in place with laboratory film to create a microtube with sample as shown in **Fig. 1B**. Affix label (*see* **Note 8**).
9. Be prepared for some changes in your NMR data acquisition (*see* **Note 10**).

### 3.4.3. Introducing Other Chemicals or Agents

If other chemicals or agents are added, follow the same method as outlined for TFE in **Subheading 3.4.2.** Add either the agent to the NMR buffer before solubilization of the peptide, or add the agent sparingly and over time.

### 3.4.4. Freeze-Drying to Remove $^1H_2O$

If an NMR sample is required that has a low $^1H_2O$ content and thus a high $^2H_2O$ content, exchange the water molecules in your system. This event is particularly useful for obtaining high-resolution DQFCOSY or NOESY NMR data, as well as for ascertaining which labile protons are in solvent exchange. The following procedure will create such a solvent exchanged system.

1. Remove sample from the NMR tube with an extended-tip NMR Pasteur pipet, and place sample in a 1.5-mL microcentrifuge tube. If it is a fresh sample, make up as for a standard NMR sample as outlined previously.
2. Freeze-dry the sample within the microcentrifuge tube. The best method for freeze-drying such samples is to rapidly cool the sample with liquid nitrogen, rather than with carbon dioxide/acetone, before placing in the freeze-dryer, under vacuum.
3. Once freeze-dried, resuspend the sample within the microcentrifuge tube in an amount of deuterium oxide ($^2H_2O$) that is equal to the initial sample volume as dictated by the tube type.
4. Freeze-dry the sample again, keeping the sample within the microcentrifuge tube. As before, use liquid nitrogen.

5. Resuspend the sample for a second time in deuterium oxide using the same microcentrifuge tube. Once again, resuspend the amount of deuterium oxide that is equal to the initial sample volume as dictated by the tube type (*see also* **Note 11**).

6. Check the pH, noting that the pH meter reads 0.4 U lower than the actual $p^2H$ reading owing to deuterium unless you have calibrated the pH meter using $^2H_2O$ standards. *See* **Note 3** for more details.

7. Transfer to an NMR sample tube, and seal as before as in **Subheading 3.4.1**.

## 4. Notes

1. The correct Shigemi tube type must be purchased that matches your spectrometer probe manufacturer. (For details, *see* http://www.geocities.com/~shigemi/.)

2. The peptide will have some solubility close to the *pI*, but it will be reduced in comparison to when it is ± 0.5 pH units away from the estimated pI. The phosphate buffer range of 5.8–8.0 should be sufficient to adopt a pH far enough from the *pI* in most cases. If the peptide has an estimated *pI* outside the phosphate-buffering range, opt for a buffer pH of 6.4 (a common buffer pH used for aqueous peptide NMR studies—the preparation of this system is shown in the procedure). Avoid using any buffering system at its buffering limits (with phosphate, pH 5.8 or 8.0), because there is no buffering potential beyond the edge of the range. However, potential problems can arise if the peptide is sensitive to phosphate or a pH outside this range is needed. In such situations, refer to **Note 3**.

3. For peptide systems that are unstable in phosphate or require buffering outside the phosphate buffer range of 5.8–8.0, you should source a suitable buffer system (for examples, *see* ref. *6*) that is available in deuterated form. For low pH (3.7–5.6), obtain acetic-$d_3$ acid (99 + % atom $^2H$), and create a sodium acetate-acetic acid buffer by carefully titrating with 0.01 *M* sodium hydroxide to obtain the desired pH. A pH value of 7.1–8.9 can be achieved using deuterated TRIS acid buffer system. High-pH (between 11.0 and 11.9) sodium monohydrogen phosphate can be titrated with 0.01 *M* sodium hydroxide. Aim for 25 m*M* buffer and 100 m*M* salt (usually as NaCl) for whatever buffer system is adopted. This is similar to that proposed for the phosphate-based system. Sodium hydroxide and hydrochloric acid can be substituted in these buffer systems with sodium deuteroxide (99 + % atom D) or deuterium chloride (99 + % atom $^2H$) solutions if low-solvent proton signal ($^1H_2O$) is desired. Note that for such systems, $p^2H = pH + 0.4$.

4. DTT is available both as standard reagent and as deuterated (DTT-$d_{10}$ 98% atom D). It is useful to incubate the peptide with NMR buffer containing DTT (or mercaptoethanol) at room temperature for 30 min in a 1.5-mL microcentrifuge tube.

5. If the sample requires an inert environment, degas the NMR sample buffer prior to use. To achieve this, place the buffer in a small Buchner or Hirsch flask, and bung the neck. Chill the buffer at 4°C for 2 h before using a magnetic flea and stirrer plate to agitate the buffer while the flask is placed under vacuum via the flask side arm for 30 min.

6. Always check with your NMR facility for their preferred NMR tube types. If fields higher than 600 MHz are used, you may have to purchase specialist tubes that are superior to those listed in **Subheading 3.3.1**.

7. Microtubes have an extended base (*see* **Fig. 1A**) that protrudes from many sample washers, preventing the washer cap (*see* **Fig. 2**) from being inserted. In this case, seal the head of the washer with laboratory film to create the vacuum, and enable the washer to work.

8. Some samples can be air-sensitive, and the NMR tube complete with sample should be purged with either dry nitrogen or argon gas prior to sealing the tube. Samples in this category benefit further from being prepared in microcell NMR tubes that create a sealed environment.

9. If the spectrometer operates well with 7% or even 5% (v/v) deuterium oxide, then adjust the figures, and use the lowest workable percentage. The examples given in **Subheading 3.4.1.** use 10% (v/v) deuterium oxide, which should work on all NMR systems.

10. If TFE-$d_3$ or another deuterated species is added, the $^2H$ lock signal on the NMR spectrometer can be more challenging to lock on to because of multiple resonances in the $^2H$ NMR spectrum. If the amount of TFE used is greater than 30% (v/v), then lock onto TFE-$d_3$, rather than $^2H_2O$ (if present). If your spectrometer does not account for the change in solvent lock from $^2H_2O$ to TFE-$d_3$, the $^1H$ carrier position (center of the NMR spectrum) will move by approx 0.93 ppm downfield at 600 MHz.

11. To remove as much $^1H_2O$ as possible, use 100% atom $^2H$ deuterium oxide to resuspend the peptide, and iteratively run through the freeze-drying and resuspension in $^2H_2O$ process three or four times.

## References

1. Roberts, G. C. K. (2000) Applications of NMR in drug discovery. *Drug Discov. Today* **5,** 230–240.
2. Shapiro, M. (2001) Applications of NMR screening in the pharmaceutical industry. *Il Farmaco* **56,** 141–143.
3. Pellecchia, M., Sem, D. S., and Wüthrich, K. (2002) NMR in drug discovery. *Nature Rev. Drug Discov.* **1,** 211–219.
4. Mayer, M. and Meyer, B. (2001) Group epitope mapping by saturation transfer difference NMR to identify segments of a ligand in direct contact with a protein. *J. Am. Chem. Soc.* **123,** 6108–6117.
5. Howard, M. J. (1998) Protein NMR spectroscopy. *Current Biol.* **8,** R331–R333.
6. Dawson, R. M. C., Elliott, D. C., Elliott, W. H., and Jones, K. M. *pH*, buffers and physiological media, in *Data for Biochemical Research*, 3rd ed. Oxford University Press, 2002, pp. 417–448.

# Index